Jim Hefferon
http://joshua.smcvt.edu/linearalgebra

Answers to exercises

Linear Algebra

Notation

$\mathbb{R}$, $\mathbb{R}^+$, $\mathbb{R}^n$	real numbers, reals greater than 0, n-tuples of reals
$\mathbb{N}$, $\mathbb{C}$	natural numbers: $\{0, 1, 2, \ldots\}$, complex numbers
$(a\,..\,b)$, $[a\,..\,b]$	interval (open, closed) of reals between a and b
$\langle \ldots \rangle$	sequence; like a set but order matters
V, W, U	vector spaces
$\vec{v}, \vec{w}, \vec{0}, \vec{0}_V$	vectors, zero vector, zero vector of V
$B, D, \vec{\beta}, \vec{\delta}$	bases, basis vectors
$\mathcal{E}_n = \langle \vec{e}_1, \ldots, \vec{e}_n \rangle$	standard basis for $\mathbb{R}^n$
$\text{Rep}_B(\vec{v})$	matrix representing the vector
$\mathcal{P}_n$	set of degree n polynomials
$\mathcal{M}_{n\times m}$	set of $n\times m$ matrices
$[S]$	span of the set S
$M \oplus N$	direct sum of subspaces
$V \cong W$	isomorphic spaces
h, g	homomorphisms, linear maps
H, G	matrices
t, s	transformations; maps from a space to itself
T, S	square matrices
$\text{Rep}_{B,D}(h)$	matrix representing the map h
$h_{i,j}$	matrix entry from row i, column j
$Z_{n\times m}, Z, I_{n\times n}, I$	zero matrix, identity matrix
$\lvert T \rvert$	determinant of the matrix T
$\mathscr{R}(h), \mathscr{N}(h)$	range space and null space of the map h
$\mathscr{R}_\infty(h), \mathscr{N}_\infty(h)$	generalized range space and null space

Lower case Greek alphabet, with pronounciation

character	name	character	name
α	alpha *AL-fuh*	ν	nu *NEW*
β	beta *BAY-tuh*	ξ	xi *KSIGH*
γ	gamma *GAM-muh*	o	omicron *OM-uh-CRON*
δ	delta *DEL-tuh*	π	pi *PIE*
ϵ	epsilon *EP-suh-lon*	ρ	rho *ROW*
ζ	zeta *ZAY-tuh*	σ	sigma *SIG-muh*
η	eta *AY-tuh*	τ	tau *TOW as in cow*
θ	theta *THAY-tuh*	υ	upsilon *OOP-suh-LON*
ι	iota *eye-OH-tuh*	ϕ	phi *FEE, or FI as in hi*
κ	kappa *KAP-uh*	χ	chi *KI as in hi*
λ	lambda *LAM-duh*	ψ	psi *SIGH, or PSIGH*
μ	mu *MEW*	ω	omega *oh-MAY-guh*

Preface

These are answers to the exercises in *Linear Algebra* by J Hefferon.

An answer labeled here as, for instance, One.II.3.4, matches the question numbered 4 from the first chapter, second section, and third subsection. The Topics are numbered separately.

Save this file in the same directory as the book so that clicking on the question number in the book takes you to its answer and clicking on the answer number takes you to the associated question, provided that you don't change the names of the saved files.*

Bug reports or comments are very welcome. For contact information see the book's home page `http://joshua.smcvt.edu/linearalgebra`.

Jim Hefferon
Saint Michael's College, Colchester VT USA

2012-Oct-12

*Yes, I once got a report of the links not working that proved to be due to the person saving the files with changed names.

Contents

Chapter Three: Maps Between Spaces

Chapter Four: Determinants

Chapter Five: Similarity

Chapter One: Linear Systems

Solving Linear Systems

One.I.1: Gauss's Method

One.I.1.17 We can perform Gauss's Method in different ways so these exhibit one possible way to get the answer.

(a) Gauss's Method

$$\xrightarrow{-(1/2)\rho_1+\rho_2} \begin{array}{rcl} 2x + & 3y = & 13 \\ & -(5/2)y = & -15/2 \end{array}$$

gives that the solution is $y = 3$ and $x = 2$.

(b) Gauss's Method here

$$\underset{\rho_1+\rho_3}{\xrightarrow{-3\rho_1+\rho_2}} \begin{array}{rcl} x & - z = & 0 \\ y + 3z & = & 1 \\ y & = & 4 \end{array} \xrightarrow{-\rho_2+\rho_3} \begin{array}{rcl} x & - \quad z = & 0 \\ y + & 3z = & 1 \\ & -3z = & 3 \end{array}$$

gives $x = -1$, $y = 4$, and $z = -1$.

One.I.1.18 **(a)** Gaussian reduction

$$\xrightarrow{-(1/2)\rho_1+\rho_2} \begin{array}{rcl} 2x + & 2y = & 5 \\ & -5y = & -5/2 \end{array}$$

shows that $y = 1/2$ and $x = 2$ is the unique solution.

(b) Gauss's Method

$$\xrightarrow{\rho_1+\rho_2} \begin{array}{rcl} -x + & y = & 1 \\ & 2y = & 3 \end{array}$$

gives $y = 3/2$ and $x = 1/2$ as the only solution.

(c) Row reduction

$$\xrightarrow{-\rho_1+\rho_2} \begin{array}{rcl} x - 3y + z = & 1 \\ 4y + z = & 13 \end{array}$$

shows, because the variable z is not a leading variable in any row, that there are many solutions.

(d) Row reduction

$$\xrightarrow{-3\rho_1+\rho_2} \begin{array}{rcl} -x - y = & 1 \\ 0 = & -1 \end{array}$$

shows that there is no solution.

(e) Gauss's Method

$$\begin{array}{rcl} & x+\ y-z=10 & & x+\ \ y-\ z=\ 10 & & x+\ \ y-\ \ z=\ 10 \\ \xrightarrow{\rho_1\leftrightarrow\rho_4} & 2x-2y+z=\ 0 & \xrightarrow[-\rho_1+\rho_3]{-2\rho_1+\rho_2} & -4y+3z=-20 & \xrightarrow[\rho_2+\rho_4]{-(1/4)\rho_2+\rho_3} & -4y+\ 3z=-20 \\ & x\qquad +z=\ 5 & & -y+2z=\ -5 & & (5/4)z=\ 0 \\ & 4y+z=20 & & 4y+\ z=\ 20 & & 4z=\ 0 \end{array}$$

gives the unique solution $(x,y,z)=(5,5,0)$.

(f) Here Gauss's Method gives

$$\begin{array}{rcl} & 2x\ +\ z+\ w=\ 5 & & 2x\ +\ z+\ w=\ 5 \\ \xrightarrow[-2\rho_1+\rho_4]{-(3/2)\rho_1+\rho_3} & y\ -\ w=\ -1 & \xrightarrow{-\rho_2+\rho_4} & y\ -\ w=\ -1 \\ & -(5/2)z-(5/2)w=-15/2 & & -(5/2)z-(5/2)w=-15/2 \\ & y\ -\ w=\ -1 & & 0=\ 0 \end{array}$$

which shows that there are many solutions.

One.I.1.19 **(a)** From $x=1-3y$ we get that $2(1-3y)+y=-3$, giving $y=1$.

(b) From $x=1-3y$ we get that $2(1-3y)+2y=0$, leading to the conclusion that $y=1/2$.

Users of this method must check any potential solutions by substituting back into all the equations.

One.I.1.20 Do the reduction

$$\xrightarrow{-3\rho_1+\rho_2}\quad \begin{array}{rcl} x-y&=&1 \\ 0&=&-3+k \end{array}$$

to conclude this system has no solutions if $k\neq 3$ and if $k=3$ then it has infinitely many solutions. It never has a unique solution.

One.I.1.21 Let $x=\sin\alpha$, $y=\cos\beta$, and $z=\tan\gamma$:

$$\begin{array}{rcl} 2x-\ y+3z=\ 3 & & 2x-\ y+\ 3z=3 \\ 4x+2y-2z=10 & \xrightarrow[-3\rho_1+\rho_3]{-2\rho_1+\rho_2} & 4y-\ 8z=4 \\ 6x-3y+\ z=\ 9 & & -8z=0 \end{array}$$

gives $z=0$, $y=1$, and $x=2$. Note that no α satisfies that requirement.

One.I.1.22 **(a)** Gauss's Method

$$\begin{array}{rcl} & x-\ 3y=\ b_1 & & x-\ 3y=\ b_1 \\ \xrightarrow[\substack{-\rho_1+\rho_3\\-2\rho_1+\rho_4}]{-3\rho_1+\rho_2} & 10y=-3b_1+b_2 & \xrightarrow[-\rho_2+\rho_4]{-\rho_2+\rho_3} & 10y=\ -3b_1+b_2 \\ & 10y=\ -b_1+b_3 & & 0=2b_1-b_2+b_3 \\ & 10y=-2b_1+b_4 & & 0=\ b_1-b_2+b_4 \end{array}$$

shows that this system is consistent if and only if both $b_3=-2b_1+b_2$ and $b_4=-b_1+b_2$.

(b) Reduction

$$\begin{array}{rcl} & x_1+\ 2x_2+3x_3=\ b_1 & & x_1+2x_2+\ 3x_3=\ b_1 \\ \xrightarrow[-\rho_1+\rho_3]{-2\rho_1+\rho_2} & x_2-3x_3=-2b_1+b_2 & \xrightarrow{2\rho_2+\rho_3} & x_2-\ 3x_3=\ -2b_1+b_2 \\ & -2x_2+5x_3=\ -b_1+b_3 & & -x_3=-5b_1+2b_2+b_3 \end{array}$$

shows that each of b_1, b_2, and b_3 can be any real number—this system always has a unique solution.

One.I.1.23 This system with more unknowns than equations

$$\begin{array}{c} x+y+z=0 \\ x+y+z=1 \end{array}$$

has no solution.

One.I.1.24 Yes. For example, the fact that we can have the same reaction in two different flasks shows that twice any solution is another, different, solution (if a physical reaction occurs then there must be at least one nonzero solution).

One.I.1.25 Because $f(1)=2$, $f(-1)=6$, and $f(2)=3$ we get a linear system.

$$\begin{aligned} 1a+1b+c&=2\\ 1a-1b+c&=6\\ 4a+2b+c&=3 \end{aligned}$$

Gauss's Method

$$\xrightarrow[-4\rho_1+\rho_3]{-\rho_1+\rho_2}\ \begin{aligned} a+\ b+\ c&=2\\ -2b\qquad&=4\\ -2b-3c&=-5 \end{aligned}\ \xrightarrow{-\rho_2+\rho_3}\ \begin{aligned} a+\ b+\ c&=2\\ -2b\qquad&=4\\ -3c&=-9 \end{aligned}$$

shows that the solution is $f(x)=1x^2-2x+3$.

One.I.1.26 Here $S_0=\{(1,1)\}$

$$\begin{aligned} x+y&=2\\ x-y&=0 \end{aligned}\ \xrightarrow{0\rho_2}\ \begin{aligned} x+y&=2\\ 0&=0 \end{aligned}$$

while S_1 is a proper superset because it contains at least two points: $(1,1)$ and $(2,0)$. In this example the solution set does not change.

$$\begin{aligned} x+\ y&=2\\ 2x+2y&=4 \end{aligned}\ \xrightarrow{0\rho_2}\ \begin{aligned} x+y&=2\\ 0&=0 \end{aligned}$$

One.I.1.27 **(a)** Yes, by inspection the given equation results from $-\rho_1+\rho_2$.

(b) No. The pair $(1,1)$ satisfies the given equation. However, that pair does not satisfy the first equation in the system.

(c) Yes. To see if the given row is $c_1\rho_1+c_2\rho_2$, solve the system of equations relating the coefficients of x, y, z, and the constants:

$$\begin{aligned} 2c_1+6c_2&=6\\ c_1-3c_2&=-9\\ -c_1+\ c_2&=5\\ 4c_1+5c_2&=-2 \end{aligned}$$

and get $c_1=-3$ and $c_2=2$, so the given row is $-3\rho_1+2\rho_2$.

One.I.1.28 If $a\neq 0$ then the solution set of the first equation is $\{(x,y)\mid x=(c-by)/a\}$. Taking $y=0$ gives the solution $(c/a,0)$, and since the second equation is supposed to have the same solution set, substituting into it gives that $a(c/a)+d\cdot 0=e$, so $c=e$. Then taking $y=1$ in $x=(c-by)/a$ gives that $a((c-b)/a)+d\cdot 1=e$, which gives that $b=d$. Hence they are the same equation.

When $a=0$ the equations can be different and still have the same solution set: e.g., $0x+3y=6$ and $0x+6y=12$.

One.I.1.29 We take three cases: that $a\neq 0$, that $a=0$ and $c\neq 0$, and that both $a=0$ and $c=0$.

For the first, we assume that $a\neq 0$. Then the reduction

$$\xrightarrow{-(c/a)\rho_1+\rho_2}\ \begin{aligned} ax+\qquad\qquad by&=j\\ (-\tfrac{cb}{a}+d)y&=-\tfrac{cj}{a}+k \end{aligned}$$

shows that this system has a unique solution if and only if $-(cb/a)+d\neq 0$; remember that $a\neq 0$ so that back substitution yields a unique x (observe, by the way, that j and k play no role in the conclusion that there is a unique solution, although if there is a unique solution then they contribute to its value). But $-(cb/a)+d=(ad-bc)/a$ and a fraction is not equal to 0 if and only if its numerator is not equal to 0. Thus, in this first case, there is a unique solution if and only if $ad-bc\neq 0$.

In the second case, if $a=0$ but $c\neq 0$, then we swap

$$\begin{aligned} cx+dy&=k\\ by&=j \end{aligned}$$

to conclude that the system has a unique solution if and only if $b\neq 0$ (we use the case assumption that $c\neq 0$ to get a unique x in back substitution). But—where $a=0$ and $c\neq 0$—the condition "$b\neq 0$" is equivalent to the condition "$ad-bc\neq 0$". That finishes the second case.

Finally, for the third case, if both a and c are 0 then the system

$$\begin{aligned} 0x + by &= j \\ 0x + dy &= k \end{aligned}$$

might have no solutions (if the second equation is not a multiple of the first) or it might have infinitely many solutions (if the second equation is a multiple of the first then for each y satisfying both equations, any pair (x, y) will do), but it never has a unique solution. Note that $a = 0$ and $c = 0$ gives that $ad - bc = 0$.

One.I.1.30 Recall that if a pair of lines share two distinct points then they are the same line. That's because two points determine a line, so these two points determine each of the two lines, and so they are the same line.

Thus the lines can share one point (giving a unique solution), share no points (giving no solutions), or share at least two points (which makes them the same line).

One.I.1.31 For the reduction operation of multiplying ρ_i by a nonzero real number k, we have that $(s_1, \ldots, s_n)$ satisfies this system

$$\begin{array}{rcl} a_{1,1}x_1 + a_{1,2}x_2 + \cdots + a_{1,n}x_n &=& d_1 \\ &\vdots& \\ ka_{i,1}x_1 + ka_{i,2}x_2 + \cdots + ka_{i,n}x_n &=& kd_i \\ &\vdots& \\ a_{m,1}x_1 + a_{m,2}x_2 + \cdots + a_{m,n}x_n &=& d_m \end{array}$$

if and only if

$$\begin{array}{rl} a_{1,1}s_1 + a_{1,2}s_2 + \cdots + a_{1,n}s_n &= d_1 \\ &\vdots \\ \text{and } ka_{i,1}s_1 + ka_{i,2}s_2 + \cdots + ka_{i,n}s_n &= kd_i \\ &\vdots \\ \text{and } a_{m,1}s_1 + a_{m,2}s_2 + \cdots + a_{m,n}s_n &= d_m \end{array}$$

by the definition of 'satisfies'. But, because $k \neq 0$, that's true if and only if

$$\begin{array}{rl} a_{1,1}s_1 + a_{1,2}s_2 + \cdots + a_{1,n}s_n &= d_1 \\ &\vdots \\ \text{and } a_{i,1}s_1 + a_{i,2}s_2 + \cdots + a_{i,n}s_n &= d_i \\ &\vdots \\ \text{and } a_{m,1}s_1 + a_{m,2}s_2 + \cdots + a_{m,n}s_n &= d_m \end{array}$$

(this is straightforward canceling on both sides of the i-th equation), which says that $(s_1, \ldots, s_n)$ solves

$$\begin{array}{rcl} a_{1,1}x_1 + a_{1,2}x_2 + \cdots + a_{1,n}x_n &=& d_1 \\ &\vdots& \\ a_{i,1}x_1 + a_{i,2}x_2 + \cdots + a_{i,n}x_n &=& d_i \\ &\vdots& \\ a_{m,1}x_1 + a_{m,2}x_2 + \cdots + a_{m,n}x_n &=& d_m \end{array}$$

as required.

For the combination operation $k\rho_i + \rho_j$, we have that $(s_1, \ldots, s_n)$ satisfies

$$\begin{array}{rcl} a_{1,1}x_1 & + \cdots + & a_{1,n}x_n = d_1 \\ & \vdots & \\ a_{i,1}x_1 & + \cdots + & a_{i,n}x_n = d_i \\ & \vdots & \\ (ka_{i,1} + a_{j,1})x_1 & + \cdots + & (ka_{i,n} + a_{j,n})x_n = kd_i + d_j \\ & \vdots & \\ a_{m,1}x_1 & + \cdots + & a_{m,n}x_n = d_m \end{array}$$

if and only if

$$\begin{aligned} a_{1,1}s_1 + \cdots + a_{1,n}s_n &= d_1 \\ &\vdots \\ \text{and } a_{i,1}s_1 + \cdots + a_{i,n}s_n &= d_i \\ &\vdots \\ \text{and } (ka_{i,1} + a_{j,1})s_1 + \cdots + (ka_{i,n} + a_{j,n})s_n &= kd_i + d_j \\ &\vdots \\ \text{and } a_{m,1}s_1 + a_{m,2}s_2 + \cdots + a_{m,n}s_n &= d_m \end{aligned}$$

again by the definition of 'satisfies'. Subtract k times the i-th equation from the j-th equation (remark: here is where we need $i \neq j$; if $i = j$ then the two d_i's above are not equal) to get that the previous compound statement holds if and only if

$$\begin{aligned} a_{1,1}s_1 + \cdots + a_{1,n}s_n &= d_1 \\ &\vdots \\ \text{and } a_{i,1}s_1 + \cdots + a_{i,n}s_n &= d_i \\ &\vdots \\ \text{and } (ka_{i,1} + a_{j,1})s_1 + \cdots + (ka_{i,n} + a_{j,n})s_n & \\ - (ka_{i,1}s_1 + \cdots + ka_{i,n}s_n) &= kd_i + d_j - kd_i \\ &\vdots \\ \text{and } a_{m,1}s_1 + \cdots + a_{m,n}s_n &= d_m \end{aligned}$$

which, after cancellation, says that $(s_1, \ldots, s_n)$ solves

$$\begin{aligned}
a_{1,1}x_1 + \cdots + a_{1,n}x_n &= d_1 \\
&\vdots \\
a_{i,1}x_1 + \cdots + a_{i,n}x_n &= d_i \\
&\vdots \\
a_{j,1}x_1 + \cdots + a_{j,n}x_n &= d_j \\
&\vdots \\
a_{m,1}x_1 + \cdots + a_{m,n}x_n &= d_m
\end{aligned}$$

as required.

One.I.1.32 Yes, this one-equation system:

$$0x + 0y = 0$$

is satisfied by every $(x, y) \in \mathbb{R}^2$.

One.I.1.33 Yes. This sequence of operations swaps rows i and j

$$\xrightarrow{\rho_i+\rho_j} \quad \xrightarrow{-\rho_j+\rho_i} \quad \xrightarrow{\rho_i+\rho_j} \quad \xrightarrow{-1\rho_i}$$

so the row-swap operation is redundant in the presence of the other two.

One.I.1.34 Reverse a row swap by swapping back.

$$\begin{array}{c} a_{1,1}x_1 + \cdots + a_{1,n}x_n = d_1 \\ \vdots \\ a_{m,1}x_1 + \cdots + a_{m,n}x_n = d_m \end{array} \xrightarrow{\rho_i\leftrightarrow\rho_j} \xrightarrow{\rho_j\leftrightarrow\rho_i} \begin{array}{c} a_{1,1}x_1 + \cdots + a_{1,n}x_n = d_1 \\ \vdots \\ a_{m,1}x_1 + \cdots + a_{m,n}x_n = d_m \end{array}$$

Multiplying both sides of a row by $k \neq 0$ is reversed by dividing by k.

$$\begin{array}{c} a_{1,1}x_1 + \cdots + a_{1,n}x_n = d_1 \\ \vdots \\ a_{m,1}x_1 + \cdots + a_{m,n}x_n = d_m \end{array} \xrightarrow{k\rho_i} \xrightarrow{(1/k)\rho_i} \begin{array}{c} a_{1,1}x_1 + \cdots + a_{1,n}x_n = d_1 \\ \vdots \\ a_{m,1}x_1 + \cdots + a_{m,n}x_n = d_m \end{array}$$

Adding k times a row to another is reversed by adding $-k$ times that row.

$$\begin{array}{c} a_{1,1}x_1 + \cdots + a_{1,n}x_n = d_1 \\ \vdots \\ a_{m,1}x_1 + \cdots + a_{m,n}x_n = d_m \end{array} \xrightarrow{k\rho_i+\rho_j} \xrightarrow{-k\rho_i+\rho_j} \begin{array}{c} a_{1,1}x_1 + \cdots + a_{1,n}x_n = d_1 \\ \vdots \\ a_{m,1}x_1 + \cdots + a_{m,n}x_n = d_m \end{array}$$

Remark: observe for the third case that if we were to allow $i = j$ then the result wouldn't hold.

$$3x + 2y = 7 \xrightarrow{2\rho_1+\rho_1} 9x + 6y = 21 \xrightarrow{-2\rho_1+\rho_1} -9x - 6y = -21$$

One.I.1.35 Let p, n, and d be the number of pennies, nickels, and dimes. For variables that are real numbers, this system

$$\begin{array}{rcl} p + n + d = 13 \\ p + 5n + 10d = 83 \end{array} \xrightarrow{-\rho_1+\rho_2} \begin{array}{rcl} p + n + d = 13 \\ 4n + 9d = 70 \end{array}$$

has more than one solution, in fact, infinitely many of them. However, it has a limited number of solutions in which p, n, and d are non-negative integers. Running through $d = 0, \ldots, d = 8$ shows that $(p, n, d) = (3, 4, 6)$ is the only solution using natural numbers.

One.I.1.36 Solving the system

$$\begin{aligned}(1/3)(a+b+c)+d&=29\\(1/3)(b+c+d)+a&=23\\(1/3)(c+d+a)+b&=21\\(1/3)(d+a+b)+c&=17\end{aligned}$$

we obtain $a=12$, $b=9$, $c=3$, $d=21$. Thus the second item, 21, is the correct answer.

One.I.1.37 *This is how the answer was given in the cited source.* A comparison of the units and hundreds columns of this addition shows that there must be a carry from the tens column. The tens column then tells us that $A<H$, so there can be no carry from the units or hundreds columns. The five columns then give the following five equations.

$$\begin{aligned}A+E&=W\\2H&=A+10\\H&=W+1\\H+T&=E+10\\A+1&=T\end{aligned}$$

The five linear equations in five unknowns, if solved simultaneously, produce the unique solution: $A=4$, $T=5$, $H=7$, $W=6$ and $E=2$, so that the original example in addition was $47474+5272=52746$.

One.I.1.38 *This is how the answer was given in the cited source.* Eight commissioners voted for B. To see this, we will use the given information to study how many voters chose each order of A, B, C.

The six orders of preference are ABC, ACB, BAC, BCA, CAB, CBA; assume they receive a, b, c, d, e, f votes respectively. We know that

$$\begin{aligned}a+b+e&=11\\d+e+f&=12\\a+c+d&=14\end{aligned}$$

from the number preferring A over B, the number preferring C over A, and the number preferring B over C. Because 20 votes were cast, we also know that

$$\begin{aligned}c+d+f&=9\\a+b+c&=8\\b+e+f&=6\end{aligned}$$

from the preferences for B over A, for A over C, and for C over B.

The solution is $a=6$, $b=1$, $c=1$, $d=7$, $e=4$, and $f=1$. The number of commissioners voting for B as their first choice is therefore $c+d=1+7=8$.

Comments. The answer to this question would have been the same had we known only that *at least* 14 commissioners preferred B over C.

The seemingly paradoxical nature of the commissioner's preferences (A is preferred to B, and B is preferred to C, and C is preferred to A), an example of "non-transitive dominance", is not uncommon when individual choices are pooled.

One.I.1.39 *This is how the answer was given in the cited source. We have not used "dependent" yet; it means here that Gauss's Method shows that there is not a unique solution.* If $n\geqslant 3$ the system is dependent and the solution is not unique. Hence $n<3$. But the term "system" implies $n>1$. Hence $n=2$. If the equations are

$$\begin{aligned}ax+(a+d)y&=a+2d\\(a+3d)x+(a+4d)y&=a+5d\end{aligned}$$

then $x=-1$, $y=2$.

One.I.2: Describing the Solution Set

One.I.2.15 **(a)** 2 **(b)** 3 **(c)** −1 **(d)** Not defined.

One.I.2.16 **(a)** 2×3 **(b)** 3×2 **(c)** 2×2

One.I.2.17 **(a)** $\begin{pmatrix}5\\1\\5\end{pmatrix}$ **(b)** $\begin{pmatrix}20\\-5\end{pmatrix}$ **(c)** $\begin{pmatrix}-2\\4\\0\end{pmatrix}$ **(d)** $\begin{pmatrix}41\\52\end{pmatrix}$ **(e)** Not defined. **(f)** $\begin{pmatrix}12\\8\\4\end{pmatrix}$

One.I.2.18 **(a)** This reduction

$$\left(\begin{array}{cc|c}3&6&18\\1&2&6\end{array}\right)\xrightarrow{(-1/3)\rho_1+\rho_2}\left(\begin{array}{cc|c}3&6&18\\0&0&0\end{array}\right)$$

leaves x leading and y free. Making y the parameter, we have $x=6-2y$ so the solution set is

$$\{\begin{pmatrix}6\\0\end{pmatrix}+\begin{pmatrix}-2\\1\end{pmatrix}y \mid y\in\mathbb{R}\}.$$

(b) This reduction

$$\left(\begin{array}{cc|c}1&1&1\\1&-1&-1\end{array}\right)\xrightarrow{-\rho_1+\rho_2}\left(\begin{array}{cc|c}1&1&1\\0&-2&-2\end{array}\right)$$

gives the unique solution $y=1$, $x=0$. The solution set is

$$\{\begin{pmatrix}0\\1\end{pmatrix}\}.$$

(c) This use of Gauss's Method

$$\left(\begin{array}{ccc|c}1&0&1&4\\1&-1&2&5\\4&-1&5&17\end{array}\right)\xrightarrow[-4\rho_1+\rho_3]{-\rho_1+\rho_2}\left(\begin{array}{ccc|c}1&0&1&4\\0&-1&1&1\\0&-1&1&1\end{array}\right)\xrightarrow{-\rho_2+\rho_3}\left(\begin{array}{ccc|c}1&0&1&4\\0&-1&1&1\\0&0&0&0\end{array}\right)$$

leaves x_1 and x_2 leading with x_3 free. The solution set is

$$\{\begin{pmatrix}4\\-1\\0\end{pmatrix}+\begin{pmatrix}-1\\1\\1\end{pmatrix}x_3 \mid x_3\in\mathbb{R}\}.$$

(d) This reduction

$$\left(\begin{array}{ccc|c}2&1&-1&2\\2&0&1&3\\1&-1&0&0\end{array}\right)\xrightarrow[-(1/2)\rho_1+\rho_3]{-\rho_1+\rho_2}\left(\begin{array}{ccc|c}2&1&-1&2\\0&-1&2&1\\0&-3/2&1/2&-1\end{array}\right)\xrightarrow{(-3/2)\rho_2+\rho_3}\left(\begin{array}{ccc|c}2&1&-1&2\\0&-1&2&1\\0&0&-5/2&-5/2\end{array}\right)$$

shows that the solution set is a singleton set.

$$\{\begin{pmatrix}1\\1\\1\end{pmatrix}\}$$

(e) This reduction is easy

$$\left(\begin{array}{cccc|c}1&2&-1&0&3\\2&1&0&1&4\\1&-1&1&1&1\end{array}\right)\xrightarrow[-\rho_1+\rho_3]{-2\rho_1+\rho_2}\left(\begin{array}{cccc|c}1&2&-1&0&3\\0&-3&2&1&-2\\0&-3&2&1&-2\end{array}\right)\xrightarrow{-\rho_2+\rho_3}\left(\begin{array}{cccc|c}1&2&-1&0&3\\0&-3&2&1&-2\\0&0&0&0&0\end{array}\right)$$

and ends with x and y leading, while z and w are free. Solving for y gives $y=(2+2z+w)/3$ and substitution shows that $x+2(2+2z+w)/3-z=3$ so $x=(5/3)-(1/3)z-(2/3)w$, making the solution set

$$\{\begin{pmatrix}5/3\\2/3\\0\\0\end{pmatrix}+\begin{pmatrix}-1/3\\2/3\\1\\0\end{pmatrix}z+\begin{pmatrix}-2/3\\1/3\\0\\1\end{pmatrix}w \mid z,w\in\mathbb{R}\}.$$

(f) The reduction

$$\left(\begin{array}{cccc|c}1&0&1&1&4\\2&1&0&-1&2\\3&1&1&0&7\end{array}\right)\xrightarrow[-3\rho_1+\rho_3]{-2\rho_1+\rho_2}\left(\begin{array}{cccc|c}1&0&1&1&4\\0&1&-2&-3&-6\\0&1&-2&-3&-5\end{array}\right)\xrightarrow{-\rho_2+\rho_3}\left(\begin{array}{cccc|c}1&0&1&1&4\\0&1&-2&-3&-6\\0&0&0&0&1\end{array}\right)$$

shows that there is no solution — the solution set is empty.

One.I.2.19 **(a)** This reduction

$$\left(\begin{array}{ccc|c}2&1&-1&1\\4&-1&0&3\end{array}\right)\xrightarrow{-2\rho_1+\rho_2}\left(\begin{array}{ccc|c}2&1&-1&1\\0&-3&2&1\end{array}\right)$$

ends with x and y leading while z is free. Solving for y gives $y=(1-2z)/(-3)$, and then substitution $2x+(1-2z)/(-3)-z=1$ shows that $x=((4/3)+(1/3)z)/2$. Hence the solution set is

$$\{\begin{pmatrix}2/3\\-1/3\\0\end{pmatrix}+\begin{pmatrix}1/6\\2/3\\1\end{pmatrix}z \mid z\in\mathbb{R}\}.$$

(b) This application of Gauss's Method

$$\left(\begin{array}{cccc|c}1&0&-1&0&1\\0&1&2&-1&3\\1&2&3&-1&7\end{array}\right)\xrightarrow{-\rho_1+\rho_3}\left(\begin{array}{cccc|c}1&0&-1&0&1\\0&1&2&-1&3\\0&2&4&-1&6\end{array}\right)\xrightarrow{-2\rho_2+\rho_3}\left(\begin{array}{cccc|c}1&0&-1&0&1\\0&1&2&-1&3\\0&0&0&1&0\end{array}\right)$$

leaves x, y, and w leading. The solution set is

$$\{\begin{pmatrix}1\\3\\0\\0\end{pmatrix}+\begin{pmatrix}1\\-2\\1\\0\end{pmatrix}z \mid z\in\mathbb{R}\}.$$

(c) This row reduction

$$\left(\begin{array}{cccc|c}1&-1&1&0&0\\0&1&0&1&0\\3&-2&3&1&0\\0&-1&0&-1&0\end{array}\right)\xrightarrow{-3\rho_1+\rho_3}\left(\begin{array}{cccc|c}1&-1&1&0&0\\0&1&0&1&0\\0&1&0&1&0\\0&-1&0&-1&0\end{array}\right)\xrightarrow[\rho_2+\rho_4]{-\rho_2+\rho_3}\left(\begin{array}{cccc|c}1&-1&1&0&0\\0&1&0&1&0\\0&0&0&0&0\\0&0&0&0&0\end{array}\right)$$

ends with z and w free. The solution set is

$$\{\begin{pmatrix}0\\0\\0\\0\end{pmatrix}+\begin{pmatrix}-1\\0\\1\\0\end{pmatrix}z+\begin{pmatrix}-1\\-1\\0\\1\end{pmatrix}w \mid z,w\in\mathbb{R}\}.$$

(d) Gauss's Method done in this way

$$\left(\begin{array}{ccccc|c}1&2&3&1&-1&1\\3&-1&1&1&1&3\end{array}\right)\xrightarrow{-3\rho_1+\rho_2}\left(\begin{array}{ccccc|c}1&2&3&1&-1&1\\0&-7&-8&-2&4&0\end{array}\right)$$

ends with c, d, and e free. Solving for b shows that $b=(8c+2d-4e)/(-7)$ and then substitution $a+2(8c+2d-4e)/(-7)+3c+1d-1e=1$ shows that $a=1-(5/7)c-(3/7)d-(1/7)e$ and so the solution set is

$$\{\begin{pmatrix}1\\0\\0\\0\\0\end{pmatrix}+\begin{pmatrix}-5/7\\-8/7\\1\\0\\0\end{pmatrix}c+\begin{pmatrix}-3/7\\-2/7\\0\\1\\0\end{pmatrix}d+\begin{pmatrix}-1/7\\4/7\\0\\0\\1\end{pmatrix}e \mid c,d,e\in\mathbb{R}\}.$$

One.I.2.20 For each problem we get a system of linear equations by looking at the equations of components.

(a) $k=5$

(b) The second components show that $i = 2$, the third components show that $j = 1$.

(c) $m = -4$, $n = 2$

One.I.2.21 For each problem we get a system of linear equations by looking at the equations of components.

(a) Yes; take $k = -1/2$.

(b) No; the system with equations $5 = 5 \cdot j$ and $4 = -4 \cdot j$ has no solution.

(c) Yes; take $r = 2$.

(d) No. The second components give $k = 0$. Then the third components give $j = 1$. But the first components don't check.

One.I.2.22 **(a)** Let c be the number of acres of corn, s be the number of acres of soy, and a be the number of acres of oats.

$$\begin{array}{rcl} c + s + a &=& 1200 \\ 20c + 50s + 12a &=& 40\,000 \end{array} \xrightarrow{-20\rho_1+\rho_2} \begin{array}{rcl} c + s + a &=& 1200 \\ 30s - 8a &=& 16\,000 \end{array}$$

To describe the solution set we can parametrize using a.

$$\{\begin{pmatrix} c \\ s \\ a \end{pmatrix} = \begin{pmatrix} 20\,000/30 \\ 16\,000/30 \\ 0 \end{pmatrix} + \begin{pmatrix} -38/30 \\ 8/30 \\ 1 \end{pmatrix} a \mid a \in \mathbb{R}\}$$

(b) There are many answers possible here. For instance we can take $a = 0$ to get $c = 20\,000/30 \approx 666.66$ and $s = 16000/30 \approx 533.33$. Another example is to take $a = 20\,000/38 \approx 526.32$, giving $c = 0$ and $s = 7360/38 \approx 193.68$.

(c) Plug your answers from the prior part into $100c + 300s + 80a$.

One.I.2.23 This system has 1 equation. The leading variable is x_1, the other variables are free.

$$\{\begin{pmatrix} -1 \\ 1 \\ \vdots \\ 0 \end{pmatrix} x_2 + \cdots + \begin{pmatrix} -1 \\ 0 \\ \vdots \\ 1 \end{pmatrix} x_n \mid x_2, \ldots, x_n \in \mathbb{R}\}$$

One.I.2.24 **(a)** Gauss's Method here gives

$$\left(\begin{array}{cccc|c} 1 & 2 & 0 & -1 & a \\ 2 & 0 & 1 & 0 & b \\ 1 & 1 & 0 & 2 & c \end{array}\right) \xrightarrow[-\rho_1+\rho_3]{-2\rho_1+\rho_2} \left(\begin{array}{cccc|c} 1 & 2 & 0 & -1 & a \\ 0 & -4 & 1 & 2 & -2a + b \\ 0 & -1 & 0 & 3 & -a + c \end{array}\right)$$

$$\xrightarrow{-(1/4)\rho_2+\rho_3} \left(\begin{array}{cccc|c} 1 & 2 & 0 & -1 & a \\ 0 & -4 & 1 & 2 & -2a + b \\ 0 & 0 & -1/4 & 5/2 & -(1/2)a - (1/4)b + c \end{array}\right),$$

leaving w free. Solve: $z = 2a + b - 4c + 10w$, and $-4y = -2a + b - (2a + b - 4c + 10w) - 2w$ so $y = a - c + 3w$, and $x = a - 2(a - c + 3w) + w = -a + 2c - 5w$. Therefore the solution set is this.

$$\{\begin{pmatrix} -a + 2c \\ a - c \\ 2a + b - 4c \\ 0 \end{pmatrix} + \begin{pmatrix} -5 \\ 3 \\ 10 \\ 1 \end{pmatrix} w \mid w \in \mathbb{R}\}$$

(b) Plug in with $a = 3$, $b = 1$, and $c = -2$.

$$\{\begin{pmatrix} -7 \\ 5 \\ 15 \\ 0 \end{pmatrix} + \begin{pmatrix} -5 \\ 3 \\ 10 \\ 1 \end{pmatrix} w \mid w \in \mathbb{R}\}$$

One.I.2.25 Leaving the comma out, say by writing a_{123}, is ambiguous because it could mean $a_{1,23}$ or $a_{12,3}$.

One.I.2.26 **(a)** $\begin{pmatrix} 2 & 3 & 4 & 5 \\ 3 & 4 & 5 & 6 \\ 4 & 5 & 6 & 7 \\ 5 & 6 & 7 & 8 \end{pmatrix}$ **(b)** $\begin{pmatrix} 1 & -1 & 1 & -1 \\ -1 & 1 & -1 & 1 \\ 1 & -1 & 1 & -1 \\ -1 & 1 & -1 & 1 \end{pmatrix}$

One.I.2.27 **(a)** $\begin{pmatrix} 1 & 4 \\ 2 & 5 \\ 3 & 6 \end{pmatrix}$ **(b)** $\begin{pmatrix} 2 & 1 \\ -3 & 1 \end{pmatrix}$ **(c)** $\begin{pmatrix} 5 & 10 \\ 10 & 5 \end{pmatrix}$ **(d)** $\begin{pmatrix} 1 & 1 & 0 \end{pmatrix}$

One.I.2.28 **(a)** Plugging in $x = 1$ and $x = -1$ gives

$$\begin{aligned} a + b + c &= 2 \\ a - b + c &= 6 \end{aligned} \xrightarrow{-\rho_1+\rho_2} \begin{aligned} a + b + c &= 2 \\ -2b &= 4 \end{aligned}$$

so the set of functions is $\{f(x) = (4-c)x^2 - 2x + c \mid c \in \mathbb{R}\}$.

(b) Putting in $x = 1$ gives

$$a + b + c = 2$$

so the set of functions is $\{f(x) = (2-b-c)x^2 + bx + c \mid b, c \in \mathbb{R}\}$.

One.I.2.29 On plugging in the five pairs (x, y) we get a system with the five equations and six unknowns a, ..., f. Because there are more unknowns than equations, if no inconsistency exists among the equations then there are infinitely many solutions (at least one variable will end up free).

But no inconsistency can exist because $a = 0$, ..., $f = 0$ is a solution (we are only using this zero solution to show that the system is consistent — the prior paragraph shows that there are nonzero solutions).

One.I.2.30 **(a)** Here is one — the fourth equation is redundant but still OK.

$$\begin{aligned} x + y - z + w &= 0 \\ y - z &= 0 \\ 2z + 2w &= 0 \\ z + w &= 0 \end{aligned}$$

(b) Here is one.

$$\begin{aligned} x + y - z + w &= 0 \\ w &= 0 \\ w &= 0 \\ w &= 0 \end{aligned}$$

(c) This is one.

$$\begin{aligned} x + y - z + w &= 0 \\ x + y - z + w &= 0 \\ x + y - z + w &= 0 \\ x + y - z + w &= 0 \end{aligned}$$

One.I.2.31 *This is how the answer was given in the cited source.* My solution was to define the numbers of arbuzoids as 3-dimensional vectors, and express all possible elementary transitions as such vectors, too:

$$\begin{array}{ll} \text{R:} & 13 \\ \text{G:} & 15 \\ \text{B:} & 17 \end{array} \qquad \text{Operations: } \begin{pmatrix} -1 \\ -1 \\ 2 \end{pmatrix}, \begin{pmatrix} -1 \\ 2 \\ -1 \end{pmatrix}, \text{ and } \begin{pmatrix} 2 \\ -1 \\ -1 \end{pmatrix}$$

Now, it is enough to check whether the solution to one of the following systems of linear equations exists:

$$\begin{pmatrix} 13 \\ 15 \\ 17 \end{pmatrix} + x \begin{pmatrix} -1 \\ -1 \\ 2 \end{pmatrix} + y \begin{pmatrix} -1 \\ 2 \\ -1 \end{pmatrix} + \begin{pmatrix} 2 \\ -1 \\ -1 \end{pmatrix} = \begin{pmatrix} 0 \\ 0 \\ 45 \end{pmatrix} \quad (\text{or } \begin{pmatrix} 0 \\ 45 \\ 0 \end{pmatrix} \text{ or } \begin{pmatrix} 45 \\ 0 \\ 0 \end{pmatrix})$$

Solving

$$\left(\begin{array}{ccc|c} -1 & -1 & 2 & -13 \\ -1 & 2 & -1 & -15 \\ 2 & -1 & -1 & 28 \end{array}\right) \xrightarrow[2\rho_1+\rho_3]{-\rho_1+\rho_2} \xrightarrow{\rho_2+\rho_3} \left(\begin{array}{ccc|c} -1 & -1 & 2 & -13 \\ 0 & 3 & -3 & -2 \\ 0 & 0 & 0 & 0 \end{array}\right)$$

gives $y + 2/3 = z$ so if the number of transformations z is an integer then y is not. The other two systems give similar conclusions so there is no solution.

One.I.2.32 *This is how the answer was given in the cited source.*

(a) Formal solution of the system yields

$$x = \frac{a^3 - 1}{a^2 - 1} \qquad y = \frac{-a^2 + a}{a^2 - 1}.$$

If $a + 1 \neq 0$ and $a - 1 \neq 0$, then the system has the single solution

$$x = \frac{a^2 + a + 1}{a + 1} \qquad y = \frac{-a}{a + 1}.$$

If $a = -1$, or if $a = +1$, then the formulas are meaningless; in the first instance we arrive at the system

$$\begin{cases} -x + y = 1 \\ x - y = 1 \end{cases}$$

which is a contradictory system. In the second instance we have

$$\begin{cases} x + y = 1 \\ x + y = 1 \end{cases}$$

which has an infinite number of solutions (for example, for x arbitrary, $y = 1 - x$).

(b) Solution of the system yields

$$x = \frac{a^4 - 1}{a^2 - 1} \qquad y = \frac{-a^3 + a}{a^2 - 1}.$$

Here, is $a^2 - 1 \neq 0$, the system has the single solution $x = a^2 + 1$, $y = -a$. For $a = -1$ and $a = 1$, we obtain the systems

$$\begin{cases} -x + y = -1 \\ x - y = 1 \end{cases} \qquad \begin{cases} x + y = 1 \\ x + y = 1 \end{cases}$$

both of which have an infinite number of solutions.

One.I.2.33 *This is how the answer was given in the cited source.* Let u, v, x, y, z be the volumes in cm^3 of Al, Cu, Pb, Ag, and Au, respectively, contained in the sphere, which we assume to be not hollow. Since the loss of weight in water (specific gravity 1.00) is 1000 grams, the volume of the sphere is 1000 cm^3. Then the data, some of which is superfluous, though consistent, leads to only 2 independent equations, one relating volumes and the other, weights.

$$\begin{aligned} u + v + x + y + z &= 1000 \\ 2.7u + 8.9v + 11.3x + 10.5y + 19.3z &= 7558 \end{aligned}$$

Clearly the sphere must contain some aluminum to bring its mean specific gravity below the specific gravities of all the other metals. There is no unique result to this part of the problem, for the amounts of three metals may be chosen arbitrarily, provided that the choices will not result in negative amounts of any metal.

If the ball contains only aluminum and gold, there are 294.5 cm^3 of gold and 705.5 cm^3 of aluminum. Another possibility is 124.7 cm^3 each of Cu, Au, Pb, and Ag and 501.2 cm^3 of Al.

One.I.3: General = Particular + Homogeneous

One.I.3.14 For the arithmetic to these, see the answers from the prior subsection.

(a) This is the solution set

$$S = \{ \begin{pmatrix} 6 \\ 0 \end{pmatrix} + \begin{pmatrix} -2 \\ 1 \end{pmatrix} y \mid y \in \mathbb{R} \}$$

Here are the particular solution and the solution set for the associated homogeneous system.

$$\begin{pmatrix}6\\0\end{pmatrix} \quad\text{and}\quad \{\begin{pmatrix}-2\\1\end{pmatrix}y \mid y\in\mathbb{R}\}$$

Comment. Students are sometimes confused on two points here. First, the set S given above is equal to this set

$$T=\{\begin{pmatrix}4\\1\end{pmatrix}+\begin{pmatrix}-2\\1\end{pmatrix}y \mid y\in\mathbb{R}\}$$

because the two sets contain the same members. All of these are correct answers to "What is a particular solution?"

$$\begin{pmatrix}6\\0\end{pmatrix},\ \begin{pmatrix}4\\1\end{pmatrix},\ \begin{pmatrix}2\\2\end{pmatrix},\ \begin{pmatrix}1\\2.5\end{pmatrix}$$

The second point of confustion is that the letter we use in the set doesn't matter. This set also equals S.

$$U=\{\begin{pmatrix}6\\0\end{pmatrix}+\begin{pmatrix}-2\\1\end{pmatrix}u \mid u\in\mathbb{R}\}$$

(b) The solution set is

$$\{\begin{pmatrix}0\\1\end{pmatrix}\}.$$

The particular solution and the solution set for the associated homogeneous system are

$$\begin{pmatrix}0\\1\end{pmatrix} \quad\text{and}\quad \{\begin{pmatrix}0\\0\end{pmatrix}\}$$

(c) The solution set is

$$\{\begin{pmatrix}4\\-1\\0\end{pmatrix}+\begin{pmatrix}-1\\1\\1\end{pmatrix}x_3 \mid x_3\in\mathbb{R}\}.$$

A particular solution and the solution set for the associated homogeneous system are

$$\begin{pmatrix}4\\-1\\0\end{pmatrix} \quad\text{and}\quad \{\begin{pmatrix}-1\\1\\1\end{pmatrix}x_3 \mid x_3\in\mathbb{R}\}.$$

(d) The solution set is a singleton

$$\{\begin{pmatrix}1\\1\\1\end{pmatrix}\}.$$

A particular solution and the solution set for the associated homogeneous system are here.

$$\begin{pmatrix}1\\1\\1\end{pmatrix} \quad \{\begin{pmatrix}0\\0\\0\end{pmatrix}\}$$

(e) The solution set is

$$\{\begin{pmatrix}5/3\\2/3\\0\\0\end{pmatrix}+\begin{pmatrix}-1/3\\2/3\\1\\0\end{pmatrix}z+\begin{pmatrix}-2/3\\1/3\\0\\1\end{pmatrix}w \mid z,w\in\mathbb{R}\}.$$

A particular solution and the solution set for the associated homogeneous system are

$$\begin{pmatrix}5/3\\2/3\\0\\0\end{pmatrix} \quad\text{and}\quad \{\begin{pmatrix}-1/3\\2/3\\1\\0\end{pmatrix}z+\begin{pmatrix}-2/3\\1/3\\0\\1\end{pmatrix}w \mid z,w\in\mathbb{R}\}.$$

(f) This system's solution set is empty. Thus, there is no particular solution. The solution set of the associated homogeneous system is

$$\{\begin{pmatrix}-1\\2\\1\\0\end{pmatrix}z+\begin{pmatrix}-1\\3\\0\\1\end{pmatrix}w \mid z,w\in\mathbb{R}\}.$$

One.I.3.15 The answers from the prior subsection show the row operations.

(a) The solution set is

$$\{\begin{pmatrix}2/3\\-1/3\\0\end{pmatrix}+\begin{pmatrix}1/6\\2/3\\1\end{pmatrix}z \mid z\in\mathbb{R}\}.$$

A particular solution and the solution set for the associated homogeneous system are

$$\begin{pmatrix}2/3\\-1/3\\0\end{pmatrix}\quad\text{and}\quad\{\begin{pmatrix}1/6\\2/3\\1\end{pmatrix}z \mid z\in\mathbb{R}\}.$$

(b) The solution set is

$$\{\begin{pmatrix}1\\3\\0\\0\end{pmatrix}+\begin{pmatrix}1\\-2\\1\\0\end{pmatrix}z \mid z\in\mathbb{R}\}.$$

A particular solution and the solution set for the associated homogeneous system are

$$\begin{pmatrix}1\\3\\0\\0\end{pmatrix}\quad\text{and}\quad\{\begin{pmatrix}1\\-2\\1\\0\end{pmatrix}z \mid z\in\mathbb{R}\}.$$

(c) The solution set is

$$\{\begin{pmatrix}0\\0\\0\\0\end{pmatrix}+\begin{pmatrix}-1\\0\\1\\0\end{pmatrix}z+\begin{pmatrix}-1\\-1\\0\\1\end{pmatrix}w \mid z,w\in\mathbb{R}\}.$$

A particular solution and the solution set for the associated homogeneous system are

$$\begin{pmatrix}0\\0\\0\\0\end{pmatrix}\quad\text{and}\quad\{\begin{pmatrix}-1\\0\\1\\0\end{pmatrix}z+\begin{pmatrix}-1\\-1\\0\\1\end{pmatrix}w \mid z,w\in\mathbb{R}\}.$$

(d) The solution set is

$$\{\begin{pmatrix}1\\0\\0\\0\\0\end{pmatrix}+\begin{pmatrix}-5/7\\-8/7\\1\\0\\0\end{pmatrix}c+\begin{pmatrix}-3/7\\-2/7\\0\\1\\0\end{pmatrix}d+\begin{pmatrix}-1/7\\4/7\\0\\0\\1\end{pmatrix}e \mid c,d,e\in\mathbb{R}\}.$$

A particular solution and the solution set for the associated homogeneous system are

$$\begin{pmatrix}1\\0\\0\\0\\0\end{pmatrix}\quad\text{and}\quad\{\begin{pmatrix}-5/7\\-8/7\\1\\0\\0\end{pmatrix}c+\begin{pmatrix}-3/7\\-2/7\\0\\1\\0\end{pmatrix}d+\begin{pmatrix}-1/7\\4/7\\0\\0\\1\end{pmatrix}e \mid c,d,e\in\mathbb{R}\}.$$

One.I.3.16 Just plug them in and see if they satisfy all three equations.

(a) No.

(b) Yes.

(c) Yes.

One.I.3.17 Gauss's Method on the associated homogeneous system gives

$$\left(\begin{array}{cccc|c}1 & -1 & 0 & 1 & 0\\ 2 & 3 & -1 & 0 & 0\\ 0 & 1 & 1 & 1 & 0\end{array}\right) \xrightarrow{-2\rho_1+\rho_2} \left(\begin{array}{cccc|c}1 & -1 & 0 & 1 & 0\\ 0 & 5 & -1 & -2 & 0\\ 0 & 1 & 1 & 1 & 0\end{array}\right) \xrightarrow{-(1/5)\rho_2+\rho_3} \left(\begin{array}{cccc|c}1 & -1 & 0 & 1 & 0\\ 0 & 5 & -1 & -2 & 0\\ 0 & 0 & 6/5 & 7/5 & 0\end{array}\right)$$

so this is the solution to the homogeneous problem:

$$\{\begin{pmatrix}-5/6\\ 1/6\\ -7/6\\ 1\end{pmatrix} w \mid w \in \mathbb{R}\}.$$

(a) That vector is indeed a particular solution, so the required general solution is

$$\{\begin{pmatrix}0\\ 0\\ 0\\ 4\end{pmatrix} + \begin{pmatrix}-5/6\\ 1/6\\ -7/6\\ 1\end{pmatrix} w \mid w \in \mathbb{R}\}.$$

(b) That vector is a particular solution so the required general solution is

$$\{\begin{pmatrix}-5\\ 1\\ -7\\ 10\end{pmatrix} + \begin{pmatrix}-5/6\\ 1/6\\ -7/6\\ 1\end{pmatrix} w \mid w \in \mathbb{R}\}.$$

(c) That vector is not a solution of the system since it does not satisfy the third equation. No such general solution exists.

One.I.3.18 The first is nonsingular while the second is singular. Just do Gauss's Method and see if the echelon form result has non-0 numbers in each entry on the diagonal.

One.I.3.19 **(a)** Nonsingular:

$$\xrightarrow{-\rho_1+\rho_2} \begin{pmatrix}1 & 2\\ 0 & 1\end{pmatrix}$$

ends with each row containing a leading entry.

(b) Singular:

$$\xrightarrow{3\rho_1+\rho_2} \begin{pmatrix}1 & 2\\ 0 & 0\end{pmatrix}$$

ends with row 2 without a leading entry.

(c) Neither. A matrix must be square for either word to apply.

(d) Singular.

(e) Nonsingular.

One.I.3.20 In each case we must decide if the vector is a linear combination of the vectors in the set.

(a) Yes. Solve

$$c_1 \begin{pmatrix}1\\ 4\end{pmatrix} + c_2 \begin{pmatrix}1\\ 5\end{pmatrix} = \begin{pmatrix}2\\ 3\end{pmatrix}$$

with

$$\left(\begin{array}{cc|c}1 & 1 & 2\\ 4 & 5 & 3\end{array}\right) \xrightarrow{-4\rho_1+\rho_2} \left(\begin{array}{cc|c}1 & 1 & 2\\ 0 & 1 & -5\end{array}\right)$$

to conclude that there are c_1 and c_2 giving the combination.

(b) No. The reduction

$$\left(\begin{array}{cc|c}2&1&-1\\1&0&0\\0&1&1\end{array}\right)\xrightarrow{-(1/2)\rho_1+\rho_2}\left(\begin{array}{cc|c}2&1&-1\\0&-1/2&1/2\\0&1&1\end{array}\right)\xrightarrow{2\rho_2+\rho_3}\left(\begin{array}{cc|c}2&1&-1\\0&-1/2&1/2\\0&0&2\end{array}\right)$$

shows that

$$c_1\begin{pmatrix}2\\1\\0\end{pmatrix}+c_2\begin{pmatrix}1\\0\\1\end{pmatrix}=\begin{pmatrix}-1\\0\\1\end{pmatrix}$$

has no solution.

(c) Yes. The reduction

$$\left(\begin{array}{cccc|c}1&2&3&4&1\\0&1&3&2&3\\4&5&0&1&0\end{array}\right)\xrightarrow{-4\rho_1+\rho_3}\left(\begin{array}{cccc|c}1&2&3&4&1\\0&1&3&2&3\\0&-3&-12&-15&-4\end{array}\right)\xrightarrow{3\rho_2+\rho_3}\left(\begin{array}{cccc|c}1&2&3&4&1\\0&1&3&2&3\\0&0&-3&-9&5\end{array}\right)$$

shows that there are infinitely many ways

$$\{\begin{pmatrix}c_1\\c_2\\c_3\\c_4\end{pmatrix}=\begin{pmatrix}-10\\8\\-5/3\\0\end{pmatrix}+\begin{pmatrix}-9\\7\\-3\\1\end{pmatrix}c_4\mid c_4\in\mathbb{R}\}$$

to write

$$\begin{pmatrix}1\\3\\0\end{pmatrix}=c_1\begin{pmatrix}1\\0\\4\end{pmatrix}+c_2\begin{pmatrix}2\\1\\5\end{pmatrix}+c_3\begin{pmatrix}3\\3\\0\end{pmatrix}+c_4\begin{pmatrix}4\\2\\1\end{pmatrix}.$$

(d) No. Look at the third components.

One.I.3.21 Because the matrix of coefficients is nonsingular, Gauss's Method ends with an echelon form where each variable leads an equation. Back substitution gives a unique solution.

(Another way to see the solution is unique is to note that with a nonsingular matrix of coefficients the associated homogeneous system has a unique solution, by definition. Since the general solution is the sum of a particular solution with each homogeneous solution, the general solution has (at most) one element.)

One.I.3.22 In this case the solution set is all of $\mathbb{R}^n$ and we can express it in the required form

$$\{c_1\begin{pmatrix}1\\0\\\vdots\\0\end{pmatrix}+c_2\begin{pmatrix}0\\1\\\vdots\\0\end{pmatrix}+\cdots+c_n\begin{pmatrix}0\\0\\\vdots\\1\end{pmatrix}\mid c_1,\ldots,c_n\in\mathbb{R}\}.$$

One.I.3.23 Assume $\vec{s},\vec{t}\in\mathbb{R}^n$ and write

$$\vec{s}=\begin{pmatrix}s_1\\\vdots\\s_n\end{pmatrix}\quad\text{and}\quad\vec{t}=\begin{pmatrix}t_1\\\vdots\\t_n\end{pmatrix}.$$

Also let $a_{i,1}x_1+\cdots+a_{i,n}x_n=0$ be the i-th equation in the homogeneous system.

(a) The check is easy:

$$\begin{aligned}a_{i,1}(s_1+t_1)+\cdots+a_{i,n}(s_n+t_n)&=(a_{i,1}s_1+\cdots+a_{i,n}s_n)+(a_{i,1}t_1+\cdots+a_{i,n}t_n)\\&=0+0.\end{aligned}$$

(b) This one is similar:

$$a_{i,1}(3s_1)+\cdots+a_{i,n}(3s_n)=3(a_{i,1}s_1+\cdots+a_{i,n}s_n)=3\cdot0=0.$$

(c) This one is not much harder:

$$\begin{aligned}a_{i,1}(ks_1+mt_1)+\cdots+a_{i,n}(ks_n+mt_n)&=k(a_{i,1}s_1+\cdots+a_{i,n}s_n)+m(a_{i,1}t_1+\cdots+a_{i,n}t_n)\\&=k\cdot0+m\cdot0.\end{aligned}$$

What is wrong with that argument is that any linear combination of the zero vector yields the zero vector again.

One.I.3.24 First the proof.

Gauss's Method will use only rationals (e.g., $-(m/n)\rho_i + \rho_j$). Thus we can express the solution set using only rational numbers as the components of each vector. Now the particular solution is all rational.

There are infinitely many (rational vector) solutions if and only if the associated homogeneous system has infinitely many (real vector) solutions. That's because setting any parameters to be rationals will produce an all-rational solution.

Linear Geometry

One.II.1: Vectors in Space

One.II.1.1 (a) $\begin{pmatrix}2\\1\end{pmatrix}$ (b) $\begin{pmatrix}-1\\2\end{pmatrix}$ (c) $\begin{pmatrix}4\\0\\-3\end{pmatrix}$ (d) $\begin{pmatrix}0\\0\\0\end{pmatrix}$

One.II.1.2 (a) No, their canonical positions are different.

$$\begin{pmatrix}1\\-1\end{pmatrix} \qquad \begin{pmatrix}0\\3\end{pmatrix}$$

(b) Yes, their canonical positions are the same.

$$\begin{pmatrix}1\\-1\\3\end{pmatrix}$$

One.II.1.3 That line is this set.

$$\{\begin{pmatrix}-2\\1\\1\\0\end{pmatrix} + \begin{pmatrix}7\\9\\-2\\4\end{pmatrix}t \mid t \in \mathbb{R}\}$$

Note that this system

$$\begin{aligned}-2+7t&=1\\1+9t&=0\\1-2t&=2\\0+4t&=1\end{aligned}$$

has no solution. Thus the given point is not in the line.

One.II.1.4 (a) Note that

$$\begin{pmatrix}2\\2\\2\\0\end{pmatrix} - \begin{pmatrix}1\\1\\5\\-1\end{pmatrix} = \begin{pmatrix}1\\1\\-3\\1\end{pmatrix} \qquad \begin{pmatrix}3\\1\\0\\4\end{pmatrix} - \begin{pmatrix}1\\1\\5\\-1\end{pmatrix} = \begin{pmatrix}2\\0\\-5\\5\end{pmatrix}$$

and so the plane is this set.

$$\{\begin{pmatrix}1\\1\\5\\-1\end{pmatrix} + \begin{pmatrix}1\\1\\-3\\1\end{pmatrix}t + \begin{pmatrix}2\\0\\-5\\5\end{pmatrix}s \mid t, s \in \mathbb{R}\}$$

(b) No; this system

$$\begin{aligned} 1+1t+2s&=0\\ 1+1t&=0\\ 5-3t-5s&=0\\ -1+1t+5s&=0 \end{aligned}$$

has no solution.

One.II.1.5 The vector

$$\begin{pmatrix}2\\0\\3\end{pmatrix}$$

is not in the line. Because

$$\begin{pmatrix}2\\0\\3\end{pmatrix}-\begin{pmatrix}-1\\0\\-4\end{pmatrix}=\begin{pmatrix}3\\0\\7\end{pmatrix}$$

we can describe that plane in this way.

$$\{\begin{pmatrix}-1\\0\\-4\end{pmatrix}+m\begin{pmatrix}1\\1\\2\end{pmatrix}+n\begin{pmatrix}3\\0\\7\end{pmatrix}\mid m,n\in\mathbb{R}\}$$

One.II.1.6 The points of coincidence are solutions of this system.

$$\begin{aligned} t&=1+2m\\ t+s&=1+3k\\ t+3s&=4m \end{aligned}$$

Gauss's Method

$$\left(\begin{array}{cccc|c}1&0&0&-2&1\\1&1&-3&0&1\\1&3&0&-4&0\end{array}\right)\xrightarrow[-\rho_1+\rho_3]{-\rho_1+\rho_2}\left(\begin{array}{cccc|c}1&0&0&-2&1\\0&1&-3&2&0\\0&3&0&-2&-1\end{array}\right)\xrightarrow{-3\rho_2+\rho_3}\left(\begin{array}{cccc|c}1&0&0&-2&1\\0&1&-3&2&0\\0&0&9&-8&-1\end{array}\right)$$

gives $k=-(1/9)+(8/9)m$, so $s=-(1/3)+(2/3)m$ and $t=1+2m$. The intersection is this.

$$\{\begin{pmatrix}1\\1\\0\end{pmatrix}+\begin{pmatrix}0\\3\\0\end{pmatrix}(-\tfrac{1}{9}+\tfrac{8}{9}m)+\begin{pmatrix}2\\0\\4\end{pmatrix}m\mid m\in\mathbb{R}\}=\{\begin{pmatrix}1\\2/3\\0\end{pmatrix}+\begin{pmatrix}2\\8/3\\4\end{pmatrix}m\mid m\in\mathbb{R}\}$$

One.II.1.7 **(a)** The system

$$\begin{aligned} 1&=1\\ 1+t&=3+s\\ 2+t&=-2+2s \end{aligned}$$

gives $s=6$ and $t=8$, so this is the solution set.

$$\{\begin{pmatrix}1\\9\\10\end{pmatrix}\}$$

(b) This system

$$\begin{aligned} 2+t&=0\\ t&=s+4w\\ 1-t&=2s+w \end{aligned}$$

gives $t=-2$, $w=-1$, and $s=2$ so their intersection is this point.

$$\begin{pmatrix}0\\-2\\3\end{pmatrix}$$

One.II.1.8 (a) The vector shown

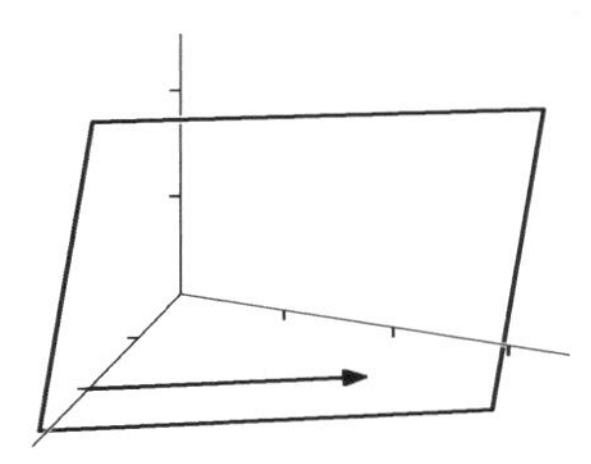

is not the result of doubling

$$\begin{pmatrix}2\\0\\0\end{pmatrix}+\begin{pmatrix}-0.5\\1\\0\end{pmatrix}\cdot 1$$

instead it is

$$\begin{pmatrix}2\\0\\0\end{pmatrix}+\begin{pmatrix}-0.5\\1\\0\end{pmatrix}\cdot 2=\begin{pmatrix}1\\2\\0\end{pmatrix}$$

which has a parameter twice as large.

(b) The vector

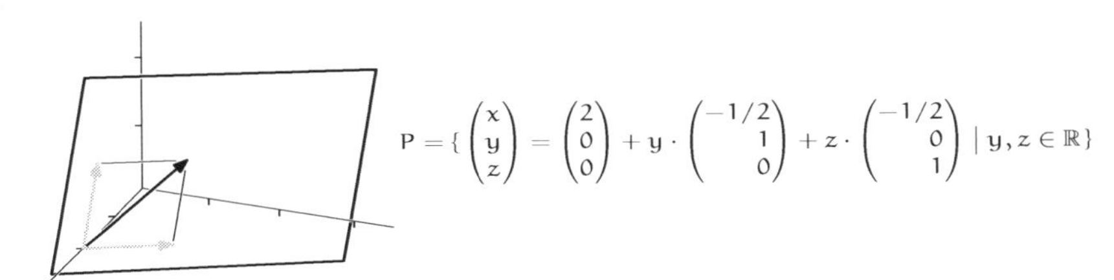

is not the result of adding

$$(\begin{pmatrix}2\\0\\0\end{pmatrix}+\begin{pmatrix}-0.5\\1\\0\end{pmatrix}\cdot 1)+(\begin{pmatrix}2\\0\\0\end{pmatrix}+\begin{pmatrix}-0.5\\0\\1\end{pmatrix}\cdot 1)$$

instead it is

$$\begin{pmatrix}2\\0\\0\end{pmatrix}+\begin{pmatrix}-0.5\\1\\0\end{pmatrix}\cdot 1+\begin{pmatrix}-0.5\\0\\1\end{pmatrix}\cdot 1=\begin{pmatrix}1\\1\\1\end{pmatrix}$$

which adds the parameters.

One.II.1.9 The "if" half is straightforward. If $b_1-a_1=d_1-c_1$ and $b_2-a_2=d_2-c_2$ then

$$\sqrt{(b_1-a_1)^2+(b_2-a_2)^2}=\sqrt{(d_1-c_1)^2+(d_2-c_2)^2}$$

so they have the same lengths, and the slopes are just as easy:

$$\frac{b_2-a_2}{b_1-a_1}=\frac{d_2-c_2}{d_1-a_1}$$

(if the denominators are 0 they both have undefined slopes).

For "only if", assume that the two segments have the same length and slope (the case of undefined slopes is easy; we will do the case where both segments have a slope m). Also assume, without loss of generality, that $a_1<b_1$ and that $c_1<d_1$. The first segment is $\overline{(a_1,a_2)(b_1,b_2)}=\{(x,y)\mid y=mx+n_1,\ x\in[a_1..b_1]\}$ (for some intercept n_1) and the second segment is $\overline{(c_1,c_2)(d_1,d_2)}=\{(x,y)\mid y=mx+n_2,\ x\in[c_1..d_1]\}$ (for some n_2). Then the lengths of those segments are

$$\sqrt{(b_1-a_1)^2+((mb_1+n_1)-(ma_1+n_1))^2}=\sqrt{(1+m^2)(b_1-a_1)^2}$$

and, similarly, $\sqrt{(1+m^2)(d_1-c_1)^2}$. Therefore, $|b_1-a_1|=|d_1-c_1|$. Thus, as we assumed that $a_1<b_1$ and $c_1<d_1$, we have that $b_1-a_1=d_1-c_1$.

The other equality is similar.

One.II.1.10 We shall later define it to be a set with one element — an "origin".

One.II.1.11 *This is how the answer was given in the cited source.* The vector triangle is as follows, so $\vec{w} = 3\sqrt{2}$ from the north west.

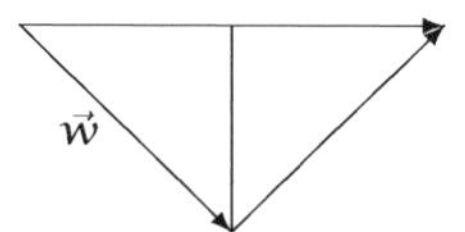

One.II.1.12 Euclid no doubt is picturing a plane inside of $\mathbb{R}^3$. Observe, however, that both $\mathbb{R}^1$ and $\mathbb{R}^2$ also satisfy that definition.

One.II.2: Length and Angle Measures

One.II.2.11 (a) $\sqrt{3^2+1^2} = \sqrt{10}$ (b) $\sqrt{5}$ (c) $\sqrt{18}$ (d) 0 (e) $\sqrt{3}$

One.II.2.12 (a) $\arccos(9/\sqrt{85}) \approx 0.22$ radians (b) $\arccos(8/\sqrt{85}) \approx 0.52$ radians (c) Not defined.

One.II.2.13 We express each displacement as a vector, rounded to one decimal place because that's the accuracy of the problem's statement, and add to find the total displacement (ignoring the curvature of the earth).

$$\begin{pmatrix}0.0\\1.2\end{pmatrix} + \begin{pmatrix}3.8\\-4.8\end{pmatrix} + \begin{pmatrix}4.0\\0.1\end{pmatrix} + \begin{pmatrix}3.3\\5.6\end{pmatrix} = \begin{pmatrix}11.1\\2.1\end{pmatrix}$$

The distance is $\sqrt{11.1^2+2.1^2} \approx 11.3$.

One.II.2.14 Solve $(k)(4)+(1)(3)=0$ to get $k=-3/4$.

One.II.2.15 We could describe the set

$$\{\begin{pmatrix}x\\y\\z\end{pmatrix} \mid 1x+3y-1z=0\}$$

with parameters in this way.

$$\{\begin{pmatrix}-3\\1\\0\end{pmatrix}y + \begin{pmatrix}1\\0\\1\end{pmatrix}z \mid y,z\in\mathbb{R}\}$$

One.II.2.16 (a) We can use the x-axis.

$$\arccos(\frac{(1)(1)+(0)(1)}{\sqrt{1}\sqrt{2}}) \approx 0.79 \text{ radians}$$

(b) Again, use the x-axis.

$$\arccos(\frac{(1)(1)+(0)(1)+(0)(1)}{\sqrt{1}\sqrt{3}}) \approx 0.96 \text{ radians}$$

(c) The x-axis worked before and it will work again.

$$\arccos(\frac{(1)(1)+\cdots+(0)(1)}{\sqrt{1}\sqrt{n}}) = \arccos(\frac{1}{\sqrt{n}})$$

(d) Using the formula from the prior item, $\lim_{n\to\infty}\arccos(1/\sqrt{n}) = \pi/2$ radians.

One.II.2.17 Clearly $u_1u_1+\cdots+u_nu_n$ is zero if and only if each u_i is zero. So only $\vec{0}\in\mathbb{R}^n$ is perpendicular to itself.

One.II.2.18 Assume that $\vec{u},\vec{v},\vec{w}\in\mathbb{R}^n$ have components $u_1,\ldots,u_n,v_1,\ldots,w_n$.

(a) Dot product is right-distributive.

$$\begin{aligned}(\vec{u}+\vec{v})\cdot\vec{w} &= [\begin{pmatrix}u_1\\ \vdots\\ u_n\end{pmatrix}+\begin{pmatrix}v_1\\ \vdots\\ v_n\end{pmatrix}]\cdot\begin{pmatrix}w_1\\ \vdots\\ w_n\end{pmatrix}\\ &= \begin{pmatrix}u_1+v_1\\ \vdots\\ u_n+v_n\end{pmatrix}\cdot\begin{pmatrix}w_1\\ \vdots\\ w_n\end{pmatrix}\\ &= (u_1+v_1)w_1+\cdots+(u_n+v_n)w_n\\ &= (u_1w_1+\cdots+u_nw_n)+(v_1w_1+\cdots+v_nw_n)\\ &= \vec{u}\cdot\vec{w}+\vec{v}\cdot\vec{w}\end{aligned}$$

(b) Dot product is also left distributive: $\vec{w}\cdot(\vec{u}+\vec{v})=\vec{w}\cdot\vec{u}+\vec{w}\cdot\vec{v}$. The proof is just like the prior one.

(c) Dot product commutes.

$$\begin{pmatrix}u_1\\ \vdots\\ u_n\end{pmatrix}\cdot\begin{pmatrix}v_1\\ \vdots\\ v_n\end{pmatrix}=u_1v_1+\cdots+u_nv_n=v_1u_1+\cdots+v_nu_n=\begin{pmatrix}v_1\\ \vdots\\ v_n\end{pmatrix}\cdot\begin{pmatrix}u_1\\ \vdots\\ u_n\end{pmatrix}$$

(d) Because $\vec{u}\cdot\vec{v}$ is a scalar, not a vector, the expression $(\vec{u}\cdot\vec{v})\cdot\vec{w}$ makes no sense; the dot product of a scalar and a vector is not defined.

(e) This is a vague question so it has many answers. Some are (1) $k(\vec{u}\cdot\vec{v})=(k\vec{u})\cdot\vec{v}$ and $k(\vec{u}\cdot\vec{v})=\vec{u}\cdot(k\vec{v})$, (2) $k(\vec{u}\cdot\vec{v})\neq(k\vec{u})\cdot(k\vec{v})$ (in general; an example is easy to produce), and (3) $\|k\vec{v}\|=|k|\|\vec{v}\|$ (the connection between norm and dot product is that the square of the norm is the dot product of a vector with itself).

One.II.2.19 **(a)** Verifying that $(k\vec{x})\cdot\vec{y}=k(\vec{x}\cdot\vec{y})=\vec{x}\cdot(k\vec{y})$ for $k\in\mathbb{R}$ and $\vec{x},\vec{y}\in\mathbb{R}^n$ is easy. Now, for $k\in\mathbb{R}$ and $\vec{v},\vec{w}\in\mathbb{R}^n$, if $\vec{u}=k\vec{v}$ then $\vec{u}\cdot\vec{v}=(k\vec{v})\cdot\vec{v}=k(\vec{v}\cdot\vec{v})$, which is k times a nonnegative real.

The $\vec{v}=k\vec{u}$ half is similar (actually, taking the k in this paragraph to be the reciprocal of the k above gives that we need only worry about the $k=0$ case).

(b) We first consider the $\vec{u}\cdot\vec{v}\geqslant 0$ case. From the Triangle Inequality we know that $\vec{u}\cdot\vec{v}=\|\vec{u}\|\,\|\vec{v}\|$ if and only if one vector is a nonnegative scalar multiple of the other. But that's all we need because the first part of this exercise shows that, in a context where the dot product of the two vectors is positive, the two statements 'one vector is a scalar multiple of the other' and 'one vector is a nonnegative scalar multiple of the other', are equivalent.

We finish by considering the $\vec{u}\cdot\vec{v}<0$ case. Because $0<|\vec{u}\cdot\vec{v}|=-(\vec{u}\cdot\vec{v})=(-\vec{u})\cdot\vec{v}$ and $\|\vec{u}\|\,\|\vec{v}\|=\|-\vec{u}\|\,\|\vec{v}\|$, we have that $0<(-\vec{u})\cdot\vec{v}=\|-\vec{u}\|\,\|\vec{v}\|$. Now the prior paragraph applies to give that one of the two vectors $-\vec{u}$ and $\vec{v}$ is a scalar multiple of the other. But that's equivalent to the assertion that one of the two vectors $\vec{u}$ and $\vec{v}$ is a scalar multiple of the other, as desired.

One.II.2.20 No. These give an example.

$$\vec{u}=\begin{pmatrix}1\\0\end{pmatrix}\quad\vec{v}=\begin{pmatrix}1\\0\end{pmatrix}\quad\vec{w}=\begin{pmatrix}1\\1\end{pmatrix}$$

One.II.2.21 We prove that a vector has length zero if and only if all its components are zero.

Let $\vec{u}\in\mathbb{R}^n$ have components $u_1,\ldots,u_n$. Recall that the square of any real number is greater than or equal to zero, with equality only when that real is zero. Thus $\|\vec{u}\|^2={u_1}^2+\cdots+{u_n}^2$ is a sum of numbers greater than or equal to zero, and so is itself greater than or equal to zero, with equality if and only if each u_i is zero. Hence $\|\vec{u}\|=0$ if and only if all the components of $\vec{u}$ are zero.

One.II.2.22 We can easily check that

$$(\frac{x_1+x_2}{2},\frac{y_1+y_2}{2})$$

is on the line connecting the two, and is equidistant from both. The generalization is obvious.

One.II.2.23 Assume that $\vec{v} \in \mathbb{R}^n$ has components $v_1, \ldots, v_n$. If $\vec{v} \neq \vec{0}$ then we have this.

$$\begin{aligned}\sqrt{\left(\frac{v_1}{\sqrt{v_1^2 + \cdots + v_n^2}}\right)^2 + \cdots + \left(\frac{v_n}{\sqrt{v_1^2 + \cdots + v_n^2}}\right)^2} &= \sqrt{\left(\frac{v_1^2}{v_1^2 + \cdots + v_n^2}\right) + \cdots + \left(\frac{v_n^2}{v_1^2 + \cdots + v_n^2}\right)} \\ &= 1\end{aligned}$$

If $\vec{v} = \vec{0}$ then $\vec{v}/\|\vec{v}\|$ is not defined.

One.II.2.24 For the first question, assume that $\vec{v} \in \mathbb{R}^n$ and $r \geqslant 0$, take the root, and factor.

$$\|r\vec{v}\| = \sqrt{(rv_1)^2 + \cdots + (rv_n)^2} = \sqrt{r^2(v_1^2 + \cdots + v_n^2} = r\|\vec{v}\|$$

For the second question, the result is r times as long, but it points in the opposite direction in that $r\vec{v} + (-r)\vec{v} = \vec{0}$.

One.II.2.25 Assume that $\vec{u}, \vec{v} \in \mathbb{R}^n$ both have length 1. Apply Cauchy-Schwartz: $|\vec{u} \cdot \vec{v}| \leqslant \|\vec{u}\|\,\|\vec{v}\| = 1$.

To see that 'less than' can happen, in $\mathbb{R}^2$ take

$$\vec{u} = \begin{pmatrix}1\\0\end{pmatrix} \quad \vec{v} = \begin{pmatrix}0\\1\end{pmatrix}$$

and note that $\vec{u} \cdot \vec{v} = 0$. For 'equal to', note that $\vec{u} \cdot \vec{u} = 1$.

One.II.2.26 Write

$$\vec{u} = \begin{pmatrix}u_1\\ \vdots \\ u_n\end{pmatrix} \quad \vec{v} = \begin{pmatrix}v_1\\ \vdots \\ v_n\end{pmatrix}$$

and then this computation works.

$$\begin{aligned}\|\vec{u} + \vec{v}\|^2 + \|\vec{u} - \vec{v}\|^2 &= (u_1 + v_1)^2 + \cdots + (u_n + v_n)^2 \\ &\quad + (u_1 - v_1)^2 + \cdots + (u_n - v_n)^2 \\ &= u_1^2 + 2u_1v_1 + v_1^2 + \cdots + u_n^2 + 2u_nv_n + v_n^2 \\ &\quad + u_1^2 - 2u_1v_1 + v_1^2 + \cdots + u_n^2 - 2u_nv_n + v_n^2 \\ &= 2(u_1^2 + \cdots + u_n^2) + 2(v_1^2 + \cdots + v_n^2) \\ &= 2\|\vec{u}\|^2 + 2\|\vec{v}\|^2\end{aligned}$$

One.II.2.27 We will prove this demonstrating that the contrapositive statement holds: if $\vec{x} \neq \vec{0}$ then there is a $\vec{y}$ with $\vec{x} \cdot \vec{y} \neq 0$.

Assume that $\vec{x} \in \mathbb{R}^n$. If $\vec{x} \neq \vec{0}$ then it has a nonzero component, say the i-th one x_i. But the vector $\vec{y} \in \mathbb{R}^n$ that is all zeroes except for a one in component i gives $\vec{x} \cdot \vec{y} = x_i$. (A slicker proof just considers $\vec{x} \cdot \vec{x}$.)

One.II.2.28 Yes; we can prove this by induction.

Assume that the vectors are in some $\mathbb{R}^k$. Clearly the statement applies to one vector. The Triangle Inequality is this statement applied to two vectors. For an inductive step assume the statement is true for n or fewer vectors. Then this

$$\|\vec{u}_1 + \cdots + \vec{u}_n + \vec{u}_{n+1}\| \leqslant \|\vec{u}_1 + \cdots + \vec{u}_n\| + \|\vec{u}_{n+1}\|$$

follows by the Triangle Inequality for two vectors. Now the inductive hypothesis, applied to the first summand on the right, gives that as less than or equal to $\|\vec{u}_1\| + \cdots + \|\vec{u}_n\| + \|\vec{u}_{n+1}\|$.

One.II.2.29 By definition

$$\frac{\vec{u} \cdot \vec{v}}{\|\vec{u}\|\,\|\vec{v}\|} = \cos\theta$$

where θ is the angle between the vectors. Thus the ratio is $|\cos\theta|$.

One.II.2.30 So that the statement 'vectors are orthogonal iff their dot product is zero' has no exceptions.

One.II.2.31 We can find the angle between (a) and (b) (for $a, b \neq 0$) with

$$\arccos(\frac{ab}{\sqrt{a^2}\sqrt{b^2}}).$$

If a or b is zero then the angle is $\pi/2$ radians. Otherwise, if a and b are of opposite signs then the angle is π radians, else the angle is zero radians.

One.II.2.32 The angle between $\vec{u}$ and $\vec{v}$ is acute if $\vec{u} \cdot \vec{v} > 0$, is right if $\vec{u} \cdot \vec{v} = 0$, and is obtuse if $\vec{u} \cdot \vec{v} < 0$. That's because, in the formula for the angle, the denominator is never negative.

One.II.2.33 Suppose that $\vec{u}, \vec{v} \in \mathbb{R}^n$. If $\vec{u}$ and $\vec{v}$ are perpendicular then

$$\|\vec{u} + \vec{v}\|^2 = (\vec{u} + \vec{v}) \cdot (\vec{u} + \vec{v}) = \vec{u} \cdot \vec{u} + 2\,\vec{u} \cdot \vec{v} + \vec{v} \cdot \vec{v} = \vec{u} \cdot \vec{u} + \vec{v} \cdot \vec{v} = \|\vec{u}\|^2 + \|\vec{v}\|^2$$

(the third equality holds because $\vec{u} \cdot \vec{v} = 0$).

One.II.2.34 Where $\vec{u}, \vec{v} \in \mathbb{R}^n$, the vectors $\vec{u}+\vec{v}$ and $\vec{u}-\vec{v}$ are perpendicular if and only if $0 = (\vec{u}+\vec{v}) \cdot (\vec{u}-\vec{v}) = \vec{u} \cdot \vec{u} - \vec{v} \cdot \vec{v}$, which shows that those two are perpendicular if and only if $\vec{u} \cdot \vec{u} = \vec{v} \cdot \vec{v}$. That holds if and only if $\|\vec{u}\| = \|\vec{v}\|$.

One.II.2.35 Suppose $\vec{u} \in \mathbb{R}^n$ is perpendicular to both $\vec{v} \in \mathbb{R}^n$ and $\vec{w} \in \mathbb{R}^n$. Then, for any $k, m \in \mathbb{R}$ we have this.

$$\vec{u} \cdot (k\vec{v} + m\vec{w}) = k(\vec{u} \cdot \vec{v}) + m(\vec{u} \cdot \vec{w}) = k(0) + m(0) = 0$$

One.II.2.36 We will show something more general: if $\|\vec{z}_1\| = \|\vec{z}_2\|$ for $\vec{z}_1, \vec{z}_2 \in \mathbb{R}^n$, then $\vec{z}_1 + \vec{z}_2$ bisects the angle between $\vec{z}_1$ and $\vec{z}_2$

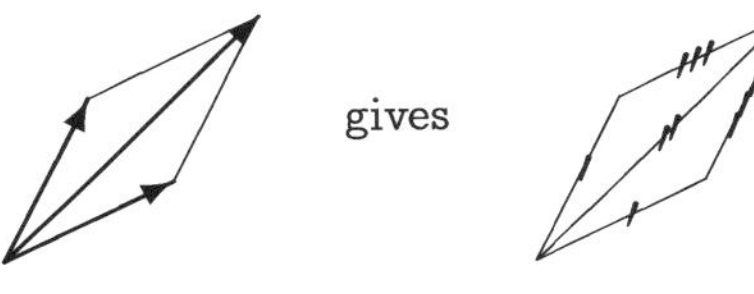

(we ignore the case where $\vec{z}_1$ and $\vec{z}_2$ are the zero vector).

The $\vec{z}_1 + \vec{z}_2 = \vec{0}$ case is easy. For the rest, by the definition of angle, we will be finished if we show this.

$$\frac{\vec{z}_1 \cdot (\vec{z}_1 + \vec{z}_2)}{\|\vec{z}_1\|\,\|\vec{z}_1 + \vec{z}_2\|} = \frac{\vec{z}_2 \cdot (\vec{z}_1 + \vec{z}_2)}{\|\vec{z}_2\|\,\|\vec{z}_1 + \vec{z}_2\|}$$

But distributing inside each expression gives

$$\frac{\vec{z}_1 \cdot \vec{z}_1 + \vec{z}_1 \cdot \vec{z}_2}{\|\vec{z}_1\|\,\|\vec{z}_1 + \vec{z}_2\|} \qquad \frac{\vec{z}_2 \cdot \vec{z}_1 + \vec{z}_2 \cdot \vec{z}_2}{\|\vec{z}_2\|\,\|\vec{z}_1 + \vec{z}_2\|}$$

and $\vec{z}_1 \cdot \vec{z}_1 = \|\vec{z}_1\|^2 = \|\vec{z}_2\|^2 = \vec{z}_2 \cdot \vec{z}_2$, so the two are equal.

One.II.2.37 We can show the two statements together. Let $\vec{u}, \vec{v} \in \mathbb{R}^n$, write

$$\vec{u} = \begin{pmatrix} u_1 \\ \vdots \\ u_n \end{pmatrix} \qquad \vec{v} = \begin{pmatrix} v_1 \\ \vdots \\ v_n \end{pmatrix}$$

and calculate.

$$\cos\theta = \frac{ku_1v_1 + \cdots + ku_nv_n}{\sqrt{(ku_1)^2 + \cdots + (ku_n)^2}\sqrt{{b_1}^2 + \cdots + {b_n}^2}} = \frac{k}{|k|}\frac{\vec{u} \cdot \vec{v}}{\|\vec{u}\|\,\|\vec{v}\|} = \pm\frac{\vec{u} \cdot \vec{v}}{\|\vec{u}\|\,\|\vec{v}\|}$$

One.II.2.38 Let

$$\vec{u} = \begin{pmatrix} u_1 \\ \vdots \\ u_n \end{pmatrix}, \quad \vec{v} = \begin{pmatrix} v_1 \\ \vdots \\ v_n \end{pmatrix} \quad \vec{w} = \begin{pmatrix} w_1 \\ \vdots \\ w_n \end{pmatrix}$$

and then

$$\begin{aligned}\vec{u}\cdot(k\vec{v}+m\vec{w}) &= \begin{pmatrix}u_1\\ \vdots\\ u_n\end{pmatrix}\cdot\Bigl(\begin{pmatrix}kv_1\\ \vdots\\ kv_n\end{pmatrix}+\begin{pmatrix}mw_1\\ \vdots\\ mw_n\end{pmatrix}\Bigr)\\
&= \begin{pmatrix}u_1\\ \vdots\\ u_n\end{pmatrix}\cdot\begin{pmatrix}kv_1+mw_1\\ \vdots\\ kv_n+mw_n\end{pmatrix}\\
&= u_1(kv_1+mw_1)+\cdots+u_n(kv_n+mw_n)\\
&= ku_1v_1+mu_1w_1+\cdots+ku_nv_n+mu_nw_n\\
&= (ku_1v_1+\cdots+ku_nv_n)+(mu_1w_1+\cdots+mu_nw_n)\\
&= k(\vec{u}\cdot\vec{v})+m(\vec{u}\cdot\vec{w})\end{aligned}$$

as required.

One.II.2.39 For $x, y\in\mathbb{R}^+$, set

$$\vec{u}=\begin{pmatrix}\sqrt{x}\\ \sqrt{y}\end{pmatrix}\qquad\vec{v}=\begin{pmatrix}\sqrt{y}\\ \sqrt{x}\end{pmatrix}$$

so that the Cauchy-Schwartz inequality asserts that (after squaring)

$$\begin{aligned}(\sqrt{x}\sqrt{y}+\sqrt{y}\sqrt{x})^2 &\leqslant (\sqrt{x}\sqrt{x}+\sqrt{y}\sqrt{y})(\sqrt{y}\sqrt{y}+\sqrt{x}\sqrt{x})\\
(2\sqrt{x}\sqrt{y})^2 &\leqslant (x+y)^2\\
\sqrt{xy} &\leqslant \frac{x+y}{2}\end{aligned}$$

as desired.

One.II.2.40 (a) For instance, a birthday of October 12 gives this.

$$\theta=\arccos\Bigl(\frac{\begin{pmatrix}7\\12\end{pmatrix}\cdot\begin{pmatrix}10\\12\end{pmatrix}}{\|\begin{pmatrix}7\\12\end{pmatrix}\|\cdot\|\begin{pmatrix}10\\12\end{pmatrix}\|}\Bigr)=\arccos\Bigl(\frac{214}{\sqrt{244}\sqrt{193}}\Bigr)\approx 0.17\text{ rad}$$

(b) Applying the same equation to $(9\quad 19)$ gives about 0.09 radians.

(c) The angle will measure 0 radians if the other person is born on the same day. It will also measure 0 if one birthday is a scalar multiple of the other. For instance, a person born on Mar 6 would be harmonious with a person born on Feb 4.

Given a birthday, we can get Sage to plot the angle for other dates. This example shows the relationship of all dates with July 12.

```
sage: plot3d(lambda x, y: math.acos((x*7+y*12)/(math.sqrt(7**2+12**2)*math.sqrt(x**2+y**2))),
                                                                    (1,12),(1,31))
```

The result looks like this.

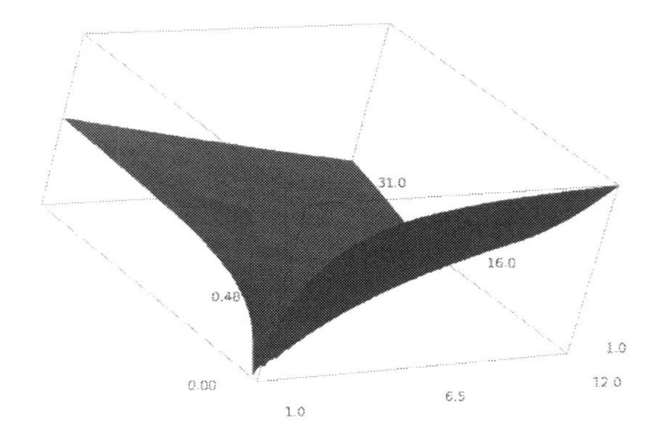

(d) We want to maximize this.

$$\theta = \arccos\left(\frac{\begin{pmatrix} 7 \\ 12 \end{pmatrix} \cdot \begin{pmatrix} m \\ d \end{pmatrix}}{\| \begin{pmatrix} 7 \\ 12 \end{pmatrix} \| \cdot \| \begin{pmatrix} m \\ d \end{pmatrix} \|} \right)$$

Of course, we cannot take m or d negative and so we cannot get a vector orthogonal to the given one. This Python script finds the largest angle by brute force.

```
import math
days={1:31,  # Jan
      2:29, 3:31, 4:30, 5:31, 6:30, 7:31, 8:31, 9:30, 10:31, 11:30, 12:31}
BDAY=(7,12)
max_res=0
max_res_date=(-1,-1)
for month in range(1,13):
    for day in range(1,days[month]+1):
        num=BDAY[0]*month+BDAY[1]*day
        denom=math.sqrt(BDAY[0]**2+BDAY[1]**2)*math.sqrt(month**2+day**2)
        if denom>0:
            res=math.acos(min(num*1.0/denom,1))
            print "day:",str(month),str(day)," angle:",str(res)
            if res>max_res:
                max_res=res
                max_res_date=(month,day)
print "For ",str(BDAY),"the worst case is",str(max_res),"radians on date",str(max_res_date)
print "  That is ",180*max_res/math.pi,"degrees"
```

The result is

```
For  (7, 12) the worst case is 0.95958064648 radians on date (12, 1)
  That is  54.9799211457 degrees
```

A more conceptual approach is to consider the relation of all points (month, day) to the point $(7,12)$. The picture below makes clear that the answer is either Dec 1 or Jan 31, depending on which is further from the birthdate. The dashed line bisects the angle between the line from the origin to Dec 1, and the line from the origin to Jan 31. Birthdays above the line are furthest from Dec 1 and birthdays below the line are furthest from Jan 31.

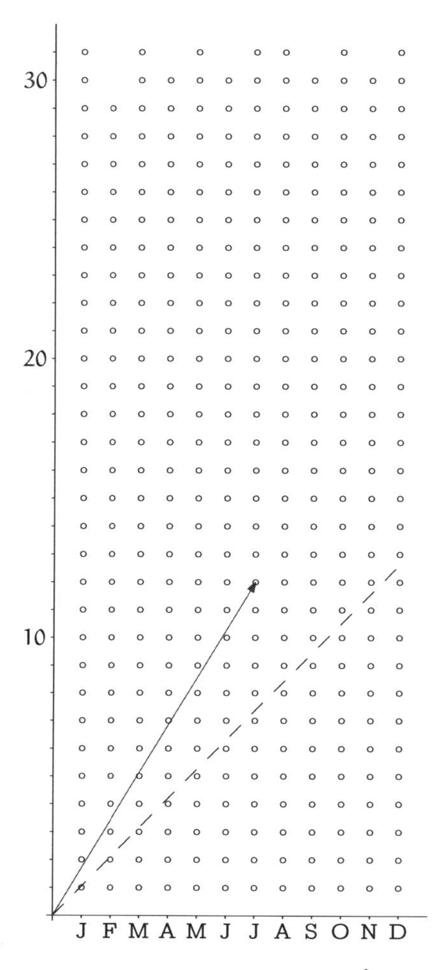

One.II.2.41 *This is how the answer was given in the cited source.* The actual velocity $\vec{v}$ of the wind is the sum of the ship's velocity and the apparent velocity of the wind. Without loss of generality we may assume $\vec{a}$ and $\vec{b}$ to be unit vectors, and may write

$$\vec{v} = \vec{v}_1 + s\vec{a} = \vec{v}_2 + t\vec{b}$$

where s and t are undetermined scalars. Take the dot product first by $\vec{a}$ and then by $\vec{b}$ to obtain

$$\begin{aligned} s - t\vec{a}\cdot\vec{b} &= \vec{a}\cdot(\vec{v}_2 - \vec{v}_1) \\ s\vec{a}\cdot\vec{b} - t &= \vec{b}\cdot(\vec{v}_2 - \vec{v}_1) \end{aligned}$$

Multiply the second by $\vec{a}\cdot\vec{b}$, subtract the result from the first, and find

$$s = \frac{[\vec{a} - (\vec{a}\cdot\vec{b})\vec{b}]\cdot(\vec{v}_2 - \vec{v}_1)}{1 - (\vec{a}\cdot\vec{b})^2}.$$

Substituting in the original displayed equation, we get

$$\vec{v} = \vec{v}_1 + \frac{[\vec{a} - (\vec{a}\cdot\vec{b})\vec{b}]\cdot(\vec{v}_2 - \vec{v}_1)\vec{a}}{1 - (\vec{a}\cdot\vec{b})^2}.$$

One.II.2.42 We use induction on n.

In the $n = 1$ base case the identity reduces to

$$(a_1 b_1)^2 = (a_1{}^2)(b_1{}^2) - 0$$

and clearly holds.

For the inductive step assume that the formula holds for the $0, \ldots, n$ cases. We will show that it then holds in the $n+1$ case. Start with the right-hand side

$$
\begin{aligned}
(\sum_{1\leqslant j\leqslant n+1} a_j^2)(\sum_{1\leqslant j\leqslant n+1} b_j^2) - \sum_{1\leqslant k<j\leqslant n+1} (a_k b_j - a_j b_k)^2 \\
&= [(\sum_{1\leqslant j\leqslant n} a_j^2) + a_{n+1}^2][(\sum_{1\leqslant j\leqslant n} b_j^2) + b_{n+1}^2] \\
&\quad - [\sum_{1\leqslant k<j\leqslant n} (a_k b_j - a_j b_k)^2 + \sum_{1\leqslant k\leqslant n} (a_k b_{n+1} - a_{n+1} b_k)^2] \\
&= (\sum_{1\leqslant j\leqslant n} a_j^2)(\sum_{1\leqslant j\leqslant n} b_j^2) + \sum_{1\leqslant j\leqslant n} b_j^2 a_{n+1}^2 + \sum_{1\leqslant j\leqslant n} a_j^2 b_{n+1}^2 + a_{n+1}^2 b_{n+1}^2 \\
&\quad - [\sum_{1\leqslant k<j\leqslant n} (a_k b_j - a_j b_k)^2 + \sum_{1\leqslant k\leqslant n} (a_k b_{n+1} - a_{n+1} b_k)^2] \\
&= (\sum_{1\leqslant j\leqslant n} a_j^2)(\sum_{1\leqslant j\leqslant n} b_j^2) - \sum_{1\leqslant k<j\leqslant n} (a_k b_j - a_j b_k)^2 \\
&\quad + \sum_{1\leqslant j\leqslant n} b_j^2 a_{n+1}^2 + \sum_{1\leqslant j\leqslant n} a_j^2 b_{n+1}^2 + a_{n+1}^2 b_{n+1}^2 \\
&\quad - \sum_{1\leqslant k\leqslant n} (a_k b_{n+1} - a_{n+1} b_k)^2
\end{aligned}
$$

and apply the inductive hypothesis

$$
\begin{aligned}
&= (\sum_{1\leqslant j\leqslant n} a_j b_j)^2 + \sum_{1\leqslant j\leqslant n} b_j^2 a_{n+1}^2 + \sum_{1\leqslant j\leqslant n} a_j^2 b_{n+1}^2 + a_{n+1}^2 b_{n+1}^2 \\
&\quad - [\sum_{1\leqslant k\leqslant n} a_k^2 b_{n+1}^2 - 2\sum_{1\leqslant k\leqslant n} a_k b_{n+1} a_{n+1} b_k + \sum_{1\leqslant k\leqslant n} a_{n+1}^2 b_k^2] \\
&= (\sum_{1\leqslant j\leqslant n} a_j b_j)^2 + 2(\sum_{1\leqslant k\leqslant n} a_k b_{n+1} a_{n+1} b_k) + a_{n+1}^2 b_{n+1}^2 \\
&= [(\sum_{1\leqslant j\leqslant n} a_j b_j) + a_{n+1} b_{n+1}]^2
\end{aligned}
$$

to derive the left-hand side.

Reduced Echelon Form

One.III.1: Gauss-Jordan Reduction

One.III.1.8 These answers show only the Gauss-Jordan reduction. With it, describing the solution set is easy.

(a) $\left(\begin{array}{cc|c} 1 & 1 & 2 \\ 1 & -1 & 0 \end{array}\right) \xrightarrow{-\rho_1+\rho_2} \left(\begin{array}{cc|c} 1 & 1 & 2 \\ 0 & -2 & -2 \end{array}\right) \xrightarrow{-(1/2)\rho_2} \left(\begin{array}{cc|c} 1 & 1 & 2 \\ 0 & 1 & 1 \end{array}\right) \xrightarrow{-\rho_2+\rho_1} \left(\begin{array}{cc|c} 1 & 0 & 1 \\ 0 & 1 & 1 \end{array}\right)$

(b) $\left(\begin{array}{ccc|c} 1 & 0 & -1 & 4 \\ 2 & 2 & 0 & 1 \end{array}\right) \xrightarrow{-2\rho_1+\rho_2} \left(\begin{array}{ccc|c} 1 & 0 & -1 & 4 \\ 0 & 2 & 2 & -7 \end{array}\right) \xrightarrow{(1/2)\rho_2} \left(\begin{array}{ccc|c} 1 & 0 & -1 & 4 \\ 0 & 1 & 1 & -7/2 \end{array}\right)$

(c)

$$\left(\begin{array}{cc|c} 3 & -2 & 1 \\ 6 & 1 & 1/2 \end{array}\right) \xrightarrow{-2\rho_1+\rho_2} \left(\begin{array}{cc|c} 3 & -2 & 1 \\ 0 & 5 & -3/2 \end{array}\right) \xrightarrow[(1/5)\rho_2]{(1/3)\rho_1} \left(\begin{array}{cc|c} 1 & -2/3 & 1/3 \\ 0 & 1 & -3/10 \end{array}\right) \xrightarrow{(2/3)\rho_2+\rho_1} \left(\begin{array}{cc|c} 1 & 0 & 2/15 \\ 0 & 1 & -3/10 \end{array}\right)$$

(d) A row swap here makes the arithmetic easier.

$$\left(\begin{array}{ccc|c}2&-1&0&-1\\1&3&-1&5\\0&1&2&5\end{array}\right)\xrightarrow{-(1/2)\rho_1+\rho_2}\left(\begin{array}{ccc|c}2&-1&0&-1\\0&7/2&-1&11/2\\0&1&2&5\end{array}\right)\xrightarrow{\rho_2\leftrightarrow\rho_3}\left(\begin{array}{ccc|c}2&-1&0&-1\\0&1&2&5\\0&7/2&-1&11/2\end{array}\right)$$

$$\xrightarrow{-(7/2)\rho_2+\rho_3}\left(\begin{array}{ccc|c}2&-1&0&-1\\0&1&2&5\\0&0&-8&-12\end{array}\right)\xrightarrow[-(1/8)\rho_2]{(1/2)\rho_1}\left(\begin{array}{ccc|c}1&-1/2&0&-1/2\\0&1&2&5\\0&0&1&3/2\end{array}\right)$$

$$\xrightarrow{-2\rho_3+\rho_2}\left(\begin{array}{ccc|c}1&-1/2&0&-1/2\\0&1&0&2\\0&0&1&3/2\end{array}\right)\xrightarrow{(1/2)\rho_2+\rho_1}\left(\begin{array}{ccc|c}1&0&0&1/2\\0&1&0&2\\0&0&1&3/2\end{array}\right)$$

One.III.1.9 Use Gauss-Jordan reduction.

(a) $\xrightarrow{-(1/2)\rho_1+\rho_2}\begin{pmatrix}2&1\\0&5/2\end{pmatrix}\xrightarrow[(2/5)\rho_2]{(1/2)\rho_1}\begin{pmatrix}1&1/2\\0&1\end{pmatrix}\xrightarrow{-(1/2)\rho_2+\rho_1}\begin{pmatrix}1&0\\0&1\end{pmatrix}$

(b) $\xrightarrow[\rho_1+\rho_3]{-2\rho_1+\rho_2}\begin{pmatrix}1&3&1\\0&-6&2\\0&0&-2\end{pmatrix}\xrightarrow[-(1/2)\rho_3]{-(1/6)\rho_2}\begin{pmatrix}1&3&1\\0&1&-1/3\\0&0&1\end{pmatrix}\xrightarrow[-\rho_3+\rho_1]{(1/3)\rho_3+\rho_2}\begin{pmatrix}1&3&0\\0&1&0\\0&0&1\end{pmatrix}\xrightarrow{-3\rho_2+\rho_1}\begin{pmatrix}1&0&0\\0&1&0\\0&0&1\end{pmatrix}$

(c)

$$\xrightarrow[-3\rho_1+\rho_3]{-\rho_1+\rho_2}\begin{pmatrix}1&0&3&1&2\\0&4&-1&0&3\\0&4&-1&-2&-4\end{pmatrix}\xrightarrow{-\rho_2+\rho_3}\begin{pmatrix}1&0&3&1&2\\0&4&-1&0&3\\0&0&0&-2&-7\end{pmatrix}$$

$$\xrightarrow[-(1/2)\rho_3]{(1/4)\rho_2}\begin{pmatrix}1&0&3&1&2\\0&1&-1/4&0&3/4\\0&0&0&1&7/2\end{pmatrix}\xrightarrow{-\rho_3+\rho_1}\begin{pmatrix}1&0&3&0&-3/2\\0&1&-1/4&0&3/4\\0&0&0&1&7/2\end{pmatrix}$$

(d)

$$\xrightarrow{\rho_1\leftrightarrow\rho_3}\begin{pmatrix}1&5&1&5\\0&0&5&6\\0&1&3&2\end{pmatrix}\xrightarrow{\rho_2\leftrightarrow\rho_3}\begin{pmatrix}1&5&1&5\\0&1&3&2\\0&0&5&6\end{pmatrix}\xrightarrow{(1/5)\rho_3}\begin{pmatrix}1&5&1&5\\0&1&3&2\\0&0&1&6/5\end{pmatrix}$$

$$\xrightarrow[-\rho_3+\rho_1]{-3\rho_3+\rho_2}\begin{pmatrix}1&5&0&19/5\\0&1&0&-8/5\\0&0&1&6/5\end{pmatrix}\xrightarrow{-5\rho_2+\rho_1}\begin{pmatrix}1&0&0&59/5\\0&1&0&-8/5\\0&0&1&6/5\end{pmatrix}$$

One.III.1.10 For the Gauss's halves, see the answers to Chapter One's section I.2 question Exercise 19.

(a) The "Jordan" half goes this way.

$$\xrightarrow[-(1/3)\rho_2]{(1/2)\rho_1}\left(\begin{array}{ccc|c}1&1/2&-1/2&1/2\\0&1&-2/3&-1/3\end{array}\right)\xrightarrow{-(1/2)\rho_2+\rho_1}\left(\begin{array}{ccc|c}1&0&-1/6&2/3\\0&1&-2/3&-1/3\end{array}\right)$$

The solution set is this

$$\{\begin{pmatrix}2/3\\-1/3\\0\end{pmatrix}+\begin{pmatrix}1/6\\2/3\\1\end{pmatrix}z \mid z\in\mathbb{R}\}$$

(b) The second half is

$$\xrightarrow{\rho_3+\rho_2}\left(\begin{array}{cccc|c}1&0&-1&0&1\\0&1&2&0&3\\0&0&0&1&0\end{array}\right)$$

so the solution is this.

$$\{\begin{pmatrix}1\\3\\0\\0\end{pmatrix}+\begin{pmatrix}1\\-2\\1\\0\end{pmatrix}z \mid z\in\mathbb{R}\}$$

(c) This Jordan half

$$\xrightarrow{\rho_2+\rho_1}\left(\begin{array}{cccc|c}1&0&1&1&0\\0&1&0&1&0\\0&0&0&0&0\\0&0&0&0&0\end{array}\right)$$

gives

$$\{\begin{pmatrix}0\\0\\0\\0\end{pmatrix}+\begin{pmatrix}-1\\0\\1\\0\end{pmatrix}z+\begin{pmatrix}-1\\-1\\0\\1\end{pmatrix}w \mid z,w\in\mathbb{R}\}$$

(of course, we could omit the zero vector from the description).

(d) The "Jordan" half

$$\xrightarrow{-(1/7)\rho_2}\left(\begin{array}{ccccc|c}1&2&3&1&-1&1\\0&1&8/7&2/7&-4/7&0\end{array}\right)\xrightarrow{-2\rho_2+\rho_1}\left(\begin{array}{ccccc|c}1&0&5/7&3/7&1/7&1\\0&1&8/7&2/7&-4/7&0\end{array}\right)$$

ends with this solution set.

$$\{\begin{pmatrix}1\\0\\0\\0\\0\end{pmatrix}+\begin{pmatrix}-5/7\\-8/7\\1\\0\\0\end{pmatrix}c+\begin{pmatrix}-3/7\\-2/7\\0\\1\\0\end{pmatrix}d+\begin{pmatrix}-1/7\\4/7\\0\\0\\1\end{pmatrix}e \mid c,d,e\in\mathbb{R}\}$$

One.III.1.11 Routine Gauss's Method gives one:

$$\xrightarrow[-(1/2)\rho_1+\rho_3]{-3\rho_1+\rho_2}\begin{pmatrix}2&1&1&3\\0&1&-2&-7\\0&9/2&1/2&7/2\end{pmatrix}\xrightarrow{-(9/2)\rho_2+\rho_3}\begin{pmatrix}2&1&1&3\\0&1&-2&-7\\0&0&19/2&35\end{pmatrix}$$

and any cosmetic change, like multiplying the bottom row by 2,

$$\begin{pmatrix}2&1&1&3\\0&1&-2&-7\\0&0&19&70\end{pmatrix}$$

gives another.

One.III.1.12 In the cases listed below, we take $a,b\in\mathbb{R}$. Thus, some canonical forms listed below actually include infinitely many cases. In particular, they includes the cases $a=0$ and $b=0$.

(a) $\begin{pmatrix}0&0\\0&0\end{pmatrix}$, $\begin{pmatrix}1&a\\0&0\end{pmatrix}$, $\begin{pmatrix}0&1\\0&0\end{pmatrix}$, $\begin{pmatrix}1&0\\0&1\end{pmatrix}$

(b) $\begin{pmatrix}0&0&0\\0&0&0\end{pmatrix}$, $\begin{pmatrix}1&a&b\\0&0&0\end{pmatrix}$, $\begin{pmatrix}0&1&a\\0&0&0\end{pmatrix}$, $\begin{pmatrix}0&0&1\\0&0&0\end{pmatrix}$, $\begin{pmatrix}1&0&a\\0&1&b\end{pmatrix}$, $\begin{pmatrix}1&a&0\\0&0&1\end{pmatrix}$, $\begin{pmatrix}0&1&0\\0&0&1\end{pmatrix}$

(c) $\begin{pmatrix}0&0\\0&0\\0&0\end{pmatrix}$, $\begin{pmatrix}1&a\\0&0\\0&0\end{pmatrix}$, $\begin{pmatrix}0&1\\0&0\\0&0\end{pmatrix}$, $\begin{pmatrix}1&0\\0&1\\0&0\end{pmatrix}$

(d) $\begin{pmatrix}0&0&0\\0&0&0\\0&0&0\end{pmatrix}$, $\begin{pmatrix}1&a&b\\0&0&0\\0&0&0\end{pmatrix}$, $\begin{pmatrix}0&1&a\\0&0&0\\0&0&0\end{pmatrix}$, $\begin{pmatrix}0&0&1\\0&0&0\\0&0&0\end{pmatrix}$, $\begin{pmatrix}1&0&a\\0&1&b\\0&0&0\end{pmatrix}$, $\begin{pmatrix}1&a&0\\0&0&1\\0&0&0\end{pmatrix}$, $\begin{pmatrix}1&0&0\\0&1&0\\0&0&1\end{pmatrix}$

One.III.1.13 A nonsingular homogeneous linear system has a unique solution. So a nonsingular matrix must reduce to a (square) matrix that is all 0's except for 1's down the upper-left to lower-right diagonal, e.g.,

$$\begin{pmatrix}1&0\\0&1\end{pmatrix},\quad\text{or}\quad\begin{pmatrix}1&0&0\\0&1&0\\0&0&1\end{pmatrix},\quad\text{etc.}$$

One.III.1.14 It is an equivalence relation. To prove that we must check that the relation is reflexive, symmetric, and transitive.

Assume that all matrices are 2×2. For reflexive, we note that a matrix has the same sum of entries as itself. For symmetric, we assume A has the same sum of entries as B and obviously then B has the same sum of entries as A. Transitivity is no harder — if A has the same sum of entries as B and B has the same sum of entries as C then A has the same as C.

One.III.1.15 To be an equivalence, each relation must be reflexive, symmetric, and transitive.

(a) This relation is not symmetric because if x has taken 4 classes and y has taken 3 then x is related to y but y is not related to x.

(b) This is reflexive because x's name starts with the same letter as does x's. It is symmetric because if x's name starts with the same letter as y's then y's starts with the same letter as does x's. And it is transitive because if x's name starts with the same letter as does y's and y's name starts with the same letter as does z's then x's starts with the same letter as does z's. So it is an equivalence.

One.III.1.16 **(a)** The $\rho_i \leftrightarrow \rho_i$ operation does not change A.

(b) For instance,

$$\begin{pmatrix}1&2\\3&4\end{pmatrix}\xrightarrow{-\rho_1+\rho_1}\begin{pmatrix}0&0\\3&4\end{pmatrix}\xrightarrow{\rho_1+\rho_1}\begin{pmatrix}0&0\\3&4\end{pmatrix}$$

leaves the matrix changed.

(c) If $i\neq j$ then

$$\begin{pmatrix}&\vdots&\\a_{i,1}&\cdots&a_{i,n}\\&\vdots&\\a_{j,1}&\cdots&a_{j,n}\\&\vdots&\end{pmatrix}\xrightarrow{k\rho_i+\rho_j}\begin{pmatrix}&\vdots&\\a_{i,1}&\cdots&a_{i,n}\\&\vdots&\\ka_{i,1}+a_{j,1}&\cdots&ka_{i,n}+a_{j,n}\\&\vdots&\end{pmatrix}$$

$$\xrightarrow{-k\rho_i+\rho_j}\begin{pmatrix}&\vdots&\\a_{i,1}&\cdots&a_{i,n}\\&\vdots&\\-ka_{i,1}+ka_{i,1}+a_{j,1}&\cdots&-ka_{i,n}+ka_{i,n}+a_{j,n}\\&\vdots&\end{pmatrix}$$

does indeed give A back. (Of course, if $i=j$ then the third matrix would have entries of the form $-k(ka_{i,j}+a_{i,j})+ka_{i,j}+a_{i,j}$.)

One.III.2: The Linear Combination Lemma

One.III.2.10 Bring each to reduced echelon form and compare.

(a) The first gives

$$\xrightarrow{-4\rho_1+\rho_2}\begin{pmatrix}1&2\\0&0\end{pmatrix}$$

while the second gives

$$\xrightarrow{\rho_1\leftrightarrow\rho_2}\begin{pmatrix}1&2\\0&1\end{pmatrix}\xrightarrow{-2\rho_2+\rho_1}\begin{pmatrix}1&0\\0&1\end{pmatrix}$$

The two reduced echelon form matrices are not identical, and so the original matrices are not row equivalent.

(b) The first is this.

$$\xrightarrow[-5\rho_1+\rho_3]{-3\rho_1+\rho_2}\begin{pmatrix}1&0&2\\0&-1&-5\\0&-1&-5\end{pmatrix}\xrightarrow{-\rho_2+\rho_3}\begin{pmatrix}1&0&2\\0&-1&-5\\0&0&0\end{pmatrix}\xrightarrow{-\rho_2}\begin{pmatrix}1&0&2\\0&1&5\\0&0&0\end{pmatrix}$$

The second is this.

$$\xrightarrow{-2\rho_1+\rho_3}\begin{pmatrix}1&0&2\\0&2&10\\0&0&0\end{pmatrix}\xrightarrow{(1/2)\rho_2}\begin{pmatrix}1&0&2\\0&1&5\\0&0&0\end{pmatrix}$$

These two are row equivalent.

(c) These two are not row equivalent because they have different sizes.

(d) The first,

$$\xrightarrow{\rho_1+\rho_2}\begin{pmatrix}1&1&1\\0&3&3\end{pmatrix}\xrightarrow{(1/3)\rho_2}\begin{pmatrix}1&1&1\\0&1&1\end{pmatrix}\xrightarrow{-\rho_2+\rho_1}\begin{pmatrix}1&0&0\\0&1&1\end{pmatrix}$$

and the second.

$$\xrightarrow{\rho_1\leftrightarrow\rho_2}\begin{pmatrix}2&2&5\\0&3&-1\end{pmatrix}\xrightarrow[(1/3)\rho_2]{(1/2)\rho_1}\begin{pmatrix}1&1&5/2\\0&1&-1/3\end{pmatrix}\xrightarrow{-\rho_2+\rho_1}\begin{pmatrix}1&0&17/6\\0&1&-1/3\end{pmatrix}$$

These are not row equivalent.

(e) Here the first is

$$\xrightarrow{(1/3)\rho_2}\begin{pmatrix}1&1&1\\0&0&1\end{pmatrix}\xrightarrow{-\rho_2+\rho_1}\begin{pmatrix}1&1&0\\0&0&1\end{pmatrix}$$

while this is the second.

$$\xrightarrow{\rho_1\leftrightarrow\rho_2}\begin{pmatrix}1&-1&1\\0&1&2\end{pmatrix}\xrightarrow{\rho_2+\rho_1}\begin{pmatrix}1&0&3\\0&1&2\end{pmatrix}$$

These are not row equivalent.

One.III.2.11 First, the only matrix row equivalent to the matrix of all 0's is itself (since row operations have no effect).

Second, the matrices that reduce to

$$\begin{pmatrix}1&a\\0&0\end{pmatrix}$$

have the form

$$\begin{pmatrix}b&ba\\c&ca\end{pmatrix}$$

(where $a, b, c \in \mathbb{R}$, and b and c are not both zero).

Next, the matrices that reduce to

$$\begin{pmatrix}0&1\\0&0\end{pmatrix}$$

have the form

$$\begin{pmatrix}0&a\\0&b\end{pmatrix}$$

(where $a, b \in \mathbb{R}$, and not both are zero).

Finally, the matrices that reduce to

$$\begin{pmatrix} 1 & 0 \\ 0 & 1 \end{pmatrix}$$

are the nonsingular matrices. That's because a linear system for which this is the matrix of coefficients will have a unique solution, and that is the definition of nonsingular. (Another way to say the same thing is to say that they fall into none of the above classes.)

One.III.2.12 **(a)** They have the form

$$\begin{pmatrix} a & 0 \\ b & 0 \end{pmatrix}$$

where $a, b \in \mathbb{R}$.

(b) They have this form (for $a, b \in \mathbb{R}$).

$$\begin{pmatrix} 1a & 2a \\ 1b & 2b \end{pmatrix}$$

(c) They have the form

$$\begin{pmatrix} a & b \\ c & d \end{pmatrix}$$

(for $a, b, c, d \in \mathbb{R}$) where $ad - bc \neq 0$. (This is the formula that determines when a 2×2 matrix is nonsingular.)

One.III.2.13 Infinitely many. For instance, in

$$\begin{pmatrix} 1 & k \\ 0 & 0 \end{pmatrix}$$

each $k \in \mathbb{R}$ gives a different class.

One.III.2.14 No. Row operations do not change the size of a matrix.

One.III.2.15 **(a)** A row operation on a matrix of zeros has no effect. Thus each such matrix is alone in its row equivalence class.

(b) No. Any nonzero entry can be rescaled.

One.III.2.16 Here are two.

$$\begin{pmatrix} 1 & 1 & 0 \\ 0 & 0 & 1 \end{pmatrix} \quad\text{and}\quad \begin{pmatrix} 1 & 0 & 0 \\ 0 & 0 & 1 \end{pmatrix}$$

One.III.2.17 Any two $n\times n$ nonsingular matrices have the same reduced echelon form, namely the matrix with all 0's except for 1's down the diagonal.

$$\begin{pmatrix} 1 & 0 & & 0 \\ 0 & 1 & & 0 \\ & & \ddots & \\ 0 & 0 & & 1 \end{pmatrix}$$

Two same-sized singular matrices need not be row equivalent. For example, these two 2×2 singular matrices are not row equivalent.

$$\begin{pmatrix} 1 & 1 \\ 0 & 0 \end{pmatrix} \quad\text{and}\quad \begin{pmatrix} 1 & 0 \\ 0 & 0 \end{pmatrix}$$

One.III.2.18 Since there is one and only one reduced echelon form matrix in each class, we can just list the possible reduced echelon form matrices.

For that list, see the answer for Exercise 12.

One.III.2.19 **(a)** If there is a linear relationship where c_0 is not zero then we can subtract $c_0\vec{\beta}_0$ from both sides and divide by $-c_0$ to get $\vec{\beta}_0$ as a linear combination of the others. (Remark: if there are no other vectors in the set — if the relationship is, say, $\vec{0} = 3 \cdot \vec{0}$ — then the statement is still true because the zero vector is by definition the sum of the empty set of vectors.)

Conversely, if $\vec{\beta}_0$ is a combination of the others $\vec{\beta}_0 = c_1\vec{\beta}_1 + \cdots + c_n\vec{\beta}_n$ then subtracting $\vec{\beta}_0$ from both sides gives a relationship where at least one of the coefficients is nonzero; namely, the -1 in front of $\vec{\beta}_0$.

(b) The first row is not a linear combination of the others for the reason given in the proof: in the equation of components from the column containing the leading entry of the first row, the only nonzero entry is the leading entry from the first row, so its coefficient must be zero. Thus, from the prior part of this exercise, the first row is in no linear relationship with the other rows.

Thus, when considering whether the second row can be in a linear relationship with the other rows, we can leave the first row out. But now the argument just applied to the first row will apply to the second row. (That is, we are arguing here by induction.)

One.III.2.20 We know that $4s + c + 10d = 8.45$ and that $3s + c + 7d = 6.30$, and we'd like to know what $s + c + d$ is. Fortunately, $s + c + d$ is a linear combination of $4s + c + 10d$ and $3s + c + 7d$. Calling the unknown price p, we have this reduction.

$$\left(\begin{array}{ccc|c} 4 & 1 & 10 & 8.45 \\ 3 & 1 & 7 & 6.30 \\ 1 & 1 & 1 & p \end{array}\right) \xrightarrow[-(1/4)\rho_1+\rho_3]{-(3/4)\rho_1+\rho_2} \left(\begin{array}{ccc|c} 4 & 1 & 10 & 8.45 \\ 0 & 1/4 & -1/2 & -0.0375 \\ 0 & 3/4 & -3/2 & p-2.1125 \end{array}\right) \xrightarrow{-3\rho_2+\rho_3} \left(\begin{array}{ccc|c} 4 & 1 & 10 & 8.45 \\ 0 & 1/4 & -1/2 & -0.0375 \\ 0 & 0 & 0 & p-2.00 \end{array}\right)$$

The price paid is \$2.00.

One.III.2.21 (1) An easy answer is this:

$$0 = 3.$$

For a less wise-guy-ish answer, solve the system:

$$\left(\begin{array}{cc|c} 3 & -1 & 8 \\ 2 & 1 & 3 \end{array}\right) \xrightarrow{-(2/3)\rho_1+\rho_2} \left(\begin{array}{cc|c} 3 & -1 & 8 \\ 0 & 5/3 & -7/3 \end{array}\right)$$

gives $y = -7/5$ and $x = 11/5$. Now any equation not satisfied by $(-7/5, 11/5)$ will do, e.g., $5x+5y = 3$.

(2) Every equation can be derived from an inconsistent system. For instance, here is how to derive "$3x + 2y = 4$" from "$0 = 5$". First,

$$0 = 5 \xrightarrow{(3/5)\rho_1} 0 = 3 \xrightarrow{x\rho_1} 0 = 3x$$

(validity of the $x = 0$ case is separate but clear). Similarly, $0 = 2y$. Ditto for $0 = 4$. But now, $0+0 = 0$ gives $3x + 2y = 4$.

One.III.2.22 Define linear systems to be equivalent if their augmented matrices are row equivalent. The proof that equivalent systems have the same solution set is easy.

One.III.2.23 **(a)** The three possible row swaps are easy, as are the three possible rescalings. One of the six possible row combinations is $k\rho_1 + \rho_2$:

$$\begin{pmatrix} 1 & 2 & 3 \\ k\cdot 1+3 & k\cdot 2+0 & k\cdot 3+3 \\ 1 & 4 & 5 \end{pmatrix}$$

and again the first and second columns add to the third. The other five combinations are similar.

(b) The obvious conjecture is that row operations do not change linear relationships among columns.

(c) A case-by-case proof follows the sketch given in the first item.

Topic: Computer Algebra Systems

1 **(a)** The commands

```
> A:=array( [[40,15],
             [-50,25]] );
> u:=array([100,50]);
> linsolve(A,u);
```

yield the answer $[1,4]$.

(b) Here there is a free variable:

```
> A:=array( [[7,0,-7,0],
             [8,1,-5,2],
             [0,1,-3,0],
             [0,3,-6,-1]] );
> u:=array([0,0,0,0]);
> linsolve(A,u);
```

prompts the reply $[\_t_1, 3\_t_1, \_t_1, 3\_t_1]$.

2 These are easy to type in. For instance, the first

```
> A:=array( [[2,2],
             [1,-4]] );
> u:=array([5,0]);
> linsolve(A,u);
```

gives the expected answer of $[2,1/2]$. The others are similar.

(a) The answer is $x = 2$ and $y = 1/2$.

(b) The answer is $x = 1/2$ and $y = 3/2$.

(c) This system has infinitely many solutions. In the first subsection, with z as a parameter, we got $x = (43-7z)/4$ and $y = (13-z)/4$. Maple responds with $[-12+7\_t_1, \_t_1, 13-4\_t_1]$, for some reason preferring y as a parameter.

(d) There is no solution to this system. When the array A and vector u are given to Maple and it is asked to `linsolve(A,u)`, it returns no result at all, that is, it responds with no solutions.

(e) The solutions is $(x,y,z) = (5,5,0)$.

(f) There are many solutions. Maple gives $[1, -1+\_t_1, 3-\_t_1, \_t_1]$.

3 As with the prior question, entering these is easy.

(a) This system has infinitely many solutions. In the second subsection we gave the solution set as

$$\{\begin{pmatrix}6\\0\end{pmatrix}+\begin{pmatrix}-2\\1\end{pmatrix}y \mid y \in \mathbb{R}\}$$

and Maple responds with $[6-2\_t_1, \_t_1]$.

(b) The solution set has only one member

$$\{\begin{pmatrix}0\\1\end{pmatrix}\}$$

and Maple has no trouble finding it $[0,1]$.

(c) This system's solution set is infinite

$$\{\begin{pmatrix}4\\-1\\0\end{pmatrix}+\begin{pmatrix}-1\\1\\1\end{pmatrix}x_3 \mid x_3 \in \mathbb{R}\}$$

and Maple gives $[\_t_1, -\_t_1+3, -\_t_1+4]$.

(d) There is a unique solution

$$\{\begin{pmatrix}1\\1\\1\end{pmatrix}\}$$

and Maple gives $[1, 1, 1]$.

(e) This system has infinitely many solutions; in the second subsection we described the solution set with two parameters

$$\{\begin{pmatrix}5/3\\2/3\\0\\0\end{pmatrix}+\begin{pmatrix}-1/3\\2/3\\1\\0\end{pmatrix}z+\begin{pmatrix}-2/3\\1/3\\0\\1\end{pmatrix}w \mid z, w \in \mathbb{R}\}$$

as does Maple $[3-2\_t_1+\_t_2, \_t_1, \_t_2, -2+3\_t_1-2\_t_2]$.

(f) The solution set is empty and Maple replies to the `linsolve(A,u)` command with no returned solutions.

4 In response to this prompting

```
> A:=array( [[a,c],
             [b,d]] );
> u:=array([p,q]);
> linsolve(A,u);
```

Maple thought for perhaps twenty seconds and gave this reply.

$$[-\frac{-dp+qc}{-bc+ad}, \frac{-bp+aq}{-bc+ad}]$$

Topic: Input-Output Analysis

1 These answers are from *Octave*.

(a) With the external use of steel as 17 789 and the external use of autos as 21 243, we get $s = 25\,952$, $a = 30\,312$.

(b) $s = 25\,857$, $a = 30\,596$

(c) $s = 25\,984$, $a = 30\,597$

2 Octave gives these answers.

(a) $s = 24\,244$, $a = 30\,307$

(b) $s = 24\,267$, $a = 30\,673$

3 **(a)** These are the equations.

$$\begin{aligned}(11.79/18.69)s-(1.28/4.27)a&=11.56\\ -(0/18.69)s+(9.87/4.27)a&=11.35\end{aligned}$$

Octave gives $s = 20.66$ and $a = 16.41$.

(b) These are the ratios.

1947	*by steel*	*by autos*
use of steel	0.63	0.09
use of autos	0.00	0.69

1958	*by steel*	*by autos*
use of steel	0.79	0.09
use of autos	0.00	0.70

(c) Octave gives (in billions of 1947 dollars) $s = 24.82$ and $a = 23.63$. In billions of 1958 dollars that is $s = 32.26$ and $a = 30.71$.

Topic: Accuracy of Computations

1 Scientific notation is convenient to express the two-place restriction. We have $.25\times 10^2+.67\times 10^0 = .25\times 10^2$. The $2/3$ has no apparent effect.

2 The reduction

$$\xrightarrow{-3\rho_1+\rho_2} \begin{array}{rl} x+2y= & 3 \\ -8= & -7.992 \end{array}$$

gives a solution of $(x,y)=(1.002, 0.999)$.

3 **(a)** The fully accurate solution is that $x=10$ and $y=0$.

(b) The four-digit conclusion is quite different.

$$\xrightarrow{-(.3454/.0003)\rho_1+\rho_2} \left(\begin{array}{cc|c} .0003 & 1.556 & 1.569 \\ 0 & 1789 & -1805 \end{array}\right) \Longrightarrow x=10460,\ y=-1.009$$

4 **(a)** For the first one, first, $(2/3)-(1/3)$ is $.666\,666\,67-.333\,333\,33=.333\,333\,34$ and so $(2/3)+((2/3)-(1/3))=.666\,666\,67+.333\,333\,34=1.000\,000\,0$. For the other one, first $((2/3)+(2/3))=.666\,666\,67+.666\,666\,67=1.333\,333\,3$ and so $((2/3)+(2/3))-(1/3)=1.333\,333\,3-.333\,333\,33=.999\,999\,97$.

(b) The first equation is $.333\,333\,33\cdot x+1.000\,000\,0\cdot y=0$ while the second is $.666\,666\,67\cdot x+2.000\,000\,0\cdot y=0$.

5 **(a)** This calculation

$$\xrightarrow[-(1/3)\rho_1+\rho_3]{-(2/3)\rho_1+\rho_2} \left(\begin{array}{ccc|c} 3 & 2 & 1 & 6 \\ 0 & -(4/3)+2\varepsilon & -(2/3)+2\varepsilon & -2+4\varepsilon \\ 0 & -(2/3)+2\varepsilon & -(1/3)-\varepsilon & -1+\varepsilon \end{array}\right)$$

$$\xrightarrow{-(1/2)\rho_2+\rho_3} \left(\begin{array}{ccc|c} 3 & 2 & 1 & 6 \\ 0 & -(4/3)+2\varepsilon & -(2/3)+2\varepsilon & -2+4\varepsilon \\ 0 & \varepsilon & -2\varepsilon & -\varepsilon \end{array}\right)$$

gives a third equation of $y-2z=-1$. Substituting into the second equation gives $((-10/3)+6\varepsilon)\cdot z=(-10/3)+6\varepsilon$ so $z=1$ and thus $y=1$. With those, the first equation says that $x=1$.

(b) The solution with two digits kept

$$\left(\begin{array}{ccc|c} .30\times 10^1 & .20\times 10^1 & .10\times 10^1 & .60\times 10^1 \\ .10\times 10^1 & .20\times 10^{-3} & .20\times 10^{-3} & .20\times 10^1 \\ .30\times 10^1 & .20\times 10^{-3} & -.10\times 10^{-3} & .10\times 10^1 \end{array}\right)$$

$$\xrightarrow[-(1/3)\rho_1+\rho_3]{-(2/3)\rho_1+\rho_2} \left(\begin{array}{ccc|c} .30\times 10^1 & .20\times 10^1 & .10\times 10^1 & .60\times 10^1 \\ 0 & -.13\times 10^1 & -.67\times 10^0 & -.20\times 10^1 \\ 0 & -.67\times 10^0 & -.33\times 10^0 & -.10\times 10^1 \end{array}\right)$$

$$\xrightarrow{-(.67/1.3)\rho_2+\rho_3} \left(\begin{array}{ccc|c} .30\times 10^1 & .20\times 10^1 & .10\times 10^1 & .60\times 10^1 \\ 0 & -.13\times 10^1 & -.67\times 10^0 & -.20\times 10^1 \\ 0 & 0 & .15\times 10^{-2} & .31\times 10^{-2} \end{array}\right)$$

comes out to be $z=2.1$, $y=2.6$, and $x=-.43$.

Topic: Analyzing Networks

1 **(a)** The total resistance is 7 ohms. With a 9 volt potential, the flow will be $9/7$ amperes. Incidentally, the voltage drops will then be: $27/7$ volts across the 3 ohm resistor, and $18/7$ volts across each of the

two 2 ohm resistors.

(b) One way to do this network is to note that the 2 ohm resistor on the left has a voltage drop of 9 volts (and hence the flow through it is 9/2 amperes), and the remaining portion on the right also has a voltage drop of 9 volts, and so we can analyze it as in the prior item. We can also use linear systems.

Using the variables from the diagram we get a linear system

$$\begin{array}{rcrcrcl} i_0 & - & i_1 & - & i_2 & & =0 \\ & & i_1 & + & i_2 & - i_3 & =0 \\ & & 2i_1 & & & & =9 \\ & & & & 7i_2 & & =9 \end{array}$$

which yields the unique solution $i_1 = 81/14$, $i_1 = 9/2$, $i_2 = 9/7$, and $i_3 = 81/14$.

Of course, the first and second paragraphs yield the same answer. Essentially, in the first paragraph we solved the linear system by a method less systematic than Gauss's Method, solving for some of the variables and then substituting.

(c) Using these variables

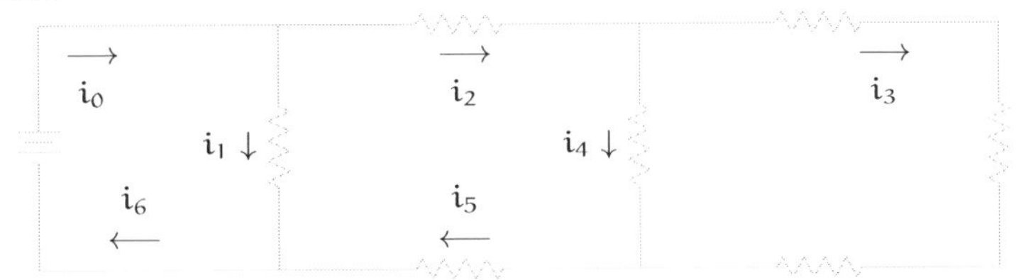

one linear system that suffices to yield a unique solution is this.

$$\begin{array}{rrrrrrrl} i_0 - & i_1 - & i_2 & & & & & =0 \\ & & i_2 - & i_3 - & i_4 & & & =0 \\ & & & i_3 + & i_4 - & i_5 & & =0 \\ & i_1 & & & & + \; i_5 & - i_6 & =0 \\ & 3i_1 & & & & & & =9 \\ & & 3i_2 & & + 2i_4 & + 2i_5 & & =9 \\ & & 3i_2 & + 9i_3 & & + 2i_5 & & =9 \end{array}$$

(The last three equations come from the circuit involving i_0-i_1-i_6, the circuit involving i_0-i_2-i_4-i_5-i_6, and the circuit with i_0-i_2-i_3-i_5-i_6.) Octave gives $i_0 = 4.35616$, $i_1 = 3.00000$, $i_2 = 1.35616$, $i_3 = 0.24658$, $i_4 = 1.10959$, $i_5 = 1.35616$, $i_6 = 4.35616$.

2 **(a)** Using the variables from the earlier analysis,

$$\begin{array}{rrrl} i_0 - & i_1 - & i_2 = & 0 \\ -i_0 + & i_1 + & i_2 = & 0 \\ & 5i_1 & = & 20 \\ & & 8i_2 = & 20 \\ & -5i_1 + & 8i_2 = & 0 \end{array}$$

The current flowing in each branch is then is $i_2 = 20/8 = 2.5$, $i_1 = 20/5 = 4$, and $i_0 = 13/2 = 6.5$, all in amperes. Thus the parallel portion is acting like a single resistor of size $20/(13/2) \approx 3.08$ ohms.

(b) A similar analysis gives that is $i_2 = i_1 = 20/8 = 4$ and $i_0 = 40/8 = 5$ amperes. The equivalent resistance is $20/5 = 4$ ohms.

(c) Another analysis like the prior ones gives is $i_2 = 20/r_2$, $i_1 = 20/r_1$, and $i_0 = 20(r_1 + r_2)/(r_1 r_2)$, all in amperes. So the parallel portion is acting like a single resistor of size $20/i_1 = r_1 r_2/(r_1 + r_2)$ ohms. (This equation is often stated as: the equivalent resistance r satisfies $1/r = (1/r_1) + (1/r_2)$.)

3 **(a)** The circuit looks like this.

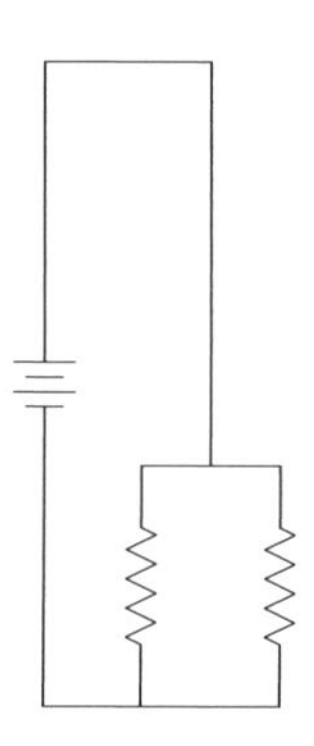

(b) The circuit looks like this.

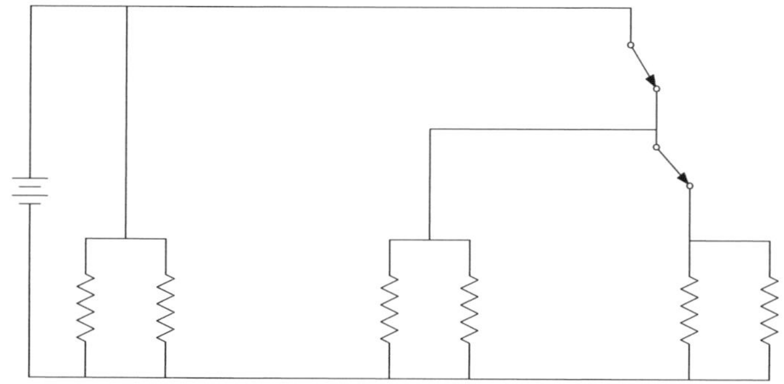

4 Not yet done.

5 **(a)** An adaptation is: in any intersection the flow in equals the flow out. It does seem reasonable in this case, unless cars are stuck at an intersection for a long time.

(b) We can label the flow in this way.

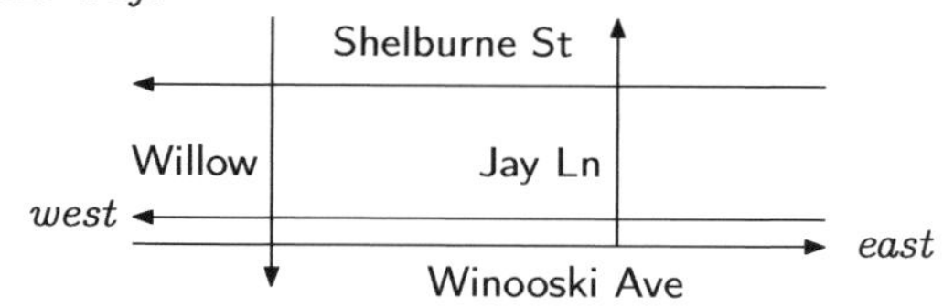

Because 50 cars leave via Main while 25 cars enter, $i_1 - 25 = i_2$. Similarly Pier's in/out balance means that $i_2 = i_3$ and North gives $i_3 + 25 = i_1$. We have this system.

$$\begin{array}{rcl} i_1 - i_2 & = & 25 \\ i_2 - i_3 & = & 0 \\ -i_1 + i_3 & = & -25 \end{array}$$

(c) The row operations $\rho_1 + \rho_2$ and $rho_2 + \rho_3$ lead to the conclusion that there are infinitely many solutions. With i_3 as the parameter,

$$\{\begin{pmatrix} 25 + i_3 \\ i_3 \\ i_3 \end{pmatrix} \mid i_3 \in \mathbb{R}\}$$

of course, since the problem is stated in number of cars, we might restrict i_3 to be a natural number.

(d) If we picture an initially-empty circle with the given input/output behavior, we can superimpose a z_3-many cars circling endlessly to get a new solution.

(e) A suitable restatement might be: the number of cars entering the circle must equal the number of cars leaving. The reasonableness of this one is not as clear. Over the five minute time period we could find that a half dozen more cars entered than left, although the problem statement's into/out table does satisfy this property. In any event, it is of no help in getting a unique solution since for that we would need to know the number of cars circling endlessly.

6 **(a)** Here is a variable for each unknown block; each known block has the flow shown.

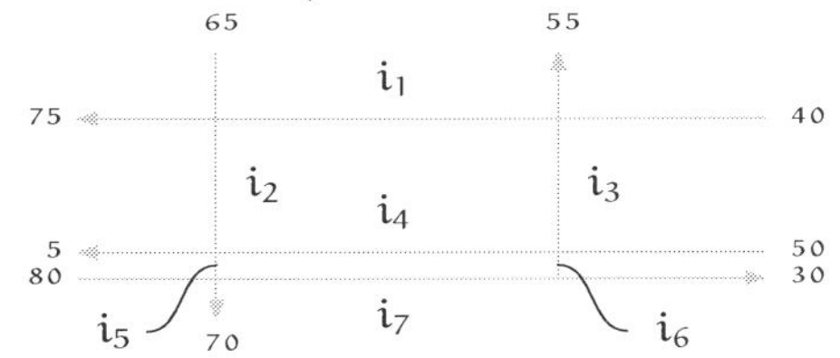

We apply Kirchhoff's principle that the flow into the intersection of Willow and Shelburne must equal the flow out to get $i_1 + 25 = i_2 + 125$. Doing the intersections from right to left and top to bottom gives these equations.

$$\begin{array}{rcrcrcrcrcrcrcr}
i_1 & - & i_2 & & & & & & & & & & & = & 10 \\
-i_1 & & & + & i_3 & & & & & & & & & = & 15 \\
& & i_2 & & & + & i_4 & & & & & & & = & 5 \\
& & & & -i_3 & - & i_4 & & & + & i_6 & & & = & -50 \\
& & & & & & & & i_5 & & & - & i_7 & = & -10 \\
& & & & & & & & & & -i_6 & + & i_7 & = & 30
\end{array}$$

The row operation $\rho_1 + \rho_2$ followed by $\rho_2 + \rho_3$ then $\rho_3 + \rho_4$ and $\rho_4 + \rho_5$ and finally $\rho_5 + \rho_6$ result in this system.

$$\begin{array}{rcrcrcrcrcrcrcr}
i_1 & - & i_2 & & & & & & & & & & & = & 10 \\
& & -i_2 & + & i_3 & & & & & & & & & = & 25 \\
& & & & i_3 & + & i_4 & - & i_5 & & & & & = & 30 \\
& & & & & & & & -i_5 & + & i_6 & & & = & -20 \\
& & & & & & & & & & -i_6 & + & i_7 & = & -30 \\
& & & & & & & & & & & & 0 & = & 0
\end{array}$$

Since the free variables are i_4 and i_7 we take them as parameters.

$$\begin{aligned}
i_6 &= i_7 - 30 \\
i_5 &= i_6 + 20 = (i_7 - 30) + 20 = i_7 - 10 \\
i_3 &= -i_4 + i_5 + 30 = -i_4 + (i_7 - 10) + 30 = -i_4 + i_7 + 20 \qquad () \\
i_2 &= i_3 - 25 = (-i_4 + i_7 + 20) - 25 = -i_4 + i_7 - 5 \\
i_1 &= i_2 + 10 = (-i_4 + i_7 - 5) + 10 = -i_4 + i_7 + 5
\end{aligned}$$

Obviously i_4 and i_7 have to be positive, and in fact the first equation shows that i_7 must be at least 30. If we start with i_7, then the i_2 equation shows that $0 \leqslant i_4 \leqslant i_7 - 5$.

(b) We cannot take i_7 to be zero or else i_6 will be negative (this would mean cars going the wrong way on the one-way street Jay). We can, however, take i_7 to be as small as 30, and then there are many suitable i_4's. For instance, the solution

$$(i_1, i_2, i_3, i_4, i_5, i_6, i_7) = (35, 25, 50, 0, 20, 0, 30)$$

results from choosing $i_4 = 0$.

Chapter Two

Chapter Two: Vector Spaces

Definition of Vector Space

Two.I.1: Definition and Examples

Two.I.1.17 **(a)** $0 + 0x + 0x^2 + 0x^3$
(b) $\begin{pmatrix} 0 & 0 & 0 & 0 \\ 0 & 0 & 0 & 0 \end{pmatrix}$
(c) The constant function $f(x) = 0$
(d) The constant function $f(n) = 0$

Two.I.1.18 **(a)** $3 + 2x - x^2$ **(b)** $\begin{pmatrix} -1 & +1 \\ 0 & -3 \end{pmatrix}$ **(c)** $-3e^x + 2e^{-x}$

Two.I.1.19 **(a)** Three elements are: $1 + 2x$, $2 - 1x$, and x. (Of course, many answers are possible.)

The verification is just like Example 1.3. We first do conditions 1-5, from the paragraph of Definition 1.1 having to do with addition. For closure under addition, condition (1), note that where $a + bx, c + dx \in \mathcal{P}_1$ we have that $(a + bx) + (c + dx) = (a + c) + (b + d)x$ is a linear polynomial with real coefficients and so is an element of $\mathcal{P}_1$. Condition (2) is verified with: where $a + bx, c + dx \in \mathcal{P}_1$ then $(a+bx)+(c+dx) = (a+c)+(b+d)x$, while in the other order they are $(c+dx)+(a+bx) = (c+a)+(d+b)x$, and both $a + c = c + a$ and $b + d = d + b$ as these are real numbers. Condition (3) is similar: suppose $a + bx, c + dx, e + fx \in \mathcal{P}$ then $((a + bx) + (c + dx)) + (e + fx) = (a + c + e) + (b + d + f)x$ while $(a + bx) + ((c + dx) + (e + fx)) = (a + c + e) + (b + d + f)x$, and the two are equal (that is, real number addition is associative so $(a + c) + e = a + (c + e)$ and $(b + d) + f = b + (d + f)$). For condition (4) observe that the linear polynomial $0 + 0x \in \mathcal{P}_1$ has the property that $(a + bx) + (0 + 0x) = a + bx$ and $(0 + 0x) + (a + bx) = a + bx$. For the last condition in this paragraph, condition (5), note that for any $a+bx \in \mathcal{P}_1$ the additive inverse is $-a-bx \in \mathcal{P}_1$ since $(a+bx)+(-a-bx) = (-a-bx)+(a+bx) = 0+0x$.

We next also check conditions (6)-(10), involving scalar multiplication. For (6), the condition that the space be closed under scalar multiplication, suppose that r is a real number and $a + bx$ is an element of $\mathcal{P}_1$, and then $r(a + bx) = (ra) + (rb)x$ is an element of $\mathcal{P}_1$ because it is a linear polynomial with real number coefficients. Condition (7) holds because $(r + s)(a + bx) = r(a + bx) + s(a + bx)$ is true from the distributive property for real number multiplication. Condition (8) is similar: $r((a + bx) + (c + dx)) = r((a + c) + (b + d)x) = r(a + c) + r(b + d)x = (ra + rc) + (rb + rd)x = r(a + bx) + r(c + dx)$. For (9) we have $(rs)(a + bx) = (rsa) + (rsb)x = r(sa + sbx) = r(s(a + bx))$. Finally, condition (10) is $1(a + bx) = (1a) + (1b)x = a + bx$.

(b) Call the set P. In the prior item in this exercise there was no restriction on the coefficients but here we are restricting attention to those linear polynomials where $a_0 - 2a_1 = 0$, that is, where the constant term minus twice the coefficient of the linear term is zero. Thus, three typical elements of P are $2+1x$, $6+3x$, and $-4-2x$.

For condition (1) we must show that if we add two linear polynomials that satisfy the restriction then we get a linear polynomial also satisfying the restriction: here that argument is that if $a+bx, c+dx \in P$ then $(a+bx)+(c+dx) = (a+c)+(b+d)x$ is an element of P because $(a+c)-2(b+d) = (a-2b)+(c-2d) = 0+0 = 0$. We can verify condition (2) with: where $a+bx, c+dx \in \mathcal{P}_1$ then $(a+bx)+(c+dx) = (a+c)+(b+d)x$, while in the other order they are $(c+dx)+(a+bx) = (c+a)+(d+b)x$, and both $a+c = c+a$ and $b+d = d+b$ as these are real numbers. (That is, this condition is not affected by the restriction and the verification is the same as the verification in the first item of this exercise). Condition (3) is also not affected by the extra restriction: suppose that $a+bx, c+dx, e+fx \in \mathcal{P}$ then $((a+bx)+(c+dx))+(e+fx) = (a+c+e)+(b+d+f)x$ while $(a+bx)+((c+dx)+(e+fx)) = (a+c+e)+(b+d+f)x$, and the two are equal. For condition (4) observe that the linear polynomial satisfies the restriction $0+0x \in P$ because its constant term minus twice the coefficient of its linear term is zero, and then the verification from the first item of this question applies: $0+0x \in \mathcal{P}_1$ has the property that $(a+bx)+(0+0x) = a+bx$ and $(0+0x)+(a+bx) = a+bx$. To check condition (5), note that for any $a+bx \in P$ the additive inverse is $-a-bx$ since it is an element of P (because $a+bx \in P$ we know that $a-2b = 0$ and multiplying both sides by -1 gives that $-a+2b = 0$), and as in the first item it acts as the additive inverse $(a+bx)+(-a-bx) = (-a-bx)+(a+bx) = 0+0x$.

We must also check conditions (6)-(10), those for scalar multiplication. For (6), the condition that the space be closed under scalar multiplication, suppose that r is a real number and $a+bx \in P$ (so that $a-2b = 0$), then $r(a+bx) = (ra)+(rb)x$ is an element of P because it is a linear polynomial with real number coefficients satisfying that $(ra)-2(rb) = r(a-2b) = 0$. Condition (7) holds for the same reason that it holds in the first item of this exercise, because $(r+s)(a+bx) = r(a+bx)+s(a+bx)$ is true from the distributive property for real number multiplication. Condition (8) is also unchanged from the first item: $r((a+bx)+(c+dx)) = r((a+c)+(b+d)x) = r(a+c)+r(b+d)x = (ra+rc)+(rb+rd)x = r(a+bx)+r(c+dx)$. So is (9): $(rs)(a+bx) = (rsa)+(rsb)x = r(sa+sbx) = r(s(a+bx))$. Finally, so is condition (10): $1(a+bx) = (1a)+(1b)x = a+bx$.

Two.I.1.20 Use Example 1.3 as a guide. (*Comment.* Because many of the conditions are quite easy to check, sometimes a person can be left with the sense that they must have missed something. But easy or routine to do is different from not necessary to do.)

(a) Here are three elements.

$$\begin{pmatrix}1&2\\3&4\end{pmatrix},\ \begin{pmatrix}-1&-2\\-3&-4\end{pmatrix},\ \begin{pmatrix}0&0\\0&0\end{pmatrix}$$

For (1), the sum of 2×2 real matrices is a 2×2 real matrix. For (2) we consider the sum of two matrices

$$\begin{pmatrix}a&b\\c&d\end{pmatrix}+\begin{pmatrix}e&f\\g&h\end{pmatrix}=\begin{pmatrix}a+e&b+f\\c+g&d+h\end{pmatrix}$$

and apply commutativity of real number addition

$$=\begin{pmatrix}e+a&f+b\\g+c&h+d\end{pmatrix}=\begin{pmatrix}e&f\\g&h\end{pmatrix}+\begin{pmatrix}a&b\\c&d\end{pmatrix}$$

to verify that the addition of the matrices is commutative. The verification for condition (3), associativity of matrix addition, is similar to the prior verification:

$$\left(\begin{pmatrix}a&b\\c&d\end{pmatrix}+\begin{pmatrix}e&f\\g&h\end{pmatrix}\right)+\begin{pmatrix}i&j\\k&l\end{pmatrix}=\begin{pmatrix}(a+e)+i&(b+f)+j\\(c+g)+k&(d+h)+l\end{pmatrix}$$

while

$$\begin{pmatrix}a&b\\c&d\end{pmatrix}+\left(\begin{pmatrix}e&f\\g&h\end{pmatrix}+\begin{pmatrix}i&j\\k&l\end{pmatrix}\right)=\begin{pmatrix}a+(e+i)&b+(f+j)\\c+(g+k)&d+(h+l)\end{pmatrix}$$

and the two are the same entry-by-entry because real number addition is associative. For (4), the zero element of this space is the 2×2 matrix of zeroes. Condition (5) holds because for any 2×2 matrix A the additive inverse is the matix whose entries are the negative of A's, the matrix $-1\cdot A$.

Condition 6 holds because a scalar multiple of a 2×2 matrix is a 2×2 matrix. For condition (7) we have this.

$$(r+s)\begin{pmatrix}a&b\\c&d\end{pmatrix}=\begin{pmatrix}(r+s)a&(r+s)b\\(r+s)c&(r+s)d\end{pmatrix}=\begin{pmatrix}ra+sa&rb+sb\\rc+sc&rd+sd\end{pmatrix}=r\begin{pmatrix}a&b\\c&d\end{pmatrix}+s\begin{pmatrix}a&b\\c&d\end{pmatrix}$$

Condition (8) goes the same way.

$$\begin{aligned}r(\begin{pmatrix}a&b\\c&d\end{pmatrix}+\begin{pmatrix}e&f\\g&h\end{pmatrix})=r\begin{pmatrix}a+e&b+f\\c+g&d+h\end{pmatrix}&=\begin{pmatrix}ra+re&rb+rf\\rc+rg&rd+rh\end{pmatrix}\\&=r\begin{pmatrix}a&b\\c&d\end{pmatrix}+r\begin{pmatrix}e&f\\g&h\end{pmatrix}=r(\begin{pmatrix}a&b\\c&d\end{pmatrix}+\begin{pmatrix}e&f\\g&h\end{pmatrix})\end{aligned}$$

For (9) we have this.

$$(rs)\begin{pmatrix}a&b\\c&d\end{pmatrix}=\begin{pmatrix}rsa&rsb\\rsc&rsd\end{pmatrix}=r\begin{pmatrix}sa&sb\\sc&sd\end{pmatrix}=r(s\begin{pmatrix}a&b\\c&d\end{pmatrix})$$

Condition (10) is just as easy.

$$1\begin{pmatrix}a&b\\c&d\end{pmatrix}=\begin{pmatrix}1\cdot a&1\cdot b\\1\cdot c&1\cdot d\end{pmatrix}=\begin{pmatrix}sa&sb\\sc&sd\end{pmatrix}$$

(b) This differs from the prior item in this exercise only in that we are restricting to the set T of matrices with a zero in the second row and first column. Here are three elements of T.

$$\begin{pmatrix}1&2\\0&4\end{pmatrix},\begin{pmatrix}-1&-2\\0&-4\end{pmatrix},\begin{pmatrix}0&0\\0&0\end{pmatrix}$$

Some of the verifications for this item are the same as for the first item in this exercise, and below we'll just do the ones that are different.

For (1), the sum of 2×2 real matrices with a zero in the $2,1$ entry is also a 2×2 real matrix with a zero in the $2,1$ entry.

$$\begin{pmatrix}a&b\\0&d\end{pmatrix}+\begin{pmatrix}e&f\\0&h\end{pmatrix}\begin{pmatrix}a+e&b+f\\0&d+h\end{pmatrix}$$

The verification for condition (2) given in the prior item works in this item also. The same holds for condition (3). For (4), note that the 2×2 matrix of zeroes is an element of T. Condition (5) holds because for any 2×2 matrix A the additive inverse is the matrix $-1\cdot A$ and so the additive inverse of a matrix with a zero in the $2,1$ entry is also a matris with a zero in the $2,1$ entry.

Condition 6 holds because a scalar multiple of a 2×2 matrix with a zero in the $2,1$ entry is a 2×2 matrix with a zero in the $2,1$ entry. Condition (7)'s verification is the same as in the prior item. So are condition (8)'s, (9)'s, and (10)'s.

Two.I.1.21 Most of the conditions are easy to check; use Example 1.3 as a guide.

(a) Three elements are $(1\ \ 2\ \ 3)$, $(2\ \ 1\ \ 3)$, and $(0\ \ 0\ \ 0)$.

We must check conditions (1)-(10) in Definition 1.1. Conditions (1)-(5) concern addition. For condition (1) recall that the sum of two three-component row vectors

$$(a\ \ b\ \ c)+(d\ \ e\ \ f)=(a+d\ \ b+e\ \ c+f)$$

is also a three-component row vector (all of the letters $a,\ldots,f$ represent real numbers). Verification of (2) is routine

$$(a\ \ b\ \ c)+(d\ \ e\ \ f)=(a+d\ \ b+e\ \ c+f)=(d+a\ \ e+b\ \ f+c)=(d\ \ e\ \ f)+(a\ \ b\ \ c)$$

(the second equality holds because the three entries are real numbers and real number addition commutes). Condition (3)'s verification is similar.

$$\begin{aligned}\big((a\ \ b\ \ c)+(d\ \ e\ \ f)\big)+(g\ \ h\ \ i) &= ((a+d)+g\ \ \ (b+e)+h\ \ \ (c+f)+i) \\ &= (a+(d+g)\ \ \ b+(e+h)\ \ \ c+(f+i)) = (a\ \ b\ \ c)+\big((d\ \ e\ \ f)+(g\ \ h\ \ i)\big)\end{aligned}$$

For (4), observe that the three-component row vector $(0\ \ 0\ \ 0)$ is the additive identity: $(a\ \ b\ \ c)+(0\ \ 0\ \ 0)=(a\ \ b\ \ c)$. To verify condition (5), assume we are given the element $(a\ \ b\ \ c)$ of the set and note that $(-a\ \ -b\ \ -c)$ is also in the set and has the desired property: $(a\ \ b\ \ c)+(-a\ \ -b\ \ -c)=(0\ \ 0\ \ 0)$.

Conditions (6)-(10) involve scalar multiplication. To verify (6), that the space is closed under the scalar multiplication operation that was given, note that $r(a\ \ b\ \ c)=(ra\ \ rb\ \ rc)$ is a three-component row vector with real entries. For (7) we compute $(r+s)(a\ \ b\ \ c)=((r+s)a\ \ (r+s)b\ \ (r+s)c)=(ra+sa\ \ rb+sb\ \ rc+sc)=(ra\ \ rb\ \ rc)+(sa\ \ sb\ \ sc)=r(a\ \ b\ \ c)+s(a\ \ b\ \ c)$. Condition (8) is very similar: $r\big((a\ \ b\ \ c)+(d\ \ e\ \ f)\big)=r(a+d\ \ b+e\ \ c+f)=(r(a+d)\ \ r(b+e)\ \ r(c+f))=(ra+rd\ \ rb+re\ \ rc+rf)=(ra\ \ rb\ \ rc)+(rd\ \ re\ \ rf)=r(a\ \ b\ \ c)+r(d\ \ e\ \ f)$. So is the computation for condition (9): $(rs)(a\ \ b\ \ c)=(rsa\ \ rsb\ \ rsc)=r(sa\ \ sb\ \ sc)=r\big(s(a\ \ b\ \ c)\big)$. Condition (10) is just as routine $1(a\ \ b\ \ c)=(1\cdot a\ \ 1\cdot b\ \ 1\cdot c)=(a\ \ b\ \ c)$.

(b) Call the set L. Closure of addition, condition (1), involves checking that if the summands are members of L then the sum

$$\begin{pmatrix}a\\b\\c\\d\end{pmatrix}+\begin{pmatrix}e\\f\\g\\h\end{pmatrix}=\begin{pmatrix}a+e\\b+f\\c+g\\d+h\end{pmatrix}$$

is also a member of L, which is true because it satisfies the criteria for membership in L: $(a+e)+(b+f)-(c+g)+(d+h)=(a+b-c+d)+(e+f-g+h)=0+0$. The verifications for conditions (2), (3), and (5) are similar to the ones in the first part of this exercise. For condition (4) note that the vector of zeroes is a member of L because its first component plus its second, minus its third, and plus its fourth, totals to zero.

Condition (6), closure of scalar multiplication, is similar: where the vector is an element of L,

$$r\begin{pmatrix}a\\b\\c\\d\end{pmatrix}=\begin{pmatrix}ra\\rb\\rc\\rd\end{pmatrix}$$

is also an element of L because $ra+rb-rc+rd=r(a+b-c+d)=r\cdot 0=0$. The verification for conditions (7), (8), (9), and (10) are as in the prior item of this exercise.

Two.I.1.22 In each item the set is called Q. For some items, there are other correct ways to show that Q is not a vector space.

(a) It is not closed under addition; it fails to meet condition (1).

$$\begin{pmatrix}1\\0\\0\end{pmatrix},\begin{pmatrix}0\\1\\0\end{pmatrix}\in Q\qquad\begin{pmatrix}1\\1\\0\end{pmatrix}\notin Q$$

(b) It is not closed under addition.

$$\begin{pmatrix}1\\0\\0\end{pmatrix},\begin{pmatrix}0\\1\\0\end{pmatrix}\in Q\qquad\begin{pmatrix}1\\1\\0\end{pmatrix}\notin Q$$

(c) It is not closed under addition.

$$\begin{pmatrix}0&1\\0&0\end{pmatrix},\begin{pmatrix}1&1\\0&0\end{pmatrix}\in Q\qquad\begin{pmatrix}1&2\\0&0\end{pmatrix}\notin Q$$

(d) It is not closed under scalar multiplication.

$$1 + 1x + 1x^2 \in Q \qquad -1 \cdot (1 + 1x + 1x^2) \notin Q$$

(e) It is empty, violating condition (4).

Two.I.1.23 The usual operations $(v_0+v_1i)+(w_0+w_1i) = (v_0+w_0)+(v_1+w_1)i$ and $r(v_0+v_1i) = (rv_0)+(rv_1)i$ suffice. The check is easy.

Two.I.1.24 No, it is not closed under scalar multiplication since, e.g., $\pi \cdot (1)$ is not a rational number.

Two.I.1.25 The natural operations are $(v_1x + v_2y + v_3z) + (w_1x + w_2y + w_3z) = (v_1 + w_1)x + (v_2 + w_2)y + (v_3 + w_3)z$ and $r \cdot (v_1x + v_2y + v_3z) = (rv_1)x + (rv_2)y + (rv_3)z$. The check that this is a vector space is easy; use Example 1.3 as a guide.

Two.I.1.26 The '+' operation is not commutative (that is, condition (2) is not met); producing two members of the set witnessing this assertion is easy.

Two.I.1.27 **(a)** It is not a vector space.

$$(1+1) \cdot \begin{pmatrix} 1 \\ 0 \\ 0 \end{pmatrix} \neq \begin{pmatrix} 1 \\ 0 \\ 0 \end{pmatrix} + \begin{pmatrix} 1 \\ 0 \\ 0 \end{pmatrix}$$

(b) It is not a vector space.

$$1 \cdot \begin{pmatrix} 1 \\ 0 \\ 0 \end{pmatrix} \neq \begin{pmatrix} 1 \\ 0 \\ 0 \end{pmatrix}$$

Two.I.1.28 For each "yes" answer, you must give a check of all the conditions given in the definition of a vector space. For each "no" answer, give a specific example of the failure of one of the conditions.

(a) Yes.

(b) Yes.

(c) No, this set is not closed under the natural addition operation. The vector of all 1/4's is a member of this set but when added to itself the result, the vector of all 1/2's, is a nonmember.

(d) Yes.

(e) No, $f(x) = e^{-2x} + (1/2)$ is in the set but $2 \cdot f$ is not (that is, condition (6) fails).

Two.I.1.29 It is a vector space. Most conditions of the definition of vector space are routine; we here check only closure. For addition, $(f_1 + f_2)\,(7) = f_1(7) + f_2(7) = 0 + 0 = 0$. For scalar multiplication, $(r \cdot f)\,(7) = rf(7) = r0 = 0$.

Two.I.1.30 We check Definition 1.1.

First, closure under '+' holds because the product of two positive reals is a positive real. The second condition is satisfied because real multiplication commutes. Similarly, as real multiplication associates, the third checks. For the fourth condition, observe that multiplying a number by $1 \in \mathbb{R}^+$ won't change the number. Fifth, any positive real has a reciprocal that is a positive real.

The sixth, closure under '·', holds because any power of a positive real is a positive real. The seventh condition is just the rule that v^{r+s} equals the product of v^r and v^s. The eight condition says that $(vw)^r = v^r w^r$. The ninth condition asserts that $(v^r)^s = v^{rs}$. The final condition says that $v^1 = v$.

Two.I.1.31 **(a)** No: $1 \cdot (0,1) + 1 \cdot (0,1) \neq (1+1) \cdot (0,1)$.

(b) No; the same calculation as the prior answer shows a condition in the definition of a vector space that is violated. Another example of a violation of the conditions for a vector space is that $1 \cdot (0,1) \neq (0,1)$.

Two.I.1.32 It is not a vector space since it is not closed under addition, as $(x^2) + (1 + x - x^2)$ is not in the set.

Two.I.1.33 **(a)** 6

(b) nm

(c) 3

(d) To see that the answer is 2, rewrite it as

$$\{\begin{pmatrix} a & 0 \\ b & -a-b \end{pmatrix} \mid a, b \in \mathbb{R}\}$$

so that there are two parameters.

Two.I.1.34 A *vector space* (over $\mathbb{R}$) consists of a set V along with two operations '$\vec{+}$' and '$\vec{\cdot}$' subject to these conditions. Where $\vec{v}, \vec{w} \in V$, (1) their *vector sum* $\vec{v} \vec{+} \vec{w}$ is an element of V. If $\vec{u}, \vec{v}, \vec{w} \in V$ then (2) $\vec{v} \vec{+} \vec{w} = \vec{w} \vec{+} \vec{v}$ and (3) $(\vec{v} \vec{+} \vec{w}) \vec{+} \vec{u} = \vec{v} \vec{+} (\vec{w} \vec{+} \vec{u})$. (4) There is a *zero vector* $\vec{0} \in V$ such that $\vec{v} \vec{+} \vec{0} = \vec{v}$ for all $\vec{v} \in V$. (5) Each $\vec{v} \in V$ has an *additive inverse* $\vec{w} \in V$ such that $\vec{w} \vec{+} \vec{v} = \vec{0}$. If r, s are *scalars*, that is, members of $\mathbb{R}$), and $\vec{v}, \vec{w} \in V$ then (6) each *scalar multiple* $r \cdot \vec{v}$ is in V. If $r, s \in \mathbb{R}$ and $\vec{v}, \vec{w} \in V$ then (7) $(r + s) \cdot \vec{v} = r \cdot \vec{v} \vec{+} s \cdot \vec{v}$, and (8) $r \vec{\cdot} (\vec{v} + \vec{w}) = r \vec{\cdot} \vec{v} + r \vec{\cdot} \vec{w}$, and (9) $(rs) \vec{\cdot} \vec{v} = r \vec{\cdot} (s \vec{\cdot} \vec{v})$, and (10) $1 \vec{\cdot} \vec{v} = \vec{v}$.

Two.I.1.35 **(a)** Let V be a vector space, assume that $\vec{v} \in V$, and assume that $\vec{w} \in V$ is the additive inverse of $\vec{v}$ so that $\vec{w} + \vec{v} = \vec{0}$. Because addition is commutative, $\vec{0} = \vec{w} + \vec{v} = \vec{v} + \vec{w}$, so therefore $\vec{v}$ is also the additive inverse of $\vec{w}$.

(b) Let V be a vector space and suppose $\vec{v}, \vec{s}, \vec{t} \in V$. The additive inverse of $\vec{v}$ is $-\vec{v}$ so $\vec{v} + \vec{s} = \vec{v} + \vec{t}$ gives that $-\vec{v} + \vec{v} + \vec{s} = -\vec{v} + \vec{v} + \vec{t}$, which says that $\vec{0} + \vec{s} = \vec{0} + \vec{t}$ and so $\vec{s} = \vec{t}$.

Two.I.1.36 Addition is commutative, so in any vector space, for any vector $\vec{v}$ we have that $\vec{v} = \vec{v} + \vec{0} = \vec{0} + \vec{v}$.

Two.I.1.37 It is not a vector space since addition of two matrices of unequal sizes is not defined, and thus the set fails to satisfy the closure condition.

Two.I.1.38 Each element of a vector space has one and only one additive inverse.

For, let V be a vector space and suppose that $\vec{v} \in V$. If $\vec{w}_1, \vec{w}_2 \in V$ are both additive inverses of $\vec{v}$ then consider $\vec{w}_1 + \vec{v} + \vec{w}_2$. On the one hand, we have that it equals $\vec{w}_1 + (\vec{v} + \vec{w}_2) = \vec{w}_1 + \vec{0} = \vec{w}_1$. On the other hand we have that it equals $(\vec{w}_1 + \vec{v}) + \vec{w}_2 = \vec{0} + \vec{w}_2 = \vec{w}_2$. Therefore, $\vec{w}_1 = \vec{w}_2$.

Two.I.1.39 **(a)** Every such set has the form $\{r \cdot \vec{v} + s \cdot \vec{w} \mid r, s \in \mathbb{R}\}$ where either or both of $\vec{v}, \vec{w}$ may be $\vec{0}$. With the inherited operations, closure of addition $(r_1\vec{v} + s_1\vec{w}) + (r_2\vec{v} + s_2\vec{w}) = (r_1 + r_2)\vec{v} + (s_1 + s_2)\vec{w}$ and scalar multiplication $c(r\vec{v} + s\vec{w}) = (cr)\vec{v} + (cs)\vec{w}$ are easy. The other conditions are also routine.

(b) No such set can be a vector space under the inherited operations because it does not have a zero element.

Two.I.1.40 Assume that $\vec{v} \in V$ is not $\vec{0}$.

(a) One direction of the if and only if is clear: if $r = 0$ then $r \cdot \vec{v} = \vec{0}$. For the other way, let r be a nonzero scalar. If $r\vec{v} = \vec{0}$ then $(1/r) \cdot r\vec{v} = (1/r) \cdot \vec{0}$ shows that $\vec{v} = \vec{0}$, contrary to the assumption.

(b) Where r_1, r_2 are scalars, $r_1\vec{v} = r_2\vec{v}$ holds if and only if $(r_1 - r_2)\vec{v} = \vec{0}$. By the prior item, then $r_1 - r_2 = 0$.

(c) A nontrivial space has a vector $\vec{v} \neq \vec{0}$. Consider the set $\{k \cdot \vec{v} \mid k \in \mathbb{R}\}$. By the prior item this set is infinite.

(d) The solution set is either trivial, or nontrivial. In the second case, it is infinite.

Two.I.1.41 Yes. A theorem of first semester calculus says that a sum of differentiable functions is differentiable and that $(f+g)' = f'+g'$, and that a multiple of a differentiable function is differentiable and that $(r \cdot f)' = r\,f'$.

Two.I.1.42 The check is routine. Note that '1' is $1 + 0i$ and the zero elements are these.

(a) $(0 + 0i) + (0 + 0i)x + (0 + 0i)x^2$

(b) $\begin{pmatrix} 0 + 0i & 0 + 0i \\ 0 + 0i & 0 + 0i \end{pmatrix}$

Two.I.1.43 Notably absent from the definition of a vector space is a distance measure.

Two.I.1.44 (a) A small rearrangement does the trick.

$$\begin{aligned}(\vec{v}_1+(\vec{v}_2+\vec{v}_3))+\vec{v}_4 &= ((\vec{v}_1+\vec{v}_2)+\vec{v}_3)+\vec{v}_4\\ &= (\vec{v}_1+\vec{v}_2)+(\vec{v}_3+\vec{v}_4)\\ &= \vec{v}_1+(\vec{v}_2+(\vec{v}_3+\vec{v}_4))\\ &= \vec{v}_1+((\vec{v}_2+\vec{v}_3)+\vec{v}_4)\end{aligned}$$

Each equality above follows from the associativity of three vectors that is given as a condition in the definition of a vector space. For instance, the second '=' applies the rule $(\vec{w}_1+\vec{w}_2)+\vec{w}_3=\vec{w}_1+(\vec{w}_2+\vec{w}_3)$ by taking $\vec{w}_1$ to be $\vec{v}_1+\vec{v}_2$, taking $\vec{w}_2$ to be $\vec{v}_3$, and taking $\vec{w}_3$ to be $\vec{v}_4$.

(b) The base case for induction is the three vector case. This case $\vec{v}_1+(\vec{v}_2+\vec{v}_3)=(\vec{v}_1+\vec{v}_2)+\vec{v}_3$ is one of the conditions in the definition of a vector space.

For the inductive step, assume that any two sums of three vectors, any two sums of four vectors, ..., any two sums of k vectors are equal no matter how we parenthesize the sums. We will show that any sum of $k+1$ vectors equals this one $((\cdots((\vec{v}_1+\vec{v}_2)+\vec{v}_3)+\cdots)+\vec{v}_k)+\vec{v}_{k+1}$.

Any parenthesized sum has an outermost '+'. Assume that it lies between $\vec{v}_m$ and $\vec{v}_{m+1}$ so the sum looks like this.

$$(\cdots\vec{v}_1\cdots\vec{v}_m\cdots)+(\cdots\vec{v}_{m+1}\cdots\vec{v}_{k+1}\cdots)$$

The second half involves fewer than $k+1$ additions, so by the inductive hypothesis we can re-parenthesize it so that it reads left to right from the inside out, and in particular, so that its outermost '+' occurs right before $\vec{v}_{k+1}$.

$$=(\cdots\vec{v}_1\cdots\vec{v}_m\cdots)+((\cdots(\vec{v}_{m+1}+\vec{v}_{m+2})+\cdots+\vec{v}_k)+\vec{v}_{k+1})$$

Apply the associativity of the sum of three things

$$=((\cdots\vec{v}_1\cdots\vec{v}_m\cdots)+(\cdots(\vec{v}_{m+1}+\vec{v}_{m+2})+\cdots\vec{v}_k))+\vec{v}_{k+1}$$

and finish by applying the inductive hypothesis inside these outermost parenthesis.

Two.I.1.45 Let $\vec{v}$ be a member of $\mathbb{R}^2$ with components v_1 and v_2. We can abbreviate the condition that both components have the same sign or are 0 by $v_1v_2\geqslant 0$.

To show the set is closed under scalar multiplication, observe that the components of $r\vec{v}$ satisfy $(rv_1)(rv_2)=r^2(v_1v_2)$ and $r^2\geqslant 0$ so $r^2v_1v_2\geqslant 0$.

To show the set is not closed under addition we need only produce one example. The vector with components -1 and 0, when added to the vector with components 0 and 1 makes a vector with mixed-sign components of -1 and 1.

Two.I.1.46 (a) We outline the check of the conditions from Definition 1.1.

Additive closure holds because if $a_0+a_1+a_2=0$ and $b_0+b_1+b_2=0$ then

$$(a_0+a_1x+a_2x^2)+(b_0+b_1x+b_2x^2)=(a_0+b_0)+(a_1+b_1)x+(a_2+b_2)x^2$$

is in the set since $(a_0+b_0)+(a_1+b_1)+(a_2+b_2)=(a_0+a_1+a_2)+(b_0+b_1+b_2)$ is zero. The second through fifth conditions are easy.

Closure under scalar multiplication holds because if $a_0+a_1+a_2=0$ then

$$r\cdot(a_0+a_1x+a_2x^2)=(ra_0)+(ra_1)x+(ra_2)x^2$$

is in the set as $ra_0+ra_1+ra_2=r(a_0+a_1+a_2)$ is zero. The remaining conditions here are also easy.

(b) This is similar to the prior answer.

(c) Call the vector space V. We have two implications: left to right, if S is a subspace then it is closed under linear combinations of pairs of vectors and, right to left, if a nonempty subset is closed under linear combinations of pairs of vectors then it is a subspace. The left to right implication is easy; we here sketch the other one by assuming S is nonempty and closed, and checking the conditions of Definition 1.1.

First, to show closure under addition, if $\vec{s}_1,\vec{s}_2\in S$ then $\vec{s}_1+\vec{s}_2\in S$ as $\vec{s}_1+\vec{s}_2=1\cdot\vec{s}_1+1\cdot\vec{s}_2$. Second, for any $\vec{s}_1,\vec{s}_2\in S$, because addition is inherited from V, the sum $\vec{s}_1+\vec{s}_2$ in S equals the sum $\vec{s}_1+\vec{s}_2$ in

V and that equals the sum $\vec{s}_2+\vec{s}_1$ in V and that in turn equals the sum $\vec{s}_2+\vec{s}_1$ in S. The argument for the third condition is similar to that for the second. For the fourth, suppose that $\vec{s}$ is in the nonempty set S and note that $0\cdot\vec{s}=\vec{0}\in S$; showing that the $\vec{0}$ of V acts under the inherited operations as the additive identity of S is easy. The fifth condition is satisfied because for any $\vec{s}\in S$ closure under linear combinations shows that the vector $0\cdot\vec{0}+(-1)\cdot\vec{s}$ is in S; showing that it is the additive inverse of $\vec{s}$ under the inherited operations is routine.

The proofs for the remaining conditions are similar.

Two.I.2: Subspaces and Spanning Sets

Two.I.2.20 By Lemma 2.9, to see if each subset of $\mathcal{M}_{2\times 2}$ is a subspace, we need only check if it is nonempty and closed.

(a) Yes, we can easily checke that it is nonempty and closed. This is a parametrization.

$$\{a\begin{pmatrix}1&0\\0&0\end{pmatrix}+b\begin{pmatrix}0&0\\0&1\end{pmatrix}\mid a,b\in\mathbb{R}\}$$

By the way, the parametrization also shows that it is a subspace, since it is given as the span of the two-matrix set, and any span is a subspace.

(b) Yes; it is easily checked to be nonempty and closed. Alternatively, as mentioned in the prior answer, the existence of a parametrization shows that it is a subspace. For the parametrization, the condition $a+b=0$ can be rewritten as $a=-b$. Then we have this.

$$\{\begin{pmatrix}-b&0\\0&b\end{pmatrix}\mid b\in\mathbb{R}\}=\{b\begin{pmatrix}-1&0\\0&1\end{pmatrix}\mid b\in\mathbb{R}\}$$

(c) No. It is not closed under addition. For instance,

$$\begin{pmatrix}5&0\\0&0\end{pmatrix}+\begin{pmatrix}5&0\\0&0\end{pmatrix}=\begin{pmatrix}10&0\\0&0\end{pmatrix}$$

is not in the set. (This set is also not closed under scalar multiplication, for instance, it does not contain the zero matrix.)

(d) Yes.

$$\{b\begin{pmatrix}-1&0\\0&1\end{pmatrix}+c\begin{pmatrix}0&1\\0&0\end{pmatrix}\mid b,c\in\mathbb{R}\}$$

Two.I.2.21 No, it is not closed. In particular, it is not closed under scalar multiplication because it does not contain the zero polynomial.

Two.I.2.22 **(a)** Yes, solving the linear system arising from

$$r_1\begin{pmatrix}1\\0\\0\end{pmatrix}+r_2\begin{pmatrix}0\\0\\1\end{pmatrix}=\begin{pmatrix}2\\0\\1\end{pmatrix}$$

gives $r_1=2$ and $r_2=1$.

(b) Yes; the linear system arising from $r_1(x^2)+r_2(2x+x^2)+r_3(x+x^3)=x-x^3$

$$\begin{aligned}2r_2+r_3&=1\\ r_1+r_2&=0\\ r_3&=-1\end{aligned}$$

gives that $-1(x^2)+1(2x+x^2)-1(x+x^3)=x-x^3$.

(c) No; any combination of the two given matrices has a zero in the upper right.

Two.I.2.23 **(a)** Yes; it is in that span since $1 \cdot \cos^2 x + 1 \cdot \sin^2 x = f(x)$.

(b) No, since $r_1 \cos^2 x + r_2 \sin^2 x = 3 + x^2$ has no scalar solutions that work for all x. For instance, setting x to be 0 and π gives the two equations $r_1 \cdot 1 + r_2 \cdot 0 = 3$ and $r_1 \cdot 1 + r_2 \cdot 0 = 3 + \pi^2$, which are not consistent with each other.

(c) No; consider what happens on setting x to be $\pi/2$ and $3\pi/2$.

(d) Yes, $\cos(2x) = 1 \cdot \cos^2(x) - 1 \cdot \sin^2(x)$.

Two.I.2.24 **(a)** Yes, for any $x, y, z \in \mathbb{R}$ this equation

$$r_1 \begin{pmatrix} 1 \\ 0 \\ 0 \end{pmatrix} + r_2 \begin{pmatrix} 0 \\ 2 \\ 0 \end{pmatrix} + r_3 \begin{pmatrix} 0 \\ 0 \\ 3 \end{pmatrix} = \begin{pmatrix} x \\ y \\ z \end{pmatrix}$$

has the solution $r_1 = x$, $r_2 = y/2$, and $r_3 = z/3$.

(b) Yes, the equation

$$r_1 \begin{pmatrix} 2 \\ 0 \\ 1 \end{pmatrix} + r_2 \begin{pmatrix} 1 \\ 1 \\ 0 \end{pmatrix} + r_3 \begin{pmatrix} 0 \\ 0 \\ 1 \end{pmatrix} = \begin{pmatrix} x \\ y \\ z \end{pmatrix}$$

gives rise to this

$$\begin{array}{rl} 2r_1 + r_2 &= x \\ r_2 &= y \\ r_1 + r_3 &= z \end{array} \xrightarrow{-(1/2)\rho_1+\rho_3} \xrightarrow{(1/2)\rho_2+\rho_3} \begin{array}{rl} 2r_1 + r_2 &= x \\ r_2 &= y \\ r_3 &= -(1/2)x + (1/2)y + z \end{array}$$

so that, given any x, y, and z, we can compute that $r_3 = (-1/2)x + (1/2)y + z$, $r_2 = y$, and $r_1 = (1/2)x - (1/2)y$.

(c) No. In particular, we cannot get the vector

$$\begin{pmatrix} 0 \\ 0 \\ 1 \end{pmatrix}$$

as a linear combination since the two given vectors both have a third component of zero.

(d) Yes. The equation

$$r_1 \begin{pmatrix} 1 \\ 0 \\ 1 \end{pmatrix} + r_2 \begin{pmatrix} 3 \\ 1 \\ 0 \end{pmatrix} + r_3 \begin{pmatrix} -1 \\ 0 \\ 0 \end{pmatrix} + r_4 \begin{pmatrix} 2 \\ 1 \\ 5 \end{pmatrix} = \begin{pmatrix} x \\ y \\ z \end{pmatrix}$$

leads to this reduction.

$$\left(\begin{array}{cccc|c} 1 & 3 & -1 & 2 & x \\ 0 & 1 & 0 & 1 & y \\ 1 & 0 & 0 & 5 & z \end{array}\right) \xrightarrow{-\rho_1+\rho_3} \xrightarrow{3\rho_2+\rho_3} \left(\begin{array}{cccc|c} 1 & 3 & -1 & 2 & x \\ 0 & 1 & 0 & 1 & y \\ 0 & 0 & 1 & 6 & -x+3y+z \end{array}\right)$$

We have infinitely many solutions. We can, for example, set r_4 to be zero and solve for r_3, r_2, and r_1 in terms of x, y, and z by the usual methods of back-substitution.

(e) No. The equation

$$r_1 \begin{pmatrix} 2 \\ 1 \\ 1 \end{pmatrix} + r_2 \begin{pmatrix} 3 \\ 0 \\ 1 \end{pmatrix} + r_3 \begin{pmatrix} 5 \\ 1 \\ 2 \end{pmatrix} + r_4 \begin{pmatrix} 6 \\ 0 \\ 2 \end{pmatrix} = \begin{pmatrix} x \\ y \\ z \end{pmatrix}$$

leads to this reduction.

$$\left(\begin{array}{cccc|c} 2 & 3 & 5 & 6 & x \\ 1 & 0 & 1 & 0 & y \\ 1 & 1 & 2 & 2 & z \end{array}\right) \xrightarrow[-(1/2)\rho_1+\rho_3]{-(1/2)\rho_1+\rho_2} \xrightarrow{-(1/3)\rho_2+\rho_3} \left(\begin{array}{cccc|c} 2 & 3 & 5 & 6 & x \\ 0 & -3/2 & -3/2 & -3 & -(1/2)x+y \\ 0 & 0 & 0 & 0 & -(1/3)x-(1/3)y+z \end{array}\right)$$

This shows that not every three-tall vector can be so expressed. Only the vectors satisfying the restriction that $-(1/3)x - (1/3)y + z = 0$ are in the span. (To see that any such vector is indeed expressible, take r_3 and r_4 to be zero and solve for r_1 and r_2 in terms of x, y, and z by back-substitution.)

Two.I.2.25 **(a)** $\{(c\ \ b\ \ c) \mid b, c \in \mathbb{R}\} = \{b(0\ \ 1\ \ 0) + c(1\ \ 0\ \ 1) \mid b, c \in \mathbb{R}\}$ The obvious choice for the set that spans is $\{(0\ \ 1\ \ 0), (1\ \ 0\ \ 1)\}$.

(b) $\{\begin{pmatrix} -d & b \\ c & d \end{pmatrix} \mid b, c, d \in \mathbb{R}\} = \{b\begin{pmatrix} 0 & 1 \\ 0 & 0 \end{pmatrix} + c\begin{pmatrix} 0 & 0 \\ 1 & 0 \end{pmatrix} + d\begin{pmatrix} -1 & 0 \\ 0 & 1 \end{pmatrix} \mid b, c, d \in \mathbb{R}\}$ One set that spans this space consists of those three matrices.

(c) The system

$$\begin{aligned} a + 3b \qquad\quad &= 0 \\ 2a \qquad\ -c - d &= 0 \end{aligned}$$

gives $b = -(c+d)/6$ and $a = (c+d)/2$. So one description is this.

$$\{c\begin{pmatrix} 1/2 & -1/6 \\ 1 & 0 \end{pmatrix} + d\begin{pmatrix} 1/2 & -1/6 \\ 0 & 1 \end{pmatrix} \mid c, d \in \mathbb{R}\}$$

That shows that a set spanning this subspace consists of those two matrices.

(d) The $a = 2b - c$ gives $\{(2b - c) + bx + cx^3 \mid b, c \in \mathbb{R}\} = \{b(2 + x) + c(-1 + x^3) \mid b, c \in \mathbb{R}\}$. So the subspace is the span of the set $\{2 + x, -1 + x^3\}$.

(e) The set $\{a + bx + cx^2 \mid a + 7b + 49c = 0\}$ parametrized as $\{b(-7 + x) + c(-49 + x^2) \mid b, c \in \mathbb{R}\}$ has the spanning set $\{-7 + x, -49 + x^2\}$.

Two.I.2.26 Each answer given is only one out of many possible.

(a) We can parametrize in this way

$$\{\begin{pmatrix} x \\ 0 \\ z \end{pmatrix} \mid x, z \in \mathbb{R}\} = \{x\begin{pmatrix} 1 \\ 0 \\ 0 \end{pmatrix} + z\begin{pmatrix} 0 \\ 0 \\ 1 \end{pmatrix} \mid x, z \in \mathbb{R}\}$$

giving this for a spanning set.

$$\{\begin{pmatrix} 1 \\ 0 \\ 0 \end{pmatrix}, \begin{pmatrix} 0 \\ 0 \\ 1 \end{pmatrix}\}$$

(b) Parametrize it with $\{y\begin{pmatrix} -2/3 \\ 1 \\ 0 \end{pmatrix} + z\begin{pmatrix} -1/3 \\ 0 \\ 1 \end{pmatrix} \mid y, z \in \mathbb{R}\}$ to get $\{\begin{pmatrix} -2/3 \\ 1 \\ 0 \end{pmatrix}, \begin{pmatrix} -1/3 \\ 0 \\ 1 \end{pmatrix}\}$.

(c) $\{\begin{pmatrix} 1 \\ -2 \\ 1 \\ 0 \end{pmatrix}, \begin{pmatrix} -1/2 \\ 0 \\ 0 \\ 1 \end{pmatrix}\}$

(d) Parametrize the description as $\{-a_1 + a_1x + a_3x^2 + a_3x^3 \mid a_1, a_3 \in \mathbb{R}\}$ to get $\{-1 + x, x^2 + x^3\}$.

(e) $\{1, x, x^2, x^3, x^4\}$

(f) $\{\begin{pmatrix} 1 & 0 \\ 0 & 0 \end{pmatrix}, \begin{pmatrix} 0 & 1 \\ 0 & 0 \end{pmatrix}, \begin{pmatrix} 0 & 0 \\ 1 & 0 \end{pmatrix}, \begin{pmatrix} 0 & 0 \\ 0 & 1 \end{pmatrix}\}$

Two.I.2.27 Technically, no. Subspaces of $\mathbb{R}^3$ are sets of three-tall vectors, while $\mathbb{R}^2$ is a set of two-tall vectors. Clearly though, $\mathbb{R}^2$ is "just like" this subspace of $\mathbb{R}^3$.

$$\{\begin{pmatrix} x \\ y \\ 0 \end{pmatrix} \mid x, y \in \mathbb{R}\}$$

Two.I.2.28 Of course, the addition and scalar multiplication operations are the ones inherited from the enclosing space.

(a) This is a subspace. It is not empty as it contains at least the two example functions given. It is closed because if f_1, f_2 are even and c_1, c_2 are scalars then we have this.

$$(c_1f_1 + c_2f_2)(-x) = c_1\,f_1(-x) + c_2\,f_2(-x) = c_1\,f_1(x) + c_2\,f_2(x) = (c_1f_1 + c_2f_2)(x)$$

(b) This is also a subspace; the check is similar to the prior one.

Two.I.2.29 It can be improper. If $\vec{v}=\vec{0}$ then this is a trivial subspace. At the opposite extreme, if the vector space is $\mathbb{R}^1$ and $\vec{v}\neq\vec{0}$ then the subspace is all of $\mathbb{R}^1$.

Two.I.2.30 No, such a set is not closed. For one thing, it does not contain the zero vector.

Two.I.2.31 **(a)** This nonempty subset of $\mathcal{M}_{2\times2}$ is not a subspace.

$$A=\{\begin{pmatrix}1&2\\3&4\end{pmatrix},\begin{pmatrix}5&6\\7&8\end{pmatrix}\}$$

One reason that it is not a subspace of $\mathcal{M}_{2\times2}$ is that it does not contain the zero matrix. (Another reason is that it is not closed under addition, since the sum of the two is not an element of A. It is also not closed under scalar multiplication.)

(b) This set of two vectors does not span $\mathbb{R}^2$.

$$\{\begin{pmatrix}1\\1\end{pmatrix},\begin{pmatrix}3\\3\end{pmatrix}\}$$

No linear combination of these two can give a vector whose second component is unequal to its first component.

Two.I.2.32 No. The only subspaces of $\mathbb{R}^1$ are the space itself and its trivial subspace. Any subspace S of $\mathbb{R}$ that contains a nonzero member $\vec{v}$ must contain the set of all of its scalar multiples $\{r\cdot\vec{v}\mid r\in\mathbb{R}\}$. But this set is all of $\mathbb{R}$.

Two.I.2.33 Item (1) is checked in the text.

Item (2) has five conditions. First, for closure, if $c\in\mathbb{R}$ and $\vec{s}\in S$ then $c\cdot\vec{s}\in S$ as $c\cdot\vec{s}=c\cdot\vec{s}+0\cdot\vec{0}$. Second, because the operations in S are inherited from V, for $c,d\in\mathbb{R}$ and $\vec{s}\in S$, the scalar product $(c+d)\cdot\vec{s}$ in S equals the product $(c+d)\cdot\vec{s}$ in V, and that equals $c\cdot\vec{s}+d\cdot\vec{s}$ in V, which equals $c\cdot\vec{s}+d\cdot\vec{s}$ in S.

The check for the third, fourth, and fifth conditions are similar to the second condition's check just given.

Two.I.2.34 An exercise in the prior subsection shows that every vector space has only one zero vector (that is, there is only one vector that is the additive identity element of the space). But a trivial space has only one element and that element must be this (unique) zero vector.

Two.I.2.35 As the hint suggests, the basic reason is the Linear Combination Lemma from the first chapter. For the full proof, we will show mutual containment between the two sets.

The first containment $[[S]]\supseteq[S]$ is an instance of the more general, and obvious, fact that for any subset T of a vector space, $[T]\supseteq T$.

For the other containment, that $[[S]]\subseteq[S]$, take m vectors from $[S]$, namely $c_{1,1}\vec{s}_{1,1}+\cdots+c_{1,n_1}\vec{s}_{1,n_1}$, $\ldots$, $c_{1,m}\vec{s}_{1,m}+\cdots+c_{1,n_m}\vec{s}_{1,n_m}$, and note that any linear combination of those

$$r_1(c_{1,1}\vec{s}_{1,1}+\cdots+c_{1,n_1}\vec{s}_{1,n_1})+\cdots+r_m(c_{1,m}\vec{s}_{1,m}+\cdots+c_{1,n_m}\vec{s}_{1,n_m})$$

is a linear combination of elements of S

$$=(r_1c_{1,1})\vec{s}_{1,1}+\cdots+(r_1c_{1,n_1})\vec{s}_{1,n_1}+\cdots+(r_mc_{1,m})\vec{s}_{1,m}+\cdots+(r_mc_{1,n_m})\vec{s}_{1,n_m}$$

and so is in $[S]$. That is, simply recall that a linear combination of linear combinations (of members of S) is a linear combination (again of members of S).

Two.I.2.36 **(a)** It is not a subspace because these are not the inherited operations. For one thing, in this space,

$$0\cdot\begin{pmatrix}x\\y\\z\end{pmatrix}=\begin{pmatrix}1\\0\\0\end{pmatrix}$$

while this does not, of course, hold in $\mathbb{R}^3$.

(b) We can combine the argument showing closure under addition with the argument showing closure under scalar multiplication into one single argument showing closure under linear combinations of two vectors. If $r_1, r_2, x_1, x_2, y_1, y_2, z_1, z_2$ are in $\mathbb{R}$ then

$$r_1\begin{pmatrix}x_1\\y_1\\z_1\end{pmatrix}+r_2\begin{pmatrix}x_2\\y_2\\z_2\end{pmatrix}=\begin{pmatrix}r_1x_1-r_1+1\\r_1y_1\\r_1z_1\end{pmatrix}+\begin{pmatrix}r_2x_2-r_2+1\\r_2y_2\\r_2z_2\end{pmatrix}=\begin{pmatrix}r_1x_1-r_1+r_2x_2-r_2+1\\r_1y_1+r_2y_2\\r_1z_1+r_2z_2\end{pmatrix}$$

(note that the definition of addition in this space is that the first components combine as $(r_1x_1-r_1+1)+(r_2x_2-r_2+1)-1$, so the first component of the last vector does not say '$+2$'). Adding the three components of the last vector gives $r_1(x_1-1+y_1+z_1)+r_2(x_2-1+y_2+z_2)+1=r_1\cdot 0+r_2\cdot 0+1=1$.

Most of the other checks of the conditions are easy (although the oddness of the operations keeps them from being routine). Commutativity of addition goes like this.

$$\begin{pmatrix}x_1\\y_1\\z_1\end{pmatrix}+\begin{pmatrix}x_2\\y_2\\z_2\end{pmatrix}=\begin{pmatrix}x_1+x_2-1\\y_1+y_2\\z_1+z_2\end{pmatrix}=\begin{pmatrix}x_2+x_1-1\\y_2+y_1\\z_2+z_1\end{pmatrix}=\begin{pmatrix}x_2\\y_2\\z_2\end{pmatrix}+\begin{pmatrix}x_1\\y_1\\z_1\end{pmatrix}$$

Associativity of addition has

$$(\begin{pmatrix}x_1\\y_1\\z_1\end{pmatrix}+\begin{pmatrix}x_2\\y_2\\z_2\end{pmatrix})+\begin{pmatrix}x_3\\y_3\\z_3\end{pmatrix}=\begin{pmatrix}(x_1+x_2-1)+x_3-1\\(y_1+y_2)+y_3\\(z_1+z_2)+z_3\end{pmatrix}$$

while

$$\begin{pmatrix}x_1\\y_1\\z_1\end{pmatrix}+(\begin{pmatrix}x_2\\y_2\\z_2\end{pmatrix}+\begin{pmatrix}x_3\\y_3\\z_3\end{pmatrix})=\begin{pmatrix}x_1+(x_2+x_3-1)-1\\y_1+(y_2+y_3)\\z_1+(z_2+z_3)\end{pmatrix}$$

and they are equal. The identity element with respect to this addition operation works this way

$$\begin{pmatrix}x\\y\\z\end{pmatrix}+\begin{pmatrix}1\\0\\0\end{pmatrix}=\begin{pmatrix}x+1-1\\y+0\\z+0\end{pmatrix}=\begin{pmatrix}x\\y\\z\end{pmatrix}$$

and the additive inverse is similar.

$$\begin{pmatrix}x\\y\\z\end{pmatrix}+\begin{pmatrix}-x+2\\-y\\-z\end{pmatrix}=\begin{pmatrix}x+(-x+2)-1\\y-y\\z-z\end{pmatrix}=\begin{pmatrix}1\\0\\0\end{pmatrix}$$

The conditions on scalar multiplication are also easy. For the first condition,

$$(r+s)\begin{pmatrix}x\\y\\z\end{pmatrix}=\begin{pmatrix}(r+s)x-(r+s)+1\\(r+s)y\\(r+s)z\end{pmatrix}$$

while

$$r\begin{pmatrix}x\\y\\z\end{pmatrix}+s\begin{pmatrix}x\\y\\z\end{pmatrix}=\begin{pmatrix}rx-r+1\\ry\\rz\end{pmatrix}+\begin{pmatrix}sx-s+1\\sy\\sz\end{pmatrix}=\begin{pmatrix}(rx-r+1)+(sx-s+1)-1\\ry+sy\\rz+sz\end{pmatrix}$$

and the two are equal. The second condition compares

$$r\cdot(\begin{pmatrix}x_1\\y_1\\z_1\end{pmatrix}+\begin{pmatrix}x_2\\y_2\\z_2\end{pmatrix})=r\cdot\begin{pmatrix}x_1+x_2-1\\y_1+y_2\\z_1+z_2\end{pmatrix}=\begin{pmatrix}r(x_1+x_2-1)-r+1\\r(y_1+y_2)\\r(z_1+z_2)\end{pmatrix}$$

with

$$r\begin{pmatrix}x_1\\y_1\\z_1\end{pmatrix}+r\begin{pmatrix}x_2\\y_2\\z_2\end{pmatrix}=\begin{pmatrix}rx_1-r+1\\ry_1\\rz_1\end{pmatrix}+\begin{pmatrix}rx_2-r+1\\ry_2\\rz_2\end{pmatrix}=\begin{pmatrix}(rx_1-r+1)+(rx_2-r+1)-1\\ry_1+ry_2\\rz_1+rz_2\end{pmatrix}$$

and they are equal. For the third condition,

$$(rs)\begin{pmatrix}x\\y\\z\end{pmatrix}=\begin{pmatrix}rsx-rs+1\\rsy\\rsz\end{pmatrix}$$

while

$$r(s\begin{pmatrix}x\\y\\z\end{pmatrix})=r(\begin{pmatrix}sx-s+1\\sy\\sz\end{pmatrix})=\begin{pmatrix}r(sx-s+1)-r+1\\rsy\\rsz\end{pmatrix}$$

and the two are equal. For scalar multiplication by 1 we have this.

$$1\cdot\begin{pmatrix}x\\y\\z\end{pmatrix}=\begin{pmatrix}1x-1+1\\1y\\1z\end{pmatrix}=\begin{pmatrix}x\\y\\z\end{pmatrix}$$

Thus all the conditions on a vector space are met by these two operations.

Remark. A way to understand this vector space is to think of it as the plane in $\mathbb{R}^3$

$$P=\{\begin{pmatrix}x\\y\\z\end{pmatrix}\mid x+y+z=0\}$$

displaced away from the origin by 1 along the x-axis. Then addition becomes: to add two members of this space,

$$\begin{pmatrix}x_1\\y_1\\z_1\end{pmatrix},\begin{pmatrix}x_2\\y_2\\z_2\end{pmatrix}$$

(such that $x_1+y_1+z_1=1$ and $x_2+y_2+z_2=1$) move them back by 1 to place them in P and add as usual,

$$\begin{pmatrix}x_1-1\\y_1\\z_1\end{pmatrix}+\begin{pmatrix}x_2-1\\y_2\\z_2\end{pmatrix}=\begin{pmatrix}x_1+x_2-2\\y_1+y_2\\z_1+z_2\end{pmatrix}\qquad\text{(in P)}$$

and then move the result back out by 1 along the x-axis.

$$\begin{pmatrix}x_1+x_2-1\\y_1+y_2\\z_1+z_2\end{pmatrix}.$$

Scalar multiplication is similar.

(c) For the subspace to be closed under the inherited scalar multiplication, where $\vec{v}$ is a member of that subspace,

$$0\cdot\vec{v}=\begin{pmatrix}0\\0\\0\end{pmatrix}$$

must also be a member.

The converse does not hold. Here is a subset of $\mathbb{R}^3$ that contains the origin

$$\{\begin{pmatrix}0\\0\\0\end{pmatrix},\begin{pmatrix}1\\0\\0\end{pmatrix}\}$$

(this subset has only two elements) but is not a subspace.

Two.I.2.37 **(a)** $(\vec{v}_1+\vec{v}_2+\vec{v}_3)-(\vec{v}_1+\vec{v}_2)=\vec{v}_3$

(b) $(\vec{v}_1+\vec{v}_2)-(\vec{v}_1)=\vec{v}_2$

(c) Surely, $\vec{v}_1$.

(d) Taking the one-long sum and subtracting gives $(\vec{v}_1) - \vec{v}_1 = \vec{0}$.

Two.I.2.38 Yes; any space is a subspace of itself, so each space contains the other.

Two.I.2.39 (a) The union of the x-axis and the y-axis in $\mathbb{R}^2$ is one.
(b) The set of integers, as a subset of $\mathbb{R}^1$, is one.
(c) The subset $\{\vec{v}\}$ of $\mathbb{R}^2$ is one, where $\vec{v}$ is any nonzero vector.

Two.I.2.40 Because vector space addition is commutative, a reordering of summands leaves a linear combination unchanged.

Two.I.2.41 We always consider that span in the context of an enclosing space.

Two.I.2.42 It is both 'if' and 'only if'.

For 'if', let S be a subset of a vector space V and assume $\vec{v} \in S$ satisfies $\vec{v} = c_1\vec{s}_1 + \cdots + c_n\vec{s}_n$ where $c_1, \ldots, c_n$ are scalars and $\vec{s}_1, \ldots, \vec{s}_n \in S$. We must show that $[S \cup \{\vec{v}\}] = [S]$.

Containment one way, $[S] \subseteq [S \cup \{\vec{v}\}]$ is obvious. For the other direction, $[S \cup \{\vec{v}\}] \subseteq [S]$, note that if a vector is in the set on the left then it has the form $d_0\vec{v} + d_1\vec{t}_1 + \cdots + d_m\vec{t}_m$ where the d's are scalars and the $\vec{t}$'s are in S. Rewrite that as $d_0(c_1\vec{s}_1 + \cdots + c_n\vec{s}_n) + d_1\vec{t}_1 + \cdots + d_m\vec{t}_m$ and note that the result is a member of the span of S.

The 'only if' is clearly true — adding $\vec{v}$ enlarges the span to include at least $\vec{v}$.

Two.I.2.43 (a) Always.

Assume that A, B are subspaces of V. Note that their intersection is not empty as both contain the zero vector. If $\vec{w}, \vec{s} \in A \cap B$ and r, s are scalars then $r\vec{v} + s\vec{w} \in A$ because each vector is in A and so a linear combination is in A, and $r\vec{v} + s\vec{w} \in B$ for the same reason. Thus the intersection is closed. Now Lemma 2.9 applies.

(b) Sometimes (more precisely, only if $A \subseteq B$ or $B \subseteq A$).

To see the answer is not 'always', take V to be $\mathbb{R}^3$, take A to be the x-axis, and B to be the y-axis. Note that

$$\begin{pmatrix}1\\0\end{pmatrix} \in A \text{ and } \begin{pmatrix}0\\1\end{pmatrix} \in B \quad \text{but} \quad \begin{pmatrix}1\\0\end{pmatrix} + \begin{pmatrix}0\\1\end{pmatrix} \not\in A \cup B$$

as the sum is in neither A nor B.

The answer is not 'never' because if $A \subseteq B$ or $B \subseteq A$ then clearly $A \cup B$ is a subspace.

To show that $A \cup B$ is a subspace only if one subspace contains the other, we assume that $A \not\subseteq B$ and $B \not\subseteq A$ and prove that the union is not a subspace. The assumption that A is not a subset of B means that there is an $\vec{a} \in A$ with $\vec{a} \not\in B$. The other assumption gives a $\vec{b} \in B$ with $\vec{b} \not\in A$. Consider $\vec{a} + \vec{b}$. Note that sum is not an element of A or else $(\vec{a} + \vec{b}) - \vec{a}$ would be in A, which it is not. Similarly the sum is not an element of B. Hence the sum is not an element of $A \cup B$, and so the union is not a subspace.

(c) Never. As A is a subspace, it contains the zero vector, and therefore the set that is A's complement does not. Without the zero vector, the complement cannot be a vector space.

Two.I.2.44 The span of a set does not depend on the enclosing space. A linear combination of vectors from S gives the same sum whether we regard the operations as those of W or as those of V, because the operations of W are inherited from V.

Two.I.2.45 It is; apply Lemma 2.9. (You must consider the following. Suppose B is a subspace of a vector space V and suppose $A \subseteq B \subseteq V$ is a subspace. From which space does A inherit its operations? The answer is that it doesn't matter — A will inherit the same operations in either case.)

Two.I.2.46 (a) Always; if $S \subseteq T$ then a linear combination of elements of S is also a linear combination of elements of T.

(b) Sometimes (more precisely, if and only if $S \subseteq T$ or $T \subseteq S$).

The answer is not 'always' as is shown by this example from $\mathbb{R}^3$

$$S = \{ \begin{pmatrix} 1 \\ 0 \\ 0 \end{pmatrix}, \begin{pmatrix} 0 \\ 1 \\ 0 \end{pmatrix} \}, \quad T = \{ \begin{pmatrix} 1 \\ 0 \\ 0 \end{pmatrix}, \begin{pmatrix} 0 \\ 0 \\ 1 \end{pmatrix} \}$$

because of this.

$$\begin{pmatrix} 1 \\ 1 \\ 1 \end{pmatrix} \in [S \cup T] \qquad \begin{pmatrix} 1 \\ 1 \\ 1 \end{pmatrix} \not\in [S] \cup [T]$$

The answer is not 'never' because if either set contains the other then equality is clear. We can characterize equality as happening only when either set contains the other by assuming $S \not\subseteq T$ (implying the existence of a vector $\vec{s} \in S$ with $\vec{s} \not\in T$) and $T \not\subseteq S$ (giving a $\vec{t} \in T$ with $\vec{t} \not\in S$), noting $\vec{s} + \vec{t} \in [S \cup T]$, and showing that $\vec{s} + \vec{t} \not\in [S] \cup [T]$.

(c) Sometimes.

Clearly $[S \cap T] \subseteq [S] \cap [T]$ because any linear combination of vectors from $S \cap T$ is a combination of vectors from S and also a combination of vectors from T.

Containment the other way does not always hold. For instance, in $\mathbb{R}^2$, take

$$S = \{ \begin{pmatrix} 1 \\ 0 \end{pmatrix}, \begin{pmatrix} 0 \\ 1 \end{pmatrix} \}, \quad T = \{ \begin{pmatrix} 2 \\ 0 \end{pmatrix} \}$$

so that $[S] \cap [T]$ is the x-axis but $[S \cap T]$ is the trivial subspace.

Characterizing exactly when equality holds is tough. Clearly equality holds if either set contains the other, but that is not 'only if' by this example in $\mathbb{R}^3$.

$$S = \{ \begin{pmatrix} 1 \\ 0 \\ 0 \end{pmatrix}, \begin{pmatrix} 0 \\ 1 \\ 0 \end{pmatrix} \}, \quad T = \{ \begin{pmatrix} 1 \\ 0 \\ 0 \end{pmatrix}, \begin{pmatrix} 0 \\ 0 \\ 1 \end{pmatrix} \}$$

(d) Never, as the span of the complement is a subspace, while the complement of the span is not (it does not contain the zero vector).

Two.I.2.47 Call the subset S. By Lemma 2.9, we need to check that $[S]$ is closed under linear combinations. If $c_1\vec{s}_1 + \cdots + c_n\vec{s}_n, c_{n+1}\vec{s}_{n+1} + \cdots + c_m\vec{s}_m \in [S]$ then for any $p, r \in \mathbb{R}$ we have

$$p \cdot (c_1\vec{s}_1 + \cdots + c_n\vec{s}_n) + r \cdot (c_{n+1}\vec{s}_{n+1} + \cdots + c_m\vec{s}_m) = pc_1\vec{s}_1 + \cdots + pc_n\vec{s}_n + rc_{n+1}\vec{s}_{n+1} + \cdots + rc_m\vec{s}_m$$

which is an element of $[S]$. (*Remark.* If the set S is empty, then that 'if ... then ...' statement is vacuously true.)

Two.I.2.48 For this to happen, one of the conditions giving the sensibleness of the addition and scalar multiplication operations must be violated. Consider $\mathbb{R}^2$ with these operations.

$$\begin{pmatrix} x_1 \\ y_1 \end{pmatrix} + \begin{pmatrix} x_2 \\ y_2 \end{pmatrix} = \begin{pmatrix} 0 \\ 0 \end{pmatrix} \qquad r \begin{pmatrix} x \\ y \end{pmatrix} = \begin{pmatrix} 0 \\ 0 \end{pmatrix}$$

The set $\mathbb{R}^2$ is closed under these operations. But it is not a vector space.

$$1 \cdot \begin{pmatrix} 1 \\ 1 \end{pmatrix} \neq \begin{pmatrix} 1 \\ 1 \end{pmatrix}$$

Linear Independence

Two.II.1: Definition and Examples

Two.II.1.18 For each of these, when the subset is independent you must prove it, and when the subset is dependent you must give an example of a dependence.

(a) It is dependent. Considering

$$c_1 \begin{pmatrix} 1 \\ -3 \\ 5 \end{pmatrix} + c_2 \begin{pmatrix} 2 \\ 2 \\ 4 \end{pmatrix} + c_3 \begin{pmatrix} 4 \\ -4 \\ 14 \end{pmatrix} = \begin{pmatrix} 0 \\ 0 \\ 0 \end{pmatrix}$$

gives this linear system.

$$\begin{aligned} c_1 + 2c_2 + 4c_3 &= 0 \\ -3c_1 + 2c_2 - 4c_3 &= 0 \\ 5c_1 + 4c_2 + 14c_3 &= 0 \end{aligned}$$

Gauss's Method

$$\left(\begin{array}{ccc|c} 1 & 2 & 4 & 0 \\ -3 & 2 & -4 & 0 \\ 5 & 4 & 14 & 0 \end{array}\right) \xrightarrow[-5\rho_1+\rho_3]{3\rho_1+\rho_2} \xrightarrow{(3/4)\rho_2+\rho_3} \left(\begin{array}{ccc|c} 1 & 2 & 4 & 0 \\ 0 & 8 & 8 & 0 \\ 0 & 0 & 0 & 0 \end{array}\right)$$

yields a free variable, so there are infinitely many solutions. For an example of a particular dependence we can set c_3 to be, say, 1. Then we get $c_2 = -1$ and $c_1 = -2$.

(b) It is dependent. The linear system that arises here

$$\left(\begin{array}{ccc|c} 1 & 2 & 3 & 0 \\ 7 & 7 & 7 & 0 \\ 7 & 7 & 7 & 0 \end{array}\right) \xrightarrow[-7\rho_1+\rho_3]{-7\rho_1+\rho_2} \xrightarrow{-\rho_2+\rho_3} \left(\begin{array}{ccc|c} 1 & 2 & 3 & 0 \\ 0 & -7 & -14 & 0 \\ 0 & 0 & 0 & 0 \end{array}\right)$$

has infinitely many solutions. We can get a particular solution by taking c_3 to be, say, 1, and back-substituting to get the resulting c_2 and c_1.

(c) It is linearly independent. The system

$$\left(\begin{array}{cc|c} 0 & 1 & 0 \\ 0 & 0 & 0 \\ -1 & 4 & 0 \end{array}\right) \xrightarrow{\rho_1\leftrightarrow\rho_2} \xrightarrow{\rho_3\leftrightarrow\rho_1} \left(\begin{array}{cc|c} -1 & 4 & 0 \\ 0 & 1 & 0 \\ 0 & 0 & 0 \end{array}\right)$$

has only the solution $c_1 = 0$ and $c_2 = 0$. (We could also have gotten the answer by inspection — the second vector is obviously not a multiple of the first, and vice versa.)

(d) It is linearly dependent. The linear system

$$\left(\begin{array}{cccc|c} 9 & 2 & 3 & 12 & 0 \\ 9 & 0 & 5 & 12 & 0 \\ 0 & 1 & -4 & -1 & 0 \end{array}\right)$$

has more unknowns than equations, and so Gauss's Method must end with at least one variable free (there can't be a contradictory equation because the system is homogeneous, and so has at least the solution of all zeroes). To exhibit a combination, we can do the reduction

$$\xrightarrow{-\rho_1+\rho_2} \xrightarrow{(1/2)\rho_2+\rho_3} \left(\begin{array}{cccc|c} 9 & 2 & 3 & 12 & 0 \\ 0 & -2 & 2 & 0 & 0 \\ 0 & 0 & -3 & -1 & 0 \end{array}\right)$$

and take, say, $c_4 = 1$. Then we have that $c_3 = -1/3$, $c_2 = -1/3$, and $c_1 = -31/27$.

Two.II.1.19 In the cases of independence, you must prove that it is independent. Otherwise, you must exhibit a dependence. (Here we give a specific dependence but others are possible.)

(a) This set is independent. Setting up the relation $c_1(3-x+9x^2)+c_2(5-6x+3x^2)+c_3(1+1x-5x^2)=0+0x+0x^2$ gives a linear system

$$\left(\begin{array}{ccc|c} 3 & 5 & 1 & 0 \\ -1 & -6 & 1 & 0 \\ 9 & 3 & -5 & 0 \end{array}\right) \xrightarrow[-3\rho_1+\rho_3]{(1/3)\rho_1+\rho_2} \xrightarrow{3\rho_2} \xrightarrow{-(12/13)\rho_2+\rho_3} \left(\begin{array}{ccc|c} 3 & 5 & 1 & 0 \\ 0 & -13 & 4 & 0 \\ 0 & 0 & -128/13 & 0 \end{array}\right)$$

with only one solution: $c_1=0$, $c_2=0$, and $c_3=0$.

(b) This set is independent. We can see this by inspection, straight from the definition of linear independence. Obviously neither is a multiple of the other.

(c) This set is linearly independent. The linear system reduces in this way

$$\left(\begin{array}{ccc|c} 2 & 3 & 4 & 0 \\ 1 & -1 & 0 & 0 \\ 7 & 2 & -3 & 0 \end{array}\right) \xrightarrow[-(7/2)\rho_1+\rho_3]{-(1/2)\rho_1+\rho_2} \xrightarrow{-(17/5)\rho_2+\rho_3} \left(\begin{array}{ccc|c} 2 & 3 & 4 & 0 \\ 0 & -5/2 & -2 & 0 \\ 0 & 0 & -51/5 & 0 \end{array}\right)$$

to show that there is only the solution $c_1=0$, $c_2=0$, and $c_3=0$.

(d) This set is linearly dependent. The linear system

$$\left(\begin{array}{cccc|c} 8 & 0 & 2 & 8 & 0 \\ 3 & 1 & 2 & -2 & 0 \\ 3 & 2 & 2 & 5 & 0 \end{array}\right)$$

must, after reduction, end with at least one variable free (there are more variables than equations, and there is no possibility of a contradictory equation because the system is homogeneous). We can take the free variables as parameters to describe the solution set. We can then set the parameter to a nonzero value to get a nontrivial linear relation.

Two.II.1.20 Let Z be the zero function $Z(x)=0$, which is the additive identity in the vector space under discussion.

(a) This set is linearly independent. Consider $c_1\cdot f(x)+c_2\cdot g(x)=Z(x)$. Plugging in $x=1$ and $x=2$ gives a linear system

$$\begin{aligned} c_1\cdot 1+c_2\cdot 1&=0 \\ c_1\cdot 2+c_2\cdot(1/2)&=0 \end{aligned}$$

with the unique solution $c_1=0$, $c_2=0$.

(b) This set is linearly independent. Consider $c_1\cdot f(x)+c_2\cdot g(x)=Z(x)$ and plug in $x=0$ and $x=\pi/2$ to get

$$\begin{aligned} c_1\cdot 1+c_2\cdot 0&=0 \\ c_1\cdot 0+c_2\cdot 1&=0 \end{aligned}$$

which obviously gives that $c_1=0$, $c_2=0$.

(c) This set is also linearly independent. Considering $c_1\cdot f(x)+c_2\cdot g(x)=Z(x)$ and plugging in $x=1$ and $x=e$

$$\begin{aligned} c_1\cdot e+c_2\cdot 0&=0 \\ c_1\cdot e^e+c_2\cdot 1&=0 \end{aligned}$$

gives that $c_1=0$ and $c_2=0$.

Two.II.1.21 In each case, if the set is independent then you must prove that and if it is dependent then you must exhibit a dependence.

(a) This set is dependent. The familiar relation $\sin^2(x)+\cos^2(x)=1$ shows that $2=c_1\cdot(4\sin^2(x))+c_2\cdot(\cos^2(x))$ is satisfied by $c_1=1/2$ and $c_2=2$.

(b) This set is independent. Consider the relationship $c_1\cdot 1+c_2\cdot\sin(x)+c_3\cdot\sin(2x)=0$ (that '0' is the zero function). Taking three suitable points such as $x=\pi$, $x=\pi/2$, $x=\pi/4$ gives a system

$$\begin{aligned} c_1\phantom{+(\sqrt{2}/2)c_2+c_3}&=0 \\ c_1+\phantom{(\sqrt{2}/2)}c_2&=0 \\ c_1+(\sqrt{2}/2)c_2+c_3&=0 \end{aligned}$$

whose only solution is $c_1=0$, $c_2=0$, and $c_3=0$.

(c) By inspection, this set is independent. Any dependence $\cos(x) = c \cdot x$ is not possible since the cosine function is not a multiple of the identity function (we are applying Corollary 1.16).

(d) By inspection, we spot that there is a dependence. Because $(1+x)^2 = x^2 + 2x + 1$, we get that $c_1 \cdot (1+x)^2 + c_2 \cdot (x^2 + 2x) = 3$ is satisfied by $c_1 = 3$ and $c_2 = -3$.

(e) This set is dependent. The easiest way to see that is to recall the trigonometric relationship $\cos^2(x) - \sin^2(x) = \cos(2x)$. (*Remark.* A person who doesn't recall this, and tries some x's, simply never gets a system leading to a unique solution, and never gets to conclude that the set is independent. Of course, this person might wonder if they simply never tried the right set of x's, but a few tries will lead most people to look instead for a dependence.)

(f) This set is dependent, because it contains the zero object in the vector space, the zero polynomial.

Two.II.1.22 No, that equation is not a linear relationship. In fact this set is independent, as the system arising from taking x to be 0, $\pi/6$ and $\pi/4$ shows.

Two.II.1.23 No. Here are two members of the plane where the second is a multiple of the first.

$$\begin{pmatrix}1\\0\\0\end{pmatrix}, \begin{pmatrix}2\\0\\0\end{pmatrix}$$

(Another reason that the answer is "no" is the the zero vector is a member of the plane and no set containing the zero vector is linearly independent.)

Two.II.1.24 We have already showed this: the Linear Combination Lemma and its corollary state that in an echelon form matrix, no nonzero row is a linear combination of the others.

Two.II.1.25 **(a)** Assume that $\{\vec{u}, \vec{v}, \vec{w}\}$ is linearly independent, so that any relationship $d_0\vec{u} + d_1\vec{v} + d_2\vec{w} = \vec{0}$ leads to the conclusion that $d_0 = 0$, $d_1 = 0$, and $d_2 = 0$.

Consider the relationship $c_1(\vec{u}) + c_2(\vec{u}+\vec{v}) + c_3(\vec{u}+\vec{v}+\vec{w}) = \vec{0}$. Rewrite it to get $(c_1 + c_2 + c_3)\vec{u} + (c_2 + c_3)\vec{v} + (c_3)\vec{w} = \vec{0}$. Taking d_0 to be $c_1 + c_2 + c_3$, taking d_1 to be $c_2 + c_3$, and taking d_2 to be c_3 we have this system.

$$\begin{aligned} c_1 + c_2 + c_3 &= 0 \\ c_2 + c_3 &= 0 \\ c_3 &= 0 \end{aligned}$$

Conclusion: the c's are all zero, and so the set is linearly independent.

(b) The second set is dependent

$$1 \cdot (\vec{u} - \vec{v}) + 1 \cdot (\vec{v} - \vec{w}) + 1 \cdot (\vec{w} - \vec{u}) = \vec{0}$$

whether or not the first set is independent.

Two.II.1.26 **(a)** A singleton set $\{\vec{v}\}$ is linearly independent if and only if $\vec{v} \neq \vec{0}$. For the 'if' direction, with $\vec{v} \neq \vec{0}$, we can apply Lemma 1.3 by considering the relationship $c \cdot \vec{v} = \vec{0}$ and noting that the only solution is the trivial one: $c = 0$. For the 'only if' direction, just recall that Example 1.9 shows that $\{\vec{0}\}$ is linearly dependent, and so if the set $\{\vec{v}\}$ is linearly independent then $\vec{v} \neq \vec{0}$.

(*Remark.* Another answer is to say that this is the special case of Lemma 1.12 where $S = \varnothing$.)

(b) A set with two elements is linearly independent if and only if neither member is a multiple of the other (note that if one is the zero vector then it is a multiple of the other). This is an equivalent statement: a set is linearly dependent if and only if one element is a multiple of the other.

The proof is easy. A set $\{\vec{v}_1, \vec{v}_2\}$ is linearly dependent if and only if there is a relationship $c_1\vec{v}_1 + c_2\vec{v}_2 = \vec{0}$ with either $c_1 \neq 0$ or $c_2 \neq 0$ (or both). That holds if and only if $\vec{v}_1 = (-c_2/c_1)\vec{v}_2$ or $\vec{v}_2 = (-c_1/c_2)\vec{v}_1$ (or both).

Two.II.1.27 This set is linearly dependent set because it contains the zero vector.

Two.II.1.28 Lemma 1.17 gives the 'if' half. The converse (the 'only if' statement) does not hold. An example is to consider the vector space $\mathbb{R}^2$ and these vectors.

$$\vec{x} = \begin{pmatrix}1\\0\end{pmatrix}, \quad \vec{y} = \begin{pmatrix}0\\1\end{pmatrix}, \quad \vec{z} = \begin{pmatrix}1\\1\end{pmatrix}$$

Two.II.1.29 (a) The linear system arising from

$$c_1\begin{pmatrix}1\\1\\0\end{pmatrix} + c_2\begin{pmatrix}-1\\2\\0\end{pmatrix} = \begin{pmatrix}0\\0\\0\end{pmatrix}$$

has the unique solution $c_1 = 0$ and $c_2 = 0$.

(b) The linear system arising from

$$c_1\begin{pmatrix}1\\1\\0\end{pmatrix} + c_2\begin{pmatrix}-1\\2\\0\end{pmatrix} = \begin{pmatrix}3\\2\\0\end{pmatrix}$$

has the unique solution $c_1 = 8/3$ and $c_2 = -1/3$.

(c) Suppose that S is linearly independent. Suppose that we have both $\vec{v} = c_1\vec{s}_1 + \cdots + c_n\vec{s}_n$ and $\vec{v} = d_1\vec{t}_1 + \cdots + d_m\vec{t}_m$ (where the vectors are members of S). Now,

$$c_1\vec{s}_1 + \cdots + c_n\vec{s}_n = \vec{v} = d_1\vec{t}_1 + \cdots + d_m\vec{t}_m$$

can be rewritten in this way.

$$c_1\vec{s}_1 + \cdots + c_n\vec{s}_n - d_1\vec{t}_1 - \cdots - d_m\vec{t}_m = \vec{0}$$

Possibly some of the $\vec{s}$'s equal some of the $\vec{t}$'s; we can combine the associated coefficients (i.e., if $\vec{s}_i = \vec{t}_j$ then $\cdots + c_i\vec{s}_i + \cdots - d_j\vec{t}_j - \cdots$ can be rewritten as $\cdots + (c_i - d_j)\vec{s}_i + \cdots$). That equation is a linear relationship among distinct (after the combining is done) members of the set S. We've assumed that S is linearly independent, so all of the coefficients are zero. If i is such that $\vec{s}_i$ does not equal any $\vec{t}_j$ then c_i is zero. If j is such that $\vec{t}_j$ does not equal any $\vec{s}_i$ then d_j is zero. In the final case, we have that $c_i - d_j = 0$ and so $c_i = d_j$.

Therefore, the original two sums are the same, except perhaps for some $0 \cdot \vec{s}_i$ or $0 \cdot \vec{t}_j$ terms that we can neglect.

(d) This set is not linearly independent:

$$S = \{\begin{pmatrix}1\\0\end{pmatrix}, \begin{pmatrix}2\\0\end{pmatrix}\} \subset \mathbb{R}^2$$

and these two linear combinations give the same result

$$\begin{pmatrix}0\\0\end{pmatrix} = 2\cdot\begin{pmatrix}1\\0\end{pmatrix} - 1\cdot\begin{pmatrix}2\\0\end{pmatrix} = 4\cdot\begin{pmatrix}1\\0\end{pmatrix} - 2\cdot\begin{pmatrix}2\\0\end{pmatrix}$$

Thus, a linearly dependent set might have indistinct sums.

In fact, this stronger statement holds: if a set is linearly dependent then it must have the property that there are two distinct linear combinations that sum to the same vector. Briefly, where $c_1\vec{s}_1 + \cdots + c_n\vec{s}_n = \vec{0}$ then multiplying both sides of the relationship by two gives another relationship. If the first relationship is nontrivial then the second is also.

Two.II.1.30 In this 'if and only if' statement, the 'if' half is clear—if the polynomial is the zero polynomial then the function that arises from the action of the polynomial must be the zero function $x \mapsto 0$. For 'only if' we write $p(x) = c_n x^n + \cdots + c_0$. Plugging in zero $p(0) = 0$ gives that $c_0 = 0$. Taking the derivative and plugging in zero $p'(0) = 0$ gives that $c_1 = 0$. Similarly we get that each c_i is zero, and p is the zero polynomial.

Two.II.1.31 The work in this section suggests that we should define an n-dimensional non-degenerate linear surface as the span of a linearly independent set of n vectors.

Two.II.1.32 **(a)** For any $a_{1,1}, \ldots, a_{2,4}$,

$$c_1 \begin{pmatrix} a_{1,1} \\ a_{2,1} \end{pmatrix} + c_2 \begin{pmatrix} a_{1,2} \\ a_{2,2} \end{pmatrix} + c_3 \begin{pmatrix} a_{1,3} \\ a_{2,3} \end{pmatrix} + c_4 \begin{pmatrix} a_{1,4} \\ a_{2,4} \end{pmatrix} = \begin{pmatrix} 0 \\ 0 \end{pmatrix}$$

yields a linear system

$$\begin{aligned} a_{1,1}c_1 + a_{1,2}c_2 + a_{1,3}c_3 + a_{1,4}c_4 &= 0 \\ a_{2,1}c_1 + a_{2,2}c_2 + a_{2,3}c_3 + a_{2,4}c_4 &= 0 \end{aligned}$$

that has infinitely many solutions (Gauss's Method leaves at least two variables free). Hence there are nontrivial linear relationships among the given members of $\mathbb{R}^2$.

(b) Any set five vectors is a superset of a set of four vectors, and so is linearly dependent.

With three vectors from $\mathbb{R}^2$, the argument from the prior item still applies, with the slight change that Gauss's Method now only leaves at least one variable free (but that still gives infinitely many solutions).

(c) The prior item shows that no three-element subset of $\mathbb{R}^2$ is independent. We know that there are two-element subsets of $\mathbb{R}^2$ that are independent — one is

$$\{ \begin{pmatrix} 1 \\ 0 \end{pmatrix}, \begin{pmatrix} 0 \\ 1 \end{pmatrix} \}$$

and so the answer is two.

Two.II.1.33 Yes; here is one.

$$\{ \begin{pmatrix} 1 \\ 0 \\ 0 \end{pmatrix}, \begin{pmatrix} 0 \\ 1 \\ 0 \end{pmatrix}, \begin{pmatrix} 0 \\ 0 \\ 1 \end{pmatrix}, \begin{pmatrix} 1 \\ 1 \\ 1 \end{pmatrix} \}$$

Two.II.1.34 Yes. The two improper subsets, the entire set and the empty subset, serve as examples.

Two.II.1.35 In $\mathbb{R}^4$ the biggest linearly independent set has four vectors. There are many examples of such sets, this is one.

$$\{ \begin{pmatrix} 1 \\ 0 \\ 0 \\ 0 \end{pmatrix}, \begin{pmatrix} 0 \\ 1 \\ 0 \\ 0 \end{pmatrix}, \begin{pmatrix} 0 \\ 0 \\ 1 \\ 0 \end{pmatrix}, \begin{pmatrix} 0 \\ 0 \\ 0 \\ 1 \end{pmatrix} \}$$

To see that no set with five or more vectors can be independent, set up

$$c_1 \begin{pmatrix} a_{1,1} \\ a_{2,1} \\ a_{3,1} \\ a_{4,1} \end{pmatrix} + c_2 \begin{pmatrix} a_{1,2} \\ a_{2,2} \\ a_{3,2} \\ a_{4,2} \end{pmatrix} + c_3 \begin{pmatrix} a_{1,3} \\ a_{2,3} \\ a_{3,3} \\ a_{4,3} \end{pmatrix} + c_4 \begin{pmatrix} a_{1,4} \\ a_{2,4} \\ a_{3,4} \\ a_{4,4} \end{pmatrix} + c_5 \begin{pmatrix} a_{1,5} \\ a_{2,5} \\ a_{3,5} \\ a_{4,5} \end{pmatrix} = \begin{pmatrix} 0 \\ 0 \\ 0 \\ 0 \end{pmatrix}$$

and note that the resulting linear system

$$\begin{aligned} a_{1,1}c_1 + a_{1,2}c_2 + a_{1,3}c_3 + a_{1,4}c_4 + a_{1,5}c_5 &= 0 \\ a_{2,1}c_1 + a_{2,2}c_2 + a_{2,3}c_3 + a_{2,4}c_4 + a_{2,5}c_5 &= 0 \\ a_{3,1}c_1 + a_{3,2}c_2 + a_{3,3}c_3 + a_{3,4}c_4 + a_{3,5}c_5 &= 0 \\ a_{4,1}c_1 + a_{4,2}c_2 + a_{4,3}c_3 + a_{4,4}c_4 + a_{4,5}c_5 &= 0 \end{aligned}$$

has four equations and five unknowns, so Gauss's Method must end with at least one c variable free, so there are infinitely many solutions, and so the above linear relationship among the four-tall vectors has more solutions than just the trivial solution.

The smallest linearly independent set is the empty set.

The biggest linearly dependent set is $\mathbb{R}^4$. The smallest is $\{\vec{0}\}$.

Two.II.1.36 **(a)** The intersection of two linearly independent sets $S \cap T$ must be linearly independent as it is a subset of the linearly independent set S (as well as the linearly independent set T also, of course).

(b) The complement of a linearly independent set is linearly dependent as it contains the zero vector.

(c) A simple example in $\mathbb{R}^2$ is these two sets.

$$S=\{\begin{pmatrix}1\\0\end{pmatrix}\}\qquad T=\{\begin{pmatrix}0\\1\end{pmatrix}\}$$

A somewhat subtler example, again in $\mathbb{R}^2$, is these two.

$$S=\{\begin{pmatrix}1\\0\end{pmatrix}\}\qquad T=\{\begin{pmatrix}1\\0\end{pmatrix},\begin{pmatrix}0\\1\end{pmatrix}\}$$

(d) We must produce an example. One, in $\mathbb{R}^2$, is

$$S=\{\begin{pmatrix}1\\0\end{pmatrix}\}\qquad T=\{\begin{pmatrix}2\\0\end{pmatrix}\}$$

since the linear dependence of $S_1 \cup S_2$ is easy to see.

Two.II.1.37 **(a)** Lemma 1.3 requires that the vectors $\vec{s}_1,\ldots,\vec{s}_n,\vec{t}_1,\ldots,\vec{t}_m$ be distinct. But we could have that the union $S\cup T$ is linearly independent with some $\vec{s}_i$ equal to some $\vec{t}_j$.

(b) One example in $\mathbb{R}^2$ is these two.

$$S=\{\begin{pmatrix}1\\0\end{pmatrix}\}\qquad T=\{\begin{pmatrix}1\\0\end{pmatrix},\begin{pmatrix}0\\1\end{pmatrix}\}$$

(c) An example from $\mathbb{R}^2$ is these sets.

$$S=\{\begin{pmatrix}1\\0\end{pmatrix},\begin{pmatrix}0\\1\end{pmatrix}\}\qquad T=\{\begin{pmatrix}1\\0\end{pmatrix},\begin{pmatrix}1\\1\end{pmatrix}\}$$

(d) The union of two linearly independent sets $S\cup T$ is linearly independent if and only if their spans of S and $T-(S\cap T)$ have a trivial intersection $[S]\cap[T-(S\cap T)]=\{\vec{0}\}$. To prove that, assume that S and T are linearly independent subsets of some vector space.

For the 'only if' direction, assume that the intersection of the spans is trivial $[S]\cap[T-(S\cap T)]=\{\vec{0}\}$. Consider the set $S\cup(T-(S\cap T))=S\cup T$ and consider the linear relationship $c_1\vec{s}_1+\cdots+c_n\vec{s}_n+d_1\vec{t}_1+\cdots+d_m\vec{t}_m=\vec{0}$. Subtracting gives $c_1\vec{s}_1+\cdots+c_n\vec{s}_n=-d_1\vec{t}_1-\cdots-d_m\vec{t}_m$. The left side of that equation sums to a vector in $[S]$, and the right side is a vector in $[T-(S\cap T)]$. Therefore, since the intersection of the spans is trivial, both sides equal the zero vector. Because S is linearly independent, all of the c's are zero. Because T is linearly independent so also is $T-(S\cap T)$ linearly independent, and therefore all of the d's are zero. Thus, the original linear relationship among members of $S\cup T$ only holds if all of the coefficients are zero. Hence, $S\cup T$ is linearly independent.

For the 'if' half we can make the same argument in reverse. Suppose that the union $S\cup T$ is linearly independent. Consider a linear relationship among members of S and $T-(S\cap T)$. $c_1\vec{s}_1+\cdots+c_n\vec{s}_n+d_1\vec{t}_1+\cdots+d_m\vec{t}_m=\vec{0}$ Note that no $\vec{s}_i$ is equal to a $\vec{t}_j$ so that is a combination of distinct vectors, as required by Lemma 1.3. So the only solution is the trivial one $c_1=0,\ldots,d_m=0$. Since any vector $\vec{v}$ in the intersection of the spans $[S]\cap[T-(S\cap T)]$ we can write $\vec{v}=c_1\vec{s}_1+\cdots+c_n\vec{s}_n=-d_1\vec{t}_1-\cdots-d_m\vec{t}_m$, and it must be the zero vector because each scalar is zero.

Two.II.1.38 **(a)** We do induction on the number of vectors in the finite set S.

The base case is that S has no elements. In this case S is linearly independent and there is nothing to check—a subset of S that has the same span as S is S itself.

For the inductive step assume that the theorem is true for all sets of size $n=0$, $n=1$, …, $n=k$ in order to prove that it holds when S has $n=k+1$ elements. If the $k+1$-element set $S=\{\vec{s}_0,\ldots,\vec{s}_k\}$ is linearly independent then the theorem is trivial, so assume that it is dependent. By Corollary 1.16 there is an $\vec{s}_i$ that is a linear combination of other vectors in S. Define $S_1=S-\{\vec{s}_i\}$ and note that S_1 has the same span as S by Lemma 1.12. The set S_1 has k elements and so the inductive hypothesis applies to give that it has a linearly independent subset with the same span. That subset of S_1 is the desired subset of S.

(b) Here is a sketch of the argument. We have left out the induction argument details.

If the finite set S is empty then there is nothing to prove. If $S = \{\vec{0}\}$ then the empty subset will do.

Otherwise, take some nonzero vector $\vec{s}_1 \in S$ and define $S_1 = \{\vec{s}_1\}$. If $[S_1] = [S]$ then we are finished with this proof by noting that S_1 is linearly independent.

If not, then there is a nonzero vector $\vec{s}_2 \in S - [S_1]$ (if every $\vec{s} \in S$ is in $[S_1]$ then $[S_1] = [S]$). Define $S_2 = S_1 \cup \{\vec{s}_2\}$. If $[S_2] = [S]$ then we are finished by using Theorem 1.16 to show that S_2 is linearly independent.

Repeat the last paragraph until a set with a big enough span appears. That must eventually happen because S is finite, and $[S]$ will be reached at worst when we have used every vector from S.

Two.II.1.39 **(a)** Assuming first that $a \neq 0$,

$$x\begin{pmatrix}a\\c\end{pmatrix}+y\begin{pmatrix}b\\d\end{pmatrix}=\begin{pmatrix}0\\0\end{pmatrix}$$

gives

$$\begin{array}{rl}ax+by=0\\cx+dy=0\end{array}\;\xrightarrow{-(c/a)\rho_1+\rho_2}\;\begin{array}{rl}ax+\quad\quad by=0\\(-(c/a)b+d)y=0\end{array}$$

which has a solution if and only if $0 \neq -(c/a)b + d = (-cb + ad)/d$ (we've assumed in this case that $a \neq 0$, and so back substitution yields a unique solution).

The $a = 0$ case is also not hard — break it into the $c \neq 0$ and $c = 0$ subcases and note that in these cases $ad - bc = 0 \cdot d - bc$.

Comment. An earlier exercise showed that a two-vector set is linearly dependent if and only if either vector is a scalar multiple of the other. We could also use that to make the calculation.

(b) The equation

$$c_1\begin{pmatrix}a\\d\\g\end{pmatrix}+c_2\begin{pmatrix}b\\e\\h\end{pmatrix}+c_3\begin{pmatrix}c\\f\\i\end{pmatrix}=\begin{pmatrix}0\\0\\0\end{pmatrix}$$

gives rise to a homogeneous linear system. We proceed by writing it in matrix form and applying Gauss's Method.

We first reduce the matrix to upper-triangular. Assume that $a \neq 0$.

$$\xrightarrow{(1/a)\rho_1}\left(\begin{array}{ccc|c}1&b/a&c/a&0\\d&e&f&0\\g&h&i&0\end{array}\right)\;\xrightarrow[-g\rho_1+\rho_3]{-d\rho_1+\rho_2}\left(\begin{array}{ccc|c}1&b/a&c/a&0\\0&(ae-bd)/a&(af-cd)/a&0\\0&(ah-bg)/a&(ai-cg)/a&0\end{array}\right)$$

$$\xrightarrow{(a/(ae-bd))\rho_2}\left(\begin{array}{ccc|c}1&b/a&c/a&0\\0&1&(af-cd)/(ae-bd)&0\\0&(ah-bg)/a&(ai-cg)/a&0\end{array}\right)$$

(where we've assumed for the moment that $ae - bd \neq 0$ in order to do the row reduction step). Then, under the assumptions, we get this.

$$\xrightarrow{((ah-bg)/a)\rho_2+\rho_3}\left(\begin{array}{ccc|c}1&\frac{b}{a}&\frac{c}{a}&0\\0&1&\frac{af-cd}{ae-bd}&0\\0&0&\frac{aei+bgf+cdh-hfa-idb-gec}{ae-bd}&0\end{array}\right)$$

shows that the original system is nonsingular if and only if the $3,3$ entry is nonzero. This fraction is defined because of the $ae - bd \neq 0$ assumption, and it will equal zero if and only if its numerator equals zero.

We next worry about the assumptions. First, if $a \neq 0$ but $ae - bd = 0$ then we swap

$$\left(\begin{array}{ccc|c}1&b/a&c/a&0\\0&0&(af-cd)/a&0\\0&(ah-bg)/a&(ai-cg)/a&0\end{array}\right)\;\xrightarrow{\rho_2\leftrightarrow\rho_3}\left(\begin{array}{ccc|c}1&b/a&c/a&0\\0&(ah-bg)/a&(ai-cg)/a&0\\0&0&(af-cd)/a&0\end{array}\right)$$

and conclude that the system is nonsingular if and only if either $ah - bg = 0$ or $af - cd = 0$. That's the same as asking that their product be zero:

$$\begin{aligned} ahaf - ahcd - bgaf + bgcd &= 0 \\ ahaf - ahcd - bgaf + aegc &= 0 \\ a(haf - hcd - bgf + egc) &= 0 \end{aligned}$$

(in going from the first line to the second we've applied the case assumption that $ae - bd = 0$ by substituting ae for bd). Since we are assuming that $a \neq 0$, we have that $haf - hcd - bgf + egc = 0$. With $ae - bd = 0$ we can rewrite this to fit the form we need: in this $a \neq 0$ and $ae - bd = 0$ case, the given system is nonsingular when $haf - hcd - bgf + egc - i(ae - bd) = 0$, as required.

The remaining cases have the same character. Do the $a = 0$ but $d \neq 0$ case and the $a = 0$ and $d = 0$ but $g \neq 0$ case by first swapping rows and then going on as above. The $a = 0$, $d = 0$, and $g = 0$ case is easy — a set with a zero vector is linearly dependent, and the formula comes out to equal zero.

(c) It is linearly dependent if and only if either vector is a multiple of the other. That is, it is not independent iff

$$\begin{pmatrix} a \\ d \\ g \end{pmatrix} = r \cdot \begin{pmatrix} b \\ e \\ h \end{pmatrix} \quad \text{or} \quad \begin{pmatrix} b \\ e \\ h \end{pmatrix} = s \cdot \begin{pmatrix} a \\ d \\ g \end{pmatrix}$$

(or both) for some scalars r and s. Eliminating r and s in order to restate this condition only in terms of the given letters a, b, d, e, g, h, we have that it is not independent — it is dependent — iff $ae - bd = ah - gb = dh - ge$.

(d) Dependence or independence is a function of the indices, so there is indeed a formula (although at first glance a person might think the formula involves cases: "if the first component of the first vector is zero then ...", this guess turns out not to be correct).

Two.II.1.40 Recall that two vectors from $\mathbb{R}^n$ are perpendicular if and only if their dot product is zero.

(a) Assume that $\vec{v}$ and $\vec{w}$ are perpendicular nonzero vectors in $\mathbb{R}^n$, with $n > 1$. With the linear relationship $c\vec{v} + d\vec{w} = \vec{0}$, apply $\vec{v}$ to both sides to conclude that $c \cdot \|\vec{v}\|^2 + d \cdot 0 = 0$. Because $\vec{v} \neq \vec{0}$ we have that $c = 0$. A similar application of $\vec{w}$ shows that $d = 0$.

(b) Two vectors in $\mathbb{R}^1$ are perpendicular if and only if at least one of them is zero.

We define $\mathbb{R}^0$ to be a trivial space, and so both $\vec{v}$ and $\vec{w}$ are the zero vector.

(c) The right generalization is to look at a set $\{\vec{v}_1, \ldots, \vec{v}_n\} \subseteq \mathbb{R}^k$ of vectors that are *mutually orthogonal* (also called *pairwise perpendicular*): if $i \neq j$ then $\vec{v}_i$ is perpendicular to $\vec{v}_j$. Mimicking the proof of the first item above shows that such a set of nonzero vectors is linearly independent.

Two.II.1.41 **(a)** This check is routine.

(b) The summation is infinite (has infinitely many summands). The definition of linear combination involves only finite sums.

(c) No nontrivial finite sum of members of $\{g, f_0, f_1, \ldots\}$ adds to the zero object: assume that

$$c_0 \cdot (1/(1-x)) + c_1 \cdot 1 + \cdots + c_n \cdot x^n = 0$$

(any finite sum uses a highest power, here n). Multiply both sides by $1 - x$ to conclude that each coefficient is zero, because a polynomial describes the zero function only when it is the zero polynomial.

Two.II.1.42 It is both 'if' and 'only if'.

Let T be a subset of the subspace S of the vector space V. The assertion that any linear relationship $c_1\vec{t}_1 + \cdots + c_n\vec{t}_n = \vec{0}$ among members of T must be the trivial relationship $c_1 = 0, \ldots, c_n = 0$ is a statement that holds in S if and only if it holds in V, because the subspace S inherits its addition and scalar multiplication operations from V.

Basis and Dimension

Two.III.1: Basis

Two.III.1.16 By Theorem 1.12, each is a basis if and only if we can express each vector in the space in a unique way as a linear combination of the given vectors.

(a) Yes this is a basis. The relation

$$c_1\begin{pmatrix}1\\2\\3\end{pmatrix}+c_2\begin{pmatrix}3\\2\\1\end{pmatrix}+c_3\begin{pmatrix}0\\0\\1\end{pmatrix}=\begin{pmatrix}x\\y\\z\end{pmatrix}$$

gives

$$\left(\begin{array}{ccc|c}1&3&0&x\\2&2&0&y\\3&1&1&z\end{array}\right)\xrightarrow[-3\rho_1+\rho_3]{-2\rho_1+\rho_2}\xrightarrow{-2\rho_2+\rho_3}\left(\begin{array}{ccc|c}1&3&0&x\\0&-4&0&-2x+y\\0&0&1&x-2y+z\end{array}\right)$$

which has the unique solution $c_3=x-2y+z$, $c_2=x/2-y/4$, and $c_1=-x/2+3y/4$.

(b) This is not a basis. Setting it up as in the prior item

$$c_1\begin{pmatrix}1\\2\\3\end{pmatrix}+c_2\begin{pmatrix}3\\2\\1\end{pmatrix}=\begin{pmatrix}x\\y\\z\end{pmatrix}$$

gives a linear system whose solution

$$\left(\begin{array}{cc|c}1&3&x\\2&2&y\\3&1&z\end{array}\right)\xrightarrow[-3\rho_1+\rho_3]{-2\rho_1+\rho_2}\xrightarrow{-2\rho_2+\rho_3}\left(\begin{array}{cc|c}1&3&x\\0&-4&-2x+y\\0&0&x-2y+z\end{array}\right)$$

is possible if and only if the three-tall vector's components x, y, and z satisfy $x-2y+z=0$. For instance, we can find the coefficients c_1 and c_2 that work when $x=1$, $y=1$, and $z=1$. However, there are no c's that work for $x=1$, $y=1$, and $z=2$. Thus this is not a basis; it does not span the space.

(c) Yes, this is a basis. Setting up the relationship leads to this reduction

$$\left(\begin{array}{ccc|c}0&1&2&x\\2&1&5&y\\-1&1&0&z\end{array}\right)\xrightarrow{\rho_1\leftrightarrow\rho_3}\xrightarrow{2\rho_1+\rho_2}\xrightarrow{-(1/3)\rho_2+\rho_3}\left(\begin{array}{ccc|c}-1&1&0&z\\0&3&5&y+2z\\0&0&1/3&x-y/3-2z/3\end{array}\right)$$

which has a unique solution for each triple of components x, y, and z.

(d) No, this is not a basis. The reduction

$$\left(\begin{array}{ccc|c}0&1&1&x\\2&1&3&y\\-1&1&0&z\end{array}\right)\xrightarrow{\rho_1\leftrightarrow\rho_3}\xrightarrow{2\rho_1+\rho_2}\xrightarrow{(-1/3)\rho_2+\rho_3}\left(\begin{array}{ccc|c}-1&1&0&z\\0&3&3&y+2z\\0&0&0&x-y/3-2z/3\end{array}\right)$$

which does not have a solution for each triple x, y, and z. Instead, the span of the given set includes only those three-tall vectors where $x=y/3+2z/3$.

Two.III.1.17 **(a)** We solve

$$c_1\begin{pmatrix}1\\1\end{pmatrix}+c_2\begin{pmatrix}-1\\1\end{pmatrix}=\begin{pmatrix}1\\2\end{pmatrix}$$

with

$$\left(\begin{array}{cc|c}1&-1&1\\1&1&2\end{array}\right)\xrightarrow{-\rho_1+\rho_2}\left(\begin{array}{cc|c}1&-1&1\\0&2&1\end{array}\right)$$

and conclude that $c_2=1/2$ and so $c_1=3/2$. Thus, the representation is this.

$$\mathrm{Rep}_B(\begin{pmatrix}1\\2\end{pmatrix})=\begin{pmatrix}3/2\\1/2\end{pmatrix}_B$$

(b) The relationship $c_1 \cdot (1) + c_2 \cdot (1+x) + c_3 \cdot (1+x+x^2) + c_4 \cdot (1+x+x^2+x^3) = x^2+x^3$ is easily solved by eye to give that $c_4 = 1$, $c_3 = 0$, $c_2 = -1$, and $c_1 = 0$.

$$\mathrm{Rep}_D(x^2+x^3) = \begin{pmatrix} 0 \\ -1 \\ 0 \\ 1 \end{pmatrix}_D$$

(c) $\mathrm{Rep}_{\mathcal{E}_4}(\begin{pmatrix} 0 \\ -1 \\ 0 \\ 1 \end{pmatrix}) = \begin{pmatrix} 0 \\ -1 \\ 0 \\ 1 \end{pmatrix}_{\mathcal{E}_4}$

Two.III.1.18 A natural basis is $\langle 1, x, x^2 \rangle$. There are bases for $\mathcal{P}_2$ that do not contain any polynomials of degree one or degree zero. One is $\langle 1+x+x^2, x+x^2, x^2 \rangle$. (Every basis has at least one polynomial of degree two, though.)

Two.III.1.19 The reduction

$$\left(\begin{array}{cccc|c} 1 & -4 & 3 & -1 & 0 \\ 2 & -8 & 6 & -2 & 0 \end{array}\right) \xrightarrow{-2\rho_1+\rho_2} \left(\begin{array}{cccc|c} 1 & -4 & 3 & -1 & 0 \\ 0 & 0 & 0 & 0 & 0 \end{array}\right)$$

gives that the only condition is that $x_1 = 4x_2 - 3x_3 + x_4$. The solution set is

$$\{\begin{pmatrix} 4x_2 - 3x_3 + x_4 \\ x_2 \\ x_3 \\ x_4 \end{pmatrix} \mid x_2, x_3, x_4 \in \mathbb{R}\} = \{x_2 \begin{pmatrix} 4 \\ 1 \\ 0 \\ 0 \end{pmatrix} + x_3 \begin{pmatrix} -3 \\ 0 \\ 1 \\ 0 \end{pmatrix} + x_4 \begin{pmatrix} 1 \\ 0 \\ 0 \\ 1 \end{pmatrix} \mid x_2, x_3, x_4 \in \mathbb{R}\}$$

and so the obvious candidate for the basis is this.

$$\langle \begin{pmatrix} 4 \\ 1 \\ 0 \\ 0 \end{pmatrix}, \begin{pmatrix} -3 \\ 0 \\ 1 \\ 0 \end{pmatrix}, \begin{pmatrix} 1 \\ 0 \\ 0 \\ 1 \end{pmatrix} \rangle$$

We've shown that this spans the space, and showing it is also linearly independent is routine.

Two.III.1.20 There are many bases. This is a natural one.

$$\langle \begin{pmatrix} 1 & 0 \\ 0 & 0 \end{pmatrix}, \begin{pmatrix} 0 & 1 \\ 0 & 0 \end{pmatrix}, \begin{pmatrix} 0 & 0 \\ 1 & 0 \end{pmatrix}, \begin{pmatrix} 0 & 0 \\ 0 & 1 \end{pmatrix} \rangle$$

Two.III.1.21 For each item, many answers are possible.

(a) One way to proceed is to parametrize by expressing the a_2 as a combination of the other two $a_2 = 2a_1 + a_0$. Then $a_2x^2 + a_1x + a_0$ is $(2a_1 + a_0)x^2 + a_1x + a_0$ and

$$\{(2a_1 + a_0)x^2 + a_1x + a_0 \mid a_1, a_0 \in \mathbb{R}\} = \{a_1 \cdot (2x^2 + x) + a_0 \cdot (x^2 + 1) \mid a_1, a_0 \in \mathbb{R}\}$$

suggests $\langle 2x^2 + x, x^2 + 1 \rangle$. This only shows that it spans, but checking that it is linearly independent is routine.

(b) Parametrize $\{(a \;\; b \;\; c) \mid a + b = 0\}$ to get $\{(-b \;\; b \;\; c) \mid b, c \in \mathbb{R}\}$, which suggests using the sequence $\langle (-1 \;\; 1 \;\; 0), (0 \;\; 0 \;\; 1) \rangle$. We've shown that it spans, and checking that it is linearly independent is easy.

(c) Rewriting

$$\{\begin{pmatrix} a & b \\ 0 & 2b \end{pmatrix} \mid a, b \in \mathbb{R}\} = \{a \cdot \begin{pmatrix} 1 & 0 \\ 0 & 0 \end{pmatrix} + b \cdot \begin{pmatrix} 0 & 1 \\ 0 & 2 \end{pmatrix} \mid a, b \in \mathbb{R}\}$$

suggests this for the basis.

$$\langle \begin{pmatrix} 1 & 0 \\ 0 & 0 \end{pmatrix}, \begin{pmatrix} 0 & 1 \\ 0 & 2 \end{pmatrix} \rangle$$

Two.III.1.22 We will show that the second is a basis; the first is similar. We will show this straight from the definition of a basis, because this example appears before Theorem 1.12.

To see that it is linearly independent, we set up $c_1\cdot(\cos\theta-\sin\theta)+c_2\cdot(2\cos\theta+3\sin\theta)=0\cos\theta+0\sin\theta$. Taking $\theta=0$ and $\theta=\pi/2$ gives this system

$$\begin{array}{rcl} c_1\cdot 1+c_2\cdot 2=0 \\ c_1\cdot(-1)+c_2\cdot 3=0 \end{array} \xrightarrow{\rho_1+\rho_2} \begin{array}{rcl} c_1+2c_2=0 \\ +5c_2=0 \end{array}$$

which shows that $c_1=0$ and $c_2=0$.

The calculation for span is also easy; for any $x,y\in\mathbb{R}$, we have that $c_1\cdot(\cos\theta-\sin\theta)+c_2\cdot(2\cos\theta+3\sin\theta)=x\cos\theta+y\sin\theta$ gives that $c_2=x/5+y/5$ and that $c_1=3x/5-2y/5$, and so the span is the entire space.

Two.III.1.23 **(a)** Asking which $a_0+a_1x+a_2x^2$ can be expressed as $c_1\cdot(1+x)+c_2\cdot(1+2x)$ gives rise to three linear equations, describing the coefficients of x^2, x, and the constants.

$$\begin{array}{rl} c_1+c_2&=a_0 \\ c_1+2c_2&=a_1 \\ 0&=a_2 \end{array}$$

Gauss's Method with back-substitution shows, provided that $a_2=0$, that $c_2=-a_0+a_1$ and $c_1=2a_0-a_1$. Thus, with $a_2=0$, we can compute appropriate c_1 and c_2 for any a_0 and a_1. So the span is the entire set of linear polynomials $\{a_0+a_1x \mid a_0,a_1\in\mathbb{R}\}$. Parametrizing that set $\{a_0\cdot 1+a_1\cdot x \mid a_0,a_1\in\mathbb{R}\}$ suggests a basis $\langle 1,x\rangle$ (we've shown that it spans; checking linear independence is easy).

(b) With

$$a_0+a_1x+a_2x^2=c_1\cdot(2-2x)+c_2\cdot(3+4x^2)=(2c_1+3c_2)+(-2c_1)x+(4c_2)x^2$$

we get this system.

$$\begin{array}{rl} 2c_1+3c_2&=a_0 \\ -2c_1&=a_1 \\ 4c_2&=a_2 \end{array} \xrightarrow{\rho_1+\rho_2} \xrightarrow{(-4/3)\rho_2+\rho_3} \begin{array}{rl} 2c_1+3c_2&=a_0 \\ 3c_2&=a_0+a_1 \\ 0&=(-4/3)a_0-(4/3)a_1+a_2 \end{array}$$

Thus, the only quadratic polynomials $a_0+a_1x+a_2x^2$ with associated c's are the ones such that $0=(-4/3)a_0-(4/3)a_1+a_2$. Hence the span is $\{(-a_1+(3/4)a_2)+a_1x+a_2x^2 \mid a_1,a_2\in\mathbb{R}\}$. Parametrizing gives $\{a_1\cdot(-1+x)+a_2\cdot((3/4)+x^2) \mid a_1,a_2\in\mathbb{R}\}$, which suggests $\langle -1+x,(3/4)+x^2\rangle$ (checking that it is linearly independent is routine).

Two.III.1.24 **(a)** The subspace is $\{a_0+a_1x+a_2x^2+a_3x^3 \mid a_0+7a_1+49a_2+343a_3=0\}$. Rewriting $a_0=-7a_1-49a_2-343a_3$ gives $\{(-7a_1-49a_2-343a_3)+a_1x+a_2x^2+a_3x^3 \mid a_1,a_2,a_3\in\mathbb{R}\}$, which, on breaking out the parameters, suggests $\langle -7+x,-49+x^2,-343+x^3\rangle$ for the basis (it is easily verified).

(b) The given subspace is the collection of cubics $p(x)=a_0+a_1x+a_2x^2+a_3x^3$ such that $a_0+7a_1+49a_2+343a_3=0$ and $a_0+5a_1+25a_2+125a_3=0$. Gauss's Method

$$\begin{array}{r} a_0+7a_1+49a_2+343a_3=0 \\ a_0+5a_1+25a_2+125a_3=0 \end{array} \xrightarrow{-\rho_1+\rho_2} \begin{array}{r} a_0+\ 7a_1+49a_2+343a_3=0 \\ -2a_1-24a_2-218a_3=0 \end{array}$$

gives that $a_1=-12a_2-109a_3$ and that $a_0=35a_2+420a_3$. Rewriting $(35a_2+420a_3)+(-12a_2-109a_3)x+a_2x^2+a_3x^3$ as $a_2\cdot(35-12x+x^2)+a_3\cdot(420-109x+x^3)$ suggests this for a basis $\langle 35-12x+x^2,420-109x+x^3\rangle$. The above shows that it spans the space. Checking it is linearly independent is routine. (*Comment.* A worthwhile check is to verify that both polynomials in the basis have both seven and five as roots.)

(c) Here there are three conditions on the cubics, that $a_0+7a_1+49a_2+343a_3=0$, that $a_0+5a_1+25a_2+125a_3=0$, and that $a_0+3a_1+9a_2+27a_3=0$. Gauss's Method

$$\begin{array}{r} a_0+7a_1+49a_2+343a_3=0 \\ a_0+5a_1+25a_2+125a_3=0 \\ a_0+3a_1+\ 9a_2+\ 27a_3=0 \end{array} \xrightarrow[-\rho_1+\rho_3]{-\rho_1+\rho_2} \xrightarrow{-2\rho_2+\rho_3} \begin{array}{r} a_0+\ 7a_1+49a_2+343a_3=0 \\ -2a_1-24a_2-218a_3=0 \\ 8a_2+120a_3=0 \end{array}$$

yields the single free variable a_3, with $a_2 = -15a_3$, $a_1 = 71a_3$, and $a_0 = -105a_3$. The parametrization is this.

$$\{(-105a_3) + (71a_3)x + (-15a_3)x^2 + (a_3)x^3 \mid a_3 \in \mathbb{R}\} = \{a_3 \cdot (-105 + 71x - 15x^2 + x^3) \mid a_3 \in \mathbb{R}\}$$

Therefore, a natural candidate for the basis is $\langle -105 + 71x - 15x^2 + x^3 \rangle$. It spans the space by the work above. It is clearly linearly independent because it is a one-element set (with that single element not the zero object of the space). Thus, any cubic through the three points $(7,0)$, $(5,0)$, and $(3,0)$ is a multiple of this one. (*Comment.* As in the prior question, a worthwhile check is to verify that plugging seven, five, and three into this polynomial yields zero each time.)

(d) This is the trivial subspace of $\mathcal{P}_3$. Thus, the basis is empty $\langle\rangle$.

Remark. Alternatively, we could have derived the polynomial in the third item by multiplying out $(x-7)(x-5)(x-3)$.

Two.III.1.25 Yes. Linear independence and span are unchanged by reordering.

Two.III.1.26 No linearly independent set contains a zero vector.

Two.III.1.27 **(a)** To show that it is linearly independent, note that $d_1(c_1\vec{\beta}_1) + d_2(c_2\vec{\beta}_2) + d_3(c_3\vec{\beta}_3) = \vec{0}$ gives that $(d_1c_1)\vec{\beta}_1 + (d_2c_2)\vec{\beta}_2 + (d_3c_3)\vec{\beta}_3 = \vec{0}$, which in turn implies that each d_ic_i is zero. But with $c_i \neq 0$ that means that each d_i is zero. Showing that it spans the space is much the same; because $\langle \vec{\beta}_1, \vec{\beta}_2, \vec{\beta}_3 \rangle$ is a basis, and so spans the space, we can for any $\vec{v}$ write $\vec{v} = d_1\vec{\beta}_1 + d_2\vec{\beta}_2 + d_3\vec{\beta}_3$, and then $\vec{v} = (d_1/c_1)(c_1\vec{\beta}_1) + (d_2/c_2)(c_2\vec{\beta}_2) + (d_3/c_3)(c_3\vec{\beta}_3)$.

If any of the scalars are zero then the result is not a basis, because it is not linearly independent.

(b) Showing that $\langle 2\vec{\beta}_1, \vec{\beta}_1 + \vec{\beta}_2, \vec{\beta}_1 + \vec{\beta}_3 \rangle$ is linearly independent is easy. To show that it spans the space, assume that $\vec{v} = d_1\vec{\beta}_1 + d_2\vec{\beta}_2 + d_3\vec{\beta}_3$. Then, we can represent the same $\vec{v}$ with respect to $\langle 2\vec{\beta}_1, \vec{\beta}_1 + \vec{\beta}_2, \vec{\beta}_1 + \vec{\beta}_3 \rangle$ in this way $\vec{v} = (1/2)(d_1 - d_2 - d_3)(2\vec{\beta}_1) + d_2(\vec{\beta}_1 + \vec{\beta}_2) + d_3(\vec{\beta}_1 + \vec{\beta}_3)$.

Two.III.1.28 Each forms a linearly independent set if we omit $\vec{v}$. To preserve linear independence, we must expand the span of each. That is, we must determine the span of each (leaving $\vec{v}$ out), and then pick a $\vec{v}$ lying outside of that span. Then to finish, we must check that the result spans the entire given space. Those checks are routine.

(a) Any vector that is not a multiple of the given one, that is, any vector that is not on the line $y = x$ will do here. One is $\vec{v} = \vec{e}_1$.

(b) By inspection, we notice that the vector $\vec{e}_3$ is not in the span of the set of the two given vectors. The check that the resulting set is a basis for $\mathbb{R}^3$ is routine.

(c) For any member of the span $\{c_1 \cdot (x) + c_2 \cdot (1 + x^2) \mid c_1, c_2 \in \mathbb{R}\}$, the coefficient of x^2 equals the constant term. So we expand the span if we add a quadratic without this property, say, $\vec{v} = 1 - x^2$. The check that the result is a basis for $\mathcal{P}_2$ is easy.

Two.III.1.29 To show that each scalar is zero, simply subtract $c_1\vec{\beta}_1 + \cdots + c_k\vec{\beta}_k - c_{k+1}\vec{\beta}_{k+1} - \cdots - c_n\vec{\beta}_n = \vec{0}$. The obvious generalization is that in any equation involving only the $\vec{\beta}$'s, and in which each $\vec{\beta}$ appears only once, each scalar is zero. For instance, an equation with a combination of the even-indexed basis vectors (i.e., $\vec{\beta}_2$, $\vec{\beta}_4$, etc.) on the right and the odd-indexed basis vectors on the left also gives the conclusion that all of the coefficients are zero.

Two.III.1.30 No; no linearly independent set contains the zero vector.

Two.III.1.31 Here is a subset of $\mathbb{R}^2$ that is not a basis, and two different linear combinations of its elements that sum to the same vector.

$$\{\begin{pmatrix}1\\2\end{pmatrix}, \begin{pmatrix}2\\4\end{pmatrix}\} \qquad 2 \cdot \begin{pmatrix}1\\2\end{pmatrix} + 0 \cdot \begin{pmatrix}2\\4\end{pmatrix} = 0 \cdot \begin{pmatrix}1\\2\end{pmatrix} + 1 \cdot \begin{pmatrix}2\\4\end{pmatrix}$$

Thus, when a subset is not a basis, it can be the case that its linear combinations are not unique.

But just because a subset is not a basis does not imply that its combinations must be not unique. For instance, this set

$$\{\begin{pmatrix}1\\2\end{pmatrix}\}$$

does have the property that

$$c_1\cdot\begin{pmatrix}1\\2\end{pmatrix}=c_2\cdot\begin{pmatrix}1\\2\end{pmatrix}$$

implies that $c_1=c_2$. The idea here is that this subset fails to be a basis because it fails to span the space; the proof of the theorem establishes that linear combinations are unique if and only if the subset is linearly independent.

Two.III.1.32 **(a)** Describing the vector space as

$$\{\begin{pmatrix}a&b\\b&c\end{pmatrix}\mid a,b,c\in\mathbb{R}\}$$

suggests this for a basis.

$$\langle\begin{pmatrix}1&0\\0&0\end{pmatrix},\begin{pmatrix}0&0\\0&1\end{pmatrix},\begin{pmatrix}0&1\\1&0\end{pmatrix}\rangle$$

Verification is easy.

(b) This is one possible basis.

$$\langle\begin{pmatrix}1&0&0\\0&0&0\\0&0&0\end{pmatrix},\begin{pmatrix}0&0&0\\0&1&0\\0&0&0\end{pmatrix},\begin{pmatrix}0&0&0\\0&0&0\\0&0&1\end{pmatrix},\begin{pmatrix}0&1&0\\1&0&0\\0&0&0\end{pmatrix},\begin{pmatrix}0&0&1\\0&0&0\\1&0&0\end{pmatrix},\begin{pmatrix}0&0&0\\0&0&1\\0&1&0\end{pmatrix}\rangle$$

(c) As in the prior two questions, we can form a basis from two kinds of matrices. First are the matrices with a single one on the diagonal and all other entries zero (there are n of those matrices). Second are the matrices with two opposed off-diagonal entries are ones and all other entries are zeros. (That is, all entries in M are zero except that $m_{i,j}$ and $m_{j,i}$ are one.)

Two.III.1.33 **(a)** Any four vectors from $\mathbb{R}^3$ are linearly related because the vector equation

$$c_1\begin{pmatrix}x_1\\y_1\\z_1\end{pmatrix}+c_2\begin{pmatrix}x_2\\y_2\\z_2\end{pmatrix}+c_3\begin{pmatrix}x_3\\y_3\\z_3\end{pmatrix}+c_4\begin{pmatrix}x_4\\y_4\\z_4\end{pmatrix}=\begin{pmatrix}0\\0\\0\end{pmatrix}$$

gives rise to a linear system

$$\begin{aligned}x_1c_1+x_2c_2+x_3c_3+x_4c_4&=0\\y_1c_1+y_2c_2+y_3c_3+y_4c_4&=0\\z_1c_1+z_2c_2+z_3c_3+z_4c_4&=0\end{aligned}$$

that is homogeneous (and so has a solution) and has four unknowns but only three equations, and therefore has nontrivial solutions. (Of course, this argument applies to any subset of $\mathbb{R}^3$ with four or more vectors.)

(b) We shall do just the two-vector case. Given $x_1, \ldots, z_2$,

$$S=\{\begin{pmatrix}x_1\\y_1\\z_1\end{pmatrix},\begin{pmatrix}x_2\\y_2\\z_2\end{pmatrix}\}$$

to decide which vectors

$$\begin{pmatrix}x\\y\\z\end{pmatrix}$$

are in the span of S, set up

$$c_1\begin{pmatrix}x_1\\y_1\\z_1\end{pmatrix}+c_2\begin{pmatrix}x_2\\y_2\\z_2\end{pmatrix}=\begin{pmatrix}x\\y\\z\end{pmatrix}$$

and row reduce the resulting system.

$$\begin{aligned}x_1c_1+x_2c_2&=x\\y_1c_1+y_2c_2&=y\\z_1c_1+z_2c_2&=z\end{aligned}$$

There are two variables c_1 and c_2 but three equations, so when Gauss's Method finishes, on the bottom row there will be some relationship of the form $0=m_1x+m_2y+m_3z$. Hence, vectors in the span of the two-element set S must satisfy some restriction. Hence the span is not all of $\mathbb{R}^3$.

Two.III.1.34 We have (using these peculiar operations with care)

$$\{\begin{pmatrix}1-y-z\\y\\z\end{pmatrix}\mid y,z\in\mathbb{R}\}=\{\begin{pmatrix}-y+1\\y\\0\end{pmatrix}+\begin{pmatrix}-z+1\\0\\z\end{pmatrix}\mid y,z\in\mathbb{R}\}=\{y\cdot\begin{pmatrix}0\\1\\0\end{pmatrix}+z\cdot\begin{pmatrix}0\\0\\1\end{pmatrix}\mid y,z\in\mathbb{R}\}$$

and so a natural candidate for a basis is this.

$$\langle\begin{pmatrix}0\\1\\0\end{pmatrix},\begin{pmatrix}0\\0\\1\end{pmatrix}\rangle$$

To check linear independence we set up

$$c_1\begin{pmatrix}0\\1\\0\end{pmatrix}+c_2\begin{pmatrix}0\\0\\1\end{pmatrix}=\begin{pmatrix}1\\0\\0\end{pmatrix}$$

(the vector on the right is the zero object in this space). That yields the linear system

$$\begin{aligned}(-c_1+1)+(-c_2+1)-1&=1\\c_1\qquad\qquad&=0\\c_2&=0\end{aligned}$$

with only the solution $c_1=0$ and $c_2=0$. Checking the span is similar.

Two.III.2: Dimension

Two.III.2.15 One basis is $\langle 1,x,x^2\rangle$, and so the dimension is three.

Two.III.2.16 The solution set is

$$\{\begin{pmatrix}4x_2-3x_3+x_4\\x_2\\x_3\\x_4\end{pmatrix}\mid x_2,x_3,x_4\in\mathbb{R}\}$$

so a natural basis is this

$$\langle\begin{pmatrix}4\\1\\0\\0\end{pmatrix},\begin{pmatrix}-3\\0\\1\\0\end{pmatrix},\begin{pmatrix}1\\0\\0\\1\end{pmatrix}\rangle$$

(checking linear independence is easy). Thus the dimension is three.

Two.III.2.17 For this space

$$\{\begin{pmatrix} a & b \\ c & d \end{pmatrix} \mid a,b,c,d \in \mathbb{R}\} = \{a \cdot \begin{pmatrix} 1 & 0 \\ 0 & 0 \end{pmatrix} + \dots + d \cdot \begin{pmatrix} 0 & 0 \\ 0 & 1 \end{pmatrix} \mid a,b,c,d \in \mathbb{R}\}$$

this is a natural basis.

$$\langle \begin{pmatrix} 1 & 0 \\ 0 & 0 \end{pmatrix}, \begin{pmatrix} 0 & 1 \\ 0 & 0 \end{pmatrix}, \begin{pmatrix} 0 & 0 \\ 1 & 0 \end{pmatrix}, \begin{pmatrix} 0 & 0 \\ 0 & 1 \end{pmatrix} \rangle$$

The dimension is four.

Two.III.2.18 (a) As in the prior exercise, the space $\mathcal{M}_{2\times 2}$ of matrices without restriction has this basis

$$\langle \begin{pmatrix} 1 & 0 \\ 0 & 0 \end{pmatrix}, \begin{pmatrix} 0 & 1 \\ 0 & 0 \end{pmatrix}, \begin{pmatrix} 0 & 0 \\ 1 & 0 \end{pmatrix}, \begin{pmatrix} 0 & 0 \\ 0 & 1 \end{pmatrix} \rangle$$

and so the dimension is four.

(b) For this space

$$\{\begin{pmatrix} a & b \\ c & d \end{pmatrix} \mid a = b - 2c \text{ and } d \in \mathbb{R}\} = \{b \cdot \begin{pmatrix} 1 & 1 \\ 0 & 0 \end{pmatrix} + c \cdot \begin{pmatrix} -2 & 0 \\ 1 & 0 \end{pmatrix} + d \cdot \begin{pmatrix} 0 & 0 \\ 0 & 1 \end{pmatrix} \mid b,c,d \in \mathbb{R}\}$$

this is a natural basis.

$$\langle \begin{pmatrix} 1 & 1 \\ 0 & 0 \end{pmatrix}, \begin{pmatrix} -2 & 0 \\ 1 & 0 \end{pmatrix}, \begin{pmatrix} 0 & 0 \\ 0 & 1 \end{pmatrix} \rangle$$

The dimension is three.

(c) Gauss's Method applied to the two-equation linear system gives that $c = 0$ and that $a = -b$. Thus, we have this description

$$\{\begin{pmatrix} -b & b \\ 0 & d \end{pmatrix} \mid b,d \in \mathbb{R}\} = \{b \cdot \begin{pmatrix} -1 & 1 \\ 0 & 0 \end{pmatrix} + d \cdot \begin{pmatrix} 0 & 0 \\ 0 & 1 \end{pmatrix} \mid b,d \in \mathbb{R}\}$$

and so this is a natural basis.

$$\langle \begin{pmatrix} -1 & 1 \\ 0 & 0 \end{pmatrix}, \begin{pmatrix} 0 & 0 \\ 0 & 1 \end{pmatrix} \rangle$$

The dimension is two.

Two.III.2.19 The bases for these spaces are developed in the answer set of the prior subsection.

(a) One basis is $\langle -7 + x, -49 + x^2, -343 + x^3 \rangle$. The dimension is three.

(b) One basis is $\langle 35 - 12x + x^2, 420 - 109x + x^3 \rangle$ so the dimension is two.

(c) A basis is $\{-105 + 71x - 15x^2 + x^3\}$. The dimension is one.

(d) This is the trivial subspace of $\mathcal{P}_3$ and so the basis is empty. The dimension is zero.

Two.III.2.20 First recall that $\cos 2\theta = \cos^2\theta - \sin^2\theta$, and so deletion of $\cos 2\theta$ from this set leaves the span unchanged. What's left, the set $\{\cos^2\theta, \sin^2\theta, \sin 2\theta\}$, is linearly independent (consider the relationship $c_1 \cos^2\theta + c_2 \sin^2\theta + c_3 \sin 2\theta = Z(\theta)$ where Z is the zero function, and then take $\theta = 0$, $\theta = \pi/4$, and $\theta = \pi/2$ to conclude that each c is zero). It is therefore a basis for its span. That shows that the span is a dimension three vector space.

Two.III.2.21 Here is a basis

$$\langle (1 + 0i, 0 + 0i, \dots, 0 + 0i),\ (0 + 1i, 0 + 0i, \dots, 0 + 0i), (0 + 0i, 1 + 0i, \dots, 0 + 0i), \dots \rangle$$

and so the dimension is $2 \cdot 47 = 94$.

Two.III.2.22 A basis is

$$\langle \begin{pmatrix} 1 & 0 & 0 & 0 & 0 \\ 0 & 0 & 0 & 0 & 0 \\ 0 & 0 & 0 & 0 & 0 \end{pmatrix}, \begin{pmatrix} 0 & 1 & 0 & 0 & 0 \\ 0 & 0 & 0 & 0 & 0 \\ 0 & 0 & 0 & 0 & 0 \end{pmatrix}, \dots, \begin{pmatrix} 0 & 0 & 0 & 0 & 0 \\ 0 & 0 & 0 & 0 & 0 \\ 0 & 0 & 0 & 0 & 1 \end{pmatrix} \rangle$$

and thus the dimension is $3 \cdot 5 = 15$.

Two.III.2.23 In a four-dimensional space a set of four vectors is linearly independent if and only if it spans the space. The form of these vectors makes linear independence easy to show (look at the equation of fourth components, then at the equation of third components, etc.).

Two.III.2.24 **(a)** The diagram for $\mathcal{P}_2$ has four levels. The top level has the only three-dimensional subspace, $\mathcal{P}_2$ itself. The next level contains the two-dimensional subspaces (*not* just the linear polynomials; any two-dimensional subspace, like those polynomials of the form ax^2+b). Below that are the one-dimensional subspaces. Finally, of course, is the only zero-dimensional subspace, the trivial subspace.
(b) For $\mathcal{M}_{2\times 2}$, the diagram has five levels, including subspaces of dimension four through zero.

Two.III.2.25 **(a)** One **(b)** Two **(c)** n

Two.III.2.26 We need only produce an infinite linearly independent set. One is $\langle f_1, f_2, \ldots\rangle$ where $f_i\colon \mathbb{R}\to\mathbb{R}$ is

$$f_i(x) = \begin{cases} 1 & \text{if } x = i \\ 0 & \text{otherwise} \end{cases}$$

the function that has value 1 only at $x = i$.

Two.III.2.27 A function is a set of ordered pairs $(x, f(x))$. So there is only one function with an empty domain, namely the empty set. A vector space with only one element a trivial vector space and has dimension zero.

Two.III.2.28 Apply Corollary 2.10.

Two.III.2.29 A plane has the form $\{\vec{p} + t_1\vec{v}_1 + t_2\vec{v}_2 \mid t_1, t_2 \in \mathbb{R}\}$. (The first chapter also calls this a '2-flat', and contains a discussion of why this is equivalent to the description often taken in Calculus as the set of points (x, y, z) subject to a condition of the form $ax + by + cz = d$). When the plane passes through the origin we can take the particular vector $\vec{p}$ to be $\vec{0}$. Thus, in the language we have developed in this chapter, a plane through the origin is the span of a set of two vectors.

Now for the statement. Asserting that the three are not coplanar is the same as asserting that no vector lies in the span of the other two — no vector is a linear combination of the other two. That's simply an assertion that the three-element set is linearly independent. By Corollary 2.14, that's equivalent to an assertion that the set is a basis for $\mathbb{R}^3$ (more precisely, any sequence made from the set's elements is a basis).

Two.III.2.30 Let the space V be finite dimensional. Let S be a subspace of V.
(a) The empty set is a linearly independent subset of S. By Corollary 2.12, it can be expanded to a basis for the vector space S.
(b) Any basis for the subspace S is a linearly independent set in the superspace V. Hence it can be expanded to a basis for the superspace, which is finite dimensional. Therefore it has only finitely many members.

Two.III.2.31 It ensures that we exhaust the $\vec{\beta}$'s. That is, it justifies the first sentence of the last paragraph.

Two.III.2.32 Let B_U be a basis for U and let B_W be a basis for W. The set $B_U \cup B_W$ is linearly dependent as it is a six member subset of the five-dimensional space $\mathbb{R}^5$. Thus some member of B_W is in the span of B_U, and thus $U \cap W$ is more than just the trivial space $\{\vec{0}\,\}$.

Generalization: if U, W are subspaces of a vector space of dimension n and if $\dim(U) + \dim(W) > n$ then they have a nontrivial intersection.

Two.III.2.33 First, note that a set is a basis for some space if and only if it is linearly independent, because in that case it is a basis for its own span.
(a) The answer to the question in the second paragraph is "yes" (implying "yes" answers for both questions in the first paragraph). If B_U is a basis for U then B_U is a linearly independent subset of W. Apply Corollary 2.12 to expand it to a basis for W. That is the desired B_W.

The answer to the question in the third paragraph is "no", which implies a "no" answer to the question of the fourth paragraph. Here is an example of a basis for a superspace with no sub-basis forming a basis for a subspace: in $W = \mathbb{R}^2$, consider the standard basis $\mathcal{E}_2$. No sub-basis of $\mathcal{E}_2$ forms a basis for the subspace U of $\mathbb{R}^2$ that is the line $y = x$.

(b) It is a basis (for its span) because the intersection of linearly independent sets is linearly independent (the intersection is a subset of each of the linearly independent sets).

It is not, however, a basis for the intersection of the spaces. For instance, these are bases for $\mathbb{R}^2$:

$$B_1 = \langle \begin{pmatrix} 1 \\ 0 \end{pmatrix}, \begin{pmatrix} 0 \\ 1 \end{pmatrix} \rangle \quad \text{and} \quad B_2 = \langle [] r] \begin{pmatrix} 2 \\ 0 \end{pmatrix}, \begin{pmatrix} 0 \\ 2 \end{pmatrix} \rangle$$

and $\mathbb{R}^2 \cap \mathbb{R}^2 = \mathbb{R}^2$, but $B_1 \cap B_2$ is empty. All we can say is that the $\cap$ of the bases is a basis for a subset of the intersection of the spaces.

(c) The $\cup$ of bases need not be a basis: in $\mathbb{R}^2$

$$B_1 = \langle \begin{pmatrix} 1 \\ 0 \end{pmatrix}, \begin{pmatrix} 1 \\ 1 \end{pmatrix} \rangle \quad \text{and} \quad B_2 = \langle \begin{pmatrix} 1 \\ 0 \end{pmatrix}, \begin{pmatrix} 0 \\ 2 \end{pmatrix} \rangle$$

$B_1 \cup B_2$ is not linearly independent. A necessary and sufficient condition for a $\cup$ of two bases to be a basis

$$B_1 \cup B_2 \text{ is linearly independent} \quad \iff \quad [B_1 \cap B_2] = [B_1] \cap [B_2]$$

it is easy enough to prove (but perhaps hard to apply).

(d) The complement of a basis cannot be a basis because it contains the zero vector.

Two.III.2.34 **(a)** A basis for U is a linearly independent set in W and so can be expanded via Corollary 2.12 to a basis for W. The second basis has at least as many members as the first.

(b) One direction is clear: if $V = W$ then they have the same dimension. For the converse, let B_U be a basis for U. It is a linearly independent subset of W and so can be expanded to a basis for W. If $\dim(U) = \dim(W)$ then this basis for W has no more members than does B_U and so equals B_U. Since U and W have the same bases, they are equal.

(c) Let W be the space of finite-degree polynomials and let U be the subspace of polynomials that have only even-powered terms $\{a_0 + a_1x^2 + a_2x^4 + \cdots + a_nx^{2n} \mid a_0, \ldots, a_n \in \mathbb{R}\}$. Both spaces have infinite dimension, but U is a proper subspace.

Two.III.2.35 The possibilities for the dimension of V are 0, 1, $n-1$, and n.

To see this, first consider the case when all the coordinates of $\vec{v}$ are equal.

$$\vec{v} = \begin{pmatrix} z \\ z \\ \vdots \\ z \end{pmatrix}$$

Then $\sigma(\vec{v}) = \vec{v}$ for every permutation σ, so V is just the span of $\vec{v}$, which has dimension 0 or 1 according to whether $\vec{v}$ is $\vec{0}$ or not.

Now suppose not all the coordinates of $\vec{v}$ are equal; let x and y with $x \neq y$ be among the coordinates of $\vec{v}$. Then we can find permutations σ_1 and σ_2 such that

$$\sigma_1(\vec{v}) = \begin{pmatrix} x \\ y \\ a_3 \\ \vdots \\ a_n \end{pmatrix} \quad \text{and} \quad \sigma_2(\vec{v}) = \begin{pmatrix} y \\ x \\ a_3 \\ \vdots \\ a_n \end{pmatrix}$$

for some $a_3, \ldots, a_n \in \mathbb{R}$. Therefore,

$$\frac{1}{y-x}\big(\sigma_1(\vec{v}) - \sigma_2(\vec{v})\big) = \begin{pmatrix} -1 \\ 1 \\ 0 \\ \vdots \\ 0 \end{pmatrix}$$

is in V. That is, $\vec{e}_2 - \vec{e}_1 \in V$, where $\vec{e}_1$, $\vec{e}_2$, ..., $\vec{e}_n$ is the standard basis for $\mathbb{R}^n$. Similarly, $\vec{e}_3 - \vec{e}_2$, ..., $\vec{e}_n - \vec{e}_1$ are all in V. It is easy to see that the vectors $\vec{e}_2 - \vec{e}_1$, $\vec{e}_3 - \vec{e}_2$, ..., $\vec{e}_n - \vec{e}_1$ are linearly independent (that is, form a linearly independent set), so $\dim V \geqslant n-1$.

Finally, we can write

$$\begin{aligned}\vec{v} &= x_1\vec{e}_1 + x_2\vec{e}_2 + \cdots + x_n\vec{e}_n \\ &= (x_1 + x_2 + \cdots + x_n)\vec{e}_1 + x_2(\vec{e}_2 - \vec{e}_1) + \cdots + x_n(\vec{e}_n - \vec{e}_1)\end{aligned}$$

This shows that if $x_1 + x_2 + \cdots + x_n = 0$ then $\vec{v}$ is in the span of $\vec{e}_2 - \vec{e}_1$, ..., $\vec{e_n} - \vec{e}_1$ (that is, is in the span of the set of those vectors); similarly, each $\sigma(\vec{v})$ will be in this span, so V will equal this span and $\dim V = n-1$. On the other hand, if $x_1 + x_2 + \cdots + x_n \neq 0$ then the above equation shows that $\vec{e}_1 \in V$ and thus $\vec{e}_1, \ldots, \vec{e}_n \in V$, so $V = \mathbb{R}^n$ and $\dim V = n$.

Two.III.3: Vector Spaces and Linear Systems

Two.III.3.16 **(a)** $\begin{pmatrix} 2 & 3 \\ 1 & 1 \end{pmatrix}$ **(b)** $\begin{pmatrix} 2 & 1 \\ 1 & 3 \end{pmatrix}$ **(c)** $\begin{pmatrix} 1 & 6 \\ 4 & 7 \\ 3 & 8 \end{pmatrix}$ **(d)** $\begin{pmatrix} 0 & 0 & 0 \end{pmatrix}$ **(e)** $\begin{pmatrix} -1 \\ -2 \end{pmatrix}$

Two.III.3.17 **(a)** Yes. To see if there are c_1 and c_2 such that $c_1 \cdot (2 \quad 1) + c_2 \cdot (3 \quad 1) = (1 \quad 0)$ we solve

$$\begin{aligned} 2c_1 + 3c_2 &= 1 \\ c_1 + c_2 &= 0 \end{aligned}$$

and get $c_1 = -1$ and $c_2 = 1$. Thus the vector is in the row space.

(b) No. The equation $c_1(0 \quad 1 \quad 3) + c_2(-1 \quad 0 \quad 1) + c_3(-1 \quad 2 \quad 7) = (1 \quad 1 \quad 1)$ has no solution.

$$\left(\begin{array}{ccc|c} 0 & -1 & -1 & 1 \\ 1 & 0 & 2 & 1 \\ 3 & 1 & 7 & 1 \end{array}\right) \xrightarrow{\rho_1 \leftrightarrow \rho_2} \xrightarrow{-3\rho_1 + \rho_2} \xrightarrow{\rho_2 + \rho_3} \left(\begin{array}{ccc|c} 1 & 0 & 2 & 1 \\ 0 & -1 & -1 & 1 \\ 0 & 0 & 0 & -1 \end{array}\right)$$

Thus, the vector is not in the row space.

Two.III.3.18 **(a)** No. To see if there are $c_1, c_2 \in \mathbb{R}$ such that

$$c_1 \begin{pmatrix} 1 \\ 1 \end{pmatrix} + c_2 \begin{pmatrix} 1 \\ 1 \end{pmatrix} = \begin{pmatrix} 1 \\ 3 \end{pmatrix}$$

we can use Gauss's Method on the resulting linear system.

$$\begin{aligned} c_1 + c_2 &= 1 \\ c_1 + c_2 &= 3 \end{aligned} \xrightarrow{-\rho_1 + \rho_2} \begin{aligned} c_1 + c_2 &= 1 \\ 0 &= 2 \end{aligned}$$

There is no solution and so the vector is not in the column space.

(b) Yes. From this relationship

$$c_1 \begin{pmatrix} 1 \\ 2 \\ 1 \end{pmatrix} + c_2 \begin{pmatrix} 3 \\ 0 \\ -3 \end{pmatrix} + c_3 \begin{pmatrix} 1 \\ 4 \\ 3 \end{pmatrix} = \begin{pmatrix} 1 \\ 0 \\ 0 \end{pmatrix}$$

we get a linear system that, when we apply Gauss's Method,

$$\left(\begin{array}{ccc|c} 1 & 3 & 1 & 1 \\ 2 & 0 & 4 & 0 \\ 1 & -3 & -3 & 0 \end{array}\right) \xrightarrow[-\rho_1 + \rho_3]{-2\rho_1 + \rho_2} \xrightarrow{-\rho_2 + \rho_3} \left(\begin{array}{ccc|c} 1 & 3 & 1 & 1 \\ 0 & -6 & 2 & -2 \\ 0 & 0 & -6 & 1 \end{array}\right)$$

yields a solution. Thus, the vector is in the column space.

Two.III.3.19 (a) Yes; we are asking if there are scalars c_1 and c_2 such that

$$c_1\begin{pmatrix}2\\2\end{pmatrix}+c_2\begin{pmatrix}1\\5\end{pmatrix}=\begin{pmatrix}1\\-3\end{pmatrix}$$

which gives rise to a linear system

$$\begin{array}{rl}2c_1+c_2=&1\\2c_1+5c_2=&-3\end{array}\xrightarrow{-\rho_1+\rho_2}\begin{array}{rl}2c_1+c_2=&1\\4c_2=&-4\end{array}$$

and Gauss's Method produces $c_2=-1$ and $c_1=1$. That is, there is indeed such a pair of scalars and so the vector is indeed in the column space of the matrix.

(b) No; we are asking if there are scalars c_1 and c_2 such that

$$c_1\begin{pmatrix}4\\2\end{pmatrix}+c_2\begin{pmatrix}-8\\-4\end{pmatrix}=\begin{pmatrix}0\\1\end{pmatrix}$$

and one way to proceed is to consider the resulting linear system

$$\begin{array}{r}4c_1-8c_2=0\\2c_1-4c_2=1\end{array}$$

that is easily seen to have no solution. Another way to proceed is to note that any linear combination of the columns on the left has a second component half as big as its first component, but the vector on the right does not meet that criterion.

(c) Yes; we can simply observe that the vector is the first column minus the second. Or, failing that, setting up the relationship among the columns

$$c_1\begin{pmatrix}1\\1\\-1\end{pmatrix}+c_2\begin{pmatrix}-1\\1\\-1\end{pmatrix}+c_3\begin{pmatrix}1\\-1\\1\end{pmatrix}=\begin{pmatrix}2\\0\\0\end{pmatrix}$$

and considering the resulting linear system

$$\begin{array}{r}c_1-c_2+c_3=2\\c_1+c_2-c_3=0\\-c_1-c_2+c_3=0\end{array}\xrightarrow[\rho_1+\rho_3]{-\rho_1+\rho_2}\begin{array}{rl}c_1-c_2+c_3=&2\\2c_2-2c_3=&-2\\-2c_2+2c_3=&2\end{array}\xrightarrow{\rho_2+\rho_3}\begin{array}{rl}c_1-c_2+c_3=&2\\2c_2-2c_3=&-2\\0=&0\end{array}$$

gives the additional information (beyond that there is at least one solution) that there are infinitely many solutions. Parametrizing gives $c_2=-1+c_3$ and $c_1=1$, and so taking c_3 to be zero gives a particular solution of $c_1=1$, $c_2=-1$, and $c_3=0$ (which is, of course, the observation made at the start).

Two.III.3.20 A routine Gaussian reduction

$$\begin{pmatrix}2&0&3&4\\0&1&1&-1\\3&1&0&2\\1&0&-4&1\end{pmatrix}\xrightarrow[-(1/2)\rho_1+\rho_4]{-(3/2)\rho_1+\rho_3}\xrightarrow{-\rho_2+\rho_3}\xrightarrow{-\rho_3+\rho_4}\begin{pmatrix}2&0&3&4\\0&1&1&-1\\0&0&-11/2&-3\\0&0&0&0\end{pmatrix}$$

suggests this basis $\langle(2\ \ 0\ \ 3\ \ 4),(0\ \ 1\ \ 1\ \ -1),(0\ \ 0\ \ -11/2\ \ -3)\rangle$.

Another, perhaps more convenient procedure, is to swap rows first,

$$\xrightarrow{\rho_1\leftrightarrow\rho_4}\xrightarrow[-2\rho_1+\rho_4]{-3\rho_1+\rho_3}\xrightarrow{-\rho_2+\rho_3}\xrightarrow{-\rho_3+\rho_4}\begin{pmatrix}1&0&-4&-1\\0&1&1&-1\\0&0&11&6\\0&0&0&0\end{pmatrix}$$

leading to the basis $\langle(1\ \ 0\ \ -4\ \ -1),(0\ \ 1\ \ 1\ \ -1),(0\ \ 0\ \ 11\ \ 6)\rangle$.

Two.III.3.21 (a) This reduction

$$\xrightarrow[-(1/2)\rho_1+\rho_3]{-(1/2)\rho_1+\rho_2}\xrightarrow{-(1/3)\rho_2+\rho_3}\begin{pmatrix}2&1&3\\0&-3/2&1/2\\0&0&4/3\end{pmatrix}$$

shows that the row rank, and hence the rank, is three.

(b) Inspection of the columns shows that the others are multiples of the first (inspection of the rows shows the same thing). Thus the rank is one.

Alternatively, the reduction

$$\begin{pmatrix} 1 & -1 & 2 \\ 3 & -3 & 6 \\ -2 & 2 & -4 \end{pmatrix} \xrightarrow[2\rho_1+\rho_3]{-3\rho_1+\rho_2} \begin{pmatrix} 1 & -1 & 2 \\ 0 & 0 & 0 \\ 0 & 0 & 0 \end{pmatrix}$$

shows the same thing.

(c) This calculation

$$\begin{pmatrix} 1 & 3 & 2 \\ 5 & 1 & 1 \\ 6 & 4 & 3 \end{pmatrix} \xrightarrow[-6\rho_1+\rho_3]{-5\rho_1+\rho_2} \xrightarrow{-\rho_2+\rho_3} \begin{pmatrix} 1 & 3 & 2 \\ 0 & -14 & -9 \\ 0 & 0 & 0 \end{pmatrix}$$

shows that the rank is two.

(d) The rank is zero.

Two.III.3.22 **(a)** This reduction

$$\begin{pmatrix} 1 & 3 \\ -1 & 3 \\ 1 & 4 \\ 2 & 1 \end{pmatrix} \xrightarrow[\substack{-\rho_1+\rho_3 \\ -2\rho_1+\rho_4}]{\rho_1+\rho_2} \xrightarrow[(5/6)\rho_2+\rho_4]{-(1/6)\rho_2+\rho_3} \begin{pmatrix} 1 & 3 \\ 0 & 6 \\ 0 & 0 \\ 0 & 0 \end{pmatrix}$$

gives $\langle (1 \;\; 3), (0 \;\; 6) \rangle$.

(b) Transposing and reducing

$$\begin{pmatrix} 1 & 2 & 1 \\ 3 & 1 & -1 \\ 1 & -3 & -3 \end{pmatrix} \xrightarrow[-\rho_1+\rho_3]{-3\rho_1+\rho_2} \begin{pmatrix} 1 & 2 & 1 \\ 0 & -5 & -4 \\ 0 & -5 & -4 \end{pmatrix} \xrightarrow{-\rho_2+\rho_3} \begin{pmatrix} 1 & 2 & 1 \\ 0 & -5 & -4 \\ 0 & 0 & 0 \end{pmatrix}$$

and then transposing back gives this basis.

$$\langle \begin{pmatrix} 1 \\ 2 \\ 1 \end{pmatrix}, \begin{pmatrix} 0 \\ -5 \\ -4 \end{pmatrix} \rangle$$

(c) Notice first that the surrounding space is as $\mathcal{P}_3$, not $\mathcal{P}_2$. Then, taking the first polynomial $1 + 1 \cdot x + 0 \cdot x^2 + 0 \cdot x^3$ to be "the same" as the row vector $(1 \;\; 1 \;\; 0 \;\; 0)$, etc., leads to

$$\begin{pmatrix} 1 & 1 & 0 & 0 \\ 1 & 0 & -1 & 0 \\ 3 & 2 & -1 & 0 \end{pmatrix} \xrightarrow[-3\rho_1+\rho_3]{-\rho_1+\rho_2} \xrightarrow{-\rho_2+\rho_3} \begin{pmatrix} 1 & 1 & 0 & 0 \\ 0 & -1 & -1 & 0 \\ 0 & 0 & 0 & 0 \end{pmatrix}$$

which yields the basis $\langle 1 + x, -x - x^2 \rangle$.

(d) Here "the same" gives

$$\begin{pmatrix} 1 & 0 & 1 & 3 & 1 & -1 \\ 1 & 0 & 3 & 2 & 1 & 4 \\ -1 & 0 & -5 & -1 & -1 & -9 \end{pmatrix} \xrightarrow[\rho_1+\rho_3]{-\rho_1+\rho_2} \xrightarrow{2\rho_2+\rho_3} \begin{pmatrix} 1 & 0 & 1 & 3 & 1 & -1 \\ 0 & 0 & 2 & -1 & 0 & 5 \\ 0 & 0 & 0 & 0 & 0 & 0 \end{pmatrix}$$

leading to this basis.

$$\langle \begin{pmatrix} 1 & 0 & 1 \\ 3 & 1 & -1 \end{pmatrix}, \begin{pmatrix} 0 & 0 & 2 \\ -1 & 0 & 5 \end{pmatrix} \rangle$$

Two.III.3.23 Only the zero matrices have rank of zero. The only matrices of rank one have the form

$$\begin{pmatrix} k_1 \cdot \rho \\ \vdots \\ k_m \cdot \rho \end{pmatrix}$$

where ρ is some nonzero row vector, and not all of the k_i's are zero. (*Remark.* We can't simply say that all of the rows are multiples of the first because the first row might be the zero row. *Another Remark.* The above also applies with 'column' replacing 'row'.)

Two.III.3.24 If $a \neq 0$ then a choice of $d = (c/a)b$ will make the second row be a multiple of the first, specifically, c/a times the first. If $a = 0$ and $b = 0$ then any non-0 choice for d will ensure that the second row is nonzero. If $a = 0$ and $b \neq 0$ and $c = 0$ then any choice for d will do, since the matrix will automatically have rank one (even with the choice of $d = 0$). Finally, if $a = 0$ and $b \neq 0$ and $c \neq 0$ then no choice for d will suffice because the matrix is sure to have rank two.

Two.III.3.25 The column rank is two. One way to see this is by inspection — the column space consists of two-tall columns and so can have a dimension of at least two, and we can easily find two columns that together form a linearly independent set (the fourth and fifth columns, for instance). Another way to see this is to recall that the column rank equals the row rank, and to perform Gauss's Method, which leaves two nonzero rows.

Two.III.3.26 We apply Theorem 3.13. The number of columns of a matrix of coefficients A of a linear system equals the number n of unknowns. A linear system with at least one solution has at most one solution if and only if the space of solutions of the associated homogeneous system has dimension zero (recall: in the 'General = Particular + Homogeneous' equation $\vec{v} = \vec{p} + \vec{h}$, provided that such a $\vec{p}$ exists, the solution $\vec{v}$ is unique if and only if the vector $\vec{h}$ is unique, namely $\vec{h} = \vec{0}$). But that means, by the theorem, that $n = r$.

Two.III.3.27 The set of columns must be dependent because the rank of the matrix is at most five while there are nine columns.

Two.III.3.28 There is little danger of their being equal since the row space is a set of row vectors while the column space is a set of columns (unless the matrix is 1×1, in which case the two spaces must be equal).

Remark. Consider

$$A = \begin{pmatrix} 1 & 3 \\ 2 & 6 \end{pmatrix}$$

and note that the row space is the set of all multiples of $(1 \quad 3)$ while the column space consists of multiples of

$$\begin{pmatrix} 1 \\ 2 \end{pmatrix}$$

so we also cannot argue that the two spaces must be simply transposes of each other.

Two.III.3.29 First, the vector space is the set of four-tuples of real numbers, under the natural operations. Although this is not the set of four-wide row vectors, the difference is slight — it is "the same" as that set. So we will treat the four-tuples like four-wide vectors.

With that, one way to see that $(1, 0, 1, 0)$ is not in the span of the first set is to note that this reduction

$$\begin{pmatrix} 1 & -1 & 2 & -3 \\ 1 & 1 & 2 & 0 \\ 3 & -1 & 6 & -6 \end{pmatrix} \xrightarrow[-3\rho_1+\rho_3]{-\rho_1+\rho_2} \xrightarrow{-\rho_2+\rho_3} \begin{pmatrix} 1 & -1 & 2 & -3 \\ 0 & 2 & 0 & 3 \\ 0 & 0 & 0 & 0 \end{pmatrix}$$

and this one

$$\begin{pmatrix} 1 & -1 & 2 & -3 \\ 1 & 1 & 2 & 0 \\ 3 & -1 & 6 & -6 \\ 1 & 0 & 1 & 0 \end{pmatrix} \xrightarrow[\substack{-3\rho_1+\rho_3 \\ -\rho_1+\rho_4}]{-\rho_1+\rho_2} \xrightarrow[-(1/2)\rho_2+\rho_4]{-\rho_2+\rho_3} \xrightarrow{\rho_3 \leftrightarrow \rho_4} \begin{pmatrix} 1 & -1 & 2 & -3 \\ 0 & 2 & 0 & 3 \\ 0 & 0 & -1 & 3/2 \\ 0 & 0 & 0 & 0 \end{pmatrix}$$

yield matrices differing in rank. This means that addition of $(1, 0, 1, 0)$ to the set of the first three four-tuples increases the rank, and hence the span, of that set. Therefore $(1, 0, 1, 0)$ is not already in the span.

Two.III.3.30 It is a subspace because it is the column space of the matrix

$$\begin{pmatrix} 3 & 2 & 4 \\ 1 & 0 & -1 \\ 2 & 2 & 5 \end{pmatrix}$$

of coefficients. To find a basis for the column space,

$$\{c_1\begin{pmatrix}3\\1\\2\end{pmatrix}+c_2\begin{pmatrix}2\\0\\2\end{pmatrix}+c_3\begin{pmatrix}4\\-1\\5\end{pmatrix}\mid c_1,c_2,c_3\in\mathbb{R}\}$$

we take the three vectors from the spanning set, transpose, reduce,

$$\begin{pmatrix}3&1&2\\2&0&2\\4&-1&5\end{pmatrix}\xrightarrow[-(4/3)\rho_1+\rho_3]{-(2/3)\rho_1+\rho_2}\xrightarrow{-(7/2)\rho_2+\rho_3}\begin{pmatrix}3&1&2\\0&-2/3&2/3\\0&0&0\end{pmatrix}$$

and transpose back to get this.

$$\langle\begin{pmatrix}3\\1\\2\end{pmatrix},\begin{pmatrix}0\\-2/3\\2/3\end{pmatrix}\rangle$$

Two.III.3.31 We can do this as a straightforward calculation.

$$\begin{aligned}(rA+sB)^{\mathsf{T}}&=\begin{pmatrix}ra_{1,1}+sb_{1,1}&\ldots&ra_{1,n}+sb_{1,n}\\&\vdots&\\ra_{m,1}+sb_{m,1}&\ldots&ra_{m,n}+sb_{m,n}\end{pmatrix}^{\mathsf{T}}\\&=\begin{pmatrix}ra_{1,1}+sb_{1,1}&\ldots&ra_{m,1}+sb_{m,1}\\&\vdots&\\ra_{1,n}+sb_{1,n}&\ldots&ra_{m,n}+sb_{m,n}\end{pmatrix}\\&=\begin{pmatrix}ra_{1,1}&\ldots&ra_{m,1}\\&\vdots&\\ra_{1,n}&\ldots&ra_{m,n}\end{pmatrix}+\begin{pmatrix}sb_{1,1}&\ldots&sb_{m,1}\\&\vdots&\\sb_{1,n}&\ldots&sb_{m,n}\end{pmatrix}\\&=rA^{\mathsf{T}}+sB^{\mathsf{T}}\end{aligned}$$

Two.III.3.32 **(a)** These reductions give different bases.

$$\begin{pmatrix}1&2&0\\1&2&1\end{pmatrix}\xrightarrow{-\rho_1+\rho_2}\begin{pmatrix}1&2&0\\0&0&1\end{pmatrix}\qquad\begin{pmatrix}1&2&0\\1&2&1\end{pmatrix}\xrightarrow{-\rho_1+\rho_2}\xrightarrow{2\rho_2}\begin{pmatrix}1&2&0\\0&0&2\end{pmatrix}$$

(b) An easy example is this.

$$\begin{pmatrix}1&2&1\\3&1&4\end{pmatrix}\qquad\begin{pmatrix}1&2&1\\3&1&4\\0&0&0\end{pmatrix}$$

This is a less simplistic example.

$$\begin{pmatrix}1&2&1\\3&1&4\end{pmatrix}\qquad\begin{pmatrix}1&2&1\\3&1&4\\2&4&2\\4&3&5\end{pmatrix}$$

(c) Assume that A and B are matrices with equal row spaces. Construct a matrix C with the rows of A above the rows of B, and another matrix D with the rows of B above the rows of A.

$$C=\begin{pmatrix}A\\B\end{pmatrix}\qquad D=\begin{pmatrix}B\\A\end{pmatrix}$$

Observe that C and D are row-equivalent (via a sequence of row-swaps) and so Gauss-Jordan reduce to the same reduced echelon form matrix.

Because the row spaces are equal, the rows of B are linear combinations of the rows of A so Gauss-Jordan reduction on C simply turns the rows of B to zero rows and thus the nonzero rows of C are just the nonzero rows obtained by Gauss-Jordan reducing A. The same can be said for the matrix D—Gauss-Jordan reduction on D gives the same non-zero rows as are produced by reduction on B alone. Therefore, A yields the same nonzero rows as C, which yields the same nonzero rows as D, which yields the same nonzero rows as B.

Two.III.3.33 It cannot be bigger.

Two.III.3.34 The number of rows in a maximal linearly independent set cannot exceed the number of rows. A better bound (the bound that is, in general, the best possible) is the minimum of m and n, because the row rank equals the column rank.

Two.III.3.35 Because the rows of a matrix A are the columns of A^T the dimension of the row space of A equals the dimension of the column space of A^T. But the dimension of the row space of A is the rank of A and the dimension of the column space of A^T is the rank of A^T. Thus the two ranks are equal.

Two.III.3.36 False. The first is a set of columns while the second is a set of rows.

This example, however,

$$A = \begin{pmatrix} 1 & 2 & 3 \\ 4 & 5 & 6 \end{pmatrix}, \qquad A^T = \begin{pmatrix} 1 & 4 \\ 2 & 5 \\ 3 & 6 \end{pmatrix}$$

indicates that as soon as we have a formal meaning for "the same", we can apply it here:

$$\text{Columnspace}(A) = [\{\begin{pmatrix} 1 \\ 4 \end{pmatrix}, \begin{pmatrix} 2 \\ 5 \end{pmatrix}, \begin{pmatrix} 3 \\ 6 \end{pmatrix}\}]$$

while

$$\text{Rowspace}(A^T) = [\{(1 \quad 4), (2 \quad 5), (3 \quad 6)\}]$$

are "the same" as each other.

Two.III.3.37 No. Here, Gauss's Method does not change the column space.

$$\begin{pmatrix} 1 & 0 \\ 3 & 1 \end{pmatrix} \xrightarrow{-3\rho_1+\rho_2} \begin{pmatrix} 1 & 0 \\ 0 & 1 \end{pmatrix}$$

Two.III.3.38 A linear system

$$c_1\vec{a}_1 + \cdots + c_n\vec{a}_n = \vec{d}$$

has a solution if and only if $\vec{d}$ is in the span of the set $\{\vec{a}_1, \ldots, \vec{a}_n\}$. That's true if and only if the column rank of the augmented matrix equals the column rank of the matrix of coefficients. Since rank equals the column rank, the system has a solution if and only if the rank of its augmented matrix equals the rank of its matrix of coefficients.

Two.III.3.39 **(a)** Row rank equals column rank so each is at most the minimum of the number of rows and columns. Hence both can be full only if the number of rows equals the number of columns. (Of course, the converse does not hold: a square matrix need not have full row rank or full column rank.)

(b) If A has full row rank then, no matter what the right-hand side, Gauss's Method on the augmented matrix ends with a leading one in each row and none of those leading ones in the furthest right column (the "augmenting" column). Back substitution then gives a solution.

On the other hand, if the linear system lacks a solution for some right-hand side it can only be because Gauss's Method leaves some row so that it is all zeroes to the left of the "augmenting" bar and has a nonzero entry on the right. Thus, if A does not have a solution for some right-hand sides, then A does not have full row rank because some of its rows have been eliminated.

(c) The matrix A has full column rank if and only if its columns form a linearly independent set. That's equivalent to the existence of only the trivial linear relationship among the columns, so the only solution of the system is where each variable is 0.

(d) The matrix A has full column rank if and only if the set of its columns is linearly independent, and so forms a basis for its span. That's equivalent to the existence of a unique linear representation of all vectors in that span. That proves it, since any linear representation of a vector in the span is a solution of the linear system.

Two.III.3.40 Instead of the row spaces being the same, the row space of B would be a subspace (possibly equal to) the row space of A.

Two.III.3.41 Clearly $\text{rank}(A) = \text{rank}(-A)$ as Gauss's Method allows us to multiply all rows of a matrix by -1. In the same way, when $k \neq 0$ we have $\text{rank}(A) = \text{rank}(kA)$.

Addition is more interesting. The rank of a sum can be smaller than the rank of the summands.

$$\begin{pmatrix} 1 & 2 \\ 3 & 4 \end{pmatrix} + \begin{pmatrix} -1 & -2 \\ -3 & -4 \end{pmatrix} = \begin{pmatrix} 0 & 0 \\ 0 & 0 \end{pmatrix}$$

The rank of a sum can be bigger than the rank of the summands.

$$\begin{pmatrix} 1 & 2 \\ 0 & 0 \end{pmatrix} + \begin{pmatrix} 0 & 0 \\ 3 & 4 \end{pmatrix} = \begin{pmatrix} 1 & 2 \\ 3 & 4 \end{pmatrix}$$

But there is an upper bound (other than the size of the matrices). In general, $\text{rank}(A+B) \leqslant \text{rank}(A) + \text{rank}(B)$.

To prove this, note that we can perform Gaussian elimination on $A+B$ in either of two ways: we can first add A to B and then apply the appropriate sequence of reduction steps

$$(A+B) \xrightarrow{\text{step}_1} \cdots \xrightarrow{\text{step}_k} \text{echelon form}$$

or we can get the same results by performing step_1 through step_k separately on A and B, and then adding. The largest rank that we can end with in the second case is clearly the sum of the ranks. (The matrices above give examples of both possibilities, $\text{rank}(A+B) < \text{rank}(A)+\text{rank}(B)$ and $\text{rank}(A+B) = \text{rank}(A)+\text{rank}(B)$, happening.)

Two.III.4: Combining Subspaces

Two.III.4.20 With each of these we can apply Lemma 4.15.

(a) Yes. The plane is the sum of this W_1 and W_2 because for any scalars a and b

$$\begin{pmatrix} a \\ b \end{pmatrix} = \begin{pmatrix} a-b \\ 0 \end{pmatrix} + \begin{pmatrix} b \\ b \end{pmatrix}$$

shows that the general vector is a sum of vectors from the two parts. And, these two subspaces are (different) lines through the origin, and so have a trivial intersection.

(b) Yes. To see that any vector in the plane is a combination of vectors from these parts, consider this relationship.

$$\begin{pmatrix} a \\ b \end{pmatrix} = c_1 \begin{pmatrix} 1 \\ 1 \end{pmatrix} + c_2 \begin{pmatrix} 1 \\ 1.1 \end{pmatrix}$$

We could now simply note that the set

$$\{ \begin{pmatrix} 1 \\ 1 \end{pmatrix}, \begin{pmatrix} 1 \\ 1.1 \end{pmatrix} \}$$

is a basis for the space (because it is clearly linearly independent, and has size two in $\mathbb{R}^2$), and thus there is one and only one solution to the above equation, implying that all decompositions are unique. Alternatively, we can solve

$$\begin{array}{rl} c_1 + c_2 &= a \\ c_1 + 1.1c_2 &= b \end{array} \xrightarrow{-\rho_1+\rho_2} \begin{array}{rl} c_1 + c_2 &= a \\ 0.1c_2 &= -a+b \end{array}$$

to get that $c_2 = 10(-a+b)$ and $c_1 = 11a - 10b$, and so we have

$$\begin{pmatrix} a \\ b \end{pmatrix} = \begin{pmatrix} 11a - 10b \\ 11a - 10b \end{pmatrix} + \begin{pmatrix} -10a + 10b \\ 1.1 \cdot (-10a + 10b) \end{pmatrix}$$

as required. As with the prior answer, each of the two subspaces is a line through the origin, and their intersection is trivial.

(c) Yes. Each vector in the plane is a sum in this way

$$\begin{pmatrix} x \\ y \end{pmatrix} = \begin{pmatrix} x \\ y \end{pmatrix} + \begin{pmatrix} 0 \\ 0 \end{pmatrix}$$

and the intersection of the two subspaces is trivial.

(d) No. The intersection is not trivial.

(e) No. These are not subspaces.

Two.III.4.21 With each of these we can use Lemma 4.15.

(a) Any vector in $\mathbb{R}^3$ can be decomposed as this sum.

$$\begin{pmatrix} x \\ y \\ z \end{pmatrix} = \begin{pmatrix} x \\ y \\ 0 \end{pmatrix} + \begin{pmatrix} 0 \\ 0 \\ z \end{pmatrix}$$

And, the intersection of the xy-plane and the z-axis is the trivial subspace.

(b) Any vector in $\mathbb{R}^3$ can be decomposed as

$$\begin{pmatrix} x \\ y \\ z \end{pmatrix} = \begin{pmatrix} x - z \\ y - z \\ 0 \end{pmatrix} + \begin{pmatrix} z \\ z \\ z \end{pmatrix}$$

and the intersection of the two spaces is trivial.

Two.III.4.22 It is. Showing that these two are subspaces is routine. To see that the space is the direct sum of these two, just note that each member of $\mathcal{P}_2$ has the unique decomposition $m+nx+px^2 = (m+px^2)+(nx)$.

Two.III.4.23 To show that they are subspaces is routine. We will argue they are complements with Lemma 4.15. The intersection $\mathcal{E} \cap \mathcal{O}$ is trivial because the only polynomial satisfying both conditions $p(-x) = p(x)$ and $p(-x) = -p(x)$ is the zero polynomial. To see that the entire space is the sum of the subspaces $\mathcal{E} + \mathcal{O} = \mathcal{P}_n$, note that the polynomials $p_0(x) = 1$, $p_2(x) = x^2$, $p_4(x) = x^4$, etc., are in $\mathcal{E}$ and also note that the polynomials $p_1(x) = x$, $p_3(x) = x^3$, etc., are in $\mathcal{O}$. Hence any member of $\mathcal{P}_n$ is a combination of members of $\mathcal{E}$ and $\mathcal{O}$.

Two.III.4.24 Each of these is $\mathbb{R}^3$.

(a) These are broken into lines for legibility.

$W_1 + W_2 + W_3$, $W_1 + W_2 + W_3 + W_4$, $W_1 + W_2 + W_3 + W_5$, $W_1 + W_2 + W_3 + W_4 + W_5$,
$W_1 + W_2 + W_4$, $W_1 + W_2 + W_4 + W_5$, $W_1 + W_2 + W_5$,
$W_1 + W_3 + W_4$, $W_1 + W_3 + W_5$, $W_1 + W_3 + W_4 + W_5$,
$W_1 + W_4$, $W_1 + W_4 + W_5$,
$W_1 + W_5$,
$W_2 + W_3 + W_4$, $W_2 + W_3 + W_4 + W_5$,
$W_2 + W_4$, $W_2 + W_4 + W_5$,
$W_3 + W_4$, $W_3 + W_4 + W_5$,
$W_4 + W_5$

(b) $W_1 \oplus W_2 \oplus W_3$, $W_1 \oplus W_4$, $W_1 \oplus W_5$, $W_2 \oplus W_4$, $W_3 \oplus W_4$

Two.III.4.25 Clearly each is a subspace. The bases $B_i = \langle x^i \rangle$ for the subspaces, when concatenated, form a basis for the whole space.

Two.III.4.26 It is W_2.

Two.III.4.27 True by Lemma 4.8.

Two.III.4.28 Two distinct direct sum decompositions of $\mathbb{R}^4$ are easy to find. Two such are $W_1 = [\{\vec{e}_1, \vec{e}_2\}]$ and $W_2 = [\{\vec{e}_3, \vec{e}_4\}]$, and also $U_1 = [\{\vec{e}_1\}]$ and $U_2 = [\{\vec{e}_2, \vec{e}_3, \vec{e}_4\}]$. (Many more are possible, for example $\mathbb{R}^4$ and its trivial subspace.)

In contrast, any partition of $\mathbb{R}^1$'s single-vector basis will give one basis with no elements and another with a single element. Thus any decomposition involves $\mathbb{R}^1$ and its trivial subspace.

Two.III.4.29 Set inclusion one way is easy: $\{\vec{w}_1 + \cdots + \vec{w}_k \mid \vec{w}_i \in W_i\}$ is a subset of $[W_1 \cup \ldots \cup W_k]$ because each $\vec{w}_1 + \cdots + \vec{w}_k$ is a sum of vectors from the union.

For the other inclusion, to any linear combination of vectors from the union apply commutativity of vector addition to put vectors from W_1 first, followed by vectors from W_2, etc. Add the vectors from W_1 to get a $\vec{w}_1 \in W_1$, add the vectors from W_2 to get a $\vec{w}_2 \in W_2$, etc. The result has the desired form.

Two.III.4.30 One example is to take the space to be $\mathbb{R}^3$, and to take the subspaces to be the xy-plane, the xz-plane, and the yz-plane.

Two.III.4.31 Of course, the zero vector is in all of the subspaces, so the intersection contains at least that one vector.. By the definition of direct sum the set $\{W_1, \ldots, W_k\}$ is independent and so no nonzero vector of W_i is a multiple of a member of W_j, when $i \neq j$. In particular, no nonzero vector from W_i equals a member of W_j.

Two.III.4.32 It can contain a trivial subspace; this set of subspaces of $\mathbb{R}^3$ is independent: $\{\{\vec{0}\}, x\text{-axis}\}$. No nonzero vector from the trivial space $\{\vec{0}\}$ is a multiple of a vector from the x-axis, simply because the trivial space has no nonzero vectors to be candidates for such a multiple (and also no nonzero vector from the x-axis is a multiple of the zero vector from the trivial subspace).

Two.III.4.33 Yes. For any subspace of a vector space we can take any basis $\langle \vec{\omega}_1, \ldots, \vec{\omega}_k \rangle$ for that subspace and extend it to a basis $\langle \vec{\omega}_1, \ldots, \vec{\omega}_k, \vec{\beta}_{k+1}, \ldots, \vec{\beta}_n \rangle$ for the whole space. Then the complement of the original subspace has this for a basis: $\langle \vec{\beta}_{k+1}, \ldots, \vec{\beta}_n \rangle$.

Two.III.4.34 **(a)** It must. We can write any member of $W_1 + W_2$ as $\vec{w}_1 + \vec{w}_2$ where $\vec{w}_1 \in W_1$ and $\vec{w}_2 \in W_2$. As S_1 spans W_1, the vector $\vec{w}_1$ is a combination of members of S_1. Similarly $\vec{w}_2$ is a combination of members of S_2.

(b) An easy way to see that it can be linearly independent is to take each to be the empty set. On the other hand, in the space $\mathbb{R}^1$, if $W_1 = \mathbb{R}^1$ and $W_2 = \mathbb{R}^1$ and $S_1 = \{1\}$ and $S_2 = \{2\}$, then their union $S_1 \cup S_2$ is not independent.

Two.III.4.35 **(a)** The intersection and sum are

$$\{\begin{pmatrix} 0 & 0 \\ c & 0 \end{pmatrix} \mid c \in \mathbb{R}\} \qquad \{\begin{pmatrix} 0 & b \\ c & d \end{pmatrix} \mid b, c, d \in \mathbb{R}\}$$

which have dimensions one and three.

(b) We write $B_{U \cap W}$ for the basis for $U \cap W$, we write B_U for the basis for U, we write B_W for the basis for W, and we write B_{U+W} for the basis under consideration.

To see that B_{U+W} spans $U + W$, observe that we can write any vector $c\vec{u} + d\vec{w}$ from $U + W$ as a linear combination of the vectors in B_{U+W}, simply by expressing $\vec{u}$ in terms of B_U and expressing $\vec{w}$ in terms of B_W.

We finish by showing that B_{U+W} is linearly independent. Consider

$$c_1\vec{\mu}_1 + \cdots + c_{j+1}\vec{\beta}_1 + \cdots + c_{j+k+p}\vec{\omega}_p = \vec{0}$$

which can be rewritten in this way.

$$c_1\vec{\mu}_1 + \cdots + c_j\vec{\mu}_j = -c_{j+1}\vec{\beta}_1 - \cdots - c_{j+k+p}\vec{\omega}_p$$

Note that the left side sums to a vector in U while right side sums to a vector in W, and thus both sides sum to a member of $U \cap W$. Since the left side is a member of $U \cap W$, it is expressible in terms of the members of $B_{U \cap W}$, which gives the combination of $\vec{\mu}$'s from the left side above as equal to a combination of $\vec{\beta}$'s. But, the fact that the basis B_U is linearly independent shows that any such combination is trivial, and in particular, the coefficients $c_1, \ldots, c_j$ from the left side above are all zero. Similarly, the coefficients of the $\vec{\omega}$'s are all zero. This leaves the above equation as a linear relationship among the $\vec{\beta}$'s, but $B_{U\cap W}$ is linearly independent, and therefore all of the coefficients of the $\vec{\beta}$'s are also zero.

(c) Just count the basis vectors in the prior item: $\dim(U+W) = j+k+p$, and $\dim(U) = j+k$, and $\dim(W) = k+p$, and $\dim(U\cap W) = k$.

(d) We know that $\dim(W_1+W_2) = \dim(W_1)+\dim(W_2)-\dim(W_1\cap W_2)$. Because $W_1 \subseteq W_1+W_2$, we know that W_1+W_2 must have dimension greater than that of W_1, that is, must have dimension eight, nine, or ten. Substituting gives us three possibilities $8 = 8+8-\dim(W_1\cap W_2)$ or $9 = 8+8-\dim(W_1\cap W_2)$ or $10 = 8+8-\dim(W_1\cap W_2)$. Thus $\dim(W_1\cap W_2)$ must be either eight, seven, or six. (Giving examples to show that each of these three cases is possible is easy, for instance in $\mathbb{R}^{10}$.)

Two.III.4.36 Expand each S_i to a basis B_i for W_i. The concatenation of those bases $B_1{}^\frown\cdots{}^\frown B_k$ is a basis for V and thus its members form a linearly independent set. But the union $S_1\cup\cdots\cup S_k$ is a subset of that linearly independent set, and thus is itself linearly independent.

Two.III.4.37 **(a)** Two such are these.

$$\begin{pmatrix} 1 & 2 \\ 2 & 3 \end{pmatrix} \qquad \begin{pmatrix} 0 & 1 \\ -1 & 0 \end{pmatrix}$$

For the antisymmetric one, entries on the diagonal must be zero.

(b) A square symmetric matrix equals its transpose. A square antisymmetric matrix equals the negative of its transpose.

(c) Showing that the two sets are subspaces is easy. Suppose that $A \in \mathcal{M}_{n\times n}$. To express A as a sum of a symmetric and an antisymmetric matrix, we observe that

$$A = (1/2)(A+A^T) + (1/2)(A-A^T)$$

and note the first summand is symmetric while the second is antisymmetric. Thus $\mathcal{M}_{n\times n}$ is the sum of the two subspaces. To show that the sum is direct, assume a matrix A is both symmetric $A = A^T$ and antisymmetric $A = -A^T$. Then $A = -A$ and so all of A's entries are zeroes.

Two.III.4.38 Assume that $\vec{v} \in (W_1\cap W_2)+(W_1\cap W_3)$. Then $\vec{v} = \vec{w}_2+\vec{w}_3$ where $\vec{w}_2 \in W_1\cap W_2$ and $\vec{w}_3 \in W_1\cap W_3$. Note that $\vec{w}_2, \vec{w}_3 \in W_1$ and, as a subspace is closed under addition, $\vec{w}_2+\vec{w}_3 \in W_1$. Thus $\vec{v} = \vec{w}_2+\vec{w}_3 \in W_1\cap(W_2+W_3)$.

This example proves that the inclusion may be strict: in $\mathbb{R}^2$ take W_1 to be the x-axis, take W_2 to be the y-axis, and take W_3 to be the line $y = x$. Then $W_1\cap W_2$ and $W_1\cap W_3$ are trivial and so their sum is trivial. But W_2+W_3 is all of $\mathbb{R}^2$ so $W_1\cap(W_2+W_3)$ is the x-axis.

Two.III.4.39 It happens when at least one of W_1, W_2 is trivial. But that is the only way it can happen.

To prove this, assume that both are non-trivial, select nonzero vectors $\vec{w}_1, \vec{w}_2$ from each, and consider $\vec{w}_1+\vec{w}_2$. This sum is not in W_1 because $\vec{w}_1+\vec{w}_2 = \vec{v} \in W_1$ would imply that $\vec{w}_2 = \vec{v}-\vec{w}_1$ is in W_1, which violates the assumption of the independence of the subspaces. Similarly, $\vec{w}_1+\vec{w}_2$ is not in W_2. Thus there is an element of V that is not in $W_1\cup W_2$.

Two.III.4.40 Yes. The left-to-right implication is Corollary 4.13. For the other direction, assume that $\dim(V) = \dim(W_1)+\cdots+\dim(W_k)$. Let $B_1, \ldots, B_k$ be bases for $W_1, \ldots, W_k$. As V is the sum of the subspaces, we can write any $\vec{v} \in V$ as $\vec{v} = \vec{w}_1+\cdots+\vec{w}_k$ and expressing each $\vec{w}_i$ as a combination of vectors from the associated basis B_i shows that the concatenation $B_1{}^\frown\cdots{}^\frown B_k$ spans V. Now, that concatenation has $\dim(W_1)+\cdots+\dim(W_k)$ members, and so it is a spanning set of size $\dim(V)$. The concatenation is therefore a basis for V. Thus V is the direct sum.

Two.III.4.41 No. The standard basis for $\mathbb{R}^2$ does not split into bases for the complementary subspaces the line $x = y$ and the line $x = -y$.

Two.III.4.42 **(a)** Yes, $W_1+W_2 = W_2+W_1$ for all subspaces W_1, W_2 because each side is the span of $W_1\cup W_2 = W_2\cup W_1$.

(b) This one is similar to the prior one—each side of that equation is the span of $(W_1 \cup W_2) \cup W_3 = W_1 \cup (W_2 \cup W_3)$.

(c) Because this is an equality between sets, we can show that it holds by mutual inclusion. Clearly $W \subseteq W + W$. For $W + W \subseteq W$ just recall that every subset is closed under addition so any sum of the form $\vec{w}_1 + \vec{w}_2$ is in W.

(d) In each vector space, the identity element with respect to subspace addition is the trivial subspace.

(e) Neither of left or right cancellation needs to hold. For an example, in $\mathbb{R}^3$ take W_1 to be the xy-plane, take W_2 to be the x-axis, and take W_3 to be the y-axis.

Two.III.4.43 **(a)** They are equal because for each, V is the direct sum if and only if we can write each $\vec{v} \in V$ in a unique way as a sum $\vec{v} = \vec{w}_1 + \vec{w}_2$ and $\vec{v} = \vec{w}_2 + \vec{w}_1$.

(b) They are equal because for each, V is the direct sum if and only if we can write each $\vec{v} \in V$ in a unique way as a sum of a vector from each $\vec{v} = (\vec{w}_1 + \vec{w}_2) + \vec{w}_3$ and $\vec{v} = \vec{w}_1 + (\vec{w}_2 + \vec{w}_3)$.

(c) We can decompose any vector in $\mathbb{R}^3$ uniquely into the sum of a vector from each axis.

(d) No. For an example, in $\mathbb{R}^2$ take W_1 to be the x-axis, take W_2 to be the y-axis, and take W_3 to be the line $y = x$.

(e) In any vector space the trivial subspace acts as the identity element with respect to direct sum.

(f) In any vector space, only the trivial subspace has a direct-sum inverse (namely, itself). One way to see this is that dimensions add, and so increase.

Topic: Fields

2 These checks are all routine; most consist only of remarking that property is so familiar that it does not need to be proved.

3 For both of these structures, these checks are all routine. As with the prior question, most of the checks consist only of remarking that property is so familiar that it does not need to be proved.

4 There is no multiplicative inverse for 2 so the integers do not satisfy condition (5).

5 We can do these checks by listing all of the possibilities. For instance, to verify the commutativity of addition, that $a + b = b + a$, we can easily check it for all possible pairs a, b, because there are only four such pairs. Similarly, for associativity, there are only eight triples a, b, c, and so the check is not too long. (There are other ways to do the checks, in particular, a reader may recognize these operations as arithmetic 'mod 2'.)

6 These will do.

+	0	1	2
0	0	1	2
1	1	2	0
2	2	0	1

·	0	1	2
0	0	0	0
1	0	1	2
2	0	2	1

As in the prior item, we can do the check that they satisfy the conditions by listing all of the cases, although this way of checking is long (making use of commutativity is helpful in shortening the work).

Topic: Crystals

1 Each fundamental unit is 3.34×10^{-10} cm, so there are about $0.1/(3.34 \times 10^{-10})$ such units. That gives 2.99×10^{8}, so there are something like $300,000,000$ (three hundred million) regions.

2 **(a)** We solve

$$c_1 \begin{pmatrix} 1.42 \\ 0 \end{pmatrix} + c_2 \begin{pmatrix} 1.23 \\ 0.71 \end{pmatrix} = \begin{pmatrix} 5.67 \\ 3.14 \end{pmatrix} \quad \Longrightarrow \quad \begin{aligned} 1.42c_1 + 1.23c_2 &= 5.67 \\ 0.71c_2 &= 3.14 \end{aligned}$$

to get $c_2 \approx 4.42$ and $c_1 \approx 0.16$.

(b) Here is the point located in the lattice. In the picture on the left, superimposed on the unit cell are the two basis vectors $\vec{\beta}_1$ and $\vec{\beta}_2$, and a box showing the offset of $0.16\vec{\beta}_1 + 4.42\vec{\beta}_2$. The picture on the right shows where that appears inside of the crystal lattice, taking as the origin the lower left corner of the hexagon in the lower left.

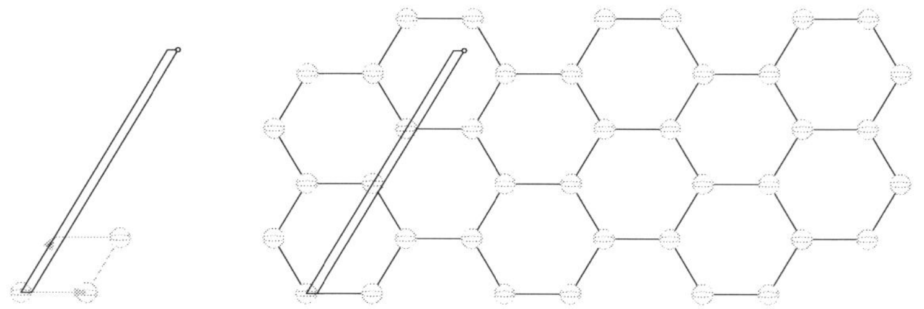

So this point is in the next column of hexagons over, and either one hexagon up or two hexagons up, depending on how you count them.

(c) This second basis

$$\langle \begin{pmatrix} 1.42 \\ 0 \end{pmatrix}, \begin{pmatrix} 0 \\ 1.42 \end{pmatrix} \rangle$$

makes the computation easier

$$c_1 \begin{pmatrix} 1.42 \\ 0 \end{pmatrix} + c_2 \begin{pmatrix} 0 \\ 1.42 \end{pmatrix} = \begin{pmatrix} 5.67 \\ 3.14 \end{pmatrix} \quad \Longrightarrow \quad \begin{aligned} 1.42c_1 \qquad &= 5.67 \\ 1.42c_2 &= 3.14 \end{aligned}$$

(we get $c_2 \approx 2.21$ and $c_1 \approx 3.99$), but it doesn't seem to have to do much with the physical structure that we are studying.

3 In terms of the basis the locations of the corner atoms are $(0,0,0)$, $(1,0,0)$, $\ldots$, $(1,1,1)$. The locations of the face atoms are $(0.5,0.5,1)$, $(1,0.5,0.5)$, $(0.5,1,0.5)$, $(0,0.5,0.5)$, $(0.5,0,0.5)$, and $(0.5,0.5,0)$. The locations of the atoms a quarter of the way down from the top are $(0.75,0.75,0.75)$ and $(0.25,0.25,0.25)$. The atoms a quarter of the way up from the bottom are at $(0.75,0.25,0.25)$ and $(0.25,0.75,0.25)$. Converting to Ångstroms is easy.

4 **(a)** $195.08/6.02 \times 10^{23} = 3.239 \times 10^{-22}$

(b) Each platinum atom in the middle of each face is split between two cubes, so that is $6/2 = 3$ atoms so far. Each atom at a corner is split among eight cubes, so that makes an additional $8/8 = 1$ atom, so the total is 4.

(c) $4 \cdot 3.239 \times 10^{-22} = 1.296 \times 10^{-21}$

(d) $1.296 \times 10^{-21}/21.45 = 6.042 \times 10^{-23}$ cubic centimeters

(e) 3.924×10^{-8} centimeters.

(f) $\langle \begin{pmatrix} 3.924 \times 10^{-8} \\ 0 \\ 0 \end{pmatrix}, \begin{pmatrix} 0 \\ 3.924 \times 10^{-8} \\ 0 \end{pmatrix}, \begin{pmatrix} 0 \\ 0 \\ 3.924 \times 10^{-8} \end{pmatrix} \rangle$

Topic: Voting Paradoxes

1 This is one example that yields a non-rational preference order for a single voter.

	character	*experience*	*policies*
Democrat	most	middle	least
Republican	middle	least	most
Third	least	most	middle

The Democrat is preferred to the Republican for character and experience. The Republican is preferred to the Third for character and policies. And, the Third is preferred to the Democrat for experience and policies.

2 First, compare the $D > R > T$ decomposition that was covered in the Topic with the decomposition of the opposite $T > R > D$ voter.

$$\begin{pmatrix}-1\\1\\1\end{pmatrix} = \frac{1}{3}\cdot\begin{pmatrix}1\\1\\1\end{pmatrix} + \frac{2}{3}\cdot\begin{pmatrix}-1\\1\\0\end{pmatrix} + \frac{2}{3}\cdot\begin{pmatrix}-1\\0\\1\end{pmatrix} \quad\text{and}\quad \begin{pmatrix}1\\-1\\-1\end{pmatrix} = d_1\cdot\begin{pmatrix}1\\1\\1\end{pmatrix} + d_2\cdot\begin{pmatrix}-1\\1\\0\end{pmatrix} + d_3\cdot\begin{pmatrix}-1\\0\\1\end{pmatrix}$$

Obviously, the second is the negative of the first, and so $d_1 = -1/3$, $d_2 = -2/3$, and $d_3 = -2/3$. This principle holds for any pair of opposite voters, and so we need only do the computation for a voter from the second row, and a voter from the third row. For a positive spin voter in the second row,

$$\begin{array}{rcl} c_1 - c_2 - c_3 &=& 1 \\ c_1 + c_2 &=& 1 \\ c_1 + c_3 &=& -1 \end{array} \xrightarrow[-\rho_1+\rho_3]{-\rho_1+\rho_2} \xrightarrow{(-1/2)\rho_2+\rho_3} \begin{array}{rcl} c_1 - c_2 - c_3 &=& 1 \\ 2c_2 + c_3 &=& 0 \\ (3/2)c_3 &=& -2 \end{array}$$

gives $c_3 = -4/3$, $c_2 = 2/3$, and $c_1 = 1/3$. For a positive spin voter in the third row,

$$\begin{array}{rcl} c_1 - c_2 - c_3 &=& 1 \\ c_1 + c_2 &=& -1 \\ c_1 + c_3 &=& 1 \end{array} \xrightarrow[-\rho_1+\rho_3]{-\rho_1+\rho_2} \xrightarrow{(-1/2)\rho_2+\rho_3} \begin{array}{rcl} c_1 - c_2 - c_3 &=& 1 \\ 2c_2 + c_3 &=& -2 \\ (3/2)c_3 &=& 1 \end{array}$$

gives $c_3 = 2/3$, $c_2 = -4/3$, and $c_1 = 1/3$.

3 The mock election corresponds to the table on page 146 in the way shown in the first table, and after cancellation the result is the second table.

positive spin	*negative spin*
$D > R > T$ 5 voters	$T > R > D$ 2 voters
$R > T > D$ 8 voters	$D > T > R$ 4 voters
$T > D > R$ 8 voters	$R > D > T$ 2 voters

positive spin	*negative spin*
$D > R > T$ 3 voters	$T > R > D$ –
$R > T > D$ 4 voters	$D > T > R$ –
$T > D > R$ 6 voters	$R > D > T$ –

All three come from the same side of the table (the left), as the result from this Topic says must happen. Tallying the election can now proceed, using the canceled numbers

3 · (D→T: −1, R→D: 1, T→R: 1) + 4 · (D→T: 1, R→D: −1, T→R: 1) + 6 · (D→T: 1, R→D: 1, T→R: −1) = (D→T: 7, R→D: 5, T→R: 1)

to get the same outcome.

4 **(a)** The two can be rewritten as $-c \leqslant a - b$ and $-c \leqslant b - a$. Either $a - b$ or $b - a$ is nonpositive and so $-c \leqslant -|a - b|$, as required.

(b) This is immediate from the supposition that $0 \leqslant a + b - c$.

(c) A trivial example starts with the zero-voter election and adds any one voter. A more interesting example is to take the Political Science mock election and add two $T > D > R$ voters (they can be added one at a time, to satisfy the "addition of one more voter" criteria in the question). Observe that the additional voters have positive spin, which is the spin of the votes remaining after cancellation in the original mock election. This is the resulting table of voters, and next to it is the result of cancellation.

positive spin	*negative spin*
$D > R > T$ 5 voters	$T > R > D$ 2 voters
$R > T > D$ 8 voters	$D > T > R$ 4 voters
$T > D > R$ 10 voters	$R > D > T$ 2 voters

positive spin	*negative spin*
$D > R > T$ 3 voters	$T > R > D$ –
$R > T > D$ 4 voters	$D > T > R$ –
$T > D > R$ 8 voters	$R > D > T$ –

The election, using the canceled numbers, is this.

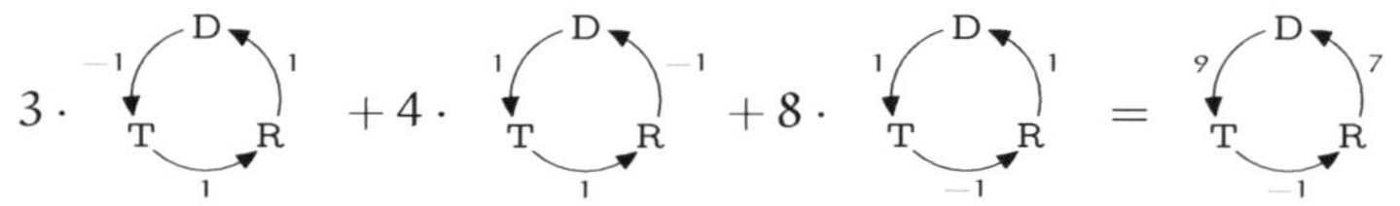

The majority cycle has indeed disappeared.

(d) One such condition is that, after cancellation, all three be nonnegative or all three be nonpositive, and: $|c| < |a + b|$ and $|b| < |a + c|$ and $|a| < |b + c|$. This follows from this diagram.

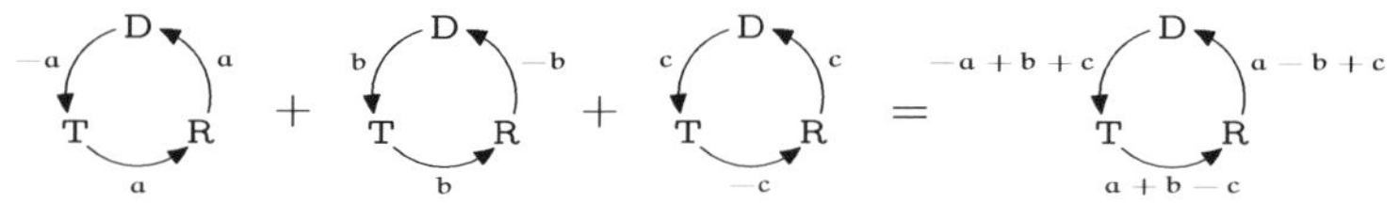

5 **(a)** A two-voter election can have a majority cycle in two ways. First, the two voters could be opposites, resulting after cancellation in the trivial election (with the majority cycle of all zeroes). Second, the two voters could have the same spin but come from different rows, as here.

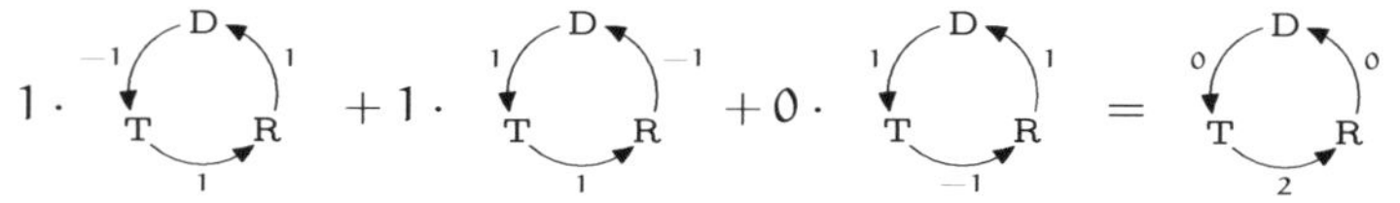

(b) There are two cases. An even number of voters can split half and half into opposites, e.g., half the voters are $D > R > T$ and half are $T > R > D$. Then cancellation gives the trivial election. If the number of voters is greater than one and odd (of the form $2k + 1$ with $k > 0$) then using the cycle diagram from the proof,

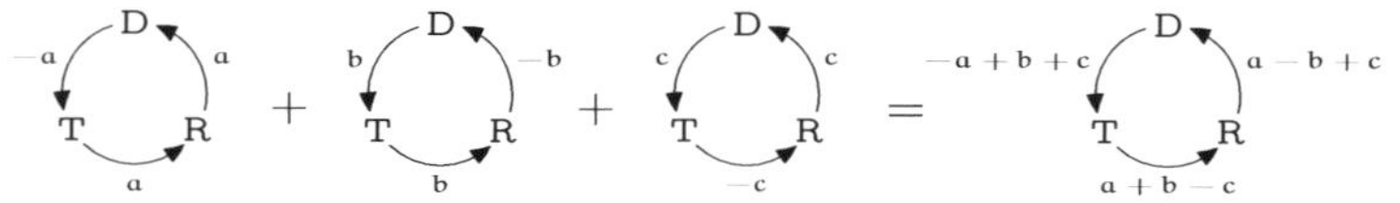

we can take $a = k$ and $b = k$ and $c = 1$. Because $k > 0$, this is a majority cycle.

6 It is nonempty because it contains the zero vector. To see that it is closed under linear combinations of two of its members, suppose that $\vec{v}_1$ and $\vec{v}_2$ are in $U^{\perp}$ and consider $c_1\vec{v}_1 + c_2\vec{v}_2$. For any $\vec{u} \in U$,

$$(c_1\vec{v}_1 + c_2\vec{v}_2) \cdot \vec{u} = c_1(\vec{v}_1 \cdot \vec{u}) + c_2(\vec{v}_2 \cdot \vec{u}) = c_1 \cdot 0 + c_2 \cdot 0 = 0$$

and so $c_1\vec{v}_1 + c_2\vec{v}_2 \in U^{\perp}$.

This holds if U is any subset, subspace or not.

Topic: Dimensional Analysis

1 **(a)** This relationship

$$(L^1M^0T^0)^{p_1}(L^1M^0T^0)^{p_2}(L^1M^0T^{-1})^{p_3}(L^0M^0T^0)^{p_4}(L^1M^0T^{-2})^{p_5}(L^0M^0T^1)^{p_6} = L^0M^0T^0$$

gives rise to this linear system

$$\begin{array}{rl} p_1 + p_2 + p_3 + p_5 &= 0 \\ 0 &= 0 \\ -p_3 - 2p_5 + p_6 &= 0 \end{array}$$

(note that there is no restriction on p_4). The natural parametrization uses the free variables to give $p_3 = -2p_5 + p_6$ and $p_1 = -p_2 + p_5 - p_6$. The resulting description of the solution set

$$\{\begin{pmatrix} p_1 \\ p_2 \\ p_3 \\ p_4 \\ p_5 \\ p_6 \end{pmatrix} = p_2\begin{pmatrix} -1 \\ 1 \\ 0 \\ 0 \\ 0 \\ 0 \end{pmatrix} + p_4\begin{pmatrix} 0 \\ 0 \\ 0 \\ 1 \\ 0 \\ 0 \end{pmatrix} + p_5\begin{pmatrix} 1 \\ 0 \\ -2 \\ 0 \\ 1 \\ 0 \end{pmatrix} + p_6\begin{pmatrix} -1 \\ 0 \\ 1 \\ 0 \\ 0 \\ 1 \end{pmatrix} \mid p_2, p_4, p_5, p_6 \in \mathbb{R}\}$$

gives $\{y/x, \theta, xt/{v_0}^2, v_0t/x\}$ as a complete set of dimensionless products (recall that "complete" in this context does not mean that there are no other dimensionless products; it simply means that the set is a basis). This is, however, not the set of dimensionless products that the question asks for.

There are two ways to proceed. The first is to fiddle with the choice of parameters, hoping to hit on the right set. For that, we can do the prior paragraph in reverse. Converting the given dimensionless products gt/v_0, gx/v_0^2, gy/v_0^2, and θ into vectors gives this description (note the *?*'s where the parameters will go).

$$\{\begin{pmatrix} p_1 \\ p_2 \\ p_3 \\ p_4 \\ p_5 \\ p_6 \end{pmatrix} = \underline{?}\begin{pmatrix} 0 \\ 0 \\ -1 \\ 0 \\ 1 \\ 1 \end{pmatrix} + \underline{?}\begin{pmatrix} 1 \\ 0 \\ -2 \\ 0 \\ 1 \\ 0 \end{pmatrix} + \underline{?}\begin{pmatrix} 0 \\ 1 \\ -2 \\ 0 \\ 1 \\ 0 \end{pmatrix} + p_4\begin{pmatrix} 0 \\ 0 \\ 0 \\ 1 \\ 0 \\ 0 \end{pmatrix} \mid p_2, p_4, p_5, p_6 \in \mathbb{R}\}$$

The p_4 is already in place. Examining the rows shows that we can also put in place p_6, p_1, and p_2.

The second way to proceed, following the hint, is to note that the given set is of size four in a four-dimensional vector space and so we need only show that it is linearly independent. That is easily done by inspection, by considering the sixth, first, second, and fourth components of the vectors.

(b) The first equation can be rewritten

$$\frac{gx}{{v_0}^2} = \frac{gt}{v_0}\cos\theta$$

so that Buckingham's function is $f_1(\Pi_1, \Pi_2, \Pi_3, \Pi_4) = \Pi_2 - \Pi_1\cos(\Pi_4)$. The second equation can be rewritten

$$\frac{gy}{{v_0}^2} = \frac{gt}{v_0}\sin\theta - \frac{1}{2}\left(\frac{gt}{v_0}\right)^2$$

and Buckingham's function here is $f_2(\Pi_1, \Pi_2, \Pi_3, \Pi_4) = \Pi_3 - \Pi_1\sin(\Pi_4) + (1/2){\Pi_1}^2$.

2 We consider

$$(L^0M^0T^{-1})^{p_1}(L^1M^{-1}T^2)^{p_2}(L^{-3}M^0T^0)^{p_3}(L^0M^1T^0)^{p_4} = (L^0M^0T^0)$$

which gives these relations among the powers.

$$\begin{array}{rl} p_2 - 3p_3 &= 0 \\ -p_2 + p_4 &= 0 \\ -p_1 + 2p_2 &= 0 \end{array} \xrightarrow{\rho_1\leftrightarrow\rho_3} \xrightarrow{\rho_2+\rho_3} \begin{array}{rl} -p_1 + 2p_2 &= 0 \\ -p_2 + p_4 &= 0 \\ -3p_3 + p_4 &= 0 \end{array}$$

This is the solution space (because we wish to express k as a function of the other quantities, we take p_2 as the parameter).

$$\{\begin{pmatrix}2\\1\\1/3\\1\end{pmatrix}p_2 \mid p_2\in\mathbb{R}\}$$

Thus, $\Pi_1 = v^2kN^{1/3}m$ is the dimensionless combination, and we have that k equals $v^{-2}N^{-1/3}m^{-1}$ times a constant (the function $\hat{f}$ is constant since it has no arguments).

3 **(a)** Setting

$$(L^2M^1T^{-2})^{p_1}(L^0M^0T^{-1})^{p_2}(L^3M^0T^0)^{p_3} = (L^0M^0T^0)$$

gives this

$$\begin{aligned}2p_1 \quad\quad + 3p_3 &= 0\\ p_1 \quad\quad\quad\quad &= 0\\ -2p_1 - p_2 \quad\quad &= 0\end{aligned}$$

which implies that $p_1 = p_2 = p_3 = 0$. That is, among quantities with these dimensional formulas, the only dimensionless product is the trivial one.

(b) Setting

$$(L^2M^1T^{-2})^{p_1}(L^0M^0T^{-1})^{p_2}(L^3M^0T^0)^{p_3}(L^{-3}M^1T^0)^{p_4} = (L^0M^0T^0)$$

gives this.

$$\begin{array}{rcl}2p_1 \quad + 3p_3 - 3p_4 &=& 0\\ p_1 \quad\quad\quad + \ p_4 &=& 0\\ -2p_1 - p_2 \quad\quad\quad &=& 0\end{array} \xrightarrow[\rho_1+\rho_3]{(-1/2)\rho_1+\rho_2} \xrightarrow{\rho_2\leftrightarrow\rho_3} \begin{array}{rcl}2p_1 \quad + \quad 3p_3 - \quad 3p_4 &=& 0\\ -p_2 + \quad 3p_3 - \quad 3p_4 &=& 0\\ (-3/2)p_3 + (5/2)p_4 &=& 0\end{array}$$

Taking p_1 as parameter to express the torque gives this description of the solution set.

$$\{\begin{pmatrix}1\\-2\\-5/3\\-1\end{pmatrix}p_1 \mid p_1\in\mathbb{R}\}$$

Denoting the torque by τ, the rotation rate by r, the volume of air by V, and the density of air by d we have that $\Pi_1 = \tau r^{-2}V^{-5/3}d^{-1}$, and so the torque is $r^2V^{5/3}d$ times a constant.

4 **(a)** These are the dimensional formulas.

quantity	*dimensional formula*
speed of the wave v	$L^1M^0T^{-1}$
separation of the dominoes d	$L^1M^0T^0$
height of the dominoes h	$L^1M^0T^0$
acceleration due to gravity g	$L^1M^0T^{-2}$

(b) The relationship

$$(L^1M^0T^{-1})^{p_1}(L^1M^0T^0)^{p_2}(L^1M^0T^0)^{p_3}(L^1M^0T^{-2})^{p_4} = (L^0M^0T^0)$$

gives this linear system.

$$\begin{array}{rcl}p_1 + p_2 + p_3 + \ p_4 &=& 0\\ 0 &=& 0\\ -p_1 \quad\quad\quad - 2p_4 &=& 0\end{array} \xrightarrow{\rho_1+\rho_4} \begin{array}{rcl}p_1 + p_2 + p_3 + p_4 &=& 0\\ p_2 + p_3 - p_4 &=& 0\end{array}$$

Taking p_3 and p_4 as parameters, we can describe the solution set in this way.

$$\{\begin{pmatrix}0\\-1\\1\\0\end{pmatrix}p_3 + \begin{pmatrix}-2\\1\\0\\1\end{pmatrix}p_4 \mid p_3,p_4\in\mathbb{R}\}$$

That gives $\{\Pi_1 = h/d, \Pi_2 = dg/v^2\}$ as a complete set.

(c) Buckingham's Theorem says that $v^2 = dg \cdot \hat{f}(h/d)$ and so, since g is a constant, if h/d is fixed then v is proportional to $\sqrt{d}$.

5 Checking the conditions in the definition of a vector space is routine.

6 (a) The dimensional formula of the circumference is L, that is, $L^1M^0T^0$. The dimensional formula of the area is L^2.

(b) One is $C + A = 2\pi r + \pi r^2$.

(c) One example is this formula relating the the length of arc subtended by an angle to the radius and the angle measure in radians: $\ell - r\theta = 0$. Both terms in that formula have dimensional formula L^1. The relationship holds for some unit systems (inches and radians, for instance) but not for all unit systems (inches and degrees, for instance).

Chapter Three: Maps Between Spaces

Isomorphisms

Three.I.1: Definition and Examples

Three.I.1.11 **(a)** Call the map f.

$$\begin{pmatrix} a & b \end{pmatrix} \stackrel{f}{\longmapsto} \begin{pmatrix} a \\ b \end{pmatrix}$$

It is one-to-one because if f sends two members of the domain to the same image, that is, if $f\left(\begin{pmatrix} a & b \end{pmatrix}\right) = f\left(\begin{pmatrix} c & d \end{pmatrix}\right)$, then the definition of f gives that

$$\begin{pmatrix} a \\ b \end{pmatrix} = \begin{pmatrix} c \\ d \end{pmatrix}$$

and since column vectors are equal only if they have equal components, we have that $a = c$ and that $b = d$. Thus, if f maps two row vectors from the domain to the same column vector then the two row vectors are equal: $\begin{pmatrix} a & b \end{pmatrix} = \begin{pmatrix} c & d \end{pmatrix}$.

To show that f is onto we must show that any member of the codomain $\mathbb{R}^2$ is the image under f of some row vector. That's easy;

$$\begin{pmatrix} x \\ y \end{pmatrix}$$

is $f\left(\begin{pmatrix} x & y \end{pmatrix}\right)$.

The computation for preservation of addition is this.

$$f\left(\begin{pmatrix} a & b \end{pmatrix} + \begin{pmatrix} c & d \end{pmatrix}\right) = f\left(\begin{pmatrix} a+c & b+d \end{pmatrix}\right) = \begin{pmatrix} a+c \\ b+d \end{pmatrix} = \begin{pmatrix} a \\ b \end{pmatrix} + \begin{pmatrix} c \\ d \end{pmatrix} = f\left(\begin{pmatrix} a & b \end{pmatrix}\right) + f\left(\begin{pmatrix} c & d \end{pmatrix}\right)$$

The computation for preservation of scalar multiplication is similar.

$$f\left(r \cdot \begin{pmatrix} a & b \end{pmatrix}\right) = f\left(\begin{pmatrix} ra & rb \end{pmatrix}\right) = \begin{pmatrix} ra \\ rb \end{pmatrix} = r \cdot \begin{pmatrix} a \\ b \end{pmatrix} = r \cdot f\left(\begin{pmatrix} a & b \end{pmatrix}\right)$$

(b) Denote the map from Example 1.2 by f. To show that it is one-to-one, assume that $f(a_0+a_1x+a_2x^2) = f(b_0 + b_1x + b_2x^2)$. Then by the definition of the function,

$$\begin{pmatrix} a_0 \\ a_1 \\ a_2 \end{pmatrix} = \begin{pmatrix} b_0 \\ b_1 \\ b_2 \end{pmatrix}$$

and so $a_0 = b_0$ and $a_1 = b_1$ and $a_2 = b_2$. Thus $a_0 + a_1x + a_2x^2 = b_0 + b_1x + b_2x^2$, and consequently f is one-to-one.

The function f is onto because there is a polynomial sent to

$$\begin{pmatrix} a \\ b \\ c \end{pmatrix}$$

by f, namely, $a + bx + cx^2$.

As for structure, this shows that f preserves addition

$$\begin{aligned}
f\big((a_0 + a_1x + a_2x^2) + (b_0 + b_1x + b_2x^2)\big) &= f\big((a_0 + b_0) + (a_1 + b_1)x + (a_2 + b_2)x^2\big) \\
&= \begin{pmatrix} a_0 + b_0 \\ a_1 + b_1 \\ a_2 + b_2 \end{pmatrix} \\
&= \begin{pmatrix} a_0 \\ a_1 \\ a_2 \end{pmatrix} + \begin{pmatrix} b_0 \\ b_1 \\ b_2 \end{pmatrix} \\
&= f(a_0 + a_1x + a_2x^2) + f(b_0 + b_1x + b_2x^2)
\end{aligned}$$

and this shows

$$\begin{aligned}
f(\,r(a_0 + a_1x + a_2x^2)\,) &= f(\,(ra_0) + (ra_1)x + (ra_2)x^2\,) \\
&= \begin{pmatrix} ra_0 \\ ra_1 \\ ra_2 \end{pmatrix} \\
&= r \cdot \begin{pmatrix} a_0 \\ a_1 \\ a_2 \end{pmatrix} \\
&= r\,f(a_0 + a_1x + a_2x^2)
\end{aligned}$$

that it preserves scalar multiplication.

Three.I.1.12 These are the images.

(a) $\begin{pmatrix} 5 \\ -2 \end{pmatrix}$ **(b)** $\begin{pmatrix} 0 \\ 2 \end{pmatrix}$ **(c)** $\begin{pmatrix} -1 \\ 1 \end{pmatrix}$

To prove that f is one-to-one, assume that it maps two linear polynomials to the same image $f(a_1 + b_1x) = f(a_2 + b_2x)$. Then

$$\begin{pmatrix} a_1 - b_1 \\ b_1 \end{pmatrix} = \begin{pmatrix} a_2 - b_2 \\ b_2 \end{pmatrix}$$

and so, since column vectors are equal only when their components are equal, $b_1 = b_2$ and $a_1 = a_2$. That shows that the two linear polynomials are equal, and so f is one-to-one.

To show that f is onto, note that this member of the codomain

$$\begin{pmatrix} s \\ t \end{pmatrix}$$

is the image of this member of the domain $(s + t) + tx$.

To check that f preserves structure, we can use item (2) of Lemma 1.10.

$$\begin{aligned} f(c_1\cdot(a_1+b_1x)+c_2\cdot(a_2+b_2x)) &= f((c_1a_1+c_2a_2)+(c_1b_1+c_2b_2)x) \\ &= \begin{pmatrix} (c_1a_1+c_2a_2)-(c_1b_1+c_2b_2) \\ c_1b_1+c_2b_2 \end{pmatrix} \\ &= c_1\cdot\begin{pmatrix} a_1-b_1 \\ b_1 \end{pmatrix}+c_2\cdot\begin{pmatrix} a_2-b_2 \\ b_2 \end{pmatrix} \\ &= c_1\cdot f(a_1+b_1x)+c_2\cdot f(a_2+b_2x) \end{aligned}$$

Three.I.1.13 To verify it is one-to-one, assume that $f_1(c_1x+c_2y+c_3z)=f_1(d_1x+d_2y+d_3z)$. Then $c_1+c_2x+c_3x^2=d_1+d_2x+d_3x^2$ by the definition of f_1. Members of $\mathcal{P}_2$ are equal only when they have the same coefficients, so this implies that $c_1=d_1$ and $c_2=d_2$ and $c_3=d_3$. Therefore $f_1(c_1x+c_2y+c_3z)=f_1(d_1x+d_2y+d_3z)$ implies that $c_1x+c_2y+c_3z=d_1x+d_2y+d_3z$, and so f_1 is one-to-one.

To verify that it is onto, consider an arbitrary member of the codomain $a_1+a_2x+a_3x^2$ and observe that it is indeed the image of a member of the domain, namely, it is $f_1(a_1x+a_2y+a_3z)$. (For instance, $0+3x+6x^2=f_1(0x+3y+6z)$.)

The computation checking that f_1 preserves addition is this.

$$\begin{aligned} f_1(\,(c_1x+c_2y+c_3z)+(d_1x+d_2y+d_3z)\,) &= f_1(\,(c_1+d_1)x+(c_2+d_2)y+(c_3+d_3)z\,) \\ &= (c_1+d_1)+(c_2+d_2)x+(c_3+d_3)x^2 \\ &= (c_1+c_2x+c_3x^2)+(d_1+d_2x+d_3x^2) \\ &= f_1(c_1x+c_2y+c_3z)+f_1(d_1x+d_2y+d_3z) \end{aligned}$$

The check that f_1 preserves scalar multiplication is this.

$$\begin{aligned} f_1(r\cdot(c_1x+c_2y+c_3z)\,) &= f_1(\,(rc_1)x+(rc_2)y+(rc_3)z\,) \\ &= (rc_1)+(rc_2)x+(rc_3)x^2 \\ &= r\cdot(c_1+c_2x+c_3x^2) \\ &= r\cdot f_1(c_1x+c_2y+c_3z) \end{aligned}$$

Three.I.1.14 **(a)** No; this map is not one-to-one. In particular, the matrix of all zeroes is mapped to the same image as the matrix of all ones.

(b) Yes, this is an isomorphism.

It is one-to-one:

$$\text{if } f(\begin{pmatrix} a_1 & b_1 \\ c_1 & d_1 \end{pmatrix})=f(\begin{pmatrix} a_2 & b_2 \\ c_2 & d_2 \end{pmatrix}) \text{ then } \begin{pmatrix} a_1+b_1+c_1+d_1 \\ a_1+b_1+c_1 \\ a_1+b_1 \\ a_1 \end{pmatrix}=\begin{pmatrix} a_2+b_2+c_2+d_2 \\ a_2+b_2+c_2 \\ a_2+b_2 \\ a_2 \end{pmatrix}$$

gives that $a_1=a_2$, and that $b_1=b_2$, and that $c_1=c_2$, and that $d_1=d_2$.

It is onto, since this shows

$$\begin{pmatrix} x \\ y \\ z \\ w \end{pmatrix}=f(\begin{pmatrix} w & z-w \\ y-z & x-y \end{pmatrix})$$

that any four-tall vector is the image of a 2×2 matrix.

Finally, it preserves combinations

$$\begin{aligned}
f(r_1\cdot\begin{pmatrix}a_1&b_1\\c_1&d_1\end{pmatrix}+r_2\cdot\begin{pmatrix}a_2&b_2\\c_2&d_2\end{pmatrix})&=f(\begin{pmatrix}r_1a_1+r_2a_2&r_1b_1+r_2b_2\\r_1c_1+r_2c_2&r_1d_1+r_2d_2\end{pmatrix})\\
&=\begin{pmatrix}r_1a_1+\cdots+r_2d_2\\r_1a_1+\cdots+r_2c_2\\r_1a_1+\cdots+r_2b_2\\r_1a_1+r_2a_2\end{pmatrix}\\
&=r_1\cdot\begin{pmatrix}a_1+\cdots+d_1\\a_1+\cdots+c_1\\a_1+b_1\\a_1\end{pmatrix}+r_2\cdot\begin{pmatrix}a_2+\cdots+d_2\\a_2+\cdots+c_2\\a_2+b_2\\a_2\end{pmatrix}\\
&=r_1\cdot f(\begin{pmatrix}a_1&b_1\\c_1&d_1\end{pmatrix})+r_2\cdot f(\begin{pmatrix}a_2&b_2\\c_2&d_2\end{pmatrix})
\end{aligned}$$

and so item (2) of Lemma 1.10 shows that it preserves structure.

(c) Yes, it is an isomorphism.

To show that it is one-to-one, we suppose that two members of the domain have the same image under f.

$$f(\begin{pmatrix}a_1&b_1\\c_1&d_1\end{pmatrix})=f(\begin{pmatrix}a_2&b_2\\c_2&d_2\end{pmatrix})$$

This gives, by the definition of f, that $c_1+(d_1+c_1)x+(b_1+a_1)x^2+a_1x^3=c_2+(d_2+c_2)x+(b_2+a_2)x^2+a_2x^3$ and then the fact that polynomials are equal only when their coefficients are equal gives a set of linear equations

$$\begin{aligned}
c_1&=c_2\\
d_1+c_1&=d_2+c_2\\
b_1+a_1&=b_2+a_2\\
a_1&=a_2
\end{aligned}$$

that has only the solution $a_1=a_2$, $b_1=b_2$, $c_1=c_2$, and $d_1=d_2$.

To show that f is onto, we note that $p+qx+rx^2+sx^3$ is the image under f of this matrix.

$$\begin{pmatrix}s&r-s\\p&q-p\end{pmatrix}$$

We can check that f preserves structure by using item (2) of Lemma 1.10.

$$\begin{aligned}
f(r_1\cdot\begin{pmatrix}a_1&b_1\\c_1&d_1\end{pmatrix}+r_2\cdot\begin{pmatrix}a_2&b_2\\c_2&d_2\end{pmatrix})&=f(\begin{pmatrix}r_1a_1+r_2a_2&r_1b_1+r_2b_2\\r_1c_1+r_2c_2&r_1d_1+r_2d_2\end{pmatrix})\\
&=(r_1c_1+r_2c_2)+(r_1d_1+r_2d_2+r_1c_1+r_2c_2)x\\
&\quad+(r_1b_1+r_2b_2+r_1a_1+r_2a_2)x^2+(r_1a_1+r_2a_2)x^3\\
&=r_1\cdot(c_1+(d_1+c_1)x+(b_1+a_1)x^2+a_1x^3)\\
&\quad+r_2\cdot(c_2+(d_2+c_2)x+(b_2+a_2)x^2+a_2x^3)\\
&=r_1\cdot f(\begin{pmatrix}a_1&b_1\\c_1&d_1\end{pmatrix})+r_2\cdot f(\begin{pmatrix}a_2&b_2\\c_2&d_2\end{pmatrix})
\end{aligned}$$

(d) No, this map does not preserve structure. For instance, it does not send the matrix of all zeroes to the zero polynomial.

Three.I.1.15 It is one-to-one and onto, a correspondence, because it has an inverse (namely, $f^{-1}(x)=\sqrt[3]{x}$). However, it is not an isomorphism. For instance, $f(1)+f(1)\neq f(1+1)$.

Three.I.1.16 Many maps are possible. Here are two.

$$\begin{pmatrix}a & b\end{pmatrix} \mapsto \begin{pmatrix}b\\a\end{pmatrix} \quad\text{and}\quad \begin{pmatrix}a & b\end{pmatrix} \mapsto \begin{pmatrix}2a\\b\end{pmatrix}$$

The verifications are straightforward adaptations of the others above.

Three.I.1.17 Here are two.

$$a_0 + a_1x + a_2x^2 \mapsto \begin{pmatrix}a_1\\a_0\\a_2\end{pmatrix} \quad\text{and}\quad a_0 + a_1x + a_2x^2 \mapsto \begin{pmatrix}a_0 + a_1\\a_1\\a_2\end{pmatrix}$$

Verification is straightforward (for the second, to show that it is onto, note that

$$\begin{pmatrix}s\\t\\u\end{pmatrix}$$

is the image of $(s - t) + tx + ux^2$).

Three.I.1.18 The space $\mathbb{R}^2$ is not a subspace of $\mathbb{R}^3$ because it is not a subset of $\mathbb{R}^3$. The two-tall vectors in $\mathbb{R}^2$ are not members of $\mathbb{R}^3$.

The natural isomorphism $\iota\colon \mathbb{R}^2 \to \mathbb{R}^3$ (called the *injection* map) is this.

$$\begin{pmatrix}x\\y\end{pmatrix} \stackrel{\iota}{\longmapsto} \begin{pmatrix}x\\y\\0\end{pmatrix}$$

This map is one-to-one because

$$f(\begin{pmatrix}x_1\\y_1\end{pmatrix}) = f(\begin{pmatrix}x_2\\y_2\end{pmatrix}) \quad\text{implies}\quad \begin{pmatrix}x_1\\y_1\\0\end{pmatrix} = \begin{pmatrix}x_2\\y_2\\0\end{pmatrix}$$

which in turn implies that $x_1 = x_2$ and $y_1 = y_2$, and therefore the initial two two-tall vectors are equal.

Because

$$\begin{pmatrix}x\\y\\0\end{pmatrix} = f(\begin{pmatrix}x\\y\end{pmatrix})$$

this map is onto the xy-plane.

To show that this map preserves structure, we will use item (2) of Lemma 1.10 and show

$$\begin{aligned} f(c_1 \cdot \begin{pmatrix}x_1\\y_1\end{pmatrix} + c_2 \cdot \begin{pmatrix}x_2\\y_2\end{pmatrix}) = f(\begin{pmatrix}c_1x_1 + c_2x_2\\c_1y_1 + c_2y_2\end{pmatrix}) &= \begin{pmatrix}c_1x_1 + c_2x_2\\c_1y_1 + c_2y_2\\0\end{pmatrix} \\ &= c_1 \cdot \begin{pmatrix}x_1\\y_1\\0\end{pmatrix} + c_2 \cdot \begin{pmatrix}x_2\\y_2\\0\end{pmatrix} = c_1 \cdot f(\begin{pmatrix}x_1\\y_1\end{pmatrix}) + c_2 \cdot f(\begin{pmatrix}x_2\\y_2\end{pmatrix}) \end{aligned}$$

that it preserves combinations of two vectors.

Three.I.1.19 Here are two:

$$\begin{pmatrix}r_1\\r_2\\\vdots\\r_{16}\end{pmatrix} \mapsto \begin{pmatrix}r_1 & r_2 & \ldots & \\ & & \ldots & r_{16}\end{pmatrix} \quad\text{and}\quad \begin{pmatrix}r_1\\r_2\\\vdots\\r_{16}\end{pmatrix} \mapsto \begin{pmatrix}r_1 & \\ r_2 & \\ \vdots & \vdots\\ & r_{16}\end{pmatrix}$$

Verification that each is an isomorphism is easy.

Three.I.1.20 When k is the product $k = mn$, here is an isomorphism.

$$\begin{pmatrix} r_1 & r_2 & \ldots & \\ \vdots & & & \\ & & \ldots & r_{m\cdot n} \end{pmatrix} \mapsto \begin{pmatrix} r_1 \\ r_2 \\ \vdots \\ r_{m\cdot n} \end{pmatrix}$$

Checking that this is an isomorphism is easy.

Three.I.1.21 If $n \geqslant 1$ then $\mathcal{P}_{n-1} \cong \mathbb{R}^n$. (If we take $\mathcal{P}_{-1}$ and $\mathbb{R}^0$ to be trivial vector spaces, then the relationship extends one dimension lower.) The natural isomorphism between them is this.

$$a_0 + a_1 x + \cdots + a_{n-1} x^{n-1} \mapsto \begin{pmatrix} a_0 \\ a_1 \\ \vdots \\ a_{n-1} \end{pmatrix}$$

Checking that it is an isomorphism is straightforward.

Three.I.1.22 This is the map, expanded.

$$\begin{aligned} f(a_0 + a_1 x + a_2 x^2 + a_3 x^3 + a_4 x^4 + a_5 x^5) &= a_0 + a_1(x-1) + a_2(x-1)^2 + a_3(x-1)^3 \\ &\quad + a_4(x-1)^4 + a_5(x-1)^5 \\ &= a_0 + a_1(x-1) + a_2(x^2 - 2x + 1) \\ &\quad + a_3(x^3 - 3x^2 + 3x - 1) \\ &\quad + a_4(x^4 - 4x^3 + 6x^2 - 4x + 1) \\ &\quad + a_5(x^5 - 5x^4 + 10x^3 - 10x^2 + 5x - 1) \\ &= (a_0 - a_1 + a_2 - a_3 + a_4 - a_5) \\ &\quad + (a_1 - 2a_2 + 3a_3 - 4a_4 + 5a_5)x \\ &\quad + (a_2 - 3a_3 + 6a_4 - 10a_5)x^2 + (a_3 - 4a_4 + 10a_5)x^3 \\ &\quad + (a_4 - 5a_5)x^4 + a_5 x^5 \end{aligned}$$

This map is a correspondence because it has an inverse, the map $p(x) \mapsto p(x+1)$.

To finish checking that it is an isomorphism, we apply item (2) of Lemma 1.10 and show that it preserves linear combinations of two polynomials. Briefly, the check goes like this.

$$\begin{aligned} &f(c \cdot (a_0 + a_1 x + \cdots + a_5 x^5) + d \cdot (b_0 + b_1 x + \cdots + b_5 x^5)) \\ &= \cdots = (ca_0 - ca_1 + ca_2 - ca_3 + ca_4 - ca_5 + db_0 - db_1 + db_2 - db_3 + db_4 - db_5) + \cdots + (ca_5 + db_5)x^5 \\ &\qquad = \cdots = c \cdot f(a_0 + a_1 x + \cdots + a_5 x^5) + d \cdot f(b_0 + b_1 x + \cdots + b_5 x^5) \end{aligned}$$

Three.I.1.23 No vector space has the empty set underlying it. We can take $\vec{v}$ to be the zero vector.

Three.I.1.24 Yes; where the two spaces are $\{\vec{a}\}$ and $\{\vec{b}\}$, the map sending $\vec{a}$ to $\vec{b}$ is clearly one-to-one and onto, and also preserves what little structure there is.

Three.I.1.25 A linear combination of $n = 0$ vectors adds to the zero vector and so Lemma 1.9 shows that the three statements are equivalent in this case.

Three.I.1.26 Consider the basis $\langle 1 \rangle$ for $\mathcal{P}_0$ and let $f(1) \in \mathbb{R}$ be k. For any $a \in \mathcal{P}_0$ we have that $f(a) = f(a \cdot 1) = af(1) = ak$ and so f's action is multiplication by k. Note that $k \neq 0$ or else the map is not one-to-one. (Incidentally, any such map $a \mapsto ka$ is an isomorphism, as is easy to check.)

Three.I.1.27 In each item, following item (2) of Lemma 1.10, we show that the map preserves structure by showing that the it preserves linear combinations of two members of the domain.

(a) The identity map is clearly one-to-one and onto. For linear combinations the check is easy.

$$\mathrm{id}(c_1 \cdot \vec{v}_1 + c_2 \cdot \vec{v}_2) = c_1 \vec{v}_1 + c_2 \vec{v}_2 = c_1 \cdot \mathrm{id}(\vec{v}_1) + c_2 \cdot \mathrm{id}(\vec{v}_2)$$

(b) The inverse of a correspondence is also a correspondence (as stated in the appendix), so we need only check that the inverse preserves linear combinations. Assume that $\vec{w}_1 = f(\vec{v}_1)$ (so $f^{-1}(\vec{w}_1) = \vec{v}_1$) and assume that $\vec{w}_2 = f(\vec{v}_2)$.

$$\begin{aligned} f^{-1}(c_1 \cdot \vec{w}_1 + c_2 \cdot \vec{w}_2) &= f^{-1}\big(c_1 \cdot f(\vec{v}_1) + c_2 \cdot f(\vec{v}_2)\big) \\ &= f^{-1}\big(f\big(c_1\vec{v}_1 + c_2\vec{v}_2\big)\big) \\ &= c_1\vec{v}_1 + c_2\vec{v}_2 \\ &= c_1 \cdot f^{-1}(\vec{w}_1) + c_2 \cdot f^{-1}(\vec{w}_2) \end{aligned}$$

(c) The composition of two correspondences is a correspondence (as stated in the appendix), so we need only check that the composition map preserves linear combinations.

$$\begin{aligned} g \circ f\,(c_1 \cdot \vec{v}_1 + c_2 \cdot \vec{v}_2) &= g\big(f(c_1\vec{v}_1 + c_2\vec{v}_2)\big) \\ &= g\big(c_1 \cdot f(\vec{v}_1) + c_2 \cdot f(\vec{v}_2)\big) \\ &= c_1 \cdot g\big(f(\vec{v}_1)\big) + c_2 \cdot g(f(\vec{v}_2)) \\ &= c_1 \cdot g \circ f\,(\vec{v}_1) + c_2 \cdot g \circ f\,(\vec{v}_2) \end{aligned}$$

Three.I.1.28 One direction is easy: by definition, if f is one-to-one then for any $\vec{w} \in W$ at most one $\vec{v} \in V$ has $f(\vec{v}) = \vec{w}$, and so in particular, at most one member of V is mapped to $\vec{0}_W$. The proof of Lemma 1.9 does not use the fact that the map is a correspondence and therefore shows that any structure-preserving map f sends $\vec{0}_V$ to $\vec{0}_W$.

For the other direction, assume that the only member of V that is mapped to $\vec{0}_W$ is $\vec{0}_V$. To show that f is one-to-one assume that $f(\vec{v}_1) = f(\vec{v}_2)$. Then $f(\vec{v}_1) - f(\vec{v}_2) = \vec{0}_W$ and so $f(\vec{v}_1 - \vec{v}_2) = \vec{0}_W$. Consequently $\vec{v}_1 - \vec{v}_2 = \vec{0}_V$, so $\vec{v}_1 = \vec{v}_2$, and so f is one-to-one.

Three.I.1.29 We will prove something stronger—not only is the existence of a dependence preserved by isomorphism, but each instance of a dependence is preserved, that is,

$$\begin{aligned} &\vec{v}_i = c_1\vec{v}_1 + \cdots + c_{i-1}\vec{v}_{i-1} + c_{i+1}\vec{v}_{i+1} + \cdots + c_k\vec{v}_k \\ &\qquad\iff f(\vec{v}_i) = c_1 f(\vec{v}_1) + \cdots + c_{i-1} f(\vec{v}_{i-1}) + c_{i+1} f(\vec{v}_{i+1}) + \cdots + c_k f(\vec{v}_k). \end{aligned}$$

The $\Longrightarrow$ direction of this statement holds by item (3) of Lemma 1.10. The $\Longleftarrow$ direction holds by regrouping

$$\begin{aligned} f(\vec{v}_i) &= c_1 f(\vec{v}_1) + \cdots + c_{i-1} f(\vec{v}_{i-1}) + c_{i+1} f(\vec{v}_{i+1}) + \cdots + c_k f(\vec{v}_k) \\ &= f(c_1\vec{v}_1 + \cdots + c_{i-1}\vec{v}_{i-1} + c_{i+1}\vec{v}_{i+1} + \cdots + c_k\vec{v}_k) \end{aligned}$$

and applying the fact that f is one-to-one, and so for the two vectors $\vec{v}_i$ and $c_1\vec{v}_1 + \cdots + c_{i-1}\vec{v}_{i-1} + c_{i+1} f\vec{v}_{i+1} + \cdots + c_k f(\vec{v}_k$ to be mapped to the same image by f, they must be equal.

Three.I.1.30 **(a)** This map is one-to-one because if $d_s(\vec{v}_1) = d_s(\vec{v}_2)$ then by definition of the map, $s \cdot \vec{v}_1 = s \cdot \vec{v}_2$ and so $\vec{v}_1 = \vec{v}_2$, as s is nonzero. This map is onto as any $\vec{w} \in \mathbb{R}^2$ is the image of $\vec{v} = (1/s) \cdot \vec{w}$ (again, note that s is nonzero). (Another way to see that this map is a correspondence is to observe that it has an inverse: the inverse of d_s is $d_{1/s}$.)

To finish, note that this map preserves linear combinations

$$d_s(c_1 \cdot \vec{v}_1 + c_2 \cdot \vec{v}_2) = s(c_1\vec{v}_1 + c_2\vec{v}_2) = c_1 s\vec{v}_1 + c_2 s\vec{v}_2 = c_1 \cdot d_s(\vec{v}_1) + c_2 \cdot d_s(\vec{v}_2)$$

and therefore is an isomorphism.

(b) As in the prior item, we can show that the map t_θ is a correspondence by noting that it has an inverse, $t_{-\theta}$.

That the map preserves structure is geometrically easy to see. For instance, adding two vectors and then rotating them has the same effect as rotating first and then adding. For an algebraic argument, consider polar coordinates: the map t_θ sends the vector with endpoint (r, ϕ) to the vector with endpoint $(r, \phi + \theta)$. Then the familiar trigonometric formulas $\cos(\phi + \theta) = \cos\phi\,\cos\theta - \sin\phi\,\sin\theta$ and

$\sin(\phi+\theta) = \sin\phi\,\cos\theta + \cos\phi\,\sin\theta$ show how to express the map's action in the usual rectangular coordinate system.

$$\begin{pmatrix}x\\y\end{pmatrix} = \begin{pmatrix}r\cos\phi\\r\sin\phi\end{pmatrix} \overset{t_\theta}{\longmapsto} \begin{pmatrix}r\cos(\phi+\theta)\\r\sin(\phi+\theta)\end{pmatrix} = \begin{pmatrix}x\cos\theta - y\sin\theta\\x\sin\theta + y\cos\theta\end{pmatrix}$$

Now the calculation for preservation of addition is routine.

$$\begin{pmatrix}x_1+x_2\\y_1+y_2\end{pmatrix} \overset{t_\theta}{\longmapsto} \begin{pmatrix}(x_1+x_2)\cos\theta-(y_1+y_2)\sin\theta\\(x_1+x_2)\sin\theta+(y_1+y_2)\cos\theta\end{pmatrix} = \begin{pmatrix}x_1\cos\theta-y_1\sin\theta\\x_1\sin\theta+y_1\cos\theta\end{pmatrix} + \begin{pmatrix}x_2\cos\theta-y_2\sin\theta\\x_2\sin\theta+y_2\cos\theta\end{pmatrix}$$

The calculation for preservation of scalar multiplication is similar.

(c) This map is a correspondence because it has an inverse (namely, itself).

As in the last item, that the reflection map preserves structure is geometrically easy to see: adding vectors and then reflecting gives the same result as reflecting first and then adding, for instance. For an algebraic proof, suppose that the line ℓ has slope k (the case of a line with undefined slope can be done as a separate, but easy, case). We can follow the hint and use polar coordinates: where the line ℓ forms an angle of ϕ with the x-axis, the action of f_ℓ is to send the vector with endpoint $(r\cos\theta, r\sin\theta)$ to the one with endpoint $(r\cos(2\phi-\theta), r\sin(2\phi-\theta))$.

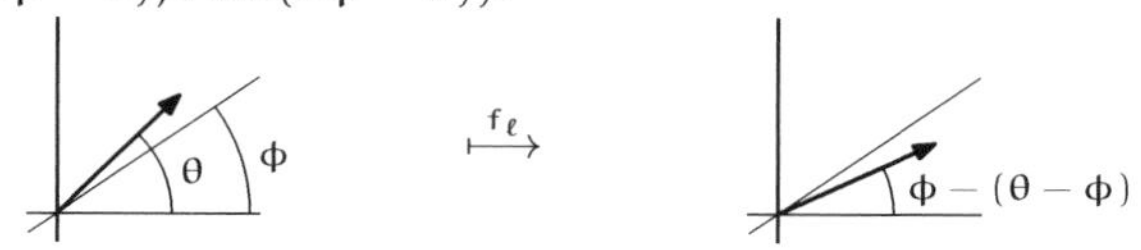

To convert to rectangular coordinates, we will use some trigonometric formulas, as we did in the prior item. First observe that $\cos\phi$ and $\sin\phi$ can be determined from the slope k of the line. This picture

gives that $\cos\phi = 1/\sqrt{1+k^2}$ and $\sin\phi = k/\sqrt{1+k^2}$. Now,

$$\begin{aligned}\cos(2\phi-\theta) &= \cos(2\phi)\,\cos\theta + \sin(2\phi)\,\sin\theta\\ &= \left(\cos^2\phi - \sin^2\phi\right)\cos\theta + (2\sin\phi\cos\phi)\,\sin\theta\\ &= \left((\frac{1}{\sqrt{1+k^2}})^2 - (\frac{k}{\sqrt{1+k^2}})^2\right)\cos\theta + \left(2\frac{k}{\sqrt{1+k^2}}\frac{1}{\sqrt{1+k^2}}\right)\sin\theta\\ &= \left(\frac{1-k^2}{1+k^2}\right)\cos\theta + \left(\frac{2k}{1+k^2}\right)\sin\theta\end{aligned}$$

and thus the first component of the image vector is this.

$$r\cdot\cos(2\phi-\theta) = \frac{1-k^2}{1+k^2}\cdot x + \frac{2k}{1+k^2}\cdot y$$

A similar calculation shows that the second component of the image vector is this.

$$r\cdot\sin(2\phi-\theta) = \frac{2k}{1+k^2}\cdot x - \frac{1-k^2}{1+k^2}\cdot y$$

With this algebraic description of the action of f_ℓ

$$\begin{pmatrix}x\\y\end{pmatrix} \overset{f_\ell}{\longmapsto} \begin{pmatrix}(1-k^2/1+k^2)\cdot x + (2k/1+k^2)\cdot y\\(2k/1+k^2)\cdot x - (1-k^2/1+k^2)\cdot y\end{pmatrix}$$

checking that it preserves structure is routine.

Three.I.1.31 First, the map $p(x) \mapsto p(x+k)$ doesn't count because it is a version of $p(x) \mapsto p(x-k)$. Here is a correct answer (many others are also correct): $a_0 + a_1x + a_2x^2 \mapsto a_2 + a_0x + a_1x^2$. Verification that this is an isomorphism is straightforward.

Three.I.1.32 (a) For the 'only if' half, let $f\colon \mathbb{R}^1 \to \mathbb{R}^1$ to be an isomorphism. Consider the basis $\langle 1\rangle \subseteq \mathbb{R}^1$. Designate $f(1)$ by k. Then for any x we have that $f(x) = f(x\cdot 1) = x\cdot f(1) = xk$, and so f's action is multiplication by k. To finish this half, just note that $k \neq 0$ or else f would not be one-to-one.

For the 'if' half we only have to check that such a map is an isomorphism when $k \neq 0$. To check that it is one-to-one, assume that $f(x_1) = f(x_2)$ so that $kx_1 = kx_2$ and divide by the nonzero factor k to conclude that $x_1 = x_2$. To check that it is onto, note that any $y \in \mathbb{R}^1$ is the image of $x = y/k$ (again, $k \neq 0$). Finally, to check that such a map preserves combinations of two members of the domain, we have this.

$$f(c_1x_1 + c_2x_2) = k(c_1x_1 + c_2x_2) = c_1kx_1 + c_2kx_2 = c_1f(x_1) + c_2f(x_2)$$

(b) By the prior item, f's action is $x \mapsto (7/3)x$. Thus $f(-2) = -14/3$.

(c) For the 'only if' half, assume that $f\colon \mathbb{R}^2 \to \mathbb{R}^2$ is an automorphism. Consider the standard basis $\mathcal{E}_2$ for $\mathbb{R}^2$. Let

$$f(\vec{e}_1) = \begin{pmatrix} a \\ c \end{pmatrix} \quad \text{and} \quad f(\vec{e}_2) = \begin{pmatrix} b \\ d \end{pmatrix}.$$

Then the action of f on any vector is determined by by its action on the two basis vectors.

$$f(\begin{pmatrix} x \\ y \end{pmatrix}) = f(x \cdot \vec{e}_1 + y \cdot \vec{e}_2) = x \cdot f(\vec{e}_1) + y \cdot f(\vec{e}_2) = x \cdot \begin{pmatrix} a \\ c \end{pmatrix} + y \cdot \begin{pmatrix} b \\ d \end{pmatrix} = \begin{pmatrix} ax + by \\ cx + dy \end{pmatrix}$$

To finish this half, note that if $ad - bc = 0$, that is, if $f(\vec{e}_2)$ is a multiple of $f(\vec{e}_1)$, then f is not one-to-one.

For 'if' we must check that the map is an isomorphism, under the condition that $ad - bc \neq 0$. The structure-preservation check is easy; we will here show that f is a correspondence. For the argument that the map is one-to-one, assume this.

$$f(\begin{pmatrix} x_1 \\ y_1 \end{pmatrix}) = f(\begin{pmatrix} x_2 \\ y_2 \end{pmatrix}) \quad \text{and so} \quad \begin{pmatrix} ax_1 + by_1 \\ cx_1 + dy_1 \end{pmatrix} = \begin{pmatrix} ax_2 + by_2 \\ cx_2 + dy_2 \end{pmatrix}$$

Then, because $ad - bc \neq 0$, the resulting system

$$\begin{aligned} a(x_1 - x_2) + b(y_1 - y_2) &= 0 \\ c(x_1 - x_2) + d(y_1 - y_2) &= 0 \end{aligned}$$

has a unique solution, namely the trivial one $x_1 - x_2 = 0$ and $y_1 - y_2 = 0$ (this follows from the hint).

The argument that this map is onto is closely related—this system

$$\begin{aligned} ax_1 + by_1 &= x \\ cx_1 + dy_1 &= y \end{aligned}$$

has a solution for any x and y if and only if this set

$$\{\begin{pmatrix} a \\ c \end{pmatrix}, \begin{pmatrix} b \\ d \end{pmatrix}\}$$

spans $\mathbb{R}^2$, i.e., if and only if this set is a basis (because it is a two-element subset of $\mathbb{R}^2$), i.e., if and only if $ad - bc \neq 0$.

(d)

$$f(\begin{pmatrix} 0 \\ -1 \end{pmatrix}) = f(\begin{pmatrix} 1 \\ 3 \end{pmatrix} - \begin{pmatrix} 1 \\ 4 \end{pmatrix}) = f(\begin{pmatrix} 1 \\ 3 \end{pmatrix}) - f(\begin{pmatrix} 1 \\ 4 \end{pmatrix}) = \begin{pmatrix} 2 \\ -1 \end{pmatrix} - \begin{pmatrix} 0 \\ 1 \end{pmatrix} = \begin{pmatrix} 2 \\ -2 \end{pmatrix}$$

Three.I.1.33 There are many answers; two are linear independence and subspaces.

To show that if a set $\{\vec{v}_1, \ldots, \vec{v}_n\}$ is linearly independent then its image $\{f(\vec{v}_1), \ldots, f(\vec{v}_n)\}$ is also linearly independent, consider a linear relationship among members of the image set.

$$0 = c_1f(\vec{v}_1) + \cdots + c_nf(\vec{v_n}) = f(c_1\vec{v}_1) + \cdots + f(c_n\vec{v_n}) = f(c_1\vec{v}_1 + \cdots + c_n\vec{v_n})$$

Because this map is an isomorphism, it is one-to-one. So f maps only one vector from the domain to the zero vector in the range, that is, $c_1\vec{v}_1 + \cdots + c_n\vec{v}_n$ equals the zero vector (in the domain, of course). But, if $\{\vec{v}_1, \ldots, \vec{v}_n\}$ is linearly independent then all of the c's are zero, and so $\{f(\vec{v}_1), \ldots, f(\vec{v}_n)\}$ is linearly independent also. (*Remark.* There is a small point about this argument that is worth mention. In a set, repeats collapse, that is, strictly speaking, this is a one-element set: $\{\vec{v}, \vec{v}\}$, because the things listed as in it are the same thing. Observe, however, the use of the subscript n in the above argument. In moving from the domain set $\{\vec{v}_1,\dots,\vec{v}_n\}$ to the image set $\{f(\vec{v}_1),\dots,f(\vec{v}_n)\}$, there is no collapsing, because the image set does not have repeats, because the isomorphism f is one-to-one.)

To show that if $f\colon V\to W$ is an isomorphism and if U is a subspace of the domain V then the set of image vectors $f(U)=\{\vec{w}\in W \mid \vec{w}=f(\vec{u}) \text{ for some } \vec{u}\in U\}$ is a subspace of W, we need only show that it is closed under linear combinations of two of its members (it is nonempty because it contains the image of the zero vector). We have

$$c_1\cdot f(\vec{u}_1)+c_2\cdot f(\vec{u}_2)=f(c_1\vec{u}_1)+f(c_2\vec{u}_2)=f(c_1\vec{u}_1+c_2\vec{u}_2)$$

and $c_1\vec{u}_1+c_2\vec{u}_2$ is a member of U because of the closure of a subspace under combinations. Hence the combination of $f(\vec{u}_1)$ and $f(\vec{u}_2)$ is a member of $f(U)$.

Three.I.1.34 (a) The association

$$\vec{p}=c_1\vec{\beta}_1+c_2\vec{\beta}_2+c_3\vec{\beta}_3 \overset{\text{Rep}_B(\cdot)}{\longmapsto} \begin{pmatrix}c_1\\c_2\\c_3\end{pmatrix}$$

is a function if every member $\vec{p}$ of the domain is associated with at least one member of the codomain, and if every member $\vec{p}$ of the domain is associated with at most one member of the codomain. The first condition holds because the basis B spans the domain — every $\vec{p}$ can be written as at least one linear combination of $\vec{\beta}$'s. The second condition holds because the basis B is linearly independent — every member $\vec{p}$ of the domain can be written as at most one linear combination of the $\vec{\beta}$'s.

(b) For the one-to-one argument, if $\text{Rep}_B(\vec{p})=\text{Rep}_B(\vec{q})$, that is, if $\text{Rep}_B(p_1\vec{\beta}_1+p_2\vec{\beta}_2+p_3\vec{\beta}_3)=\text{Rep}_B(q_1\vec{\beta}_1+q_2\vec{\beta}_2+q_3\vec{\beta}_3)$ then

$$\begin{pmatrix}p_1\\p_2\\p_3\end{pmatrix}=\begin{pmatrix}q_1\\q_2\\q_3\end{pmatrix}$$

and so $p_1=q_1$ and $p_2=q_2$ and $p_3=q_3$, which gives the conclusion that $\vec{p}=\vec{q}$. Therefore this map is one-to-one.

For onto, we can just note that

$$\begin{pmatrix}a\\b\\c\end{pmatrix}$$

equals $\text{Rep}_B(a\vec{\beta}_1+b\vec{\beta}_2+c\vec{\beta}_3)$, and so any member of the codomain $\mathbb{R}^3$ is the image of some member of the domain $\mathcal{P}_2$.

(c) This map respects addition and scalar multiplication because it respects combinations of two members of the domain (that is, we are using item (2) of Lemma 1.10): where $\vec{p}=p_1\vec{\beta}_1+p_2\vec{\beta}_2+p_3\vec{\beta}_3$ and $\vec{q}=q_1\vec{\beta}_1+q_2\vec{\beta}_2+q_3\vec{\beta}_3$, we have this.

$$\begin{aligned}\text{Rep}_B(c\cdot\vec{p}+d\cdot\vec{q})&=\text{Rep}_B(\,(cp_1+dq_1)\vec{\beta}_1+(cp_2+dq_2)\vec{\beta}_2+(cp_3+dq_3)\vec{\beta}_3\,)\\&=\begin{pmatrix}cp_1+dq_1\\cp_2+dq_2\\cp_3+dq_3\end{pmatrix}\\&=c\cdot\begin{pmatrix}p_1\\p_2\\p_3\end{pmatrix}+d\cdot\begin{pmatrix}q_1\\q_2\\q_3\end{pmatrix}\\&=\text{Rep}_B(\vec{p})+\text{Rep}_B(\vec{q})\end{aligned}$$

(d) Use any basis B for $\mathcal{P}_2$ whose first two members are $x+x^2$ and $1-x$, say $B=\langle x+x^2,1-x,1\rangle$.

Three.I.1.35 See the next subsection.

Three.I.1.36 **(a)** Most of the conditions in the definition of a vector space are routine. We here sketch the verification of part (1) of that definition.

For closure of $U \times W$, note that because U and W are closed, we have that $\vec{u}_1 + \vec{u}_2 \in U$ and $\vec{w}_1 + \vec{w}_2 \in W$ and so $(\vec{u}_1 + \vec{u}_2, \vec{w}_1 + \vec{w}_2) \in U \times W$. Commutativity of addition in $U \times W$ follows from commutativity of addition in U and W.

$$(\vec{u}_1, \vec{w}_1) + (\vec{u}_2, \vec{w}_2) = (\vec{u}_1 + \vec{u}_2, \vec{w}_1 + \vec{w}_2) = (\vec{u}_2 + \vec{u}_1, \vec{w}_2 + \vec{w}_1) = (\vec{u}_2, \vec{w}_2) + (\vec{u}_1, \vec{w}_1)$$

The check for associativity of addition is similar. The zero element is $(\vec{0}_U, \vec{0}_W) \in U \times W$ and the additive inverse of $(\vec{u}, \vec{w})$ is $(-\vec{u}, -\vec{w})$.

The checks for the second part of the definition of a vector space are also straightforward.

(b) This is a basis

$$\langle (1, \begin{pmatrix} 0 \\ 0 \end{pmatrix}), (x, \begin{pmatrix} 0 \\ 0 \end{pmatrix}), (x^2, \begin{pmatrix} 0 \\ 0 \end{pmatrix}), (1, \begin{pmatrix} 1 \\ 0 \end{pmatrix}), (1, \begin{pmatrix} 0 \\ 1 \end{pmatrix}) \rangle$$

because there is one and only one way to represent any member of $\mathcal{P}_2 \times \mathbb{R}^2$ with respect to this set; here is an example.

$$(3 + 2x + x^2, \begin{pmatrix} 5 \\ 4 \end{pmatrix}) = 3 \cdot (1, \begin{pmatrix} 0 \\ 0 \end{pmatrix}) + 2 \cdot (x, \begin{pmatrix} 0 \\ 0 \end{pmatrix}) + (x^2, \begin{pmatrix} 0 \\ 0 \end{pmatrix}) + 5 \cdot (1, \begin{pmatrix} 1 \\ 0 \end{pmatrix}) + 4 \cdot (1, \begin{pmatrix} 0 \\ 1 \end{pmatrix})$$

The dimension of this space is five.

(c) We have $\dim(U \times W) = \dim(U) + \dim(W)$ as this is a basis.

$$\langle (\vec{\mu}_1, \vec{0}_W), \ldots, (\vec{\mu}_{\dim(U)}, \vec{0}_W), (\vec{0}_U, \vec{\omega}_1), \ldots, (\vec{0}_U, \vec{\omega}_{\dim(W)}) \rangle$$

(d) We know that if $V = U \oplus W$ then each $\vec{v} \in V$ can be written as $\vec{v} = \vec{u} + \vec{w}$ in one and only one way. This is just what we need to prove that the given function an isomorphism.

First, to show that f is one-to-one we can show that if $f((\vec{u}_1, \vec{w}_1)) = ((\vec{u}_2, \vec{w}_2))$, that is, if $\vec{u}_1 + \vec{w}_1 = \vec{u}_2 + \vec{w}_2$ then $\vec{u}_1 = \vec{u}_2$ and $\vec{w}_1 = \vec{w}_2$. But the statement 'each $\vec{v}$ is such a sum in only one way' is exactly what is needed to make this conclusion. Similarly, the argument that f is onto is completed by the statement that 'each $\vec{v}$ is such a sum in at least one way'.

This map also preserves linear combinations

$$\begin{aligned}
f(\, c_1 \cdot (\vec{u}_1, \vec{w}_1) + c_2 \cdot (\vec{u}_2, \vec{w}_2)\,) &= f(\, (c_1\vec{u}_1 + c_2\vec{u}_2, c_1\vec{w}_1 + c_2\vec{w}_2)\,) \\
&= c_1\vec{u}_1 + c_2\vec{u}_2 + c_1\vec{w}_1 + c_2\vec{w}_2 \\
&= c_1\vec{u}_1 + c_1\vec{w}_1 + c_2\vec{u}_2 + c_2\vec{w}_2 \\
&= c_1 \cdot f(\,(\vec{u}_1, \vec{w}_1)\,) + c_2 \cdot f(\,(\vec{u}_2, \vec{w}_2)\,)
\end{aligned}$$

and so it is an isomorphism.

Three.I.2: Dimension Characterizes Isomorphism

Three.I.2.9 Each pair of spaces is isomorphic if and only if the two have the same dimension. We can, when there is an isomorphism, state a map, but it isn't strictly necessary.

(a) No, they have different dimensions.

(b) No, they have different dimensions.

(c) Yes, they have the same dimension. One isomorphism is this.

$$\begin{pmatrix} a & b & c \\ d & e & f \end{pmatrix} \mapsto \begin{pmatrix} a \\ \vdots \\ f \end{pmatrix}$$

(d) Yes, they have the same dimension. This is an isomorphism.

$$a + bx + \cdots + fx^5 \mapsto \begin{pmatrix} a & b & c \\ d & e & f \end{pmatrix}$$

(e) Yes, both have dimension $2k$.

Three.I.2.10 **(a)** $\mathrm{Rep}_B(3-2x) = \begin{pmatrix} 5 \\ -2 \end{pmatrix}$ **(b)** $\begin{pmatrix} 0 \\ 2 \end{pmatrix}$ **(c)** $\begin{pmatrix} -1 \\ 1 \end{pmatrix}$

Three.I.2.11 They have different dimensions.

Three.I.2.12 Yes, both are mn-dimensional.

Three.I.2.13 Yes, any two (nondegenerate) planes are both two-dimensional vector spaces.

Three.I.2.14 There are many answers, one is the set of $\mathcal{P}_k$ (taking $\mathcal{P}_{-1}$ to be the trivial vector space).

Three.I.2.15 False (except when $n = 0$). For instance, if $f\colon V \to \mathbb{R}^n$ is an isomorphism then multiplying by any nonzero scalar, gives another, different, isomorphism. (Between trivial spaces the isomorphisms are unique; the only map possible is $\vec{0}_V \mapsto 0_W$.)

Three.I.2.16 No. A proper subspace has a strictly lower dimension than it's superspace; if U is a proper subspace of V then any linearly independent subset of U must have fewer than $\dim(V)$ members or else that set would be a basis for V, and U wouldn't be proper.

Three.I.2.17 Where $B = \langle \vec{\beta}_1, \ldots, \vec{\beta}_n \rangle$, the inverse is this.

$$\begin{pmatrix} c_1 \\ \vdots \\ c_n \end{pmatrix} \mapsto c_1\vec{\beta}_1 + \cdots + c_n\vec{\beta}_n$$

Three.I.2.18 All three spaces have dimension equal to the rank of the matrix.

Three.I.2.19 We must show that if $\vec{a} = \vec{b}$ then $f(\vec{a}) = f(\vec{b})$. So suppose that $a_1\vec{\beta}_1 + \cdots + a_n\vec{\beta}_n = b_1\vec{\beta}_1 + \cdots + b_n\vec{\beta}_n$. Each vector in a vector space (here, the domain space) has a unique representation as a linear combination of basis vectors, so we can conclude that $a_1 = b_1$, ..., $a_n = b_n$. Thus,

$$f(\vec{a}) = \begin{pmatrix} a_1 \\ \vdots \\ a_n \end{pmatrix} = \begin{pmatrix} b_1 \\ \vdots \\ b_n \end{pmatrix} = f(\vec{b})$$

and so the function is well-defined.

Three.I.2.20 Yes, because a zero-dimensional space is a trivial space.

Three.I.2.21 **(a)** No, this collection has no spaces of odd dimension.

(b) Yes, because $\mathcal{P}_k \cong \mathbb{R}^{k+1}$.

(c) No, for instance, $\mathcal{M}_{2\times3} \cong \mathcal{M}_{3\times2}$.

Three.I.2.22 One direction is easy: if the two are isomorphic via f then for any basis $B \subseteq V$, the set $D = f(B)$ is also a basis (this is shown in Lemma 2.4). The check that corresponding vectors have the same coordinates: $f(c_1\vec{\beta}_1 + \cdots + c_n\vec{\beta}_n) = c_1f(\vec{\beta}_1) + \cdots + c_nf(\vec{\beta}_n) = c_1\vec{\delta}_1 + \cdots + c_n\vec{\delta}_n$ is routine.

For the other half, assume that there are bases such that corresponding vectors have the same coordinates with respect to those bases. Because f is a correspondence, to show that it is an isomorphism, we need only show that it preserves structure. Because $\mathrm{Rep}_B(\vec{v}) = \mathrm{Rep}_D(f(\vec{v}))$, the map f preserves structure if and only if representations preserve addition: $\mathrm{Rep}_B(\vec{v}_1 + \vec{v}_2) = \mathrm{Rep}_B(\vec{v}_1) + \mathrm{Rep}_B(\vec{v}_2)$ and scalar multiplication: $\mathrm{Rep}_B(r\cdot\vec{v}) = r\cdot\mathrm{Rep}_B(\vec{v})$ The addition calculation is this: $(c_1+d_1)\vec{\beta}_1 + \cdots + (c_n+d_n)\vec{\beta}_n = c_1\vec{\beta}_1 + \cdots + c_n\vec{\beta}_n + d_1\vec{\beta}_1 + \cdots + d_n\vec{\beta}_n$, and the scalar multiplication calculation is similar.

Three.I.2.23 **(a)** Pulling the definition back from $\mathbb{R}^4$ to $\mathcal{P}_3$ gives that $a_0 + a_1x + a_2x^2 + a_3x^3$ is orthogonal to $b_0 + b_1x + b_2x^2 + b_3x^3$ if and only if $a_0b_0 + a_1b_1 + a_2b_2 + a_3b_3 = 0$.

(b) A natural definition is this.

$$D(\begin{pmatrix} a_0 \\ a_1 \\ a_2 \\ a_3 \end{pmatrix}) = \begin{pmatrix} a_1 \\ 2a_2 \\ 3a_3 \\ 0 \end{pmatrix}$$

Three.I.2.24 Yes.

Assume that V is a vector space with basis $B = \langle \vec{\beta}_1, \ldots, \vec{\beta}_n \rangle$ and that W is another vector space such that the map $f\colon B \to W$ is a correspondence. Consider the extension $\hat{f}\colon V \to W$ of f.

$$\hat{f}(c_1\vec{\beta}_1 + \cdots + c_n\vec{\beta}_n) = c_1 f(\vec{\beta}_1) + \cdots + c_n f(\vec{\beta}_n).$$

The map $\hat{f}$ is an isomorphism.

First, $\hat{f}$ is well-defined because every member of V has one and only one representation as a linear combination of elements of B.

Second, $\hat{f}$ is one-to-one because every member of W has only one representation as a linear combination of elements of $\langle f(\vec{\beta}_1), \ldots, f(\vec{\beta}_n) \rangle$. That map $\hat{f}$ is onto because every member of W has at least one representation as a linear combination of members of $\langle f(\vec{\beta}_1), \ldots, f(\vec{\beta}_n) \rangle$.

Finally, preservation of structure is routine to check. For instance, here is the preservation of addition calculation.

$$\begin{aligned} \hat{f}(\,(c_1\vec{\beta}_1 + \cdots + c_n\vec{\beta}_n) + (d_1\vec{\beta}_1 + \cdots + d_n\vec{\beta}_n)\,) &= \hat{f}(\,(c_1 + d_1)\vec{\beta}_1 + \cdots + (c_n + d_n)\vec{\beta}_n\,) \\ &= (c_1 + d_1)f(\vec{\beta}_1) + \cdots + (c_n + d_n)f(\vec{\beta}_n) \\ &= c_1 f(\vec{\beta}_1) + \cdots + c_n f(\vec{\beta}_n) + d_1 f(\vec{\beta}_1) + \cdots + d_n f(\vec{\beta}_n) \\ &= \hat{f}(c_1\vec{\beta}_1 + \cdots + c_n\vec{\beta}_n) + +\hat{f}(d_1\vec{\beta}_1 + \cdots + d_n\vec{\beta}_n). \end{aligned}$$

Preservation of scalar multiplication is similar.

Three.I.2.25 Because $V_1 \cap V_2 = \{\vec{0}_V\}$ and f is one-to-one we have that $f(V_1) \cap f(V_2) = \{\vec{0}_U\}$. To finish, count the dimensions: $\dim(U) = \dim(V) = \dim(V_1) + \dim(V_2) = \dim(f(V_1)) + \dim(f(V_2))$, as required.

Three.I.2.26 Rational numbers have many representations, e.g., $1/2 = 3/6$, and the numerators can vary among representations.

Homomorphisms

Three.II.1: Definition

Three.II.1.17 **(a)** Yes. The verification is straightforward.

$$\begin{aligned} h(c_1 \cdot \begin{pmatrix} x_1 \\ y_1 \\ z_1 \end{pmatrix} + c_2 \cdot \begin{pmatrix} x_2 \\ y_2 \\ z_2 \end{pmatrix}) &= h(\begin{pmatrix} c_1x_1 + c_2x_2 \\ c_1y_1 + c_2y_2 \\ c_1z_1 + c_2z_2 \end{pmatrix}) \\ &= \begin{pmatrix} c_1x_1 + c_2x_2 \\ c_1x_1 + c_2x_2 + c_1y_1 + c_2y_2 + c_1z_1 + c_2z_2 \end{pmatrix} \\ &= c_1 \cdot \begin{pmatrix} x_1 \\ x_1 + y_1 + z_1 \end{pmatrix} + c_2 \cdot \begin{pmatrix} x_2 \\ c_2 + y_2 + z_2 \end{pmatrix} \\ &= c_1 \cdot h(\begin{pmatrix} x_1 \\ y_1 \\ z_1 \end{pmatrix}) + c_2 \cdot h(\begin{pmatrix} x_2 \\ y_2 \\ z_2 \end{pmatrix}) \end{aligned}$$

(b) Yes. The verification is easy.

$$h(c_1\cdot\begin{pmatrix}x_1\\y_1\\z_1\end{pmatrix}+c_2\cdot\begin{pmatrix}x_2\\y_2\\z_2\end{pmatrix})=h(\begin{pmatrix}c_1x_1+c_2x_2\\c_1y_1+c_2y_2\\c_1z_1+c_2z_2\end{pmatrix})$$
$$=\begin{pmatrix}0\\0\end{pmatrix}$$
$$=c_1\cdot h(\begin{pmatrix}x_1\\y_1\\z_1\end{pmatrix})+c_2\cdot h(\begin{pmatrix}x_2\\y_2\\z_2\end{pmatrix})$$

(c) No. An example of an addition that is not respected is this.

$$h(\begin{pmatrix}0\\0\\0\end{pmatrix}+\begin{pmatrix}0\\0\\0\end{pmatrix})=\begin{pmatrix}1\\1\end{pmatrix}\neq h(\begin{pmatrix}0\\0\\0\end{pmatrix})+h(\begin{pmatrix}0\\0\\0\end{pmatrix})$$

(d) Yes. The verification is straightforward.

$$h(c_1\cdot\begin{pmatrix}x_1\\y_1\\z_1\end{pmatrix}+c_2\cdot\begin{pmatrix}x_2\\y_2\\z_2\end{pmatrix})=h(\begin{pmatrix}c_1x_1+c_2x_2\\c_1y_1+c_2y_2\\c_1z_1+c_2z_2\end{pmatrix})$$
$$=\begin{pmatrix}2(c_1x_1+c_2x_2)+(c_1y_1+c_2y_2)\\3(c_1y_1+c_2y_2)-4(c_1z_1+c_2z_2)\end{pmatrix}$$
$$=c_1\cdot\begin{pmatrix}2x_1+y_1\\3y_1-4z_1\end{pmatrix}+c_2\cdot\begin{pmatrix}2x_2+y_2\\3y_2-4z_2\end{pmatrix}$$
$$=c_1\cdot h(\begin{pmatrix}x_1\\y_1\\z_1\end{pmatrix})+c_2\cdot h(\begin{pmatrix}x_2\\y_2\\z_2\end{pmatrix})$$

Three.II.1.18 For each, we must either check that the map preserves linear combinations or give an example of a linear combination that is not.

(a) Yes. The check that it preserves combinations is routine.

$$h(r_1\cdot\begin{pmatrix}a_1&b_1\\c_1&d_1\end{pmatrix}+r_2\cdot\begin{pmatrix}a_2&b_2\\c_2&d_2\end{pmatrix})=h(\begin{pmatrix}r_1a_1+r_2a_2&r_1b_1+r_2b_2\\r_1c_1+r_2c_2&r_1d_1+r_2d_2\end{pmatrix})$$
$$=(r_1a_1+r_2a_2)+(r_1d_1+r_2d_2)$$
$$=r_1(a_1+d_1)+r_2(a_2+d_2)$$
$$=r_1\cdot h(\begin{pmatrix}a_1&b_1\\c_1&d_1\end{pmatrix})+r_2\cdot h(\begin{pmatrix}a_2&b_2\\c_2&d_2\end{pmatrix})$$

(b) No. For instance, not preserved is multiplication by the scalar 2.

$$h(2\cdot\begin{pmatrix}1&0\\0&1\end{pmatrix})=h(\begin{pmatrix}2&0\\0&2\end{pmatrix})=4\quad\text{while}\quad 2\cdot h(\begin{pmatrix}1&0\\0&1\end{pmatrix})=2\cdot 1=2$$

(c) Yes. This is the check that it preserves combinations of two members of the domain.

$$h(r_1\cdot\begin{pmatrix}a_1&b_1\\c_1&d_1\end{pmatrix}+r_2\cdot\begin{pmatrix}a_2&b_2\\c_2&d_2\end{pmatrix})=h(\begin{pmatrix}r_1a_1+r_2a_2&r_1b_1+r_2b_2\\r_1c_1+r_2c_2&r_1d_1+r_2d_2\end{pmatrix})$$
$$=2(r_1a_1+r_2a_2)+3(r_1b_1+r_2b_2)+(r_1c_1+r_2c_2)-(r_1d_1+r_2d_2)$$
$$=r_1(2a_1+3b_1+c_1-d_1)+r_2(2a_2+3b_2+c_2-d_2)$$
$$=r_1\cdot h(\begin{pmatrix}a_1&b_1\\c_1&d_1\end{pmatrix})+r_2\cdot h(\begin{pmatrix}a_2&b_2\\c_2&d_2\end{pmatrix})$$

(d) No. An example of a combination that is not preserved is this.

$$h(\begin{pmatrix}1&0\\0&0\end{pmatrix}+\begin{pmatrix}1&0\\0&0\end{pmatrix})=h(\begin{pmatrix}2&0\\0&0\end{pmatrix})=4\quad\text{while}\quad h(\begin{pmatrix}1&0\\0&0\end{pmatrix})+h(\begin{pmatrix}1&0\\0&0\end{pmatrix})=1+1=2$$

Three.II.1.19 The check that each is a homomorphisms is routine. Here is the check for the differentiation map.

$$\begin{aligned}\frac{d}{dx}&(r\cdot(a_0+a_1x+a_2x^2+a_3x^3)+s\cdot(b_0+b_1x+b_2x^2+b_3x^3))\\&=\frac{d}{dx}((ra_0+sb_0)+(ra_1+sb_1)x+(ra_2+sb_2)x^2+(ra_3+sb_3)x^3)\\&=(ra_1+sb_1)+2(ra_2+sb_2)x+3(ra_3+sb_3)x^2\\&=r\cdot(a_1+2a_2x+3a_3x^2)+s\cdot(b_1+2b_2x+3b_3x^2)\\&=r\cdot\frac{d}{dx}(a_0+a_1x+a_2x^2+a_3x^3)+s\cdot\frac{d}{dx}(b_0+b_1x+b_2x^2+b_3x^3)\end{aligned}$$

(An alternate proof is to simply note that this is a property of differentiation that is familiar from calculus.)

These two maps are not inverses as this composition does not act as the identity map on this element of the domain.

$$1\in\mathcal{P}_3\overset{d/dx}{\longmapsto}0\in\mathcal{P}_2\overset{\int}{\longmapsto}0\in\mathcal{P}_3$$

Three.II.1.20 Each of these projections is a homomorphism. Projection to the xz-plane and to the yz-plane are these maps.

$$\begin{pmatrix}x\\y\\z\end{pmatrix}\mapsto\begin{pmatrix}x\\0\\z\end{pmatrix}\qquad\begin{pmatrix}x\\y\\z\end{pmatrix}\mapsto\begin{pmatrix}0\\y\\z\end{pmatrix}$$

Projection to the x-axis, to the y-axis, and to the z-axis are these maps.

$$\begin{pmatrix}x\\y\\z\end{pmatrix}\mapsto\begin{pmatrix}x\\0\\0\end{pmatrix}\qquad\begin{pmatrix}x\\y\\z\end{pmatrix}\mapsto\begin{pmatrix}0\\y\\0\end{pmatrix}\qquad\begin{pmatrix}x\\y\\z\end{pmatrix}\mapsto\begin{pmatrix}0\\0\\z\end{pmatrix}$$

And projection to the origin is this map.

$$\begin{pmatrix}x\\y\\z\end{pmatrix}\mapsto\begin{pmatrix}0\\0\\0\end{pmatrix}$$

Verification that each is a homomorphism is straightforward. (The last one, of course, is the zero transformation on $\mathbb{R}^3$.)

Three.II.1.21 The first is not onto; for instance, there is no polynomial that is sent the constant polynomial $p(x)=1$. The second is not one-to-one; both of these members of the domain

$$\begin{pmatrix}1&0\\0&0\end{pmatrix}\quad\text{and}\quad\begin{pmatrix}0&0\\0&1\end{pmatrix}$$

map to the same member of the codomain, $1\in\mathbb{R}$.

Three.II.1.22 Yes; in any space $\mathrm{id}(c\cdot\vec{v}+d\cdot\vec{w})=c\cdot\vec{v}+d\cdot\vec{w}=c\cdot\mathrm{id}(\vec{v})+d\cdot\mathrm{id}(\vec{w})$.

Three.II.1.23 **(a)** This map does not preserve structure since $f(1+1)=3$, while $f(1)+f(1)=2$.

(b) The check is routine.

$$\begin{aligned}f(r_1\cdot\begin{pmatrix}x_1\\y_1\end{pmatrix}+r_2\cdot\begin{pmatrix}x_2\\y_2\end{pmatrix})&=f(\begin{pmatrix}r_1x_1+r_2x_2\\r_1y_1+r_2y_2\end{pmatrix})\\&=(r_1x_1+r_2x_2)+2(r_1y_1+r_2y_2)\\&=r_1\cdot(x_1+2y_1)+r_2\cdot(x_2+2y_2)\\&=r_1\cdot f(\begin{pmatrix}x_1\\y_1\end{pmatrix})+r_2\cdot f(\begin{pmatrix}x_2\\y_2\end{pmatrix})\end{aligned}$$

Three.II.1.24 Yes. Where $h\colon V \to W$ is linear, $h(\vec{u}-\vec{v}) = h(\vec{u}+(-1)\cdot\vec{v}) = h(\vec{u})+(-1)\cdot h(\vec{v}) = h(\vec{u})-h(\vec{v})$.

Three.II.1.25 **(a)** Let $\vec{v} \in V$ be represented with respect to the basis as $\vec{v} = c_1\vec{\beta}_1 + \cdots + c_n\vec{\beta}_n$. Then $h(\vec{v}) = h(c_1\vec{\beta}_1 + \cdots + c_n\vec{\beta}_n) = c_1h(\vec{\beta}_1) + \cdots + c_nh(\vec{\beta}_n) = c_1\cdot\vec{0} + \cdots + c_n\cdot\vec{0} = \vec{0}$.

(b) This argument is similar to the prior one. Let $\vec{v} \in V$ be represented with respect to the basis as $\vec{v} = c_1\vec{\beta}_1 + \cdots + c_n\vec{\beta}_n$. Then $h(c_1\vec{\beta}_1 + \cdots + c_n\vec{\beta}_n) = c_1h(\vec{\beta}_1) + \cdots + c_nh(\vec{\beta}_n) = c_1\vec{\beta}_1 + \cdots + c_n\vec{\beta}_n = \vec{v}$.

(c) As above, only $c_1h(\vec{\beta}_1) + \cdots + c_nh(\vec{\beta}_n) = c_1r\vec{\beta}_1 + \cdots + c_nr\vec{\beta}_n = r(c_1\vec{\beta}_1 + \cdots + c_n\vec{\beta}_n) = r\vec{v}$.

Three.II.1.26 That it is a homomorphism follows from the familiar rules that the logarithm of a product is the sum of the logarithms $\ln(ab) = \ln(a) + \ln(b)$ and that the logarithm of a power is the multiple of the logarithm $\ln(a^r) = r\ln(a)$. This map is an isomorphism because it has an inverse, namely, the exponential map, so it is a correspondence, and therefore it is an isomorphism.

Three.II.1.27 Where $\hat{x} = x/2$ and $\hat{y} = y/3$, the image set is

$$\{\begin{pmatrix}\hat{x}\\ \hat{y}\end{pmatrix} \mid \frac{(2\hat{x})^2}{4} + \frac{(3\hat{y})^2}{9} = 1\} = \{\begin{pmatrix}\hat{x}\\ \hat{y}\end{pmatrix} \mid \hat{x}^2 + \hat{y}^2 = 1\}$$

the unit circle in the $\hat{x}\hat{y}$-plane.

Three.II.1.28 The circumference function $r \mapsto 2\pi r$ is linear. Thus we have $2\pi\cdot(r_{\text{earth}}+6) - 2\pi\cdot(r_{\text{earth}}) = 12\pi$. Observe that it takes the same amount of extra rope to raise the circle from tightly wound around a basketball to six feet above that basketball as it does to raise it from tightly wound around the earth to six feet above the earth.

Three.II.1.29 Verifying that it is linear is routine.

$$\begin{aligned}
h(c_1\cdot\begin{pmatrix}x_1\\ y_1\\ z_1\end{pmatrix} + c_2\cdot\begin{pmatrix}x_2\\ y_2\\ z_2\end{pmatrix}) &= h(\begin{pmatrix}c_1x_1+c_2x_2\\ c_1y_1+c_2y_2\\ c_1z_1+c_2z_2\end{pmatrix}) \\
&= 3(c_1x_1+c_2x_2) - (c_1y_1+c_2y_2) - (c_1z_1+c_2z_2) \\
&= c_1\cdot(3x_1-y_1-z_1) + c_2\cdot(3x_2-y_2-z_2) \\
&= c_1\cdot h(\begin{pmatrix}x_1\\ y_1\\ z_1\end{pmatrix}) + c_2\cdot h(\begin{pmatrix}x_2\\ y_2\\ z_2\end{pmatrix})
\end{aligned}$$

The natural guess at a generalization is that for any fixed $\vec{k} \in \mathbb{R}^3$ the map $\vec{v} \mapsto \vec{v}\cdot\vec{k}$ is linear. This statement is true. It follows from properties of the dot product we have seen earlier: $(\vec{v}+\vec{u})\cdot\vec{k} = \vec{v}\cdot\vec{k} + \vec{u}\cdot\vec{k}$ and $(r\vec{v})\cdot\vec{k} = r(\vec{v}\cdot\vec{k})$. (The natural guess at a generalization of this generalization, that the map from $\mathbb{R}^n$ to $\mathbb{R}$ whose action consists of taking the dot product of its argument with a fixed vector $\vec{k} \in \mathbb{R}^n$ is linear, is also true.)

Three.II.1.30 Let $h\colon \mathbb{R}^1 \to \mathbb{R}^1$ be linear. A linear map is determined by its action on a basis, so fix the basis $\langle 1\rangle$ for $\mathbb{R}^1$. For any $r \in \mathbb{R}^1$ we have that $h(r) = h(r\cdot 1) = r\cdot h(1)$ and so h acts on any argument r by multiplying it by the constant $h(1)$. If $h(1)$ is not zero then the map is a correspondence — its inverse is division by $h(1)$ — so any nontrivial transformation of $\mathbb{R}^1$ is an isomorphism.

This projection map is an example that shows that not every transformation of $\mathbb{R}^n$ acts via multiplication by a constant when $n > 1$, including when $n = 2$.

$$\begin{pmatrix}x_1\\ x_2\\ \vdots\\ x_n\end{pmatrix} \mapsto \begin{pmatrix}x_1\\ 0\\ \vdots\\ 0\end{pmatrix}$$

Three.II.1.31 **(a)** Where c and d are scalars, we have this.

$$\begin{aligned}
h(c\cdot\begin{pmatrix}x_1\\ \vdots\\ x_n\end{pmatrix}+d\cdot\begin{pmatrix}y_1\\ \vdots\\ y_n\end{pmatrix}) &= h(\begin{pmatrix}cx_1+dy_1\\ \vdots\\ cx_n+dy_n\end{pmatrix}) \\
&= \begin{pmatrix}a_{1,1}(cx_1+dy_1)+\cdots+a_{1,n}(cx_n+dy_n)\\ \vdots\\ a_{m,1}(cx_1+dy_1)+\cdots+a_{m,n}(cx_n+dy_n)\end{pmatrix} \\
&= c\cdot\begin{pmatrix}a_{1,1}x_1+\cdots+a_{1,n}x_n\\ \vdots\\ a_{m,1}x_1+\cdots+a_{m,n}x_n\end{pmatrix}+d\cdot\begin{pmatrix}a_{1,1}y_1+\cdots+a_{1,n}y_n\\ \vdots\\ a_{m,1}y_1+\cdots+a_{m,n}y_n\end{pmatrix} \\
&= c\cdot h(\begin{pmatrix}x_1\\ \vdots\\ x_n\end{pmatrix})+d\cdot h(\begin{pmatrix}y_1\\ \vdots\\ y_n\end{pmatrix})
\end{aligned}$$

(b) Each power i of the derivative operator is linear because of these rules familiar from calculus.

$$\frac{d^i}{dx^i}(\,f(x)+g(x)\,)=\frac{d^i}{dx^i}f(x)+\frac{d^i}{dx^i}g(x)\quad\text{and}\quad\frac{d^i}{dx^i}r\cdot f(x)=r\cdot\frac{d^i}{dx^i}f(x)$$

Thus the given map is a linear transformation of $\mathcal{P}_n$ because any linear combination of linear maps is also a linear map.

Three.II.1.32 (This argument has already appeared, as part of the proof that isomorphism is an equivalence.) Let $f\colon U\to V$ and $g\colon V\to W$ be linear. The composition preserves linear combinations

$$\begin{aligned}
g\circ f(c_1\vec{u}_1+c_2\vec{u}_2) &= g(\,f(c_1\vec{u}_1+c_2\vec{u}_2)\,)=g(\,c_1f(\vec{u}_1)+c_2f(\vec{u}_2)\,) \\
&= c_1\cdot g(f(\vec{u}_1))+c_2\cdot g(f(\vec{u}_2))=c_1\cdot g\circ f(\vec{u}_1)+c_2\cdot g\circ f(\vec{u}_2)
\end{aligned}$$

where $\vec{u}_1,\vec{u}_2\in U$ and scalars c_1,c_2

Three.II.1.33 **(a)** Yes. The set of $\vec{w}$'s cannot be linearly independent if the set of $\vec{v}$'s is linearly dependent because any nontrivial relationship in the domain $\vec{0}_V=c_1\vec{v}_1+\cdots+c_n\vec{v}_n$ would give a nontrivial relationship in the range $f(\vec{0}_V)=\vec{0}_W=f(c_1\vec{v}_1+\cdots+c_n\vec{v}_n)=c_1f(\vec{v}_1)+\cdots+c_nf(\vec{v}_n)=c_1\vec{w}+\cdots+c_n\vec{w}_n$.

(b) Not necessarily. For instance, the transformation of $\mathbb{R}^2$ given by

$$\begin{pmatrix}x\\ y\end{pmatrix}\mapsto\begin{pmatrix}x+y\\ x+y\end{pmatrix}$$

sends this linearly independent set in the domain to a linearly dependent image.

$$\{\vec{v}_1,\vec{v}_2\}=\{\begin{pmatrix}1\\ 0\end{pmatrix},\begin{pmatrix}1\\ 1\end{pmatrix}\}\ \mapsto\ \{\begin{pmatrix}1\\ 1\end{pmatrix},\begin{pmatrix}2\\ 2\end{pmatrix}\}=\{\vec{w}_1,\vec{w}_2\}$$

(c) Not necessarily. An example is the projection map $\pi\colon\mathbb{R}^3\to\mathbb{R}^2$

$$\begin{pmatrix}x\\ y\\ z\end{pmatrix}\mapsto\begin{pmatrix}x\\ y\end{pmatrix}$$

and this set that does not span the domain but maps to a set that does span the codomain.

$$\{\begin{pmatrix}1\\ 0\\ 0\end{pmatrix},\begin{pmatrix}0\\ 1\\ 0\end{pmatrix}\}\stackrel{\pi}{\longmapsto}\{\begin{pmatrix}1\\ 0\end{pmatrix},\begin{pmatrix}0\\ 1\end{pmatrix}\}$$

(d) Not necessarily. For instance, the injection map $\iota\colon\mathbb{R}^2\to\mathbb{R}^3$ sends the standard basis $\mathcal{E}_2$ for the domain to a set that does not span the codomain. (*Remark.* However, the set of $\vec{w}$'s does span the range. A proof is easy.)

Three.II.1.34 Recall that the entry in row i and column j of the transpose of M is the entry $m_{j,i}$ from row j and column i of M. Now, the check is routine.

$$\begin{aligned}
[r\cdot\begin{pmatrix} & \vdots & \\ \cdots & a_{i,j} & \cdots \\ & \vdots & \end{pmatrix}+s\cdot\begin{pmatrix} & \vdots & \\ \cdots & b_{i,j} & \cdots \\ & \vdots & \end{pmatrix}]^{T} &= \begin{pmatrix} & \vdots & \\ \cdots & ra_{i,j}+sb_{i,j} & \cdots \\ & \vdots & \end{pmatrix}^{T} \\
&= \begin{pmatrix} & \vdots & \\ \cdots & ra_{j,i}+sb_{j,i} & \cdots \\ & \vdots & \end{pmatrix} \\
&= r\cdot\begin{pmatrix} & \vdots & \\ \cdots & a_{j,i} & \cdots \\ & \vdots & \end{pmatrix}+s\cdot\begin{pmatrix} & \vdots & \\ \cdots & b_{j,i} & \cdots \\ & \vdots & \end{pmatrix} \\
&= r\cdot\begin{pmatrix} & \vdots & \\ \cdots & a_{j,i} & \cdots \\ & \vdots & \end{pmatrix}^{T}+s\cdot\begin{pmatrix} & \vdots & \\ \cdots & b_{j,i} & \cdots \\ & \vdots & \end{pmatrix}^{T}
\end{aligned}$$

The domain is $\mathcal{M}_{m\times n}$ while the codomain is $\mathcal{M}_{n\times m}$.

Three.II.1.35 **(a)** For any homomorphism $h\colon \mathbb{R}^n \to \mathbb{R}^m$ we have

$$h(\ell)=\{h(t\cdot\vec{u}+(1-t)\cdot\vec{v}) \mid t\in[0..1]\}=\{t\cdot h(\vec{u})+(1-t)\cdot h(\vec{v}) \mid t\in[0..1]\}$$

which is the line segment from $h(\vec{u})$ to $h(\vec{v})$.

(b) We must show that if a subset of the domain is convex then its image, as a subset of the range, is also convex. Suppose that $C\subseteq\mathbb{R}^n$ is convex and consider its image $h(C)$. To show $h(C)$ is convex we must show that for any two of its members, $\vec{d}_1$ and $\vec{d}_2$, the line segment connecting them

$$\ell=\{t\cdot\vec{d}_1+(1-t)\cdot\vec{d}_2 \mid t\in[0..1]\}$$

is a subset of $h(C)$.

Fix any member $\hat{t}\cdot\vec{d}_1+(1-\hat{t})\cdot\vec{d}_2$ of that line segment. Because the endpoints of ℓ are in the image of C, there are members of C that map to them, say $h(\vec{c}_1)=\vec{d}_1$ and $h(\vec{c}_2)=\vec{d}_2$. Now, where $\hat{t}$ is the scalar that we fixed in the first sentence of this paragraph, observe that $h(\hat{t}\cdot\vec{c}_1+(1-\hat{t})\cdot\vec{c}_2)=\hat{t}\cdot h(\vec{c}_1)+(1-\hat{t})\cdot h(\vec{c}_2)=\hat{t}\cdot\vec{d}_1+(1-\hat{t})\cdot\vec{d}_2$ Thus, any member of ℓ is a member of $h(C)$, and so $h(C)$ is convex.

Three.II.1.36 **(a)** For $\vec{v}_0,\vec{v}_1\in\mathbb{R}^n$, the line through $\vec{v}_0$ with direction $\vec{v}_1$ is the set $\{\vec{v}_0+t\cdot\vec{v}_1 \mid t\in\mathbb{R}\}$. The image under h of that line $\{h(\vec{v}_0+t\cdot\vec{v}_1) \mid t\in\mathbb{R}\}=\{h(\vec{v}_0)+t\cdot h(\vec{v}_1) \mid t\in\mathbb{R}\}$ is the line through $h(\vec{v}_0)$ with direction $h(\vec{v}_1)$. If $h(\vec{v}_1)$ is the zero vector then this line is degenerate.

(b) A k-dimensional linear surface in $\mathbb{R}^n$ maps to a (possibly degenerate) k-dimensional linear surface in $\mathbb{R}^m$. The proof is just like that the one for the line.

Three.II.1.37 Suppose that $h\colon V\to W$ is a homomorphism and suppose that S is a subspace of V. Consider the map $\hat{h}\colon S\to W$ defined by $\hat{h}(\vec{s})=h(\vec{s})$. (The only difference between $\hat{h}$ and h is the difference in domain.) Then this new map is linear: $\hat{h}(c_1\cdot\vec{s}_1+c_2\cdot\vec{s}_2)=h(c_1\vec{s}_1+c_2\vec{s}_2)=c_1h(\vec{s}_1)+c_2h(\vec{s}_2)=c_1\cdot\hat{h}(\vec{s}_1)+c_2\cdot\hat{h}(\vec{s}_2)$.

Three.II.1.38 This will appear as a lemma in the next subsection.

(a) The range is nonempty because V is nonempty. To finish we need to show that it is closed under combinations. A combination of range vectors has the form, where $\vec{v}_1,\ldots,\vec{v}_n\in V$,

$$c_1\cdot h(\vec{v}_1)+\cdots+c_n\cdot h(\vec{v}_n)=h(c_1\vec{v}_1)+\cdots+h(c_n\vec{v}_n)=h(c_1\cdot\vec{v}_1+\cdots+c_n\cdot\vec{v}_n),$$

which is itself in the range as $c_1\cdot\vec{v}_1+\cdots+c_n\cdot\vec{v}_n$ is a member of domain V. Therefore the range is a subspace.

(b) The null space is nonempty since it contains $\vec{0}_V$, as $\vec{0}_V$ maps to $\vec{0}_W$. It is closed under linear combinations because, where $\vec{v}_1,\ldots,\vec{v}_n\in V$ are elements of the inverse image $\{\vec{v}\in V \mid h(\vec{v})=\vec{0}_W\}$, for $c_1,\ldots,c_n\in\mathbb{R}$

$$\vec{0}_W = c_1\cdot h(\vec{v}_1)+\cdots+c_n\cdot h(\vec{v}_n) = h(c_1\cdot\vec{v}_1+\cdots+c_n\cdot\vec{v}_n)$$

and so $c_1\cdot\vec{v}_1+\cdots+c_n\cdot\vec{v}_n$ is also in the inverse image of $\vec{0}_W$.

(c) This image of U nonempty because U is nonempty. For closure under combinations, where $\vec{u}_1,\ldots,\vec{u}_n\in U$,

$$c_1\cdot h(\vec{u}_1)+\cdots+c_n\cdot h(\vec{u}_n) = h(c_1\cdot\vec{u}_1)+\cdots+h(c_n\cdot\vec{u}_n) = h(c_1\cdot\vec{u}_1+\cdots+c_n\cdot\vec{u}_n)$$

which is itself in $h(U)$ as $c_1\cdot\vec{u}_1+\cdots+c_n\cdot\vec{u}_n$ is in U. Thus this set is a subspace.

(d) The natural generalization is that the inverse image of a subspace of is a subspace.

Suppose that X is a subspace of W. Note that $\vec{0}_W\in X$ so the set $\{\vec{v}\in V \mid h(\vec{v})\in X\}$ is not empty. To show that this set is closed under combinations, let $\vec{v}_1,\ldots,\vec{v}_n$ be elements of V such that $h(\vec{v}_1)=\vec{x}_1$, …, $h(\vec{v}_n)=\vec{x}_n$ and note that

$$h(c_1\cdot\vec{v}_1+\cdots+c_n\cdot\vec{v}_n) = c_1\cdot h(\vec{v}_1)+\cdots+c_n\cdot h(\vec{v}_n) = c_1\cdot\vec{x}_1+\cdots+c_n\cdot\vec{x}_n$$

so a linear combination of elements of $h^{-1}(X)$ is also in $h^{-1}(X)$.

Three.II.1.39 No; the set of isomorphisms does not contain the zero map (unless the space is trivial).

Three.II.1.40 If $\langle\vec{\beta}_1,\ldots,\vec{\beta}_n\rangle$ doesn't span the space then the map needn't be unique. For instance, if we try to define a map from $\mathbb{R}^2$ to itself by specifying only that $\vec{e}_1$ maps to itself, then there is more than one homomorphism possible; both the identity map and the projection map onto the first component fit this condition.

If we drop the condition that $\langle\vec{\beta}_1,\ldots,\vec{\beta}_n\rangle$ is linearly independent then we risk an inconsistent specification (i.e, there could be no such map). An example is if we consider $\langle\vec{e}_2,\vec{e}_1,2\vec{e}_1\rangle$, and try to define a map from $\mathbb{R}^2$ to itself that sends $\vec{e}_2$ to itself, and sends both $\vec{e}_1$ and $2\vec{e}_1$ to $\vec{e}_1$. No homomorphism can satisfy these three conditions.

Three.II.1.41 **(a)** Briefly, the check of linearity is this.

$$F(r_1\cdot\vec{v}_1+r_2\cdot\vec{v}_2) = \begin{pmatrix} f_1(r_1\vec{v}_1+r_2\vec{v}_2) \\ f_2(r_1\vec{v}_1+r_2\vec{v}_2) \end{pmatrix} = r_1\begin{pmatrix} f_1(\vec{v}_1) \\ f_2(\vec{v}_1) \end{pmatrix} + r_2\begin{pmatrix} f_1(\vec{v}_2) \\ f_2(\vec{v}_2) \end{pmatrix} = r_1\cdot F(\vec{v}_1)+r_2\cdot F(\vec{v}_2)$$

(b) Yes. Let $\pi_1\colon\mathbb{R}^2\to\mathbb{R}^1$ and $\pi_2\colon\mathbb{R}^2\to\mathbb{R}^1$ be the projections

$$\begin{pmatrix} x \\ y \end{pmatrix} \stackrel{\pi_1}{\longmapsto} x \quad\text{and}\quad \begin{pmatrix} x \\ y \end{pmatrix} \stackrel{\pi_2}{\longmapsto} y$$

onto the two axes. Now, where $f_1(\vec{v})=\pi_1(F(\vec{v}))$ and $f_2(\vec{v})=\pi_2(F(\vec{v}))$ we have the desired component functions.

$$F(\vec{v}) = \begin{pmatrix} f_1(\vec{v}) \\ f_2(\vec{v}) \end{pmatrix}$$

They are linear because they are the composition of linear functions, and the fact that the composition of linear functions is linear was part of the proof that isomorphism is an equivalence relation (alternatively, the check that they are linear is straightforward).

(c) In general, a map from a vector space V to an $\mathbb{R}^n$ is linear if and only if each of the component functions is linear. The verification is as in the prior item.

Three.II.2: Range space and Null space

Three.II.2.22 First, to answer whether a polynomial is in the null space, we have to consider it as a member of the domain $\mathcal{P}_3$. To answer whether it is in the range space, we consider it as a member of the codomain

$\mathcal{P}_4$. That is, for $p(x) = x^4$, the question of whether it is in the range space is sensible but the question of whether it is in the null space is not because it is not even in the domain.

(a) The polynomial $x^3 \in \mathcal{P}_3$ is not in the null space because $h(x^3) = x^4$ is not the zero polynomial in $\mathcal{P}_4$. The polynomial $x^3 \in \mathcal{P}_4$ is in the range space because $x^2 \in \mathcal{P}_3$ is mapped by h to x^3.

(b) The answer to both questions is, "Yes, because $h(0) = 0$." The polynomial $0 \in \mathcal{P}_3$ is in the null space because it is mapped by h to the zero polynomial in $\mathcal{P}_4$. The polynomial $0 \in \mathcal{P}_4$ is in the range space because it is the image, under h, of $0 \in \mathcal{P}_3$.

(c) The polynomial $7 \in \mathcal{P}_3$ is not in the null space because $h(7) = 7x$ is not the zero polynomial in $\mathcal{P}_4$. The polynomial $7 \in \mathcal{P}_4$ is not in the range space because there is no member of the domain that when multiplied by x gives the constant polynomial $p(x) = 7$.

(d) The polynomial $12x - 0.5x^3 \in \mathcal{P}_3$ is not in the null space because $h(12x - 0.5x^3) = 12x^2 - 0.5x^4$. The polynomial $12x - 0.5x^3 \in \mathcal{P}_4$ is in the range space because it is the image of $12 - 0.5x^2$.

(e) The polynomial $1 + 3x^2 - x^3 \in \mathcal{P}_3$ is not in the null space because $h(1 + 3x^2 - x^3) = x + 3x^3 - x^4$. The polynomial $1 + 3x^2 - x^3 \in \mathcal{P}_4$ is not in the range space because of the constant term.

Three.II.2.23 **(a)** The null space is

$$\mathcal{N}(h) = \{\begin{pmatrix} a \\ b \end{pmatrix} \in \mathbb{R}^2 \mid a + ax + ax^2 + 0x^3 = 0 + 0x + 0x^2 + 0x^3\} = \{\begin{pmatrix} 0 \\ b \end{pmatrix} \mid b \in \mathbb{R}\}$$

while the range space is

$$\mathcal{R}(h) = \{a + ax + ax^2 \in \mathcal{P}_3 \mid a, b \in \mathbb{R}\} = \{a \cdot (1 + x + x^2) \mid a \in \mathbb{R}\}$$

and so the nullity is one and the rank is one.

(b) The null space is this.

$$\mathcal{N}(h) = \{\begin{pmatrix} a & b \\ c & d \end{pmatrix} \mid a + d = 0\} = \{\begin{pmatrix} -d & b \\ c & d \end{pmatrix} \mid b, c, d \in \mathbb{R}\}$$

The range space

$$\mathcal{R}(h) = \{a + d \mid a, b, c, d \in \mathbb{R}\}$$

is all of $\mathbb{R}$ (we can get any real number by taking d to be 0 and taking a to be the desired number). Thus, the nullity is three and the rank is one.

(c) The null space is

$$\mathcal{N}(h) = \{\begin{pmatrix} a & b \\ c & d \end{pmatrix} \mid a + b + c = 0 \text{ and } d = 0\} = \{\begin{pmatrix} -b - c & b \\ c & 0 \end{pmatrix} \mid b, c \in \mathbb{R}\}$$

while the range space is $\mathcal{R}(h) = \{r + sx^2 \mid r, s \in \mathbb{R}\}$. Thus, the nullity is two and the rank is two.

(d) The null space is all of $\mathbb{R}^3$ so the nullity is three. The range space is the trivial subspace of $\mathbb{R}^4$ so the rank is zero.

Three.II.2.24 For each, use the result that the rank plus the nullity equals the dimension of the domain.

(a) 0 **(b)** 3 **(c)** 3 **(d)** 0

Three.II.2.25 Because

$$\frac{d}{dx}(a_0 + a_1x + \cdots + a_nx^n) = a_1 + 2a_2x + 3a_3x^2 + \cdots + na_nx^{n-1}$$

we have this.

$$\begin{aligned}\mathcal{N}(\frac{d}{dx}) &= \{a_0 + \cdots + a_nx^n \mid a_1 + 2a_2x + \cdots + na_nx^{n-1} = 0 + 0x + \cdots + 0x^{n-1}\} \\ &= \{a_0 + \cdots + a_nx^n \mid a_1 = 0, \text{ and } a_2 = 0, \ldots, a_n = 0\} \\ &= \{a_0 + 0x + 0x^2 + \cdots + 0x^n \mid a_0 \in \mathbb{R}\}\end{aligned}$$

In the same way,

$$\mathcal{N}(\frac{d^k}{dx^k}) = \{a_0 + a_1x + \cdots + a_nx^n \mid a_0, \ldots, a_{k-1} \in \mathbb{R}\}$$

for $k \leqslant n$.

Three.II.2.26 The shadow of a scalar multiple is the scalar multiple of the shadow.

Three.II.2.27 (a) Setting $a_0 + (a_0 + a_1)x + (a_2 + a_3)x^3 = 0 + 0x + 0x^2 + 0x^3$ gives $a_0 = 0$ and $a_0 + a_1 = 0$ and $a_2 + a_3 = 0$, so the null space is $\{-a_3x^2 + a_3x^3 \mid a_3 \in \mathbb{R}\}$.

(b) Setting $a_0 + (a_0 + a_1)x + (a_2 + a_3)x^3 = 2 + 0x + 0x^2 - x^3$ gives that $a_0 = 2$, and $a_1 = -2$, and $a_2 + a_3 = -1$. Taking a_3 as a parameter, and renaming it $a_3 = a$ gives this set description $\{2 - 2x + (-1 - a)x^2 + ax^3 \mid a \in \mathbb{R}\} = \{(2 - 2x - x^2) + a \cdot (-x^2 + x^3) \mid a \in \mathbb{R}\}$.

(c) This set is empty because the range of h includes only those polynomials with a $0x^2$ term.

Three.II.2.28 All inverse images are lines with slope -2.

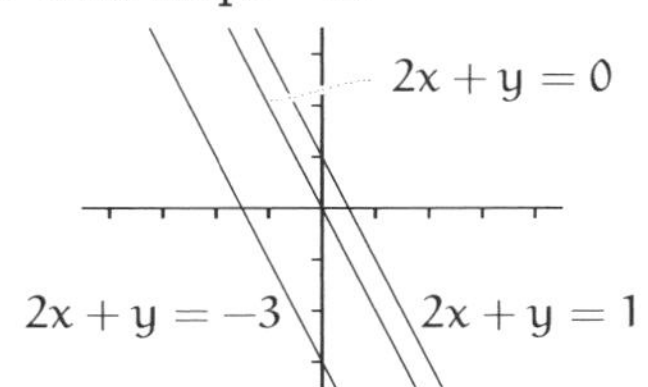

Three.II.2.29 These are the inverses.

(a) $a_0 + a_1x + a_2x^2 + a_3x^3 \mapsto a_0 + a_1x + (a_2/2)x^2 + (a_3/3)x^3$

(b) $a_0 + a_1x + a_2x^2 + a_3x^3 \mapsto a_0 + a_2x + a_1x^2 + a_3x^3$

(c) $a_0 + a_1x + a_2x^2 + a_3x^3 \mapsto a_3 + a_0x + a_1x^2 + a_2x^3$

(d) $a_0 + a_1x + a_2x^2 + a_3x^3 \mapsto a_0 + (a_1 - a_0)x + (a_2 - a_1)x^2 + (a_3 - a_2)x^3$

For instance, for the second one, the map given in the question sends $0 + 1x + 2x^2 + 3x^3 \mapsto 0 + 2x + 1x^2 + 3x^3$ and then the inverse above sends $0 + 2x + 1x^2 + 3x^3 \mapsto 0 + 1x + 2x^2 + 3x^3$. So this map is actually self-inverse.

Three.II.2.30 For any vector space V, the null space

$$\{\vec{v} \in V \mid 2\vec{v} = \vec{0}\}$$

is trivial, while the range space

$$\{\vec{w} \in V \mid \vec{w} = 2\vec{v} \text{ for some } \vec{v} \in V\}$$

is all of V, because every vector $\vec{w}$ is twice some other vector, specifically, it is twice $(1/2)\vec{w}$. (Thus, this transformation is actually an automorphism.)

Three.II.2.31 Because the rank plus the nullity equals the dimension of the domain (here, five), and the rank is at most three, the possible pairs are: $(3,2)$, $(2,3)$, $(1,4)$, and $(0,5)$. Coming up with linear maps that show that each pair is indeed possible is easy.

Three.II.2.32 No (unless $\mathcal{P}_n$ is trivial), because the two polynomials $f_0(x) = 0$ and $f_1(x) = 1$ have the same derivative; a map must be one-to-one to have an inverse.

Three.II.2.33 The null space is this.

$$\{a_0 + a_1x + \cdots + a_nx^n \mid a_0(1) + \frac{a_1}{2}(1^2) + \cdots + \frac{a_n}{n+1}(1^{n+1}) = 0\}$$
$$= \{a_0 + a_1x + \cdots + a_nx^n \mid a_0 + (a_1/2) + \cdots + (a_{n+1}/n+1) = 0\}$$

Thus the nullity is n.

Three.II.2.34 (a) One direction is obvious: if the homomorphism is onto then its range is the codomain and so its rank equals the dimension of its codomain. For the other direction assume that the map's rank equals the dimension of the codomain. Then the map's range is a subspace of the codomain, and has dimension equal to the dimension of the codomain. Therefore, the map's range must equal the codomain, and the map is onto. (The 'therefore' is because there is a linearly independent subset of the range that is of size equal to the dimension of the codomain, but any such linearly independent subset of the codomain must be a basis for the codomain, and so the range equals the codomain.)

(b) By Theorem 2.21, a homomorphism is one-to-one if and only if its nullity is zero. Because rank plus nullity equals the dimension of the domain, it follows that a homomorphism is one-to-one if and only if its rank equals the dimension of its domain. But this domain and codomain have the same dimension, so the map is one-to-one if and only if it is onto.

Three.II.2.35 We are proving that $h\colon V\to W$ is one-to-one if and only if for every linearly independent subset S of V the subset $h(S)=\{h(\vec{s})\mid \vec{s}\in S\}$ of W is linearly independent.

One half is easy — by Theorem 2.21, if h is not one-to-one then its null space is nontrivial, that is, it contains more than just the zero vector. So where $\vec{v}\neq\vec{0}_V$ is in that null space, the singleton set $\{\vec{v}\}$ is independent while its image $\{h(\vec{v})\}=\{\vec{0}_W\}$ is not.

For the other half, assume that h is one-to-one and so by Theorem 2.21 has a trivial null space. Then for any $\vec{v}_1,\dots,\vec{v}_n\in V$, the relation

$$\vec{0}_W=c_1\cdot h(\vec{v}_1)+\dots+c_n\cdot h(\vec{v}_n)=h(c_1\cdot\vec{v}_1+\dots+c_n\cdot\vec{v}_n)$$

implies the relation $c_1\cdot\vec{v}_1+\dots+c_n\cdot\vec{v}_n=\vec{0}_V$. Hence, if a subset of V is independent then so is its image in W.

Remark. The statement is that a linear map is one-to-one if and only if it preserves independence for *all* sets (that is, if a set is independent then its image is also independent). A map that is not one-to-one may well preserve some independent sets. One example is this map from $\mathbb{R}^3$ to $\mathbb{R}^2$.

$$\begin{pmatrix}x\\y\\z\end{pmatrix}\mapsto\begin{pmatrix}x+y+z\\0\end{pmatrix}$$

Linear independence is preserved for this set

$$\{\begin{pmatrix}1\\0\\0\end{pmatrix}\}\mapsto\{\begin{pmatrix}1\\0\end{pmatrix}\}$$

and (in a somewhat more tricky example) also for this set

$$\{\begin{pmatrix}1\\0\\0\end{pmatrix},\begin{pmatrix}0\\1\\0\end{pmatrix}\}\mapsto\{\begin{pmatrix}1\\0\end{pmatrix}\}$$

(recall that in a set, repeated elements do not appear twice). However, there are sets whose independence is not preserved under this map

$$\{\begin{pmatrix}1\\0\\0\end{pmatrix},\begin{pmatrix}0\\2\\0\end{pmatrix}\}\mapsto\{\begin{pmatrix}1\\0\end{pmatrix},\begin{pmatrix}2\\0\end{pmatrix}\}$$

and so not all sets have independence preserved.

Three.II.2.36 (We use the notation from Theorem 1.9.) Fix a basis $\langle\vec{\beta}_1,\dots,\vec{\beta}_n\rangle$ for V and a basis $\langle\vec{w}_1,\dots,\vec{w}_k\rangle$ for W. If the dimension k of W is less than or equal to the dimension n of V then the theorem gives a linear map from V to W determined in this way.

$$\vec{\beta}_1\mapsto\vec{w}_1,\ \dots,\ \vec{\beta}_k\mapsto\vec{w}_k\quad\text{and}\quad\vec{\beta}_{k+1}\mapsto\vec{w}_k,\ \dots,\ \vec{\beta}_n\mapsto\vec{w}_k$$

We need only to verify that this map is onto.

We can write any member of W as a linear combination of basis elements $c_1\cdot\vec{w}_1+\dots+c_k\cdot\vec{w}_k$. This vector is the image, under the map described above, of $c_1\cdot\vec{\beta}_1+\dots+c_k\cdot\vec{\beta}_k+0\cdot\vec{\beta}_{k+1}\dots+0\cdot\vec{\beta}_n$. Thus the map is onto.

Three.II.2.37 Yes. For the transformation of $\mathbb{R}^2$ given by

$$\begin{pmatrix}x\\y\end{pmatrix}\overset{h}{\longmapsto}\begin{pmatrix}0\\x\end{pmatrix}$$

we have this.

$$\mathscr{N}(h)=\{\begin{pmatrix}0\\y\end{pmatrix}\mid y\in\mathbb{R}\}=\mathscr{R}(h)$$

Remark. We will see more of this in the fifth chapter.

Three.II.2.38 This is a simple calculation.

$$\begin{aligned}h([S]) &= \{h(c_1\vec{s}_1 + \cdots + c_n\vec{s}_n) \mid c_1, \ldots, c_n \in \mathbb{R} \text{ and } \vec{s}_1, \ldots, \vec{s}_n \in S\} \\ &= \{c_1 h(\vec{s}_1) + \cdots + c_n h(\vec{s}_n) \mid c_1, \ldots, c_n \in \mathbb{R} \text{ and } \vec{s}_1, \ldots, \vec{s}_n \in S\} \\ &= [h(S)]\end{aligned}$$

Three.II.2.39 (a) We will show that the two sets are equal $h^{-1}(\vec{w}) = \{\vec{v}+\vec{n} \mid \vec{n} \in \mathscr{N}(h)\}$ by mutual inclusion. For the $\{\vec{v}+\vec{n} \mid \vec{n} \in \mathscr{N}(h)\} \subseteq h^{-1}(\vec{w})$ direction, just note that $h(\vec{v}+\vec{n}) = h(\vec{v}) + h(\vec{n})$ equals $\vec{w}$, and so any member of the first set is a member of the second. For the $h^{-1}(\vec{w}) \subseteq \{\vec{v}+\vec{n} \mid \vec{n} \in \mathscr{N}(h)\}$ direction, consider $\vec{u} \in h^{-1}(\vec{w})$. Because h is linear, $h(\vec{u}) = h(\vec{v})$ implies that $h(\vec{u}-\vec{v}) = \vec{0}$. We can write $\vec{u}-\vec{v}$ as $\vec{n}$, and then we have that $\vec{u} \in \{\vec{v}+\vec{n} \mid \vec{n} \in \mathscr{N}(h)\}$, as desired, because $\vec{u} = \vec{v} + (\vec{u}-\vec{v})$.

(b) This check is routine.

(c) This is immediate.

(d) For the linearity check, briefly, where c, d are scalars and $\vec{x}, \vec{y} \in \mathbb{R}^n$ have components $x_1, \ldots, x_n$ and $y_1, \ldots, y_n$, we have this.

$$\begin{aligned}h(c \cdot \vec{x} + d \cdot \vec{y}) &= \begin{pmatrix} a_{1,1}(cx_1 + dy_1) + \cdots + a_{1,n}(cx_n + dy_n) \\ \vdots \\ a_{m,1}(cx_1 + dy_1) + \cdots + a_{m,n}(cx_n + dy_n) \end{pmatrix} \\ &= \begin{pmatrix} a_{1,1}cx_1 + \cdots + a_{1,n}cx_n \\ \vdots \\ a_{m,1}cx_1 + \cdots + a_{m,n}cx_n \end{pmatrix} + \begin{pmatrix} a_{1,1}dy_1 + \cdots + a_{1,n}dy_n \\ \vdots \\ a_{m,1}dy_1 + \cdots + a_{m,n}dy_n \end{pmatrix} \\ &= c \cdot h(\vec{x}) + d \cdot h(\vec{y})\end{aligned}$$

The appropriate conclusion is that General = Particular + Homogeneous.

(e) Each power of the derivative is linear because of the rules

$$\frac{d^k}{dx^k}(f(x) + g(x)) = \frac{d^k}{dx^k}f(x) + \frac{d^k}{dx^k}g(x) \quad \text{and} \quad \frac{d^k}{dx^k}rf(x) = r\frac{d^k}{dx^k}f(x)$$

from calculus. Thus the given map is a linear transformation of the space because any linear combination of linear maps is also a linear map by Lemma 1.16. The appropriate conclusion is General = Particular + Homogeneous, where the associated homogeneous differential equation has a constant of 0.

Three.II.2.40 Because the rank of t is one, the range space of t is a one-dimensional set. Taking $\langle h(\vec{v})\rangle$ as a basis (for some appropriate $\vec{v}$), we have that for every $\vec{w} \in V$, the image $h(\vec{w}) \in V$ is a multiple of this basis vector — associated with each $\vec{w}$ there is a scalar $c_{\vec{w}}$ such that $t(\vec{w}) = c_{\vec{w}} t(\vec{v})$. Apply t to both sides of that equation and take r to be $c_{t(\vec{v})}$

$$t \circ t(\vec{w}) = t(c_{\vec{w}} \cdot t(\vec{v})) = c_{\vec{w}} \cdot t \circ t(\vec{v}) = c_{\vec{w}} \cdot c_{t(\vec{v})} \cdot t(\vec{v}) = c_{\vec{w}} \cdot r \cdot t(\vec{v}) = r \cdot c_{\vec{w}} \cdot t(\vec{v}) = r \cdot t(\vec{w})$$

to get the desired conclusion.

Three.II.2.41 By assumption, h is not the zero map and so a vector $\vec{v} \in V$ exists that is not in the null space. Note that $\langle h(\vec{v})\rangle$ is a basis for $\mathbb{R}$, because it is a size-one linearly independent subset of $\mathbb{R}$. Consequently h is onto, as for any $r \in \mathbb{R}$ we have $r = c \cdot h(\vec{v})$ for some scalar c, and so $r = h(c\vec{v})$.

Thus the rank of h is one. Because the nullity is n, the dimension of the domain of h, the vector space V, is $n+1$. We can finish by showing $\{\vec{v}, \vec{\beta}_1, \ldots, \vec{\beta}_n\}$ is linearly independent, as it is a size $n+1$ subset of a dimension $n+1$ space. Because $\{\vec{\beta}_1, \ldots, \vec{\beta}_n\}$ is linearly independent we need only show that $\vec{v}$ is not a linear combination of the other vectors. But $c_1\vec{\beta}_1 + \cdots + c_n\vec{\beta}_n = \vec{v}$ would give $-\vec{v} + c_1\vec{\beta}_1 + \cdots + c_n\vec{\beta}_n = \vec{0}$ and applying h to both sides would give a contradiction.

Three.II.2.42 Fix a basis $\langle \vec{\beta}_1, \ldots, \vec{\beta}_n\rangle$ for V. We shall prove that this map

$$h \stackrel{\Phi}{\longmapsto} \begin{pmatrix} h(\vec{\beta}_1) \\ \vdots \\ h(\vec{\beta}_n) \end{pmatrix}$$

is an isomorphism from V^* to $\mathbb{R}^n$.

To see that Φ is one-to-one, assume that h_1 and h_2 are members of V^* such that $\Phi(h_1) = \Phi(h_2)$. Then

$$\begin{pmatrix} h_1(\vec{\beta}_1) \\ \vdots \\ h_1(\vec{\beta}_n) \end{pmatrix} = \begin{pmatrix} h_2(\vec{\beta}_1) \\ \vdots \\ h_2(\vec{\beta}_n) \end{pmatrix}$$

and consequently, $h_1(\vec{\beta}_1) = h_2(\vec{\beta}_1)$, etc. But a homomorphism is determined by its action on a basis, so $h_1 = h_2$, and therefore Φ is one-to-one.

To see that Φ is onto, consider

$$\begin{pmatrix} x_1 \\ \vdots \\ x_n \end{pmatrix}$$

for $x_1, \ldots, x_n \in \mathbb{R}$. This function h from V to $\mathbb{R}$

$$c_1\vec{\beta}_1 + \cdots + c_n\vec{\beta}_n \overset{h}{\longmapsto} c_1x_1 + \cdots + c_nx_n$$

is linear and Φ maps it to the given vector in $\mathbb{R}^n$, so Φ is onto.

The map Φ also preserves structure: where

$$c_1\vec{\beta}_1 + \cdots + c_n\vec{\beta}_n \overset{h_1}{\longmapsto} c_1h_1(\vec{\beta}_1) + \cdots + c_nh_1(\vec{\beta}_n)$$
$$c_1\vec{\beta}_1 + \cdots + c_n\vec{\beta}_n \overset{h_2}{\longmapsto} c_1h_2(\vec{\beta}_1) + \cdots + c_nh_2(\vec{\beta}_n)$$

we have

$$\begin{aligned}(r_1h_1 + r_2h_2)(c_1\vec{\beta}_1 + \cdots + c_n\vec{\beta}_n) &= c_1(r_1h_1(\vec{\beta}_1) + r_2h_2(\vec{\beta}_1)) + \cdots + c_n(r_1h_1(\vec{\beta}_n) + r_2h_2(\vec{\beta}_n)) \\ &= r_1(c_1h_1(\vec{\beta}_1) + \cdots + c_nh_1(\vec{\beta}_n)) + r_2(c_1h_2(\vec{\beta}_1) + \cdots + c_nh_2(\vec{\beta}_n))\end{aligned}$$

so $\Phi(r_1h_1 + r_2h_2) = r_1\Phi(h_1) + r_2\Phi(h_2)$.

Three.II.2.43 Let $h\colon V \to W$ be linear and fix a basis $\langle \vec{\beta}_1, \ldots, \vec{\beta}_n \rangle$ for V. Consider these n maps from V to W

$$h_1(\vec{v}) = c_1 \cdot h(\vec{\beta}_1), \quad h_2(\vec{v}) = c_2 \cdot h(\vec{\beta}_2), \quad \ldots \quad , h_n(\vec{v}) = c_n \cdot h(\vec{\beta}_n)$$

for any $\vec{v} = c_1\vec{\beta}_1 + \cdots + c_n\vec{\beta}_n$. Clearly h is the sum of the h_i's. We need only check that each h_i is linear: where $\vec{u} = d_1\vec{\beta}_1 + \cdots + d_n\vec{\beta}_n$ we have $h_i(r\vec{v} + s\vec{u}) = rc_i + sd_i = rh_i(\vec{v}) + sh_i(\vec{u})$.

Three.II.2.44 Either yes (trivially) or no (nearly trivially).

If we take V 'is homomorphic to' W to mean there is a homomorphism from V into (but not necessarily onto) W, then every space is homomorphic to every other space as a zero map always exists.

If we take V 'is homomorphic to' W to mean there is an onto homomorphism from V to W then the relation is not an equivalence. For instance, there is an onto homomorphism from $\mathbb{R}^3$ to $\mathbb{R}^2$ (projection is one) but no homomorphism from $\mathbb{R}^2$ onto $\mathbb{R}^3$ by Corollary 2.18, so the relation is not reflexive.*

Three.II.2.45 That they form the chains is obvious. For the rest, we show here that $\mathscr{R}(t^{j+1}) = \mathscr{R}(t^j)$ implies that $\mathscr{R}(t^{j+2}) = \mathscr{R}(t^{j+1})$. Induction then applies.

Assume that $\mathscr{R}(t^{j+1}) = \mathscr{R}(t^j)$. Then $t\colon \mathscr{R}(t^{j+1}) \to \mathscr{R}(t^{j+2})$ is the same map, with the same domain, as $t\colon \mathscr{R}(t^j) \to \mathscr{R}(t^{j+1})$. Thus it has the same range: $\mathscr{R}(t^{j+2}) = \mathscr{R}(t^{j+1})$.

Computing Linear Maps

Three.III.1: Representing Linear Maps with Matrices

* More information on equivalence relations is in the appendix.

Three.III.1.12 (a) $\begin{pmatrix} 1\cdot 2+3\cdot 1+1\cdot 0 \\ 0\cdot 2+(-1)\cdot 1+2\cdot 0 \\ 1\cdot 2+1\cdot 1+0\cdot 0 \end{pmatrix} = \begin{pmatrix} 5 \\ -1 \\ 3 \end{pmatrix}$ (b) Not defined. (c) $\begin{pmatrix} 0 \\ 0 \\ 0 \end{pmatrix}$

Three.III.1.13 (a) $\begin{pmatrix} 2\cdot 4+1\cdot 2 \\ 3\cdot 4-(1/2)\cdot 2 \end{pmatrix} = \begin{pmatrix} 10 \\ 11 \end{pmatrix}$ (b) $\begin{pmatrix} 4 \\ 1 \end{pmatrix}$ (c) Not defined.

Three.III.1.14 Matrix-vector multiplication gives rise to a linear system.

$$\begin{aligned} 2x+y+z&=8 \\ y+3z&=4 \\ x-y+2z&=4 \end{aligned}$$

Gaussian reduction shows that $z=1$, $y=1$, and $x=3$.

Three.III.1.15 Here are two ways to get the answer.

First, obviously $1-3x+2x^2 = 1\cdot 1-3\cdot x+2\cdot x^2$, and so we can apply the general property of preservation of combinations to get $h(1-3x+2x^2) = h(1\cdot 1-3\cdot x+2\cdot x^2) = 1\cdot h(1)-3\cdot h(x)+2\cdot h(x^2) = 1\cdot(1+x)-3\cdot(1+2x)+2\cdot(x-x^3) = -2-3x-2x^3$.

The other way uses the computation scheme developed in this subsection. Because we know where these elements of the space go, we consider this basis $B=\langle 1,x,x^2\rangle$ for the domain. Arbitrarily, we can take $D=\langle 1,x,x^2,x^3\rangle$ as a basis for the codomain. With those choices, we have that

$$\mathrm{Rep}_{B,D}(h) = \begin{pmatrix} 1 & 1 & 0 \\ 1 & 2 & 1 \\ 0 & 0 & 0 \\ 0 & 0 & -1 \end{pmatrix}_{B,D}$$

and, as

$$\mathrm{Rep}_B(1-3x+2x^2) = \begin{pmatrix} 1 \\ -3 \\ 2 \end{pmatrix}_B$$

the matrix-vector multiplication calculation gives this.

$$\mathrm{Rep}_D(h(1-3x+2x^2)) = \begin{pmatrix} 1 & 1 & 0 \\ 1 & 2 & 1 \\ 0 & 0 & 0 \\ 0 & 0 & -1 \end{pmatrix}_{B,D} \begin{pmatrix} 1 \\ -3 \\ 2 \end{pmatrix}_B = \begin{pmatrix} -2 \\ -3 \\ 0 \\ -2 \end{pmatrix}_D$$

Thus, $h(1-3x+2x^2) = -2\cdot 1-3\cdot x+0\cdot x^2-2\cdot x^3 = -2-3x-2x^3$, as above.

Three.III.1.16 Again, as recalled in the subsection, with respect to $\mathcal{E}_i$, a column vector represents itself.

(a) To represent h with respect to $\mathcal{E}_2, \mathcal{E}_3$ we take the images of the basis vectors from the domain, and represent them with respect to the basis for the codomain.

$$\mathrm{Rep}_{\mathcal{E}_3}(\,h(\vec{e}_1)\,) = \mathrm{Rep}_{\mathcal{E}_3}(\begin{pmatrix} 2 \\ 2 \\ 0 \end{pmatrix}) = \begin{pmatrix} 2 \\ 2 \\ 0 \end{pmatrix} \qquad \mathrm{Rep}_{\mathcal{E}_3}(\,h(\vec{e}_2)\,) = \mathrm{Rep}_{\mathcal{E}_3}(\begin{pmatrix} 0 \\ 1 \\ -1 \end{pmatrix}) = \begin{pmatrix} 0 \\ 1 \\ -1 \end{pmatrix}$$

These are adjoined to make the matrix.

$$\mathrm{Rep}_{\mathcal{E}_2,\mathcal{E}_3}(h) = \begin{pmatrix} 2 & 0 \\ 2 & 1 \\ 0 & -1 \end{pmatrix}$$

(b) For any $\vec{v}$ in the domain $\mathbb{R}^2$,

$$\mathrm{Rep}_{\mathcal{E}_2}(\vec{v}) = \mathrm{Rep}_{\mathcal{E}_2}(\begin{pmatrix} v_1 \\ v_2 \end{pmatrix}) = \begin{pmatrix} v_1 \\ v_2 \end{pmatrix}$$

and so

$$\mathrm{Rep}_{\mathcal{E}_3}(\,h(\vec{v})\,) = \begin{pmatrix} 2 & 0 \\ 2 & 1 \\ 0 & -1 \end{pmatrix} \begin{pmatrix} v_1 \\ v_2 \end{pmatrix} = \begin{pmatrix} 2v_1 \\ 2v_1 + v_2 \\ -v_2 \end{pmatrix}$$

is the desired representation.

Three.III.1.17 **(a)** We must first find the image of each vector from the domain's basis, and then represent that image with respect to the codomain's basis.

$$\mathrm{Rep}_B(\frac{d\,1}{dx}) = \begin{pmatrix} 0 \\ 0 \\ 0 \\ 0 \end{pmatrix} \quad \mathrm{Rep}_B(\frac{d\,x}{dx}) = \begin{pmatrix} 1 \\ 0 \\ 0 \\ 0 \end{pmatrix} \quad \mathrm{Rep}_B(\frac{d\,x^2}{dx}) = \begin{pmatrix} 0 \\ 2 \\ 0 \\ 0 \end{pmatrix} \quad \mathrm{Rep}_B(\frac{d\,x^3}{dx}) = \begin{pmatrix} 0 \\ 0 \\ 3 \\ 0 \end{pmatrix}$$

Those representations are then adjoined to make the matrix representing the map.

$$\mathrm{Rep}_{B,B}(\frac{d}{dx}) = \begin{pmatrix} 0 & 1 & 0 & 0 \\ 0 & 0 & 2 & 0 \\ 0 & 0 & 0 & 3 \\ 0 & 0 & 0 & 0 \end{pmatrix}$$

(b) Proceeding as in the prior item, we represent the images of the domain's basis vectors

$$\mathrm{Rep}_B(\frac{d\,1}{dx}) = \begin{pmatrix} 0 \\ 0 \\ 0 \\ 0 \end{pmatrix} \quad \mathrm{Rep}_B(\frac{d\,x}{dx}) = \begin{pmatrix} 1 \\ 0 \\ 0 \\ 0 \end{pmatrix} \quad \mathrm{Rep}_B(\frac{d\,x^2}{dx}) = \begin{pmatrix} 0 \\ 1 \\ 0 \\ 0 \end{pmatrix} \quad \mathrm{Rep}_B(\frac{d\,x^3}{dx}) = \begin{pmatrix} 0 \\ 0 \\ 1 \\ 0 \end{pmatrix}$$

and adjoin to make the matrix.

$$\mathrm{Rep}_{B,D}(\frac{d}{dx}) = \begin{pmatrix} 0 & 1 & 0 & 0 \\ 0 & 0 & 1 & 0 \\ 0 & 0 & 0 & 1 \\ 0 & 0 & 0 & 0 \end{pmatrix}$$

Three.III.1.18 For each, we must find the image of each of the domain's basis vectors, represent each image with respect to the codomain's basis, and then adjoin those representations to get the matrix.

(a) The basis vectors from the domain have these images

$$1 \mapsto 0 \quad x \mapsto 1 \quad x^2 \mapsto 2x \quad \ldots$$

and these images are represented with respect to the codomain's basis in this way.

$$\mathrm{Rep}_B(0) = \begin{pmatrix} 0 \\ 0 \\ 0 \\ \vdots \\ \\ \end{pmatrix} \quad \mathrm{Rep}_B(1) = \begin{pmatrix} 1 \\ 0 \\ 0 \\ \vdots \\ \\ \end{pmatrix} \quad \mathrm{Rep}_B(2x) = \begin{pmatrix} 0 \\ 2 \\ 0 \\ \vdots \\ \\ \end{pmatrix} \quad \ldots \quad \mathrm{Rep}_B(nx^{n-1}) = \begin{pmatrix} 0 \\ 0 \\ 0 \\ \vdots \\ n \\ 0 \end{pmatrix}$$

The matrix

$$\mathrm{Rep}_{B,B}(\frac{d}{dx}) = \begin{pmatrix} 0 & 1 & 0 & \ldots & 0 \\ 0 & 0 & 2 & \ldots & 0 \\ & & \vdots & & \\ 0 & 0 & 0 & \ldots & n \\ 0 & 0 & 0 & \ldots & 0 \end{pmatrix}$$

has $n+1$ rows and columns.

(b) Once the images under this map of the domain's basis vectors are determined

$$1 \mapsto x \quad x \mapsto x^2/2 \quad x^2 \mapsto x^3/3 \quad \ldots$$

then they can be represented with respect to the codomain's basis

$$\mathrm{Rep}_{B_{n+1}}(x) = \begin{pmatrix} 0 \\ 1 \\ 0 \\ \vdots \end{pmatrix} \quad \mathrm{Rep}_{B_{n+1}}(x^2/2) = \begin{pmatrix} 0 \\ 0 \\ 1/2 \\ \vdots \end{pmatrix} \quad \ldots \quad \mathrm{Rep}_{B_{n+1}}(x^{n+1}/(n+1)) = \begin{pmatrix} 0 \\ 0 \\ 0 \\ \vdots \\ 1/(n+1) \end{pmatrix}$$

and put together to make the matrix.

$$\mathrm{Rep}_{B_n,B_{n+1}}(\int) = \begin{pmatrix} 0 & 0 & \ldots & 0 & 0 \\ 1 & 0 & \ldots & 0 & 0 \\ 0 & 1/2 & \ldots & 0 & 0 \\ & \vdots & & & \\ 0 & 0 & \ldots & 0 & 1/(n+1) \end{pmatrix}$$

(c) The images of the basis vectors of the domain are

$$1 \mapsto 1 \quad x \mapsto 1/2 \quad x^2 \mapsto 1/3 \quad \ldots$$

and they are represented with respect to the codomain's basis as

$$\mathrm{Rep}_{\mathcal{E}_1}(1) = 1 \quad \mathrm{Rep}_{\mathcal{E}_1}(1/2) = 1/2 \quad \ldots$$

so the matrix is

$$\mathrm{Rep}_{B,\mathcal{E}_1}(\int) = \begin{pmatrix} 1 & 1/2 & \cdots & 1/n & 1/(n+1) \end{pmatrix}$$

(this is an $1 \times (n+1)$ matrix).

(d) Here, the images of the domain's basis vectors are

$$1 \mapsto 1 \quad x \mapsto 3 \quad x^2 \mapsto 9 \quad \ldots$$

and they are represented in the codomain as

$$\mathrm{Rep}_{\mathcal{E}_1}(1) = 1 \quad \mathrm{Rep}_{\mathcal{E}_1}(3) = 3 \quad \mathrm{Rep}_{\mathcal{E}_1}(9) = 9 \quad \ldots$$

and so the matrix is this.

$$\mathrm{Rep}_{B,\mathcal{E}_1}(\int_0^1) = \begin{pmatrix} 1 & 3 & 9 & \cdots & 3^n \end{pmatrix}$$

(e) The images of the basis vectors from the domain are

$$1 \mapsto 1 \quad x \mapsto x+1 = 1+x \quad x^2 \mapsto (x+1)^2 = 1+2x+x^2 \quad x^3 \mapsto (x+1)^3 = 1+3x+3x^2+x^3 \quad \ldots$$

which are represented as

$$\mathrm{Rep}_B(1) = \begin{pmatrix} 1 \\ 0 \\ 0 \\ 0 \\ \vdots \\ 0 \end{pmatrix} \quad \mathrm{Rep}_B(1+x) = \begin{pmatrix} 1 \\ 1 \\ 0 \\ 0 \\ \vdots \\ 0 \end{pmatrix} \quad \mathrm{Rep}_B(1+2x+x^2) = \begin{pmatrix} 1 \\ 2 \\ 1 \\ 0 \\ \vdots \\ 0 \end{pmatrix} \quad \ldots$$

The resulting matrix

$$\mathrm{Rep}_{B,B}(\mathrm{slide}_{-1}) = \begin{pmatrix} 1 & 1 & 1 & 1 & \ldots & 1 \\ 0 & 1 & 2 & 3 & \ldots & \binom{n}{1} \\ 0 & 0 & 1 & 3 & \ldots & \binom{n}{2} \\ & \vdots & & & & \\ 0 & 0 & 0 & & \ldots & 1 \end{pmatrix}$$

is *Pascal's triangle* (recall that $\binom{n}{r}$ is the number of ways to choose r things, without order and without repetition, from a set of size n).

Three.III.1.19 Where the space is n-dimensional,

$$\text{Rep}_{B,B}(\text{id})=\begin{pmatrix}1&0\ldots&0\\0&1\ldots&0\\&\vdots&\\0&0\ldots&1\end{pmatrix}_{B,B}$$

is the $n\times n$ identity matrix.

Three.III.1.20 Taking this as the natural basis

$$B=\langle\vec{\beta}_1,\vec{\beta}_2,\vec{\beta}_3,\vec{\beta}_4\rangle=\langle\begin{pmatrix}1&0\\0&0\end{pmatrix},\begin{pmatrix}0&1\\0&0\end{pmatrix},\begin{pmatrix}0&0\\1&0\end{pmatrix},\begin{pmatrix}0&0\\0&1\end{pmatrix}\rangle$$

the transpose map acts in this way

$$\vec{\beta}_1\mapsto\vec{\beta}_1\quad\vec{\beta}_2\mapsto\vec{\beta}_3\quad\vec{\beta}_3\mapsto\vec{\beta}_2\quad\vec{\beta}_4\mapsto\vec{\beta}_4$$

so that representing the images with respect to the codomain's basis and adjoining those column vectors together gives this.

$$\text{Rep}_{B,B}(\text{trans})=\begin{pmatrix}1&0&0&0\\0&0&1&0\\0&1&0&0\\0&0&0&1\end{pmatrix}_{B,B}$$

Three.III.1.21 **(a)** With respect to the basis of the codomain, the images of the members of the basis of the domain are represented as

$$\text{Rep}_B(\vec{\beta}_2)=\begin{pmatrix}0\\1\\0\\0\end{pmatrix}\quad\text{Rep}_B(\vec{\beta}_3)=\begin{pmatrix}0\\0\\1\\0\end{pmatrix}\quad\text{Rep}_B(\vec{\beta}_4)=\begin{pmatrix}0\\0\\0\\1\end{pmatrix}\quad\text{Rep}_B(\vec{0})=\begin{pmatrix}0\\0\\0\\0\end{pmatrix}$$

and consequently, the matrix representing the transformation is this.

$$\begin{pmatrix}0&0&0&0\\1&0&0&0\\0&1&0&0\\0&0&1&0\end{pmatrix}$$

(b) $\begin{pmatrix}0&0&0&0\\1&0&0&0\\0&0&0&0\\0&0&1&0\end{pmatrix}$

(c) $\begin{pmatrix}0&0&0&0\\1&0&0&0\\0&1&0&0\\0&0&0&0\end{pmatrix}$

Three.III.1.22 **(a)** The picture of $d_s\colon\mathbb{R}^2\to\mathbb{R}^2$ is this.

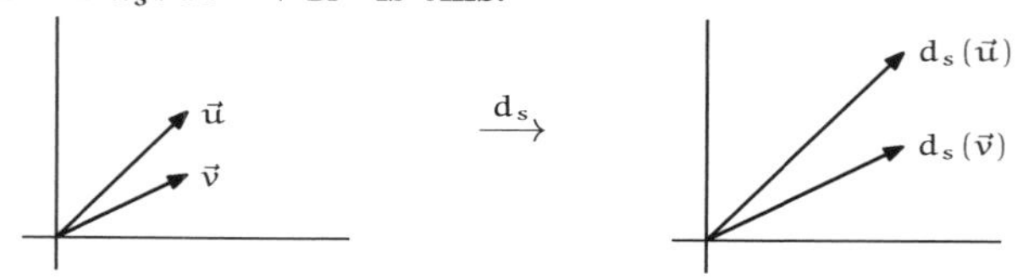

This map's effect on the vectors in the standard basis for the domain is

$$\begin{pmatrix}1\\0\end{pmatrix}\overset{d_s}{\longmapsto}\begin{pmatrix}s\\0\end{pmatrix}\qquad\begin{pmatrix}0\\1\end{pmatrix}\overset{d_s}{\longmapsto}\begin{pmatrix}0\\s\end{pmatrix}$$

and those images are represented with respect to the codomain's basis (again, the standard basis) by themselves.

$$\text{Rep}_{\mathcal{E}_2}(\begin{pmatrix} s \\ 0 \end{pmatrix}) = \begin{pmatrix} s \\ 0 \end{pmatrix} \qquad \text{Rep}_{\mathcal{E}_2}(\begin{pmatrix} 0 \\ s \end{pmatrix}) = \begin{pmatrix} 0 \\ s \end{pmatrix}$$

Thus the representation of the dilation map is this.

$$\text{Rep}_{\mathcal{E}_2,\mathcal{E}_2}(d_s) = \begin{pmatrix} s & 0 \\ 0 & s \end{pmatrix}$$

(b) The picture of $f_\ell \colon \mathbb{R}^2 \to \mathbb{R}^2$ is this.

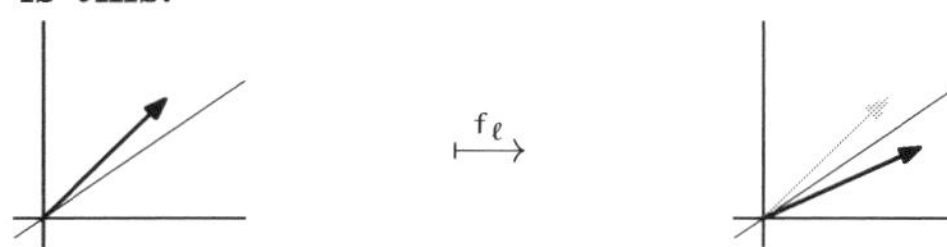

Some calculation (see Exercise I.30) shows that when the line has slope k

$$\begin{pmatrix} 1 \\ 0 \end{pmatrix} \stackrel{f_\ell}{\longmapsto} \begin{pmatrix} (1-k^2)/(1+k^2) \\ 2k/(1+k^2) \end{pmatrix} \qquad \begin{pmatrix} 0 \\ 1 \end{pmatrix} \stackrel{f_\ell}{\longmapsto} \begin{pmatrix} 2k/(1+k^2) \\ -(1-k^2)/(1+k^2) \end{pmatrix}$$

(the case of a line with undefined slope is separate but easy) and so the matrix representing reflection is this.

$$\text{Rep}_{\mathcal{E}_2,\mathcal{E}_2}(f_\ell) = \frac{1}{1+k^2} \cdot \begin{pmatrix} 1-k^2 & 2k \\ 2k & -(1-k^2) \end{pmatrix}$$

Three.III.1.23 Call the map $t \colon \mathbb{R}^2 \to \mathbb{R}^2$.

(a) To represent this map with respect to the standard bases, we must find, and then represent, the images of the vectors $\vec{e}_1$ and $\vec{e}_2$ from the domain's basis. The image of $\vec{e}_1$ is given.

One way to find the image of $\vec{e}_2$ is by eye—we can see this.

$$\begin{pmatrix} 1 \\ 1 \end{pmatrix} - \begin{pmatrix} 1 \\ 0 \end{pmatrix} = \begin{pmatrix} 0 \\ 1 \end{pmatrix} \stackrel{t}{\longmapsto} \begin{pmatrix} 2 \\ 0 \end{pmatrix} - \begin{pmatrix} -1 \\ 0 \end{pmatrix} = \begin{pmatrix} 3 \\ 0 \end{pmatrix}$$

A more systematic way to find the image of $\vec{e}_2$ is to use the given information to represent the transformation, and then use that representation to determine the image. Taking this for a basis,

$$C = \langle \begin{pmatrix} 1 \\ 1 \end{pmatrix}, \begin{pmatrix} 1 \\ 0 \end{pmatrix} \rangle$$

the given information says this.

$$\text{Rep}_{C,\mathcal{E}_2}(t) \begin{pmatrix} 2 & -1 \\ 0 & 0 \end{pmatrix}$$

As

$$\text{Rep}_C(\vec{e}_2) = \begin{pmatrix} 1 \\ -1 \end{pmatrix}_C$$

we have that

$$\text{Rep}_{\mathcal{E}_2}(t(\vec{e}_2)) = \begin{pmatrix} 2 & -1 \\ 0 & 0 \end{pmatrix}_{C,\mathcal{E}_2} \begin{pmatrix} 1 \\ -1 \end{pmatrix}_C = \begin{pmatrix} 3 \\ 0 \end{pmatrix}_{\mathcal{E}_2}$$

and consequently we know that $t(\vec{e}_2) = 3 \cdot \vec{e}_1$ (since, with respect to the standard basis, this vector is represented by itself). Therefore, this is the representation of t with respect to $\mathcal{E}_2, \mathcal{E}_2$.

$$\text{Rep}_{\mathcal{E}_2,\mathcal{E}_2}(t) = \begin{pmatrix} -1 & 3 \\ 0 & 0 \end{pmatrix}_{\mathcal{E}_2,\mathcal{E}_2}$$

(b) To use the matrix developed in the prior item, note that

$$\text{Rep}_{\mathcal{E}_2}(\begin{pmatrix} 0 \\ 5 \end{pmatrix}) = \begin{pmatrix} 0 \\ 5 \end{pmatrix}_{\mathcal{E}_2}$$

and so we have this is the representation, with respect to the codomain's basis, of the image of the given vector.

$$\mathrm{Rep}_{\mathcal{E}_2}(t(\begin{pmatrix}0\\5\end{pmatrix})) = \begin{pmatrix}-1&3\\0&0\end{pmatrix}_{\mathcal{E}_2,\mathcal{E}_2}\begin{pmatrix}0\\5\end{pmatrix}_{\mathcal{E}_2} = \begin{pmatrix}15\\0\end{pmatrix}_{\mathcal{E}_2}$$

Because the codomain's basis is the standard one, and so vectors in the codomain are represented by themselves, we have this.

$$t(\begin{pmatrix}0\\5\end{pmatrix}) = \begin{pmatrix}15\\0\end{pmatrix}$$

(c) We first find the image of each member of B, and then represent those images with respect to D. For the first step, we can use the matrix developed earlier.

$$\mathrm{Rep}_{\mathcal{E}_2}(\begin{pmatrix}1\\-1\end{pmatrix}) = \begin{pmatrix}-1&3\\0&0\end{pmatrix}_{\mathcal{E}_2,\mathcal{E}_2}\begin{pmatrix}1\\-1\end{pmatrix}_{\mathcal{E}_2} = \begin{pmatrix}-4\\0\end{pmatrix}_{\mathcal{E}_2} \quad\text{so}\quad t(\begin{pmatrix}1\\-1\end{pmatrix}) = \begin{pmatrix}-4\\0\end{pmatrix}$$

Actually, for the second member of B there is no need to apply the matrix because the problem statement gives its image.

$$t(\begin{pmatrix}1\\1\end{pmatrix}) = \begin{pmatrix}2\\0\end{pmatrix}$$

Now representing those images with respect to D is routine.

$$\mathrm{Rep}_D(\begin{pmatrix}-4\\0\end{pmatrix}) = \begin{pmatrix}-1\\2\end{pmatrix}_D \quad\text{and}\quad \mathrm{Rep}_D(\begin{pmatrix}2\\0\end{pmatrix}) = \begin{pmatrix}1/2\\-1\end{pmatrix}_D$$

Thus, the matrix is this.

$$\mathrm{Rep}_{B,D}(t) = \begin{pmatrix}-1&1/2\\2&-1\end{pmatrix}_{B,D}$$

(d) We know the images of the members of the domain's basis from the prior item.

$$t(\begin{pmatrix}1\\-1\end{pmatrix}) = \begin{pmatrix}-4\\0\end{pmatrix} \qquad t(\begin{pmatrix}1\\1\end{pmatrix}) = \begin{pmatrix}2\\0\end{pmatrix}$$

We can compute the representation of those images with respect to the codomain's basis.

$$\mathrm{Rep}_B(\begin{pmatrix}-4\\0\end{pmatrix}) = \begin{pmatrix}-2\\-2\end{pmatrix}_B \quad\text{and}\quad \mathrm{Rep}_B(\begin{pmatrix}2\\0\end{pmatrix}) = \begin{pmatrix}1\\1\end{pmatrix}_B$$

Thus this is the matrix.

$$\mathrm{Rep}_{B,B}(t) = \begin{pmatrix}-2&1\\-2&1\end{pmatrix}_{B,B}$$

Three.III.1.24 **(a)** The images of the members of the domain's basis are

$$\vec{\beta}_1 \mapsto h(\vec{\beta}_1) \quad \vec{\beta}_2 \mapsto h(\vec{\beta}_2) \quad \ldots \quad \vec{\beta}_n \mapsto h(\vec{\beta}_n)$$

and those images are represented with respect to the codomain's basis in this way.

$$\mathrm{Rep}_{h(B)}(\,h(\vec{\beta}_1)\,) = \begin{pmatrix}1\\0\\\vdots\\0\end{pmatrix} \quad \mathrm{Rep}_{h(B)}(\,h(\vec{\beta}_2)\,) = \begin{pmatrix}0\\1\\\vdots\\0\end{pmatrix} \quad \ldots \quad \mathrm{Rep}_{h(B)}(\,h(\vec{\beta}_n)\,) = \begin{pmatrix}0\\0\\\vdots\\1\end{pmatrix}$$

Hence, the matrix is the identity.

$$\mathrm{Rep}_{B,h(B)}(h) = \begin{pmatrix}1&0&\ldots&0\\0&1&&0\\&&\ddots&\\0&0&&1\end{pmatrix}$$

(b) Using the matrix in the prior item, the representation is this.

$$\mathrm{Rep}_{h(B)}(\,h(\vec{v})\,) = \begin{pmatrix} c_1 \\ \vdots \\ c_n \end{pmatrix}_{h(B)}$$

Three.III.1.25 The product

$$\begin{pmatrix} h_{1,1} & \ldots & h_{1,i} & \ldots & h_{1,n} \\ h_{2,1} & \ldots & h_{2,i} & \ldots & h_{2,n} \\ & \vdots & & & \\ h_{m,1} & \ldots & h_{m,i} & \ldots & h_{1,n} \end{pmatrix} \begin{pmatrix} 0 \\ \vdots \\ 1 \\ \vdots \\ 0 \end{pmatrix} = \begin{pmatrix} h_{1,i} \\ h_{2,i} \\ \vdots \\ h_{m,i} \end{pmatrix}$$

gives the i-th column of the matrix.

Three.III.1.26 **(a)** The images of the basis vectors for the domain are $\cos x \stackrel{d/dx}{\longmapsto} -\sin x$ and $\sin x \stackrel{d/dx}{\longmapsto} \cos x$. Representing those with respect to the codomain's basis (again, B) and adjoining the representations gives this matrix.

$$\mathrm{Rep}_{B,B}(\frac{d}{dx}) = \begin{pmatrix} 0 & 1 \\ -1 & 0 \end{pmatrix}_{B,B}$$

(b) The images of the vectors in the domain's basis are $e^x \stackrel{d/dx}{\longmapsto} e^x$ and $e^{2x} \stackrel{d/dx}{\longmapsto} 2e^{2x}$. Representing with respect to the codomain's basis and adjoining gives this matrix.

$$\mathrm{Rep}_{B,B}(\frac{d}{dx}) = \begin{pmatrix} 1 & 0 \\ 0 & 2 \end{pmatrix}_{B,B}$$

(c) The images of the members of the domain's basis are $1 \stackrel{d/dx}{\longmapsto} 0$, $x \stackrel{d/dx}{\longmapsto} 1$, $e^x \stackrel{d/dx}{\longmapsto} e^x$, and $xe^x \stackrel{d/dx}{\longmapsto} e^x + xe^x$. Representing these images with respect to B and adjoining gives this matrix.

$$\mathrm{Rep}_{B,B}(\frac{d}{dx}) = \begin{pmatrix} 0 & 1 & 0 & 0 \\ 0 & 0 & 0 & 0 \\ 0 & 0 & 1 & 1 \\ 0 & 0 & 0 & 1 \end{pmatrix}_{B,B}$$

Three.III.1.27 **(a)** It is the set of vectors of the codomain represented with respect to the codomain's basis in this way.

$$\{\begin{pmatrix} 1 & 0 \\ 0 & 0 \end{pmatrix}\begin{pmatrix} x \\ y \end{pmatrix} \mid x, y \in \mathbb{R}\} = \{\begin{pmatrix} x \\ 0 \end{pmatrix} \mid x, y \in \mathbb{R}\}$$

As the codomain's basis is $\mathcal{E}_2$, and so each vector is represented by itself, the range of this transformation is the x-axis.

(b) It is the set of vectors of the codomain represented in this way.

$$\{\begin{pmatrix} 0 & 0 \\ 3 & 2 \end{pmatrix}\begin{pmatrix} x \\ y \end{pmatrix} \mid x, y \in \mathbb{R}\} = \{\begin{pmatrix} 0 \\ 3x + 2y \end{pmatrix} \mid x, y \in \mathbb{R}\}$$

With respect to $\mathcal{E}_2$ vectors represent themselves so this range is the y axis.

(c) The set of vectors represented with respect to $\mathcal{E}_2$ as

$$\{\begin{pmatrix} a & b \\ 2a & 2b \end{pmatrix}\begin{pmatrix} x \\ y \end{pmatrix} \mid x, y \in \mathbb{R}\} = \{\begin{pmatrix} ax + by \\ 2ax + 2by \end{pmatrix} \mid x, y \in \mathbb{R}\} = \{(ax + by) \cdot \begin{pmatrix} 1 \\ 2 \end{pmatrix} \mid x, y \in \mathbb{R}\}$$

is the line $y = 2x$, provided either a or b is not zero, and is the set consisting of just the origin if both are zero.

Three.III.1.28 Yes, for two reasons.

First, the two maps h and $\hat{h}$ need not have the same domain and codomain. For instance,

$$\begin{pmatrix}1&2\\3&4\end{pmatrix}$$

represents a map $h\colon \mathbb{R}^2 \to \mathbb{R}^2$ with respect to the standard bases that sends

$$\begin{pmatrix}1\\0\end{pmatrix} \mapsto \begin{pmatrix}1\\3\end{pmatrix} \quad\text{and}\quad \begin{pmatrix}0\\1\end{pmatrix} \mapsto \begin{pmatrix}2\\4\end{pmatrix}$$

and also represents a $\hat{h}\colon \mathcal{P}_1 \to \mathbb{R}^2$ with respect to $\langle 1, x\rangle$ and $\mathcal{E}_2$ that acts in this way.

$$1 \mapsto \begin{pmatrix}1\\3\end{pmatrix} \quad\text{and}\quad x \mapsto \begin{pmatrix}2\\4\end{pmatrix}$$

The second reason is that, even if the domain and codomain of h and $\hat{h}$ coincide, different bases produce different maps. An example is the 2×2 identity matrix

$$I = \begin{pmatrix}1&0\\0&1\end{pmatrix}$$

which represents the identity map on $\mathbb{R}^2$ with respect to $\mathcal{E}_2, \mathcal{E}_2$. However, with respect to $\mathcal{E}_2$ for the domain but the basis $D = \langle \vec{e}_2, \vec{e}_1\rangle$ for the codomain, the same matrix I represents the map that swaps the first and second components

$$\begin{pmatrix}x\\y\end{pmatrix} \mapsto \begin{pmatrix}y\\x\end{pmatrix}$$

(that is, reflection about the line $y = x$).

Three.III.1.29 We mimic Example 1.1, just replacing the numbers with letters.

Write B as $\langle \vec{\beta}_1, \dots, \vec{\beta}_n\rangle$ and D as $\langle \vec{\delta}_1, \dots, \vec{\delta}_m\rangle$. By definition of representation of a map with respect to bases, the assumption that

$$\text{Rep}_{B,D}(h) = \begin{pmatrix}h_{1,1}&\dots&h_{1,n}\\ \vdots&&\vdots\\ h_{m,1}&\dots&h_{m,n}\end{pmatrix}$$

means that $h(\vec{\beta}_i) = h_{i,1}\vec{\delta}_1 + \cdots + h_{i,n}\vec{\delta}_n$. And, by the definition of the representation of a vector with respect to a basis, the assumption that

$$\text{Rep}_B(\vec{v}) = \begin{pmatrix}c_1\\ \vdots\\ c_n\end{pmatrix}$$

means that $\vec{v} = c_1\vec{\beta}_1 + \cdots + c_n\vec{\beta}_n$. Substituting gives

$$\begin{aligned}h(\vec{v}) &= h(c_1\cdot\vec{\beta}_1 + \cdots + c_n\cdot\vec{\beta}_n)\\ &= c_1\cdot h(\vec{\beta}_1) + \cdots + c_n\cdot\vec{\beta}_n\\ &= c_1\cdot(h_{1,1}\vec{\delta}_1 + \cdots + h_{m,1}\vec{\delta}_m) + \cdots + c_n\cdot(h_{1,n}\vec{\delta}_1 + \cdots + h_{m,n}\vec{\delta}_m)\\ &= (h_{1,1}c_1 + \cdots + h_{1,n}c_n)\cdot\vec{\delta}_1 + \cdots + (h_{m,1}c_1 + \cdots + h_{m,n}c_n)\cdot\vec{\delta}_m\end{aligned}$$

and so $h(\vec{v})$ is represented as required.

Three.III.1.30 (a) The picture is this.

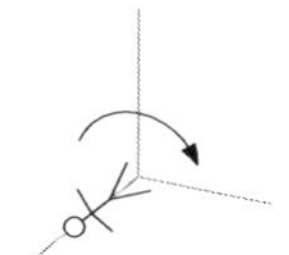

The images of the vectors from the domain's basis

$$\begin{pmatrix}1\\0\\0\end{pmatrix}\mapsto\begin{pmatrix}1\\0\\0\end{pmatrix}\quad\begin{pmatrix}0\\1\\0\end{pmatrix}\mapsto\begin{pmatrix}0\\\cos\theta\\-\sin\theta\end{pmatrix}\quad\begin{pmatrix}0\\0\\1\end{pmatrix}\mapsto\begin{pmatrix}0\\\sin\theta\\\cos\theta\end{pmatrix}$$

are represented with respect to the codomain's basis (again, $\mathcal{E}_3$) by themselves, so adjoining the representations to make the matrix gives this.

$$\mathrm{Rep}_{\mathcal{E}_3,\mathcal{E}_3}(r_\theta)=\begin{pmatrix}1&0&0\\0&\cos\theta&\sin\theta\\0&-\sin\theta&\cos\theta\end{pmatrix}$$

(b) The picture is similar to the one in the prior answer. The images of the vectors from the domain's basis

$$\begin{pmatrix}1\\0\\0\end{pmatrix}\mapsto\begin{pmatrix}\cos\theta\\0\\\sin\theta\end{pmatrix}\quad\begin{pmatrix}0\\1\\0\end{pmatrix}\mapsto\begin{pmatrix}0\\1\\0\end{pmatrix}\quad\begin{pmatrix}0\\0\\1\end{pmatrix}\mapsto\begin{pmatrix}-\sin\theta\\0\\\cos\theta\end{pmatrix}$$

are represented with respect to the codomain's basis $\mathcal{E}_3$ by themselves, so this is the matrix.

$$\begin{pmatrix}\cos\theta&0&-\sin\theta\\0&1&0\\\sin\theta&0&\cos\theta\end{pmatrix}$$

(c) To a person standing up, with the vertical z-axis, a rotation of the xy-plane that is clockwise proceeds from the positive y-axis to the positive x-axis. That is, it rotates opposite to the direction in Example 1.9. The images of the vectors from the domain's basis

$$\begin{pmatrix}1\\0\\0\end{pmatrix}\mapsto\begin{pmatrix}\cos\theta\\-\sin\theta\\0\end{pmatrix}\quad\begin{pmatrix}0\\1\\0\end{pmatrix}\mapsto\begin{pmatrix}\sin\theta\\\cos\theta\\0\end{pmatrix}\quad\begin{pmatrix}0\\0\\1\end{pmatrix}\mapsto\begin{pmatrix}0\\0\\1\end{pmatrix}$$

are represented with respect to $\mathcal{E}_3$ by themselves, so the matrix is this.

$$\begin{pmatrix}\cos\theta&\sin\theta&0\\-\sin\theta&\cos\theta&0\\0&0&1\end{pmatrix}$$

(d) $\begin{pmatrix}\cos\theta&\sin\theta&0&0\\-\sin\theta&\cos\theta&0&0\\0&0&1&0\\0&0&0&1\end{pmatrix}$

Three.III.1.31 **(a)** Write B_U as $\langle\vec{\beta}_1,\ldots,\vec{\beta}_k\rangle$ and then B_V as $\langle\vec{\beta}_1,\ldots,\vec{\beta}_k,\vec{\beta}_{k+1},\ldots,\vec{\beta}_n\rangle$. If

$$\mathrm{Rep}_{B_U}(\vec{v})=\begin{pmatrix}c_1\\\vdots\\c_k\end{pmatrix}\qquad\text{so that }\vec{v}=c_1\cdot\vec{\beta}_1+\cdots+c_k\cdot\vec{\beta}_k$$

then,

$$\mathrm{Rep}_{B_V}(\vec{v})=\begin{pmatrix}c_1\\\vdots\\c_k\\0\\\vdots\\0\end{pmatrix}$$

because $\vec{v}=c_1\cdot\vec{\beta}_1+\cdots+c_k\cdot\vec{\beta}_k+0\cdot\vec{\beta}_{k+1}+\cdots+0\cdot\vec{\beta}_n$.

(b) We must first decide what the question means. Compare $h\colon V \to W$ with its restriction to the subspace $h\restriction_U\colon U \to W$. The range space of the restriction is a subspace of W, so fix a basis $D_{h(U)}$ for this range space and extend it to a basis D_V for W. We want the relationship between these two.

$$\text{Rep}_{B_V,D_V}(h) \quad\text{and}\quad \text{Rep}_{B_U,D_{h(U)}}(h\restriction_U)$$

The answer falls right out of the prior item: if

$$\text{Rep}_{B_U,D_{h(U)}}(h\restriction_U) = \begin{pmatrix} h_{1,1} & \ldots & h_{1,k} \\ \vdots & & \vdots \\ h_{p,1} & \ldots & h_{p,k} \end{pmatrix}$$

then the extension is represented in this way.

$$\text{Rep}_{B_V,D_V}(h) = \begin{pmatrix} h_{1,1} & \ldots & h_{1,k} & h_{1,k+1} & \ldots & h_{1,n} \\ \vdots & & & & & \vdots \\ h_{p,1} & \ldots & h_{p,k} & h_{p,k+1} & \ldots & h_{p,n} \\ 0 & \ldots & 0 & h_{p+1,k+1} & \ldots & h_{p+1,n} \\ \vdots & & & & & \vdots \\ 0 & \ldots & 0 & h_{m,k+1} & \ldots & h_{m,n} \end{pmatrix}$$

(c) Take W_i to be the span of $\{h(\vec{\beta}_1),\ldots,h(\vec{\beta}_i)\}$.

(d) Apply the answer from the second item to the third item.

(e) No. For instance $\pi_x\colon \mathbb{R}^2 \to \mathbb{R}^2$, projection onto the x axis, is represented by these two upper-triangular matrices

$$\text{Rep}_{\mathcal{E}_2,\mathcal{E}_2}(\pi_x) = \begin{pmatrix} 1 & 0 \\ 0 & 0 \end{pmatrix} \quad\text{and}\quad \text{Rep}_{C,\mathcal{E}_2}(\pi_x) = \begin{pmatrix} 0 & 1 \\ 0 & 0 \end{pmatrix}$$

where $C = \langle \vec{e}_2, \vec{e}_1 \rangle$.

Three.III.2: Any Matrix Represents a Linear Map

Three.III.2.12 With respect to B the vector's representation is this.

$$\text{Rep}_B(2x-1) = \begin{pmatrix} -1 \\ 3 \end{pmatrix}$$

Using the matrix-vector product we can compute $\text{Rep}_D(h(\vec{v}))$

$$\text{Rep}_D(h(2x-1)) = \begin{pmatrix} 2 & 1 \\ 4 & 2 \end{pmatrix} \begin{pmatrix} -1 \\ 3 \end{pmatrix}_B = \begin{pmatrix} 1 \\ 2 \end{pmatrix}_D$$

From that representation we can compute $h(\vec{v})$.

$$h(2x-1) = 1 \cdot \begin{pmatrix} 1 \\ 1 \end{pmatrix} + 2 \cdot \begin{pmatrix} 1 \\ 0 \end{pmatrix} = \begin{pmatrix} 3 \\ 1 \end{pmatrix}$$

Three.III.2.13 As described in the subsection, with respect to the standard bases, representations are transparent, and so, for instance, the first matrix describes this map.

$$\begin{pmatrix} 1 \\ 0 \\ 0 \end{pmatrix} = \begin{pmatrix} 1 \\ 0 \\ 0 \end{pmatrix}_{\mathcal{E}_3} \mapsto \begin{pmatrix} 1 \\ 0 \end{pmatrix}_{\mathcal{E}_2} = \begin{pmatrix} 1 \\ 0 \end{pmatrix} \qquad \begin{pmatrix} 0 \\ 1 \\ 0 \end{pmatrix} \mapsto \begin{pmatrix} 1 \\ 1 \end{pmatrix} \qquad \begin{pmatrix} 0 \\ 0 \\ 1 \end{pmatrix} \mapsto \begin{pmatrix} 3 \\ 4 \end{pmatrix}$$

So, for this first one, we are asking whether there are scalars such that

$$c_1 \begin{pmatrix} 1 \\ 0 \end{pmatrix} + c_2 \begin{pmatrix} 1 \\ 1 \end{pmatrix} + c_3 \begin{pmatrix} 3 \\ 4 \end{pmatrix} = \begin{pmatrix} 1 \\ 3 \end{pmatrix}$$

that is, whether the vector is in the column space of the matrix.

(a) Yes. We can get this conclusion by setting up the resulting linear system and applying Gauss's Method, as usual. Another way to get it is to note by inspection of the equation of columns that taking $c_3 = 3/4$, and $c_1 = -5/4$, and $c_2 = 0$ will do. Still a third way to get this conclusion is to note that the rank of the matrix is two, which equals the dimension of the codomain, and so the map is onto — the range is all of $\mathbb{R}^2$ and in particular includes the given vector.

(b) No; note that all of the columns in the matrix have a second component that is twice the first, while the vector does not. Alternatively, the column space of the matrix is

$$\{c_1\begin{pmatrix}2\\4\end{pmatrix}+c_2\begin{pmatrix}0\\0\end{pmatrix}+c_3\begin{pmatrix}3\\6\end{pmatrix} \mid c_1,c_2,c_3\in\mathbb{R}\}=\{c\begin{pmatrix}1\\2\end{pmatrix} \mid c\in\mathbb{R}\}$$

(which is the fact already noted, but we got it by calculation rather than inspiration), and the given vector is not in this set.

Three.III.2.14 **(a)** The first member of the basis

$$\begin{pmatrix}0\\1\end{pmatrix}=\begin{pmatrix}1\\0\end{pmatrix}_B$$

maps to

$$\begin{pmatrix}1/2\\-1/2\end{pmatrix}_D$$

which is this member of the codomain.

$$\frac{1}{2}\cdot\begin{pmatrix}1\\1\end{pmatrix}-\frac{1}{2}\cdot\begin{pmatrix}1\\-1\end{pmatrix}=\begin{pmatrix}0\\1\end{pmatrix}$$

(b) The second member of the basis maps

$$\begin{pmatrix}1\\0\end{pmatrix}=\begin{pmatrix}0\\1\end{pmatrix}_B\mapsto\begin{pmatrix}(1/2\\1/2\end{pmatrix}_D$$

to this member of the codomain.

$$\frac{1}{2}\cdot\begin{pmatrix}1\\1\end{pmatrix}+\frac{1}{2}\cdot\begin{pmatrix}1\\-1\end{pmatrix}=\begin{pmatrix}1\\0\end{pmatrix}$$

(c) Because the map that the matrix represents is the identity map on the basis, it must be the identity on all members of the domain. We can come to the same conclusion in another way by considering

$$\begin{pmatrix}x\\y\end{pmatrix}=\begin{pmatrix}y\\x\end{pmatrix}_B$$

which maps to

$$\begin{pmatrix}(x+y)/2\\(x-y)/2\end{pmatrix}_D$$

which represents this member of $\mathbb{R}^2$.

$$\frac{x+y}{2}\cdot\begin{pmatrix}1\\1\end{pmatrix}+\frac{x-y}{2}\cdot\begin{pmatrix}1\\-1\end{pmatrix}=\begin{pmatrix}x\\y\end{pmatrix}$$

Three.III.2.15 A general member of the domain, represented with respect to the domain's basis as

$$a\cos\theta+b\sin\theta=\begin{pmatrix}a\\a+b\end{pmatrix}_B$$

maps to

$$\begin{pmatrix}0\\a\end{pmatrix}_D \quad \text{representing} \quad 0\cdot(\cos\theta+\sin\theta)+a\cdot(\cos\theta)$$

and so the linear map represented by the matrix with respect to these bases

$$a\cos\theta+b\sin\theta\mapsto a\cos\theta$$

is projection onto the first component.

Three.III.2.16 Denote the given basis of $\mathcal{P}_2$ by B. Application of the linear map is represented by matrix-vector multiplication. Thus the first vector in $\mathcal{E}_3$ maps to the element of $\mathcal{P}_2$ represented with respect to B by

$$\begin{pmatrix} 1 & 3 & 0 \\ 0 & 1 & 0 \\ 1 & 0 & 1 \end{pmatrix} \begin{pmatrix} 1 \\ 0 \\ 0 \end{pmatrix} = \begin{pmatrix} 1 \\ 0 \\ 1 \end{pmatrix}$$

and that element is $1 + x$. Calculate the other two images of basis vectors in the same way.

$$\begin{pmatrix} 1 & 3 & 0 \\ 0 & 1 & 0 \\ 1 & 0 & 1 \end{pmatrix} \begin{pmatrix} 0 \\ 1 \\ 0 \end{pmatrix} = \begin{pmatrix} 3 \\ 1 \\ 0 \end{pmatrix} = \mathrm{Rep}_B(4 + x^2) \qquad \begin{pmatrix} 1 & 3 & 0 \\ 0 & 1 & 0 \\ 1 & 0 & 1 \end{pmatrix} \begin{pmatrix} 0 \\ 0 \\ 1 \end{pmatrix} = \begin{pmatrix} 0 \\ 0 \\ 1 \end{pmatrix} = \mathrm{Rep}_B(x)$$

So the range of h is the span of three polynomials $1 + x$, $4 + x^2$, and x. We can thus decide if $1 + 2x$ is in the range of the map by looking for scalars c_1, c_2, and c_3 such that

$$c_1 \cdot (1 + x) + c_2 \cdot (4 + x^2) + c_3 \cdot (x) = 1 + 2x$$

and obviously $c_1 = 1$, $c_2 = 0$, and $c_3 = 1$ suffice. Thus $1 + 2x$ is in the range, since it is the image of this vector.

$$1 \cdot \begin{pmatrix} 1 \\ 0 \\ 0 \end{pmatrix} + 0 \cdot \begin{pmatrix} 0 \\ 1 \\ 0 \end{pmatrix} + 1 \cdot \begin{pmatrix} 0 \\ 0 \\ 1 \end{pmatrix}$$

Comment. A slicker argument is to note that the matrix is nonsingular, so it has rank 3, so the range has dimension 3, and since the codomain has dimension 3 the map is onto. Thus every polynomial is the image of some vector and in particular $1 + 2x$ is the image of a vector in the domain.

Three.III.2.17 Let the matrix be G, and suppose that it represents $g\colon V \to W$ with respect to bases B and D. Because G has two columns, V is two-dimensional. Because G has two rows, W is two-dimensional. The action of g on a general member of the domain is this.

$$\begin{pmatrix} x \\ y \end{pmatrix}_B \mapsto \begin{pmatrix} x + 2y \\ 3x + 6y \end{pmatrix}_D$$

(a) The only representation of the zero vector in the codomain is

$$\mathrm{Rep}_D(\vec{0}) = \begin{pmatrix} 0 \\ 0 \end{pmatrix}_D$$

and so the set of representations of members of the null space is this.

$$\{\begin{pmatrix} x \\ y \end{pmatrix}_B \mid x + 2y = 0 \text{ and } 3x + 6y = 0\} = \{y \cdot \begin{pmatrix} -1/2 \\ 1 \end{pmatrix}_D \mid y \in \mathbb{R}\}$$

(b) The representation map $\mathrm{Rep}_D\colon W \to \mathbb{R}^2$ and its inverse are isomorphisms, and so preserve the dimension of subspaces. The subspace of $\mathbb{R}^2$ that is in the prior item is one-dimensional. Therefore, the image of that subspace under the inverse of the representation map—the null space of G, is also one-dimensional.

(c) The set of representations of members of the range space is this.

$$\{\begin{pmatrix} x + 2y \\ 3x + 6y \end{pmatrix}_D \mid x, y \in \mathbb{R}\} = \{k \cdot \begin{pmatrix} 1 \\ 3 \end{pmatrix}_D \mid k \in \mathbb{R}\}$$

(d) Of course, Theorem 2.4 gives that the rank of the map equals the rank of the matrix, which is one. Alternatively, the same argument that we used above for the null space gives here that the dimension of the range space is one.

(e) One plus one equals two.

Three.III.2.18 (a) The defined map h is onto if and only if for every $\vec{w} \in W$ there is a $\vec{v} \in V$ such that $h(\vec{v}) = \vec{w}$. Since for every vector there is exactly one representation, converting to representations gives that h is onto if and only if for every representation $\mathrm{Rep}_D(\vec{w})$ there is a representation $\mathrm{Rep}_B(\vec{v})$ such that $H \cdot \mathrm{Rep}_B(\vec{v}) = \mathrm{Rep}_D(\vec{w})$.

(b) This is just like the prior part.

(c) As described at the start of this subsection, by definition the map h defined by the matrix H associates this domain vector $\vec{v}$ with this codomain vector $\vec{w}$.

$$\mathrm{Rep}_B(\vec{v}) = \begin{pmatrix} v_1 \\ \vdots \\ v_n \end{pmatrix} \qquad \mathrm{Rep}_D(\vec{w}) = H \cdot \mathrm{Rep}_B(\vec{v}) = \begin{pmatrix} h_{1,1}v_1 + \cdots + h_{1,n}v_n \\ \vdots \\ h_{m,1}v_1 + \cdots + h_{m,n}v_n \end{pmatrix}$$

Fix $\vec{w} \in W$ and consider the linear system defined by the above equation.

$$\begin{aligned} h_{1,1}v_1 + \cdots + h_{1,n}v_n &= w_1 \\ h_{2,1}v_1 + \cdots + h_{2,n}v_n &= w_2 \\ &\vdots \\ h_{n,1}v_1 + \cdots + h_{n,n}v_n &= w_n \end{aligned}$$

(Again, here the w_i are fixed and the v_j are unknowns.) Now, H is nonsingular if and only if for all w_1, ..., w_n this system has a solution and the solution is unique. By the first two parts of this exercise this is true if and only if the map h is onto and one-to-one. This in turn is true if and only if h is an isomorphism.

Three.III.2.19 No, the range spaces may differ. Example 2.3 shows this.

Three.III.2.20 Recall that the representation map

$$V \overset{\mathrm{Rep}_B}{\longmapsto} \mathbb{R}^n$$

is an isomorphism. Thus, its inverse map $\mathrm{Rep}_B^{-1} \colon \mathbb{R}^n \to V$ is also an isomorphism. The desired transformation of $\mathbb{R}^n$ is then this composition.

$$\mathbb{R}^n \overset{\mathrm{Rep}_B^{-1}}{\longmapsto} V \overset{\mathrm{Rep}_D}{\longmapsto} \mathbb{R}^n$$

Because a composition of isomorphisms is also an isomorphism, this map $\mathrm{Rep}_D \circ \mathrm{Rep}_B^{-1}$ is an isomorphism.

Three.III.2.21 Yes. Consider

$$H = \begin{pmatrix} 1 & 0 \\ 0 & 1 \end{pmatrix}$$

representing a map from $\mathbb{R}^2$ to $\mathbb{R}^2$. With respect to the standard bases $B_1 = \mathcal{E}_2, D_1 = \mathcal{E}_2$ this matrix represents the identity map. With respect to

$$B_2 = D_2 = \langle \begin{pmatrix} 1 \\ 1 \end{pmatrix}, \begin{pmatrix} 1 \\ -1 \end{pmatrix} \rangle$$

this matrix again represents the identity. In fact, as long as the starting and ending bases are equal—as long as $B_i = D_i$—then the map represented by H is the identity.

Three.III.2.22 This is immediate from Lemma 2.9.

Three.III.2.23 The first map

$$\begin{pmatrix} x \\ y \end{pmatrix} = \begin{pmatrix} x \\ y \end{pmatrix}_{\mathcal{E}_2} \mapsto \begin{pmatrix} 3x \\ 2y \end{pmatrix}_{\mathcal{E}_2} = \begin{pmatrix} 3x \\ 2y \end{pmatrix}$$

stretches vectors by a factor of three in the x direction and by a factor of two in the y direction. The second map

$$\begin{pmatrix} x \\ y \end{pmatrix} = \begin{pmatrix} x \\ y \end{pmatrix}_{\mathcal{E}_2} \mapsto \begin{pmatrix} x \\ 0 \end{pmatrix}_{\mathcal{E}_2} = \begin{pmatrix} x \\ 0 \end{pmatrix}$$

projects vectors onto the x axis. The third

$$\begin{pmatrix} x \\ y \end{pmatrix} = \begin{pmatrix} x \\ y \end{pmatrix}_{\mathcal{E}_2} \mapsto \begin{pmatrix} y \\ x \end{pmatrix}_{\mathcal{E}_2} = \begin{pmatrix} y \\ x \end{pmatrix}$$

interchanges first and second components (that is, it is a reflection about the line $y = x$). The last

$$\begin{pmatrix} x \\ y \end{pmatrix} = \begin{pmatrix} x \\ y \end{pmatrix}_{\mathcal{E}_2} \mapsto \begin{pmatrix} x+3y \\ y \end{pmatrix}_{\mathcal{E}_2} = \begin{pmatrix} x+3y \\ y \end{pmatrix}$$

stretches vectors parallel to the y axis, by an amount equal to three times their distance from that axis (this is a *skew.*)

Three.III.2.24 **(a)** This is immediate from Theorem 2.4.

(b) Yes. This is immediate from the prior item.

To give a specific example, we can start with $\mathcal{E}_3$ as the basis for the domain, and then we require a basis D for the codomain $\mathbb{R}^3$. The matrix H gives the action of the map as this

$$\begin{pmatrix} 1 \\ 0 \\ 0 \end{pmatrix} = \begin{pmatrix} 1 \\ 0 \\ 0 \end{pmatrix}_{\mathcal{E}_3} \mapsto \begin{pmatrix} 1 \\ 2 \\ 0 \end{pmatrix}_D \quad \begin{pmatrix} 0 \\ 1 \\ 0 \end{pmatrix} = \begin{pmatrix} 0 \\ 1 \\ 0 \end{pmatrix}_{\mathcal{E}_3} \mapsto \begin{pmatrix} 0 \\ 0 \\ 1 \end{pmatrix}_D \quad \begin{pmatrix} 0 \\ 0 \\ 1 \end{pmatrix} = \begin{pmatrix} 0 \\ 0 \\ 1 \end{pmatrix}_{\mathcal{E}_3} \mapsto \begin{pmatrix} 0 \\ 0 \\ 0 \end{pmatrix}_D$$

and there is no harm in finding a basis D so that

$$\mathrm{Rep}_D(\begin{pmatrix} 1 \\ 0 \\ 0 \end{pmatrix}) = \begin{pmatrix} 1 \\ 2 \\ 0 \end{pmatrix}_D \quad \text{and} \quad \mathrm{Rep}_D(\begin{pmatrix} 0 \\ 1 \\ 0 \end{pmatrix}) = \begin{pmatrix} 0 \\ 0 \\ 1 \end{pmatrix}_D$$

that is, so that the map represented by H with respect to $\mathcal{E}_3, D$ is projection down onto the xy plane. The second condition gives that the third member of D is $\vec{e}_2$. The first condition gives that the first member of D plus twice the second equals $\vec{e}_1$, and so this basis will do.

$$D = \langle \begin{pmatrix} 0 \\ -1 \\ 0 \end{pmatrix}, \begin{pmatrix} 1/2 \\ 1/2 \\ 0 \end{pmatrix}, \begin{pmatrix} 0 \\ 1 \\ 0 \end{pmatrix} \rangle$$

Three.III.2.25 **(a)** Recall that the representation map $\mathrm{Rep}_B\colon V \to \mathbb{R}^n$ is linear (it is actually an isomorphism, but we do not need that it is one-to-one or onto here). Considering the column vector x to be a $n\times 1$ matrix gives that the map from $\mathbb{R}^n$ to $\mathbb{R}$ that takes a column vector to its dot product with $\vec{x}$ is linear (this is a matrix-vector product and so Theorem 2.2 applies). Thus the map under consideration $h_{\vec{x}}$ is linear because it is the composition of two linear maps.

$$\vec{v} \mapsto \mathrm{Rep}_B(\vec{v}) \mapsto \vec{x}\cdot\mathrm{Rep}_B(\vec{v})$$

(b) Any linear map $g\colon V \to \mathbb{R}$ is represented by some matrix

$$\begin{pmatrix} g_1 & g_2 & \cdots & g_n \end{pmatrix}$$

(the matrix has n columns because V is n-dimensional and it has only one row because $\mathbb{R}$ is one-dimensional). Then taking $\vec{x}$ to be the column vector that is the transpose of this matrix

$$\vec{x} = \begin{pmatrix} g_1 \\ \vdots \\ g_n \end{pmatrix}$$

has the desired action.

$$\vec{v} = \begin{pmatrix} v_1 \\ \vdots \\ v_n \end{pmatrix} \mapsto \begin{pmatrix} g_1 \\ \vdots \\ g_n \end{pmatrix} \cdot \begin{pmatrix} v_1 \\ \vdots \\ v_n \end{pmatrix} = g_1v_1 + \cdots + g_nv_n$$

(c) No. If $\vec{x}$ has any nonzero entries then $h_{\vec{x}}$ cannot be the zero map (and if $\vec{x}$ is the zero vector then $h_{\vec{x}}$ can only be the zero map).

Three.III.2.26 See the following section.

Matrix Operations

Three.IV.1: Sums and Scalar Products

Three.IV.1.8 (a) $\begin{pmatrix} 7 & 0 & 6 \\ 9 & 1 & 6 \end{pmatrix}$ (b) $\begin{pmatrix} 12 & -6 & -6 \\ 6 & 12 & 18 \end{pmatrix}$ (c) $\begin{pmatrix} 4 & 2 \\ 0 & 6 \end{pmatrix}$ (d) $\begin{pmatrix} -1 & 28 \\ 2 & 1 \end{pmatrix}$ (e) Not defined.

Three.IV.1.9 Represent the domain vector $\vec{v} \in V$ and the maps $g, h\colon V \to W$ with respect to bases B, D in the usual way.

(a) The representation of $(g+h)(\vec{v}) = g(\vec{v}) + h(\vec{v})$

$$\begin{aligned}&\big((g_{1,1}v_1 + \cdots + g_{1,n}v_n)\vec{\delta}_1 + \cdots + (g_{m,1}v_1 + \cdots + g_{m,n}v_n)\vec{\delta}_m\big)\\ &\qquad + \big((h_{1,1}v_1 + \cdots + h_{1,n}v_n)\vec{\delta}_1 + \cdots + (h_{m,1}v_1 + \cdots + h_{m,n}v_n)\vec{\delta}_m\big)\end{aligned}$$

regroups

$$= \big((g_{1,1} + h_{1,1})v_1 + \cdots + (g_{1,1} + h_{1,n})v_n\big) \cdot \vec{\delta}_1 + \cdots + \big((g_{m,1} + h_{m,1})v_1 + \cdots + (g_{m,n} + h_{m,n})v_n\big) \cdot \vec{\delta}_m$$

to the entry-by-entry sum of the representation of $g(\vec{v})$ and the representation of $h(\vec{v})$.

(b) The representation of $(r \cdot h)(\vec{v}) = r \cdot \big(h(\vec{v})\big)$

$$\begin{aligned}&r \cdot \big((h_{1,1}v_1 + h_{1,2}v_2 + \cdots + h_{1,n}v_n)\vec{\delta}_1 + \cdots + (h_{m,1}v_1 + h_{m,2}v_2 + \cdots + h_{m,n}v_n)\vec{\delta}_m\big)\\ &\qquad = (rh_{1,1}v_1 + \cdots + rh_{1,n}v_n) \cdot \vec{\delta}_1 + \cdots + (rh_{m,1}v_1 + \cdots + rh_{m,n}v_n) \cdot \vec{\delta}_m\end{aligned}$$

is the entry-by-entry multiple of r and the representation of h.

Three.IV.1.10 First, each of these properties is easy to check in an entry-by-entry way. For example, writing

$$G = \begin{pmatrix} g_{1,1} & \ldots & g_{1,n} \\ \vdots & & \vdots \\ g_{m,1} & \ldots & g_{m,n} \end{pmatrix} \qquad H = \begin{pmatrix} h_{1,1} & \ldots & h_{1,n} \\ \vdots & & \vdots \\ h_{m,1} & \ldots & h_{m,n} \end{pmatrix}$$

then, by definition we have

$$G + H = \begin{pmatrix} g_{1,1} + h_{1,1} & \ldots & g_{1,n} + h_{1,n} \\ \vdots & & \vdots \\ g_{m,1} + h_{m,1} & \ldots & g_{m,n} + h_{m,n} \end{pmatrix} \qquad H + G = \begin{pmatrix} h_{1,1} + g_{1,1} & \ldots & h_{1,n} + g_{1,n} \\ \vdots & & \vdots \\ h_{m,1} + g_{m,1} & \ldots & h_{m,n} + g_{m,n} \end{pmatrix}$$

and the two are equal since their entries are equal $g_{i,j} + h_{i,j} = h_{i,j} + g_{i,j}$. That is, each of these is easy to check by using Definition 1.3 alone.

However, each property is also easy to understand in terms of the represented maps, by applying Theorem 1.4 as well as the definition.

(a) The two maps $g+h$ and $h+g$ are equal because $g(\vec{v}) + h(\vec{v}) = h(\vec{v}) + g(\vec{v})$, as addition is commutative in any vector space. Because the maps are the same, they must have the same representative.

(b) As with the prior answer, except that here we apply that vector space addition is associative.

(c) As before, except that here we note that $g(\vec{v}) + z(\vec{v}) = g(\vec{v}) + \vec{0} = g(\vec{v})$.

(d) Apply that $0 \cdot g(\vec{v}) = \vec{0} = z(\vec{v})$.

(e) Apply that $(r+s) \cdot g(\vec{v}) = r \cdot g(\vec{v}) + s \cdot g(\vec{v})$.

(f) Apply the prior two items with $r = 1$ and $s = -1$.

(g) Apply that $r \cdot (g(\vec{v}) + h(\vec{v})) = r \cdot g(\vec{v}) + r \cdot h(\vec{v})$.

(h) Apply that $(rs) \cdot g(\vec{v}) = r \cdot (s \cdot g(\vec{v}))$.

Three.IV.1.11 For any V, W with bases B, D, the (appropriately-sized) zero matrix represents this map.

$$\vec{\beta}_1 \mapsto 0 \cdot \vec{\delta}_1 + \cdots + 0 \cdot \vec{\delta}_m \quad \cdots \quad \vec{\beta}_n \mapsto 0 \cdot \vec{\delta}_1 + \cdots + 0 \cdot \vec{\delta}_m$$

This is the zero map.

There are no other matrices that represent only one map. For, suppose that H is not the zero matrix. Then it has a nonzero entry; assume that $h_{i,j} \neq 0$. With respect to bases B, D, it represents $h_1 \colon V \to W$ sending

$$\vec{\beta}_j \mapsto h_{1,j}\vec{\delta}_1 + \cdots + h_{i,j}\vec{\delta}_i + \cdots + h_{m,j}\vec{\delta}_m$$

and with respect to $B, 2\cdot D$ it also represents $h_2 \colon V \to W$ sending

$$\vec{\beta}_j \mapsto h_{1,j}\cdot(2\vec{\delta}_1) + \cdots + h_{i,j}\cdot(2\vec{\delta}_i) + \cdots + h_{m,j}\cdot(2\vec{\delta}_m)$$

(the notation $2\cdot D$ means to double all of the members of D). These maps are easily seen to be unequal.

Three.IV.1.12 Fix bases B and D for V and W, and consider $\mathrm{Rep}_{B,D} \colon \mathcal{L}(V,W) \to \mathcal{M}_{m\times n}$ associating each linear map with the matrix representing that map $h \mapsto \mathrm{Rep}_{B,D}(h)$. From the prior section we know that (under fixed bases) the matrices correspond to linear maps, so the representation map is one-to-one and onto. That it preserves linear operations is Theorem 1.4.

Three.IV.1.13 Fix bases and represent the transformations with 2×2 matrices. The space of matrices $\mathcal{M}_{2\times 2}$ has dimension four, and hence the above six-element set is linearly dependent. By the prior exercise that extends to a dependence of maps. (The misleading part is only that there are six transformations, not five, so that we have more than we need to give the existence of the dependence.)

Three.IV.1.14 That the trace of a sum is the sum of the traces holds because both $\mathrm{trace}(H+G)$ and $\mathrm{trace}(H)+\mathrm{trace}(G)$ are the sum of $h_{1,1}+g_{1,1}$ with $h_{2,2}+g_{2,2}$, etc. For scalar multiplication we have $\mathrm{trace}(r\cdot H) = r\cdot\mathrm{trace}(H)$; the proof is easy. Thus the trace map is a homomorphism from $\mathcal{M}_{n\times n}$ to $\mathbb{R}$.

Three.IV.1.15 **(a)** The i,j entry of $(G+H)^{\mathsf{T}}$ is $g_{j,i}+h_{j,i}$. That is also the i,j entry of $G^{\mathsf{T}}+H^{\mathsf{T}}$.
(b) The i,j entry of $(r\cdot H)^{\mathsf{T}}$ is $rh_{j,i}$, which is also the i,j entry of $r\cdot H^{\mathsf{T}}$.

Three.IV.1.16 **(a)** For $H+H^{\mathsf{T}}$, the i,j entry is $h_{i,j}+h_{j,i}$ and the j,i entry of is $h_{j,i}+h_{i,j}$. The two are equal and thus $H+H^{\mathsf{T}}$ is symmetric.

Every symmetric matrix does have that form, since we can write $H = (1/2)\cdot(H+H^{\mathsf{T}})$.

(b) The set of symmetric matrices is nonempty as it contains the zero matrix. Clearly a scalar multiple of a symmetric matrix is symmetric. A sum $H+G$ of two symmetric matrices is symmetric because $h_{i,j}+g_{i,j} = h_{j,i}+g_{j,i}$ (since $h_{i,j}=h_{j,i}$ and $g_{i,j}=g_{j,i}$). Thus the subset is nonempty and closed under the inherited operations, and so it is a subspace.

Three.IV.1.17 **(a)** Scalar multiplication leaves the rank of a matrix unchanged except that multiplication by zero leaves the matrix with rank zero. (This follows from the first theorem of the book, that multiplying a row by a nonzero scalar doesn't change the solution set of the associated linear system.)
(b) A sum of rank n matrices can have rank less than n. For instance, for any matrix H, the sum $H+(-1)\cdot H$ has rank zero.

A sum of rank n matrices can have rank greater than n. Here are rank one matrices that sum to a rank two matrix.

$$\begin{pmatrix}1&0\\0&0\end{pmatrix}+\begin{pmatrix}0&0\\0&1\end{pmatrix}=\begin{pmatrix}1&0\\0&1\end{pmatrix}$$

Three.IV.2: Matrix Multiplication

Three.IV.2.14 **(a)** $\begin{pmatrix}0&15.5\\0&-19\end{pmatrix}$ **(b)** $\begin{pmatrix}2&-1&-1\\17&-1&-1\end{pmatrix}$ **(c)** Not defined. **(d)** $\begin{pmatrix}1&0\\0&1\end{pmatrix}$

Three.IV.2.15 **(a)** $\begin{pmatrix}1&-2\\10&4\end{pmatrix}$ **(b)** $\begin{pmatrix}1&-2\\10&4\end{pmatrix}\begin{pmatrix}-2&3\\-4&1\end{pmatrix}=\begin{pmatrix}6&1\\-36&34\end{pmatrix}$ **(c)** $\begin{pmatrix}-18&17\\-24&16\end{pmatrix}$
(d) $\begin{pmatrix}1&-1\\2&0\end{pmatrix}\begin{pmatrix}-18&17\\-24&16\end{pmatrix}=\begin{pmatrix}6&1\\-36&34\end{pmatrix}$

Three.IV.2.16 **(a)** Yes. **(b)** Yes. **(c)** No. **(d)** No.

Three.IV.2.17 **(a)** 2×1 **(b)** 1×1 **(c)** Not defined. **(d)** 2×2

Three.IV.2.18 We have

$$\begin{aligned}h_{1,1}\cdot(g_{1,1}y_1+g_{1,2}y_2)+h_{1,2}\cdot(g_{2,1}y_1+g_{2,2}y_2)+h_{1,3}\cdot(g_{3,1}y_1+g_{3,2}y_2)&=d_1\\h_{2,1}\cdot(g_{1,1}y_1+g_{1,2}y_2)+h_{2,2}\cdot(g_{2,1}y_1+g_{2,2}y_2)+h_{2,3}\cdot(g_{3,1}y_1+g_{3,2}y_2)&=d_2\end{aligned}$$

which, after expanding and regrouping about the y's yields this.

$$\begin{aligned}(h_{1,1}g_{1,1}+h_{1,2}g_{2,1}+h_{1,3}g_{3,1})y_1+(h_{1,1}g_{1,2}+h_{1,2}g_{2,2}+h_{1,3}g_{3,2})y_2&=d_1\\(h_{2,1}g_{1,1}+h_{2,2}g_{2,1}+h_{2,3}g_{3,1})y_1+(h_{2,1}g_{1,2}+h_{2,2}g_{2,2}+h_{2,3}g_{3,2})y_2&=d_2\end{aligned}$$

We can express the starting system and the system used for the substitutions in matrix language.

$$\begin{pmatrix}h_{1,1}&h_{1,2}&h_{1,3}\\h_{2,1}&h_{2,2}&h_{2,3}\end{pmatrix}\begin{pmatrix}x_1\\x_2\\x_3\end{pmatrix}=H\begin{pmatrix}x_1\\x_2\\x_3\end{pmatrix}=\begin{pmatrix}d_1\\d_2\end{pmatrix}\qquad\begin{pmatrix}g_{1,1}&g_{1,2}\\g_{2,1}&g_{2,2}\\g_{3,1}&g_{3,2}\end{pmatrix}\begin{pmatrix}y_1\\y_2\end{pmatrix}=G\begin{pmatrix}y_1\\y_2\end{pmatrix}=\begin{pmatrix}x_1\\x_2\\x_3\end{pmatrix}$$

With this, the substitution is $\vec{d}=H\vec{x}=H(G\vec{y})=(HG)\vec{y}$.

Three.IV.2.19 Technically, no. The dot product operation yields a scalar while the matrix product yields a 1×1 matrix. However, we usually will ignore the distinction.

Three.IV.2.20 The action of d/dx on B is $1\mapsto 0$, $x\mapsto 1$, $x^2\mapsto 2x$, ... and so this is its $(n+1)\times(n+1)$ matrix representation.

$$\text{Rep}_{B,B}(\frac{d}{dx})=\begin{pmatrix}0&1&0&&0\\0&0&2&&0\\&&&\ddots&\\0&0&0&&n\\0&0&0&&0\end{pmatrix}$$

The product of this matrix with itself is defined because the matrix is square.

$$\begin{pmatrix}0&1&0&&0\\0&0&2&&0\\&&&\ddots&\\0&0&0&&n\\0&0&0&&0\end{pmatrix}^2=\begin{pmatrix}0&0&2&0&&0\\0&0&0&6&&0\\&&&&\ddots&\\0&0&0&&&n(n-1)\\0&0&0&&&0\\0&0&0&&&0\end{pmatrix}$$

The map so represented is the composition

$$p\overset{\frac{d}{dx}}{\longmapsto}\frac{d\,p}{dx}\overset{\frac{d}{dx}}{\longmapsto}\frac{d^2\,p}{dx^2}$$

which is the second derivative operation.

Three.IV.2.21 **(a)** iii
(b) iv
(c) None
(d) None (or (i) if we allow multiplication from the left)

Three.IV.2.22 It is true for all one-dimensional spaces. Let f and g be transformations of a one-dimensional space. We must show that $g\circ f\,(\vec{v})=f\circ g\,(\vec{v})$ for all vectors. Fix a basis B for the space and then the transformations are represented by 1×1 matrices.

$$F=\text{Rep}_{B,B}(f)=\begin{pmatrix}f_{1,1}\end{pmatrix}\qquad G=\text{Rep}_{B,B}(g)=\begin{pmatrix}g_{1,1}\end{pmatrix}$$

Therefore, the compositions can be represented as GF and FG.

$$GF = \text{Rep}_{B,B}(g \circ f) = \begin{pmatrix} g_{1,1}f_{1,1} \end{pmatrix} \qquad FG = \text{Rep}_{B,B}(f \circ g) = \begin{pmatrix} f_{1,1}g_{1,1} \end{pmatrix}$$

These two matrices are equal and so the compositions have the same effect on each vector in the space.

Three.IV.2.23 It would not represent linear map composition; Theorem 2.6 would fail.

Three.IV.2.24 Each follows easily from the associated map fact. For instance, p applications of the transformation h, following q applications, is simply $p+q$ applications.

Three.IV.2.25 Although we can do these by going through the indices, they are best understood in terms of the represented maps. That is, fix spaces and bases so that the matrices represent linear maps f, g, h.

(a) Yes; we have both $r \cdot (g \circ h)(\vec{v}) = r \cdot g(\,h(\vec{v})\,) = (r \cdot g) \circ h\,(\vec{v})$ and $g \circ (r \cdot h)\,(\vec{v}) = g(\,r \cdot h(\vec{v})\,) = r \cdot g(h(\vec{v})) = r \cdot (g \circ h)\,(\vec{v})$ (the second equality holds because of the linearity of g).

(b) Both answers are yes. First, $f \circ (rg+sh)$ and $r \cdot (f \circ g) + s \cdot (f \circ h)$ both send $\vec{v}$ to $r \cdot f(g(\vec{v})) + s \cdot f(h(\vec{v}))$; the calculation is as in the prior item (using the linearity of f for the first one). For the other, $(rf+sg) \circ h$ and $r \cdot (f \circ h) + s \cdot (g \circ h)$ both send $\vec{v}$ to $r \cdot f(h(\vec{v})) + s \cdot g(h(\vec{v}))$.

Three.IV.2.26 We have not seen a map interpretation of the transpose operation, so we will verify these by considering the entries.

(a) The i,j entry of GH^{T} is the j,i entry of GH, which is the dot product of the j-th row of G and the i-th column of H. The i,j entry of $H^{\mathsf{T}}G^{\mathsf{T}}$ is the dot product of the i-th row of H^{T} and the j-th column of G^{T}, which is the dot product of the i-th column of H and the j-th row of G. Dot product is commutative and so these two are equal.

(b) By the prior item each equals its transpose, e.g., $(HH^{\mathsf{T}})^{\mathsf{T}} = H^{\mathsf{T}^{\mathsf{T}}}H^{\mathsf{T}} = HH^{\mathsf{T}}$.

Three.IV.2.27 Consider $r_x, r_y \colon \mathbb{R}^3 \to \mathbb{R}^3$ rotating all vectors $\pi/2$ radians counterclockwise about the x and y axes (counterclockwise in the sense that a person whose head is at $\vec{e}_1$ or $\vec{e}_2$ and whose feet are at the origin sees, when looking toward the origin, the rotation as counterclockwise).

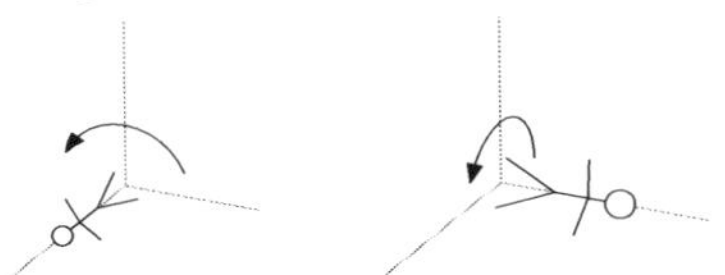

Rotating r_x first and then r_y is different than rotating r_y first and then r_x. In particular, $r_x(\vec{e}_3) = -\vec{e}_2$ so $r_y \circ r_x(\vec{e}_3) = -\vec{e}_2$, while $r_y(\vec{e}_3) = \vec{e}_1$ so $r_x \circ r_y(\vec{e}_3) = \vec{e}_1$, and hence the maps do not commute.

Three.IV.2.28 It doesn't matter (as long as the spaces have the appropriate dimensions).

For associativity, suppose that F is $m \times r$, that G is $r \times n$, and that H is $n \times k$. We can take any r dimensional space, any m dimensional space, any n dimensional space, and any k dimensional space—for instance, $\mathbb{R}^r$, $\mathbb{R}^m$, $\mathbb{R}^n$, and $\mathbb{R}^k$ will do. We can take any bases A, B, C, and D, for those spaces. Then, with respect to C, D the matrix H represents a linear map h, with respect to B, C the matrix G represents a g, and with respect to A, B the matrix F represents an f. We can use those maps in the proof.

The second half is similar, except that we add G and H and so we must take them to represent maps with the same domain and codomain.

Three.IV.2.29 **(a)** The product of rank n matrices can have rank less than or equal to n but not greater than n.

To see that the rank can fall, consider the maps $\pi_x, \pi_y \colon \mathbb{R}^2 \to \mathbb{R}^2$ projecting onto the axes. Each is rank one but their composition $\pi_x \circ \pi_y$, which is the zero map, is rank zero. That translates over to matrices representing those maps in this way.

$$\text{Rep}_{\mathcal{E}_2,\mathcal{E}_2}(\pi_x) \cdot \text{Rep}_{\mathcal{E}_2,\mathcal{E}_2}(\pi_y) = \begin{pmatrix} 1 & 0 \\ 0 & 0 \end{pmatrix} \begin{pmatrix} 0 & 0 \\ 0 & 1 \end{pmatrix} = \begin{pmatrix} 0 & 0 \\ 0 & 0 \end{pmatrix}$$

To prove that the product of rank n matrices cannot have rank greater than n, we can apply the map result that the image of a linearly dependent set is linearly dependent. That is, if $h \colon V \to W$ and

$g\colon W \to X$ both have rank n then a set in the range $\mathscr{R}(g \circ h)$ of size larger than n is the image under g of a set in W of size larger than n and so is linearly dependent (since the rank of h is n). Now, the image of a linearly dependent set is dependent, so any set of size larger than n in the range is dependent. (By the way, observe that the rank of g was not mentioned. See the next part.)

(b) Fix spaces and bases and consider the associated linear maps f and g. Recall that the dimension of the image of a map (the map's rank) is less than or equal to the dimension of the domain, and consider the arrow diagram.

$$V \stackrel{f}{\longmapsto} \mathscr{R}(f) \stackrel{g}{\longmapsto} \mathscr{R}(g \circ f)$$

First, the image of $\mathscr{R}(f)$ must have dimension less than or equal to the dimension of $\mathscr{R}(f)$, by the prior sentence. On the other hand, $\mathscr{R}(f)$ is a subset of the domain of g, and thus its image has dimension less than or equal the dimension of the domain of g. Combining those two, the rank of a composition is less than or equal to the minimum of the two ranks.

The matrix fact follows immediately.

Three.IV.2.30 The 'commutes with' relation is reflexive and symmetric. However, it is not transitive: for instance, with

$$G=\begin{pmatrix}1&2\\3&4\end{pmatrix}\quad H=\begin{pmatrix}1&0\\0&1\end{pmatrix}\quad J=\begin{pmatrix}5&6\\7&8\end{pmatrix}$$

G commutes with H and H commutes with J, but G does not commute with J.

Three.IV.2.31 **(a)** Either of these.

$$\begin{pmatrix}x\\y\\z\end{pmatrix}\stackrel{\pi_x}{\longmapsto}\begin{pmatrix}x\\0\\0\end{pmatrix}\stackrel{\pi_y}{\longmapsto}\begin{pmatrix}0\\0\\0\end{pmatrix}\qquad\begin{pmatrix}x\\y\\z\end{pmatrix}\stackrel{\pi_y}{\longmapsto}\begin{pmatrix}0\\y\\0\end{pmatrix}\stackrel{\pi_x}{\longmapsto}\begin{pmatrix}0\\0\\0\end{pmatrix}$$

(b) The composition is the fifth derivative map d^5/dx^5 on the space of fourth-degree polynomials.

(c) With respect to the natural bases,

$$\text{Rep}_{\mathcal{E}_3,\mathcal{E}_3}(\pi_x)=\begin{pmatrix}1&0&0\\0&0&0\\0&0&0\end{pmatrix}\qquad\text{Rep}_{\mathcal{E}_3,\mathcal{E}_3}(\pi_y)=\begin{pmatrix}0&0&0\\0&1&0\\0&0&0\end{pmatrix}$$

and their product (in either order) is the zero matrix.

(d) Where $B=\langle 1,x,x^2,x^3,x^4\rangle$,

$$\text{Rep}_{B,B}(\frac{d^2}{dx^2})=\begin{pmatrix}0&0&2&0&0\\0&0&0&6&0\\0&0&0&0&12\\0&0&0&0&0\\0&0&0&0&0\end{pmatrix}\qquad\text{Rep}_{B,B}(\frac{d^3}{dx^3})=\begin{pmatrix}0&0&0&6&0\\0&0&0&0&24\\0&0&0&0&0\\0&0&0&0&0\\0&0&0&0&0\end{pmatrix}$$

and their product (in either order) is the zero matrix.

Three.IV.2.32 Note that $(S+T)(S-T)=S^2-ST+TS-T^2$, so a reasonable try is to look at matrices that do not commute so that $-ST$ and TS don't cancel: with

$$S=\begin{pmatrix}1&2\\3&4\end{pmatrix}\quad T=\begin{pmatrix}5&6\\7&8\end{pmatrix}$$

we have the desired inequality.

$$(S+T)(S-T)=\begin{pmatrix}-56&-56\\-88&-88\end{pmatrix}\qquad S^2-T^2=\begin{pmatrix}-60&-68\\-76&-84\end{pmatrix}$$

Three.IV.2.33 Because the identity map acts on the basis B as $\vec{\beta}_1 \mapsto \vec{\beta}_1, \ldots, \vec{\beta}_n \mapsto \vec{\beta}_n$, the representation is this.

$$\begin{pmatrix} 1 & 0 & 0 & & 0 \\ 0 & 1 & 0 & & 0 \\ 0 & 0 & 1 & & 0 \\ & & & \ddots & \\ 0 & 0 & 0 & & 1 \end{pmatrix}$$

The second part of the question is obvious from Theorem 2.6.

Three.IV.2.34 Here are four solutions.

$$T = \begin{pmatrix} \pm 1 & 0 \\ 0 & \pm 1 \end{pmatrix}$$

Three.IV.2.35 **(a)** The vector space $\mathcal{M}_{2\times 2}$ has dimension four. The set $\{T^4, \ldots, T, I\}$ has five elements and thus is linearly dependent.

(b) Where T is $n\times n$, generalizing the argument from the prior item shows that there is such a polynomial of degree n^2 or less, since $\{T^{n^2}, \ldots, T, I\}$ is a n^2+1-member subset of the n^2-dimensional space $\mathcal{M}_{n\times n}$.

(c) First compute the powers

$$T^2 = \begin{pmatrix} 1/2 & -\sqrt{3}/2 \\ \sqrt{3}/2 & 1/2 \end{pmatrix} \qquad T^3 = \begin{pmatrix} 0 & -1 \\ 1 & 0 \end{pmatrix} \qquad T^4 = \begin{pmatrix} -1/2 & -\sqrt{3}/2 \\ \sqrt{3}/2 & -1/2 \end{pmatrix}$$

(observe that rotating by $\pi/6$ three times results in a rotation by $\pi/2$, which is indeed what T^3 represents). Then set $c_4T^4 + c_3T^3 + c_2T^2 + c_1T + c_0I$ equal to the zero matrix

$$\begin{pmatrix} -1/2 & -\sqrt{3}/2 \\ \sqrt{3}/2 & -1/2 \end{pmatrix} c_4 + \begin{pmatrix} 0 & -1 \\ 1 & 0 \end{pmatrix} c_3 + \begin{pmatrix} 1/2 & -\sqrt{3}/2 \\ \sqrt{3}/2 & 1/2 \end{pmatrix} c_2 + \begin{pmatrix} \sqrt{3}/2 & -1/2 \\ 1/2 & \sqrt{3}/2 \end{pmatrix} c_1 + \begin{pmatrix} 1 & 0 \\ 0 & 1 \end{pmatrix} c_0 = \begin{pmatrix} 0 & 0 \\ 0 & 0 \end{pmatrix}$$

to get this linear system.

$$\begin{aligned} -(1/2)c_4 \quad\quad + (1/2)c_2 + (\sqrt{3}/2)c_1 + c_0 &= 0 \\ -(\sqrt{3}/2)c_4 - c_3 - (\sqrt{3}/2)c_2 - (1/2)c_1 \quad &= 0 \\ (\sqrt{3}/2)c_4 + c_3 + (\sqrt{3}/2)c_2 + (1/2)c_1 \quad &= 0 \\ -(1/2)c_4 \quad\quad + (1/2)c_2 + (\sqrt{3}/2)c_1 + c_0 &= 0 \end{aligned}$$

Apply Gaussian reduction.

$$\xrightarrow[]{-\rho_1+\rho_4}\ \xrightarrow[]{\rho_2+\rho_3} \begin{aligned} -(1/2)c_4 \quad\quad + (1/2)c_2 + (\sqrt{3}/2)c_1 + c_0 &= 0 \\ -(\sqrt{3}/2)c_4 - c_3 - (\sqrt{3}/2)c_2 - (1/2)c_1 \quad &= 0 \\ 0 &= 0 \\ 0 &= 0 \end{aligned}$$

$$\xrightarrow[]{-\sqrt{3}\rho_1+\rho_2} \begin{aligned} -(1/2)c_4 \quad + (1/2)c_2 + (\sqrt{3}/2)c_1 + \quad c_0 &= 0 \\ -c_3 - \sqrt{3}c_2 - \quad 2c_1 - \sqrt{3}c_0 &= 0 \\ 0 &= 0 \\ 0 &= 0 \end{aligned}$$

Setting c_4, c_3, and c_2 to zero makes c_1 and c_0 also come out to be zero so no degree one or degree zero polynomial will do. Setting c_4 and c_3 to zero (and c_2 to one) gives a linear system

$$\begin{aligned} (1/2) + (\sqrt{3}/2)c_1 + \quad c_0 &= 0 \\ -\sqrt{3} - \quad 2c_1 - \sqrt{3}c_0 &= 0 \end{aligned}$$

with solution $c_1 = -\sqrt{3}$ and $c_0 = 1$. Conclusion: the polynomial $m(x) = x^2 - \sqrt{3}x + 1$ is minimal for the matrix T.

Three.IV.2.36 The check is routine:

$$a_0 + a_1x + \cdots + a_nx^n \stackrel{s}{\longmapsto} a_0x + a_1x^2 + \cdots + a_nx^{n+1} \stackrel{d/dx}{\longmapsto} a_0 + 2a_1x + \cdots + (n+1)a_nx^n$$

while

$$a_0 + a_1x + \cdots + a_nx^n \stackrel{d/dx}{\longmapsto} a_1 + \cdots + na_nx^{n-1} \stackrel{s}{\longmapsto} a_1x + \cdots + a_nx^n$$

so that under the map $(d/dx \circ s) - (s \circ d/dx)$ we have $a_0 + a_1x + \cdots + a_nx^n \mapsto a_0 + a_1x + \cdots + a_nx^n$.

Three.IV.2.37 **(a)** Tracing through the remark at the end of the subsection gives that the i,j entry of $(FG)H$ is this

$$\begin{aligned}\sum_{t=1}^{s}\Big(\sum_{k=1}^{r} f_{i,k}g_{k,t}\Big)h_{t,j} &= \sum_{t=1}^{s}\sum_{k=1}^{r}(f_{i,k}g_{k,t})h_{t,j} = \sum_{t=1}^{s}\sum_{k=1}^{r} f_{i,k}(g_{k,t}h_{t,j}) \\ &= \sum_{k=1}^{r}\sum_{t=1}^{s} f_{i,k}(g_{k,t}h_{t,j}) = \sum_{k=1}^{r} f_{i,k}\Big(\sum_{t=1}^{s} g_{k,t}h_{t,j}\Big)\end{aligned}$$

(the first equality comes from using the distributive law to multiply through the h's, the second equality is the associative law for real numbers, the third is the commutative law for reals, and the fourth equality follows on using the distributive law to factor the f's out), which is the i,j entry of $F(GH)$.

(b) The k-th component of $h(\vec{v})$ is

$$\sum_{j=1}^{n} h_{k,j}v_j$$

and so the i-th component of $g \circ h\,(\vec{v})$ is this

$$\begin{aligned}\sum_{k=1}^{r} g_{i,k}\Big(\sum_{j=1}^{n} h_{k,j}v_j\Big) &= \sum_{k=1}^{r}\sum_{j=1}^{n} g_{i,k}h_{k,j}v_j = \sum_{k=1}^{r}\sum_{j=1}^{n}(g_{i,k}h_{k,j})v_j \\ &= \sum_{j=1}^{n}\sum_{k=1}^{r}(g_{i,k}h_{k,j})v_j = \sum_{j=1}^{n}\Big(\sum_{k=1}^{r} g_{i,k}h_{k,j}\Big)\,v_j\end{aligned}$$

(the first equality holds by using the distributive law to multiply the g's through, the second equality represents the use of associativity of reals, the third follows by commutativity of reals, and the fourth comes from using the distributive law to factor the v's out).

Three.IV.3: Mechanics of Matrix Multiplication

Three.IV.3.24 **(a)** The second matrix has its first row multiplied by 3.

$$\begin{pmatrix} 3 & 6 \\ 3 & 4 \end{pmatrix}$$

(b) The second matrix has its second row multiplied by 2.

$$\begin{pmatrix} 1 & 2 \\ 6 & 8 \end{pmatrix}$$

(c) The second matrix undergoes the combination operation of replacing the second row with -2 times the first row added to the second.

$$\begin{pmatrix} 1 & 2 \\ 1 & 0 \end{pmatrix}$$

(d) The first matrix undergoes the column operation of: replace the second column by -1 times the first column plus the second.

$$\begin{pmatrix} 1 & 1 \\ 3 & 1 \end{pmatrix}$$

(e) The first matrix has its columns swapped.

$$\begin{pmatrix} 2 & 1 \\ 4 & 3 \end{pmatrix}$$

Three.IV.3.25 (a) The second matrix has its first row multiplied by 3 and its second row multiplied by 0.

$$\begin{pmatrix} 3 & 6 \\ 0 & 0 \end{pmatrix}$$

(b) The second matrix has its first row multiplied by 4 and its second row multiplied by 2.

$$\begin{pmatrix} 4 & 8 \\ 6 & 8 \end{pmatrix}$$

Three.IV.3.26 The pay due each person appears in the matrix product of the two arrays.

Three.IV.3.27 (a) The adjacency matrix is this (e.g, the first row shows that there is only one connection including Burlington, the road to Winooski).

$$\begin{pmatrix} 0 & 0 & 0 & 0 & 1 \\ 0 & 0 & 1 & 1 & 1 \\ 0 & 1 & 0 & 1 & 0 \\ 0 & 1 & 1 & 0 & 0 \\ 1 & 1 & 0 & 0 & 0 \end{pmatrix}$$

(b) Because these are two-way roads, any road connecting city i to city j gives a connection between city j and city i.

(c) The square of the adjacency matrix tells how cities are connected by trips involving two roads.

Three.IV.3.28 The product is the identity matrix (recall that $\cos^2\theta + \sin^2\theta = 1$). An explanation is that the given matrix represents, with respect to the standard bases, a rotation in $\mathbb{R}^2$ of θ radians while the transpose represents a rotation of $-\theta$ radians. The two cancel.

Three.IV.3.29 The set of diagonal matrices is nonempty as the zero matrix is diagonal. Clearly it is closed under scalar multiples and sums. Therefore it is a subspace. The dimension is n; here is a basis.

$$\{\begin{pmatrix} 1 & 0 & \ldots & \\ 0 & 0 & & \\ & & \ddots & \\ 0 & 0 & & 0 \end{pmatrix}, \ldots, \begin{pmatrix} 0 & 0 & \ldots & \\ 0 & 0 & & \\ & & \ddots & \\ 0 & 0 & & 1 \end{pmatrix}\}$$

Three.IV.3.30 No. In $\mathcal{P}_1$, with respect to the unequal bases $B = \langle 1, x \rangle$ and $D = \langle 1 + x, 1 - x \rangle$, the identity transformation is represented by this matrix.

$$\text{Rep}_{B,D}(\text{id}) = \begin{pmatrix} 1/2 & 1/2 \\ 1/2 & -1/2 \end{pmatrix}_{B,D}$$

Three.IV.3.31 For any scalar r and square matrix H we have $(rI)H = r(IH) = rH = r(HI) = (Hr)I = H(rI)$.

There are no other such matrices; here is an argument for 2×2 matrices that is easily extended to $n\times n$. If a matrix commutes with all others then it commutes with this unit matrix.

$$\begin{pmatrix} 0 & a \\ 0 & c \end{pmatrix} = \begin{pmatrix} a & b \\ c & d \end{pmatrix}\begin{pmatrix} 0 & 1 \\ 0 & 0 \end{pmatrix} = \begin{pmatrix} 0 & 1 \\ 0 & 0 \end{pmatrix}\begin{pmatrix} a & b \\ c & d \end{pmatrix} = \begin{pmatrix} c & d \\ 0 & 0 \end{pmatrix}$$

From this we first conclude that the upper left entry a must equal its lower right entry d. We also conclude that the lower left entry c is zero. The argument for the upper right entry b is similar.

Three.IV.3.32 It is false; these two don't commute.

$$\begin{pmatrix} 1 & 2 \\ 3 & 4 \end{pmatrix} \qquad \begin{pmatrix} 5 & 6 \\ 7 & 8 \end{pmatrix}$$

Three.IV.3.33 A permutation matrix has a single one in each row and column, and all its other entries are zeroes. Fix such a matrix. Suppose that the i-th row has its one in its j-th column. Then no other row has its one in the j-th column; every other row has a zero in the j-th column. Thus the dot product of the i-th row and any other row is zero.

The i-th row of the product is made up of the dot products of the i-th row of the matrix and the columns of the transpose. By the last paragraph, all such dot products are zero except for the i-th one, which is one.

Three.IV.3.34 The generalization is to go from the first and second rows to the i_1-th and i_2-th rows. Row i of GH is made up of the dot products of row i of G and the columns of H. Thus if rows i_1 and i_2 of G are equal then so are rows i_1 and i_2 of GH.

Three.IV.3.35 If the product of two diagonal matrices is defined—if both are $n\times n$—then the product of the diagonals is the diagonal of the products: where G, H are equal-sized diagonal matrices, GH is all zeros except each that i,i entry is $g_{i,i}h_{i,i}$.

Three.IV.3.36 One way to produce this matrix from the identity is to use the column operations of first multiplying the second column by three, and then adding the negative of the resulting second column to the first.

$$\begin{pmatrix}1&0\\0&1\end{pmatrix} \longrightarrow \begin{pmatrix}1&0\\0&3\end{pmatrix} \longrightarrow \begin{pmatrix}1&0\\-3&3\end{pmatrix}$$

In contrast with row operations, column operations are written from left to right, so this matrix product expresses doing the above two operations.

$$\begin{pmatrix}1&0\\0&3\end{pmatrix}\begin{pmatrix}1&0\\-1&1\end{pmatrix}$$

Remark. Alternatively, we could get the required matrix with row operations. Starting with the identity, first adding the negative of the first row to the second, and then multiplying the second row by three will work. Because we write successive row operations as matrix products from right to left, doing these two row operations is expressed with: the same matrix product.

Three.IV.3.37 The i-th row of GH is made up of the dot products of the i-th row of G with the columns of H. The dot product of a zero row with a column is zero.

It works for columns if stated correctly: if H has a column of zeros then GH (if defined) has a column of zeros. The proof is easy.

Three.IV.3.38 Perhaps the easiest way is to show that each $n\times m$ matrix is a linear combination of unit matrices in one and only one way:

$$c_1\begin{pmatrix}1&0&\ldots\\0&0\\\vdots\end{pmatrix}+\cdots+c_{n,m}\begin{pmatrix}0&0&\ldots\\\vdots\\0&\ldots&1\end{pmatrix}=\begin{pmatrix}a_{1,1}&a_{1,2}&\ldots\\\vdots\\a_{n,1}&\ldots&a_{n,m}\end{pmatrix}$$

has the unique solution $c_1=a_{1,1}$, $c_2=a_{1,2}$, etc.

Three.IV.3.39 Call that matrix F. We have

$$F^2=\begin{pmatrix}2&1\\1&1\end{pmatrix}\quad F^3=\begin{pmatrix}3&2\\2&1\end{pmatrix}\quad F^4=\begin{pmatrix}5&3\\3&2\end{pmatrix}$$

In general,

$$F^n=\begin{pmatrix}f_{n+1}&f_n\\f_n&f_{n-1}\end{pmatrix}$$

where f_i is the i-th Fibonacci number $f_i=f_{i-1}+f_{i-2}$ and $f_0=0$, $f_1=1$, which we verify by induction, based on this equation.

$$\begin{pmatrix}f_{i-1}&f_{i-2}\\f_{i-2}&f_{i-3}\end{pmatrix}\begin{pmatrix}1&1\\1&0\end{pmatrix}=\begin{pmatrix}f_i&f_{i-1}\\f_{i-1}&f_{i-2}\end{pmatrix}$$

Three.IV.3.40 *Chapter Five gives a less computational reason — the trace of a matrix is the second coefficient in its characteristic polynomial — but for now we can use indices.* We have

$$\begin{aligned}\mathrm{Tr}(GH) &= (g_{1,1}h_{1,1} + g_{1,2}h_{2,1} + \cdots + g_{1,n}h_{n,1}) \\ &\quad + (g_{2,1}h_{1,2} + g_{2,2}h_{2,2} + \cdots + g_{2,n}h_{n,2}) \\ &\quad + \cdots + (g_{n,1}h_{1,n} + g_{n,2}h_{2,n} + \cdots + g_{n,n}h_{n,n})\end{aligned}$$

while

$$\begin{aligned}\mathrm{Tr}(HG) &= (h_{1,1}g_{1,1} + h_{1,2}g_{2,1} + \cdots + h_{1,n}g_{n,1}) \\ &\quad + (h_{2,1}g_{1,2} + h_{2,2}g_{2,2} + \cdots + h_{2,n}g_{n,2}) \\ &\quad + \cdots + (h_{n,1}g_{1,n} + h_{n,2}g_{2,n} + \cdots + h_{n,n}g_{n,n})\end{aligned}$$

and the two are equal.

Three.IV.3.41 A matrix is upper triangular if and only if its i,j entry is zero whenever $i > j$. Thus, if G, H are upper triangular then $h_{i,j}$ and $g_{i,j}$ are zero when $i > j$. An entry in the product $p_{i,j} = g_{i,1}h_{1,j} + \cdots + g_{i,n}h_{n,j}$ is zero unless at least some of the terms are nonzero, that is, unless for at least some of the summands $g_{i,r}h_{r,j}$ both $i \leqslant r$ and $r \leqslant j$. Of course, if $i > j$ this cannot happen and so the product of two upper triangular matrices is upper triangular. (A similar argument works for lower triangular matrices.)

Three.IV.3.42 The sum along the i-th row of the product is this.

$$\begin{aligned}p_{i,1} + \cdots + p_{i,n} &= (h_{i,1}g_{1,1} + h_{i,2}g_{2,1} + \cdots + h_{i,n}g_{n,1}) \\ &\quad + (h_{i,1}g_{1,2} + h_{i,2}g_{2,2} + \cdots + h_{i,n}g_{n,2}) \\ &\quad + \cdots + (h_{i,1}g_{1,n} + h_{i,2}g_{2,n} + \cdots + h_{i,n}g_{n,n}) \\ &= h_{i,1}(g_{1,1} + g_{1,2} + \cdots + g_{1,n}) \\ &\quad + h_{i,2}(g_{2,1} + g_{2,2} + \cdots + g_{2,n}) \\ &\quad + \cdots + h_{i,n}(g_{n,1} + g_{n,2} + \cdots + g_{n,n}) \\ &= h_{i,1} \cdot 1 + \cdots + h_{i,n} \cdot 1 \\ &= 1\end{aligned}$$

Three.IV.3.43 Matrices representing (say, with respect to $\mathcal{E}_2, \mathcal{E}_2 \subset \mathbb{R}^2$) the maps that send

$$\vec{\beta}_1 \stackrel{h}{\longmapsto} \vec{\beta}_1 \quad \vec{\beta}_2 \stackrel{h}{\longmapsto} \vec{0}$$

and

$$\vec{\beta}_1 \stackrel{g}{\longmapsto} \vec{\beta}_2 \quad \vec{\beta}_2 \stackrel{g}{\longmapsto} \vec{0}$$

will do.

Three.IV.3.44 The combination is to have all entries of the matrix be zero except for one (possibly) nonzero entry in each row and column. We can write such a matrix as the product of a permutation matrix and a diagonal matrix, e.g.,

$$\begin{pmatrix} 0 & 4 & 0 \\ 2 & 0 & 0 \\ 0 & 0 & -5 \end{pmatrix} = \begin{pmatrix} 0 & 1 & 0 \\ 1 & 0 & 0 \\ 0 & 0 & 1 \end{pmatrix} \begin{pmatrix} 4 & 0 & 0 \\ 0 & 2 & 0 \\ 0 & 0 & -5 \end{pmatrix}$$

and its action is thus to rescale the rows and permute them.

Three.IV.3.45 **(a)** Each entry $p_{i,j} = g_{i,1}h_{1,j} + \cdots + g_{1,r}h_{r,1}$ takes r multiplications and there are $m \cdot n$ entries. Thus there are $m \cdot n \cdot r$ multiplications.

(b) Let H_1 be $5{\times}10$, let H_2 be $10{\times}20$, let H_3 be $20{\times}5$, let H_4 be $5{\times}1$. Then, using the formula from the prior part,

this association	*uses this many multiplications*
$((H_1H_2)H_3)H_4$	$1000+500+25=1525$
$(H_1(H_2H_3))H_4$	$1000+250+25=1275$
$(H_1H_2)(H_3H_4)$	$1000+100+100=1200$
$H_1(H_2(H_3H_4))$	$100+200+50=350$
$H_1((H_2H_3)H_4)$	$1000+50+50=1100$

shows which is cheapest.

(c) This is an improvement by S. Winograd of a formula due to V. Strassen: where $w = aA-(a-c-d)(A-C+D)$,

$$\begin{pmatrix}a&b\\c&d\end{pmatrix}\begin{pmatrix}A&B\\C&D\end{pmatrix}$$
$$=\begin{pmatrix}aA+bB & w+(c+d)(C-A)+(a+b-c-d)D\\ w+(a-c)(D-C)-d(A-B-C+D) & w+(a-c)(D-C)+(c+d)(C-A)\end{pmatrix}$$

takes seven multiplications and fifteen additions (save the intermediate results).

Three.IV.3.46 *This is how the answer was given in the cited source.* No, it does not. Let A and B represent, with respect to the standard bases, these transformations of $\mathbb{R}^3$.

$$\begin{pmatrix}x\\y\\z\end{pmatrix}\overset{a}{\longmapsto}\begin{pmatrix}x\\y\\0\end{pmatrix}\qquad\begin{pmatrix}x\\y\\z\end{pmatrix}\overset{a}{\longmapsto}\begin{pmatrix}0\\x\\y\end{pmatrix}$$

Observe that

$$\begin{pmatrix}x\\y\\z\end{pmatrix}\overset{abab}{\longmapsto}\begin{pmatrix}0\\0\\0\end{pmatrix}\quad\text{but}\quad\begin{pmatrix}x\\y\\z\end{pmatrix}\overset{baba}{\longmapsto}\begin{pmatrix}0\\0\\x\end{pmatrix}.$$

Three.IV.3.47 *This is how the answer was given in the cited source.*

(a) Obvious.

(b) If $A^{\mathsf{T}}A\vec{x}=\vec{0}$ then $\vec{y}\cdot\vec{y}=0$ where $\vec{y}=A\vec{x}$. Hence $\vec{y}=\vec{0}$ by (a).

The converse is obvious.

(c) By (b), $A\vec{x}_1,\ldots,A\vec{x}_n$ are linearly independent iff $A^{\mathsf{T}}A\vec{x}_1,\ldots,A^{\mathsf{T}}A\vec{v}_n$ are linearly independent.

(d) We have $\text{col rank}(A)=\text{col rank}(A^{\mathsf{T}}A)=\dim\{A^{\mathsf{T}}(A\vec{x})\mid\text{all }\vec{x}\}\leqslant\dim\{A^{\mathsf{T}}\vec{y}\mid\text{all }\vec{y}\}=\text{col rank}(A^{\mathsf{T}})$. Thus also $\text{col rank}(A^{\mathsf{T}})\leqslant\text{col rank}(A^{\mathsf{T}\mathsf{T}})$ and so we have $\text{col rank}(A)=\text{col rank}(A^{\mathsf{T}})=\text{row rank}(A)$.

Three.IV.3.48 *This is how the answer was given in the cited source.* Let $\langle\vec{z}_1,\ldots,\vec{z}_k\rangle$ be a basis for $\mathscr{R}(A)\cap\mathscr{N}(A)$ (k might be 0). Let $\vec{x}_1,\ldots,\vec{x}_k\in V$ be such that $A\vec{x}_i=\vec{z}_i$. Note $\{A\vec{x}_1,\ldots,A\vec{x}_k\}$ is linearly independent, and extend to a basis for $\mathscr{R}(A)$: $A\vec{x}_1,\ldots,A\vec{x}_k,A\vec{x}_{k+1},\ldots,A\vec{x}_{r_1}$ where $r_1=\dim(\mathscr{R}(A))$.

Now take $\vec{x}\in V$. Write

$$A\vec{x}=a_1(A\vec{x}_1)+\cdots+a_{r_1}(A\vec{x}_{r_1})$$

and so

$$A^2\vec{x}=a_1(A^2\vec{x}_1)+\cdots+a_{r_1}(A^2\vec{x}_{r_1}).$$

But $A\vec{x}_1,\ldots,A\vec{x}_k\in\mathscr{N}(A)$, so $A^2\vec{x}_1=\vec{0},\ldots,A^2\vec{x}_k=\vec{0}$ and we now know

$$A^2\vec{x}_{k+1},\ldots,A^2\vec{x}_{r_1}$$

spans $\mathscr{R}(A^2)$.

To see $\{A^2\vec{x}_{k+1},\ldots,A^2\vec{x}_{r_1}\}$ is linearly independent, write

$$b_{k+1}A^2\vec{x}_{k+1}+\cdots+b_{r_1}A^2\vec{x}_{r_1}=\vec{0}$$
$$A[b_{k+1}A\vec{x}_{k+1}+\cdots+b_{r_1}A\vec{x}_{r_1}]=\vec{0}$$

and, since $b_{k+1}A\vec{x}_{k+1}+\cdots+b_{r_1}A\vec{x}_{r_1}\in\mathscr{N}(A)$ we get a contradiction unless it is $\vec{0}$ (clearly it is in $\mathscr{R}(A)$, but $A\vec{x}_1,\ldots,A\vec{x}_k$ is a basis for $\mathscr{R}(A)\cap\mathscr{N}(A)$).

Hence $\dim(\mathscr{R}(A^2))=r_1-k=\dim(\mathscr{R}(A))-\dim(\mathscr{R}(A)\cap\mathscr{N}(A))$.

Three.IV.4: Inverses

Three.IV.4.12 Here is one way to proceed.

$$\xrightarrow{\rho_1\leftrightarrow\rho_2}\left(\begin{array}{ccc|ccc}1&0&1&0&1&0\\0&3&-1&1&0&0\\1&-1&0&0&0&1\end{array}\right)\xrightarrow{-\rho_1+\rho_3}\left(\begin{array}{ccc|ccc}1&0&1&0&1&0\\0&3&-1&1&0&0\\0&-1&-1&0&-1&1\end{array}\right)$$

$$\xrightarrow{(1/3)\rho_2+\rho_3}\left(\begin{array}{ccc|ccc}1&0&1&0&1&0\\0&3&-1&1&0&0\\0&0&-4/3&1/3&-1&1\end{array}\right)\xrightarrow[-(3/4)\rho_3]{(1/3)\rho_2}\left(\begin{array}{ccc|ccc}1&0&1&0&1&0\\0&1&-1/3&1/3&0&0\\0&0&1&-1/4&3/4&-3/4\end{array}\right)$$

$$\xrightarrow[-\rho_3+\rho_1]{(1/3)\rho_3+\rho_2}\left(\begin{array}{ccc|ccc}1&0&0&1/4&1/4&3/4\\0&1&0&1/4&1/4&-1/4\\0&0&1&-1/4&3/4&-3/4\end{array}\right)$$

Three.IV.4.13 (a) Yes, it has an inverse: $ad-bc=2\cdot 1-1\cdot(-1)\neq 0$. (b) Yes. (c) No.

Three.IV.4.14 (a) $\frac{1}{2\cdot 1-1\cdot(-1)}\cdot\begin{pmatrix}1&-1\\1&2\end{pmatrix}=\frac{1}{3}\cdot\begin{pmatrix}1&-1\\1&2\end{pmatrix}=\begin{pmatrix}1/3&-1/3\\1/3&2/3\end{pmatrix}$

(b) $\frac{1}{0\cdot(-3)-4\cdot 1}\cdot\begin{pmatrix}-3&-4\\-1&0\end{pmatrix}=\begin{pmatrix}3/4&1\\1/4&0\end{pmatrix}$

(c) The prior question shows that no inverse exists.

Three.IV.4.15 (a) The reduction is routine.

$$\left(\begin{array}{cc|cc}3&1&1&0\\0&2&0&1\end{array}\right)\xrightarrow[(1/2)\rho_2]{(1/3)\rho_1}\left(\begin{array}{cc|cc}1&1/3&1/3&0\\0&1&0&1/2\end{array}\right)\xrightarrow{-(1/3)\rho_2+\rho_1}\left(\begin{array}{cc|cc}1&0&1/3&-1/6\\0&1&0&1/2\end{array}\right)$$

This answer agrees with the answer from the check.

$$\begin{pmatrix}3&1\\0&2\end{pmatrix}^{-1}=\frac{1}{3\cdot 2-0\cdot 1}\cdot\begin{pmatrix}2&-1\\0&3\end{pmatrix}=\frac{1}{6}\cdot\begin{pmatrix}2&-1\\0&3\end{pmatrix}$$

(b) This reduction is easy.

$$\left(\begin{array}{cc|cc}2&1/2&1&0\\3&1&0&1\end{array}\right)\xrightarrow{-(3/2)\rho_1+\rho_2}\left(\begin{array}{cc|cc}2&1/2&1&0\\0&1/4&-3/2&1\end{array}\right)$$

$$\xrightarrow[4\rho_2]{(1/2)\rho_1}\left(\begin{array}{cc|cc}1&1/4&1/2&0\\0&1&-6&4\end{array}\right)\xrightarrow{-(1/4)\rho_2+\rho_1}\left(\begin{array}{cc|cc}1&0&2&-1\\0&1&-6&4\end{array}\right)$$

The check agrees.

$$\frac{1}{2\cdot 1-3\cdot(1/2)}\cdot\begin{pmatrix}1&-1/2\\-3&2\end{pmatrix}=2\cdot\begin{pmatrix}1&-1/2\\-3&2\end{pmatrix}$$

(c) Trying the Gauss-Jordan reduction

$$\left(\begin{array}{cc|cc}2&-4&1&0\\-1&2&0&1\end{array}\right)\xrightarrow{(1/2)\rho_1+\rho_2}\left(\begin{array}{cc|cc}2&-4&1&0\\0&0&1/2&1\end{array}\right)$$

shows that the left side won't reduce to the identity, so no inverse exists. The check $ad-bc=2\cdot 2-(-4)\cdot(-1)=0$ agrees.

(d) This produces an inverse.

$$\left(\begin{array}{ccc|ccc} 1 & 1 & 3 & 1 & 0 & 0 \\ 0 & 2 & 4 & 0 & 1 & 0 \\ -1 & 1 & 0 & 0 & 0 & 1 \end{array}\right) \xrightarrow{\rho_1+\rho_3} \left(\begin{array}{ccc|ccc} 1 & 1 & 3 & 1 & 0 & 0 \\ 0 & 2 & 4 & 0 & 1 & 0 \\ 0 & 2 & 3 & 1 & 0 & 1 \end{array}\right) \xrightarrow{-\rho_2+\rho_3} \left(\begin{array}{ccc|ccc} 1 & 1 & 3 & 1 & 0 & 0 \\ 0 & 2 & 4 & 0 & 1 & 0 \\ 0 & 0 & -1 & 1 & -1 & 1 \end{array}\right)$$

$$\xrightarrow[-\rho_3]{(1/2)\rho_2} \left(\begin{array}{ccc|ccc} 1 & 1 & 3 & 1 & 0 & 0 \\ 0 & 1 & 2 & 0 & 1/2 & 0 \\ 0 & 0 & 1 & -1 & 1 & -1 \end{array}\right) \xrightarrow[-3\rho_3+\rho_1]{-2\rho_3+\rho_2} \left(\begin{array}{ccc|ccc} 1 & 1 & 0 & 4 & -3 & 3 \\ 0 & 1 & 0 & 2 & -3/2 & 2 \\ 0 & 0 & 1 & -1 & 1 & -1 \end{array}\right)$$

$$\xrightarrow{-\rho_2+\rho_1} \left(\begin{array}{ccc|ccc} 1 & 0 & 0 & 2 & -3/2 & 1 \\ 0 & 1 & 0 & 2 & -3/2 & 2 \\ 0 & 0 & 1 & -1 & 1 & -1 \end{array}\right)$$

(e) This is one way to do the reduction.

$$\left(\begin{array}{ccc|ccc} 0 & 1 & 5 & 1 & 0 & 0 \\ 0 & -2 & 4 & 0 & 1 & 0 \\ 2 & 3 & -2 & 0 & 0 & 1 \end{array}\right) \xrightarrow{\rho_3\leftrightarrow\rho_1} \left(\begin{array}{ccc|ccc} 2 & 3 & -2 & 0 & 0 & 1 \\ 0 & -2 & 4 & 0 & 1 & 0 \\ 0 & 1 & 5 & 1 & 0 & 0 \end{array}\right)$$

$$\xrightarrow{(1/2)\rho_2+\rho_3} \left(\begin{array}{ccc|ccc} 2 & 3 & -2 & 0 & 0 & 1 \\ 0 & -2 & 4 & 0 & 1 & 0 \\ 0 & 0 & 7 & 1 & 1/2 & 0 \end{array}\right) \xrightarrow[\substack{-(1/2)\rho_2 \\ (1/7)\rho_3}]{(1/2)\rho_1} \left(\begin{array}{ccc|ccc} 1 & 3/2 & -1 & 0 & 0 & 1/2 \\ 0 & 1 & -2 & 0 & -1/2 & 0 \\ 0 & 0 & 1 & 1/7 & 1/14 & 0 \end{array}\right)$$

$$\xrightarrow[\rho_3+\rho_1]{2\rho_3+\rho_2} \left(\begin{array}{ccc|ccc} 1 & 3/2 & 0 & 1/7 & 1/14 & 1/2 \\ 0 & 1 & 0 & 2/7 & -5/14 & 0 \\ 0 & 0 & 1 & 1/7 & 1/14 & 0 \end{array}\right) \xrightarrow{-(3/2)\rho_2+\rho_1} \left(\begin{array}{ccc|ccc} 1 & 0 & 0 & -2/7 & 17/28 & 1/2 \\ 0 & 1 & 0 & 2/7 & -5/14 & 0 \\ 0 & 0 & 1 & 1/7 & 1/14 & 0 \end{array}\right)$$

(f) There is no inverse.

$$\left(\begin{array}{ccc|ccc} 2 & 2 & 3 & 1 & 0 & 0 \\ 1 & -2 & -3 & 0 & 1 & 0 \\ 4 & -2 & -3 & 0 & 0 & 1 \end{array}\right) \xrightarrow[-2\rho_1+\rho_3]{-(1/2)\rho_1+\rho_2} \left(\begin{array}{ccc|ccc} 2 & 2 & 3 & 1 & 0 & 0 \\ 0 & -3 & -9/2 & -1/2 & 1 & 0 \\ 0 & -6 & -9 & -2 & 0 & 1 \end{array}\right)$$

$$\xrightarrow{-2\rho_2+\rho_3} \left(\begin{array}{ccc|ccc} 2 & 2 & 3 & 1 & 0 & 0 \\ 0 & -3 & -9/2 & -1/2 & 1 & 0 \\ 0 & 0 & 0 & -1 & -2 & 1 \end{array}\right)$$

As a check, note that the third column of the starting matrix is $3/2$ times the second, and so it is indeed singular and therefore has no inverse.

Three.IV.4.16 We can use Corollary 4.11.

$$\frac{1}{1\cdot 5 - 2\cdot 3}\cdot\begin{pmatrix} 5 & -3 \\ -2 & 1 \end{pmatrix} = \begin{pmatrix} -5 & 3 \\ 2 & -1 \end{pmatrix}$$

Three.IV.4.17 **(a)** The proof that the inverse is $r^{-1}H^{-1} = (1/r)\cdot H^{-1}$ (provided, of course, that the matrix is invertible) is easy.

(b) No. For one thing, the fact that $H+G$ has an inverse doesn't imply that H has an inverse or that G has an inverse. Neither of these matrices is invertible but their sum is.

$$\begin{pmatrix} 1 & 0 \\ 0 & 0 \end{pmatrix} \qquad \begin{pmatrix} 0 & 0 \\ 0 & 1 \end{pmatrix}$$

Another point is that just because H and G each has an inverse doesn't mean $H+G$ has an inverse; here is an example.

$$\begin{pmatrix} 1 & 0 \\ 0 & 1 \end{pmatrix} \qquad \begin{pmatrix} -1 & 0 \\ 0 & -1 \end{pmatrix}$$

Still a third point is that, even if the two matrices have inverses, and the sum has an inverse, doesn't imply that the equation holds:

$$\begin{pmatrix}2&0\\0&2\end{pmatrix}^{-1}=\begin{pmatrix}1/2&0\\0&1/2\end{pmatrix}^{-1}\qquad\begin{pmatrix}3&0\\0&3\end{pmatrix}^{-1}=\begin{pmatrix}1/3&0\\0&1/3\end{pmatrix}^{-1}$$

but

$$\begin{pmatrix}5&0\\0&5\end{pmatrix}^{-1}=\begin{pmatrix}1/5&0\\0&1/5\end{pmatrix}^{-1}$$

and $(1/2)_+(1/3)$ does not equal $1/5$.

Three.IV.4.18 Yes: $T^k(T^{-1})^k=(TT\cdots T)\cdot(T^{-1}T^{-1}\cdots T^{-1})=T^{k-1}(TT^{-1})(T^{-1})^{k-1}=\cdots=I$.

Three.IV.4.19 Yes, the inverse of H^{-1} is H.

Three.IV.4.20 One way to check that the first is true is with the angle sum formulas from trigonometry.

$$\begin{aligned}\begin{pmatrix}\cos(\theta_1+\theta_2)&-\sin(\theta_1+\theta_2)\\\sin(\theta_1+\theta_2)&\cos(\theta_1+\theta_2)\end{pmatrix}&=\begin{pmatrix}\cos\theta_1\cos\theta_2-\sin\theta_1\sin\theta_2&-\sin\theta_1\cos\theta_2-\cos\theta_1\sin\theta_2\\\sin\theta_1\cos\theta_2+\cos\theta_1\sin\theta_2&\cos\theta_1\cos\theta_2-\sin\theta_1\sin\theta_2\end{pmatrix}\\&=\begin{pmatrix}\cos\theta_1&-\sin\theta_1\\\sin\theta_1&\cos\theta_1\end{pmatrix}\begin{pmatrix}\cos\theta_2&-\sin\theta_2\\\sin\theta_2&\cos\theta_2\end{pmatrix}\end{aligned}$$

Checking the second equation in this way is similar.

Of course, the equations can be not just checked but also understood by recalling that t_θ is the map that rotates vectors about the origin through an angle of θ radians.

Three.IV.4.21 There are two cases. For the first case we assume that a is nonzero. Then

$$\xrightarrow{-(c/a)\rho_1+\rho_2}\left(\begin{array}{cc|cc}a&b&1&0\\0&-(bc/a)+d&-c/a&1\end{array}\right)=\left(\begin{array}{cc|cc}a&b&1&0\\0&(ad-bc)/a&-c/a&1\end{array}\right)$$

shows that the matrix is invertible (in this $a\neq0$ case) if and only if $ad-bc\neq0$. To find the inverse, we finish with the Jordan half of the reduction.

$$\xrightarrow[(a/ad-bc)\rho_2]{(1/a)\rho_1}\left(\begin{array}{cc|cc}1&b/a&1/a&0\\0&1&-c/(ad-bc)&a/(ad-bc)\end{array}\right)\xrightarrow{-(b/a)\rho_2+\rho_1}\left(\begin{array}{cc|cc}1&0&d/(ad-bc)&-b/(ad-bc)\\0&1&-c/(ad-bc)&a/(ad-bc)\end{array}\right)$$

The other case is the $a=0$ case. We swap to get c into the $1,1$ position.

$$\xrightarrow{\rho_1\leftrightarrow\rho_2}\left(\begin{array}{cc|cc}c&d&0&1\\0&b&1&0\end{array}\right)$$

This matrix is nonsingular if and only if both b and c are nonzero (which, under the case assumption that $a=0$, holds if and only if $ad-bc\neq0$). To find the inverse we do the Jordan half.

$$\xrightarrow[(1/b)\rho_2]{(1/c)\rho_1}\left(\begin{array}{cc|cc}1&d/c&0&1/c\\0&1&1/b&0\end{array}\right)\xrightarrow{-(d/c)\rho_2+\rho_1}\left(\begin{array}{cc|cc}1&0&-d/bc&1/c\\0&1&1/b&0\end{array}\right)$$

(Note that this is what is required, since $a=0$ gives that $ad-bc=-bc$).

Three.IV.4.22 With H a 2×3 matrix, in looking for a matrix G such that the combination HG acts as the 2×2 identity we need G to be 3×2. Setting up the equation

$$\begin{pmatrix}1&0&1\\0&1&0\end{pmatrix}\begin{pmatrix}m&n\\p&q\\r&s\end{pmatrix}=\begin{pmatrix}1&0\\0&1\end{pmatrix}$$

and solving the resulting linear system

$$\begin{array}{rrrrrrl}m&&&&+r&&=1\\&n&&&&+s&=0\\&&p&&&&=0\\&&&q&&&=1\end{array}$$

gives infinitely many solutions.

$$\{\begin{pmatrix} m \\ n \\ p \\ q \\ r \\ s \end{pmatrix} = \begin{pmatrix} 1 \\ 0 \\ 0 \\ 1 \\ 0 \\ 0 \end{pmatrix} + r \cdot \begin{pmatrix} -1 \\ 0 \\ 0 \\ 0 \\ 1 \\ 0 \end{pmatrix} + s \cdot \begin{pmatrix} 0 \\ -1 \\ 0 \\ 0 \\ 0 \\ 1 \end{pmatrix} \mid r, s \in \mathbb{R}\}$$

Thus H has infinitely many right inverses.

As for left inverses, the equation

$$\begin{pmatrix} a & b \\ c & d \end{pmatrix} \begin{pmatrix} 1 & 0 & 1 \\ 0 & 1 & 0 \end{pmatrix} = \begin{pmatrix} 1 & 0 & 0 \\ 0 & 1 & 0 \\ 0 & 0 & 1 \end{pmatrix}$$

gives rise to a linear system with nine equations and four unknowns.

$$\begin{array}{rcl} a & & =1 \\ & b & =0 \\ a & & =0 \\ & c & =0 \\ & d & =1 \\ & c & =0 \\ & e & =0 \\ & f & =0 \\ & e & =1 \end{array}$$

This system is inconsistent (the first equation conflicts with the third, as do the seventh and ninth) and so there is no left inverse.

Three.IV.4.23 With respect to the standard bases we have

$$\text{Rep}_{\mathcal{E}_2,\mathcal{E}_3}(\iota) = \begin{pmatrix} 1 & 0 \\ 0 & 1 \\ 0 & 0 \end{pmatrix}$$

and setting up the equation to find the matrix inverse

$$\begin{pmatrix} a & b & c \\ d & e & f \end{pmatrix} \begin{pmatrix} 1 & 0 \\ 0 & 1 \\ 0 & 0 \end{pmatrix} = \begin{pmatrix} 1 & 0 \\ 0 & 1 \end{pmatrix} = \text{Rep}_{\mathcal{E}_2,\mathcal{E}_2}(\text{id})$$

gives rise to a linear system.

$$\begin{array}{rcl} a & & =1 \\ & b & =0 \\ & d & =0 \\ & e & =1 \end{array}$$

There are infinitely many solutions in $a, \ldots, f$ to this system because two of these variables are entirely unrestricted

$$\{\begin{pmatrix} a \\ b \\ c \\ d \\ e \\ f \end{pmatrix} = \begin{pmatrix} 1 \\ 0 \\ 0 \\ 0 \\ 1 \\ 0 \end{pmatrix} + c \cdot \begin{pmatrix} 0 \\ 0 \\ 1 \\ 0 \\ 0 \\ 0 \end{pmatrix} + f \cdot \begin{pmatrix} 0 \\ 0 \\ 0 \\ 0 \\ 0 \\ 1 \end{pmatrix} \mid c, f \in \mathbb{R}\}$$

and so there are infinitely many solutions to the matrix equation.

$$\{\begin{pmatrix} 1 & 0 & c \\ 0 & 1 & f \end{pmatrix} \mid c, f \in \mathbb{R}\}$$

With the bases still fixed at $\mathcal{E}_2, \mathcal{E}_2$, for instance taking $c = 2$ and $f = 3$ gives a matrix representing this map.

$$\begin{pmatrix} x \\ y \\ z \end{pmatrix} \overset{f_{2,3}}{\longmapsto} \begin{pmatrix} x + 2z \\ y + 3z \end{pmatrix}$$

The check that $f_{2,3} \circ \iota$ is the identity map on $\mathbb{R}^2$ is easy.

Three.IV.4.24 By Lemma 4.2 it cannot have infinitely many left inverses, because a matrix with both left and right inverses has only one of each (and that one of each is one of both—the left and right inverse matrices are equal).

Three.IV.4.25 **(a)** True, It must be linear, as the proof from Theorem II.2.21 shows.

(b) False. It may be linear, but it need not be. Consider the projection map $\pi\colon \mathbb{R}^3 \to \mathbb{R}^2$ described at the start of this subsection. Define $\eta\colon \mathbb{R}^2 \to \mathbb{R}^3$ in this way.

$$\begin{pmatrix} x \\ y \end{pmatrix} \mapsto \begin{pmatrix} x \\ y \\ 1 \end{pmatrix}$$

It is a right inverse of π because $\pi \circ \eta$ does this.

$$\begin{pmatrix} x \\ y \end{pmatrix} \mapsto \begin{pmatrix} x \\ y \\ 1 \end{pmatrix} \mapsto \begin{pmatrix} x \\ y \end{pmatrix}$$

It is not linear because it does not map the zero vector to the zero vector.

Three.IV.4.26 The associativity of matrix multiplication gives on the one hand $H^{-1}(HG) = H^{-1}Z = Z$, and on the other that $H^{-1}(HG) = (H^{-1}H)G = IG = G$.

Three.IV.4.27 Multiply both sides of the first equation by H.

Three.IV.4.28 Checking that when $I - T$ is multiplied on both sides by that expression (assuming that T^4 is the zero matrix) then the result is the identity matrix is easy. The obvious generalization is that if T^n is the zero matrix then $(I - T)^{-1} = I + T + T^2 + \cdots + T^{n-1}$; the check again is easy.

Three.IV.4.29 The powers of the matrix are formed by taking the powers of the diagonal entries. That is, D^2 is all zeros except for diagonal entries of ${d_{1,1}}^2$, ${d_{2,2}}^2$, etc. This suggests defining D^0 to be the identity matrix.

Three.IV.4.30 Assume that B is row equivalent to A and that A is invertible. Because they are row-equivalent, there is a sequence of row steps to reduce one to the other. We can do that reduction with matrices, for instance, A can change by row operations to B as $B = R_n \cdots R_1 A$. This equation gives B as a product of invertible matrices and by Lemma 4.4 then, B is also invertible.

Three.IV.4.31 **(a)** See the answer to Exercise 29.

(b) We will show that both conditions are equivalent to the condition that the two matrices be nonsingular.

As T and S are square and their product is defined, they are equal-sized, say $n \times n$. Consider the $TS = I$ half. By the prior item the rank of I is less than or equal to the minimum of the rank of T and the rank of S. But the rank of I is n, so the rank of T and the rank of S must each be n. Hence each is nonsingular.

The same argument shows that $ST = I$ implies that each is nonsingular.

Three.IV.4.32 Inverses are unique, so we need only show that it works. The check appears above as Exercise 33.

Three.IV.4.33 **(a)** See the answer for Exercise 26.

(b) See the answer for Exercise 26.

(c) Apply the first part to $I = AA^{-1}$ to get $I = I^T = (AA^{-1})^T = (A^{-1})^T A^T$.

(d) Apply the prior item with $A^T = A$, as A is symmetric.

Three.IV.4.34 For the answer to the items making up the first half, see Exercise 31. For the proof in the second half, assume that A is a zero divisor so there is a nonzero matrix B with $AB = Z$ (or else $BA = Z$; this case is similar), If A is invertible then $A^{-1}(AB) = (A^{-1}A)B = IB = B$ but also $A^{-1}(AB) = A^{-1}Z = Z$, contradicting that B is nonzero.

Three.IV.4.35 Here are four solutions to $H^2 = I$.

$$\begin{pmatrix} \pm 1 & 0 \\ 0 & \pm 1 \end{pmatrix}$$

Three.IV.4.36 It is not reflexive since, for instance,

$$H = \begin{pmatrix} 1 & 0 \\ 0 & 2 \end{pmatrix}$$

is not a two-sided inverse of itself. The same example shows that it is not transitive. That matrix has this two-sided inverse

$$G = \begin{pmatrix} 1 & 0 \\ 0 & 1/2 \end{pmatrix}$$

and while H is a two-sided inverse of G and G is a two-sided inverse of H, we know that H is not a two-sided inverse of H. However, the relation is symmetric: if G is a two-sided inverse of H then $GH = I = HG$ and therefore H is also a two-sided inverse of G.

Three.IV.4.37 *This is how the answer was given in the cited source.* Let A be $m \times m$, non-singular, with the stated property. Let B be its inverse. Then for $n \leqslant m$,

$$1 = \sum_{r=1}^{m} \delta_{nr} = \sum_{r=1}^{m} \sum_{s=1}^{m} b_{ns} a_{sr} = \sum_{s=1}^{m} \sum_{r=1}^{m} b_{ns} a_{sr} = k \sum_{s=1}^{m} b_{ns}$$

(A is singular if $k = 0$).

Change of Basis

Three.V.1: Changing Representations of Vectors

Three.V.1.7 For the matrix to change bases from D to $\mathcal{E}_2$ we need that $\text{Rep}_{\mathcal{E}_2}(\text{id}(\vec{\delta}_1)) = \text{Rep}_{\mathcal{E}_2}(\vec{\delta}_1)$ and that $\text{Rep}_{\mathcal{E}_2}(\text{id}(\vec{\delta}_2)) = \text{Rep}_{\mathcal{E}_2}(\vec{\delta}_2)$. Of course, the representation of a vector in $\mathbb{R}^2$ with respect to the standard basis is easy.

$$\text{Rep}_{\mathcal{E}_2}(\vec{\delta}_1) = \begin{pmatrix} 2 \\ 1 \end{pmatrix} \qquad \text{Rep}_{\mathcal{E}_2}(\vec{\delta}_2) = \begin{pmatrix} -2 \\ 4 \end{pmatrix}$$

Concatenating those two together to make the columns of the change of basis matrix gives this.

$$\text{Rep}_{D,\mathcal{E}_2}(\text{id}) = \begin{pmatrix} 2 & -2 \\ 1 & 4 \end{pmatrix}$$

For the change of basis matrix in the other direction we can calculate $\text{Rep}_D(\text{id}(\vec{e}_1)) = \text{Rep}_D(\vec{e}_1)$ and $\text{Rep}_D(\text{id}(\vec{e}_2)) = \text{Rep}_D(\vec{e}_2)$ (this job is routine) or we can take the inverse of the above matrix. Because of the formula for the inverse of a 2×2 matrix, this is easy.

$$\text{Rep}_{\mathcal{E}_2,D}(\text{id}) = \frac{1}{10} \cdot \begin{pmatrix} 4 & 2 \\ -1 & 2 \end{pmatrix} = \begin{pmatrix} 4/10 & 2/10 \\ -1/10 & 2/10 \end{pmatrix}$$

Three.V.1.8 In each case, concatenate the columns $\text{Rep}_D(\text{id}(\vec{\beta}_1)) = \text{Rep}_D(\vec{\beta}_1)$ and $\text{Rep}_D(\text{id}(\vec{\beta}_2)) = \text{Rep}_D(\vec{\beta}_2)$ to make the change of basis matrix $\text{Rep}_{B,D}(\text{id})$.

(a) $\begin{pmatrix}0&1\\1&0\end{pmatrix}$ (b) $\begin{pmatrix}2&-1/2\\-1&1/2\end{pmatrix}$ (c) $\begin{pmatrix}1&1\\2&4\end{pmatrix}$ (d) $\begin{pmatrix}1&-1\\-1&2\end{pmatrix}$

Three.V.1.9 One way to go is to find $\mathrm{Rep}_B(\vec{\delta}_1)$ and $\mathrm{Rep}_B(\vec{\delta}_2)$, and then concatenate them into the columns of the desired change of basis matrix. Another way is to find the inverse of the matrices that answer Exercise 8.

(a) $\begin{pmatrix}0&1\\1&0\end{pmatrix}$ (b) $\begin{pmatrix}1&1\\2&4\end{pmatrix}$ (c) $\begin{pmatrix}2&-1/2\\-1&1/2\end{pmatrix}$ (d) $\begin{pmatrix}2&1\\1&1\end{pmatrix}$

Three.V.1.10 The columns vector representations $\mathrm{Rep}_D(\mathrm{id}(\vec{\beta}_1))=\mathrm{Rep}_D(\vec{\beta}_1)$, and $\mathrm{Rep}_D(\mathrm{id}(\vec{\beta}_2))=\mathrm{Rep}_D(\vec{\beta}_2)$, and $\mathrm{Rep}_D(\mathrm{id}(\vec{\beta}_3))=\mathrm{Rep}_D(\vec{\beta}_3)$ make the change of basis matrix $\mathrm{Rep}_{B,D}(\mathrm{id})$.

(a) $\begin{pmatrix}0&0&1\\1&0&0\\0&1&0\end{pmatrix}$ (b) $\begin{pmatrix}1&-1&0\\0&1&-1\\0&0&1\end{pmatrix}$ (c) $\begin{pmatrix}1&-1&1/2\\1&1&-1/2\\0&2&0\end{pmatrix}$

E.g., for the first column of the first matrix, $1=0\cdot x^2+1\cdot 1+0\cdot x$.

Three.V.1.11 A matrix changes bases if and only if it is nonsingular.

(a) This matrix is nonsingular and so changes bases. Finding to what basis $\mathcal{E}_2$ is changed means finding D such that

$$\mathrm{Rep}_{\mathcal{E}_2,D}(\mathrm{id})=\begin{pmatrix}5&0\\0&4\end{pmatrix}$$

and by the definition of how a matrix represents a linear map, we have this.

$$\mathrm{Rep}_D(\mathrm{id}(\vec{e}_1))=\mathrm{Rep}_D(\vec{e}_1)=\begin{pmatrix}5\\0\end{pmatrix}\qquad \mathrm{Rep}_D(\mathrm{id}(\vec{e}_2))=\mathrm{Rep}_D(\vec{e}_2)=\begin{pmatrix}0\\4\end{pmatrix}$$

Where

$$D=\langle\begin{pmatrix}x_1\\y_1\end{pmatrix},\begin{pmatrix}x_2\\y_2\end{pmatrix}\rangle$$

we can either solve the system

$$\begin{pmatrix}1\\0\end{pmatrix}=5\begin{pmatrix}x_1\\y_1\end{pmatrix}+0\begin{pmatrix}x_2\\y_1\end{pmatrix}\qquad \begin{pmatrix}0\\1\end{pmatrix}=0\begin{pmatrix}x_1\\y_1\end{pmatrix}+4\begin{pmatrix}x_2\\y_1\end{pmatrix}$$

or else just spot the answer (thinking of the proof of Lemma 1.5).

$$D=\langle\begin{pmatrix}1/5\\0\end{pmatrix},\begin{pmatrix}0\\1/4\end{pmatrix}\rangle$$

(b) Yes, this matrix is nonsingular and so changes bases. To calculate D, we proceed as above with

$$D=\langle\begin{pmatrix}x_1\\y_1\end{pmatrix},\begin{pmatrix}x_2\\y_2\end{pmatrix}\rangle$$

to solve

$$\begin{pmatrix}1\\0\end{pmatrix}=2\begin{pmatrix}x_1\\y_1\end{pmatrix}+3\begin{pmatrix}x_2\\y_1\end{pmatrix}\quad\text{and}\quad \begin{pmatrix}0\\1\end{pmatrix}=1\begin{pmatrix}x_1\\y_1\end{pmatrix}+1\begin{pmatrix}x_2\\y_1\end{pmatrix}$$

and get this.

$$D=\langle\begin{pmatrix}-1\\3\end{pmatrix},\begin{pmatrix}1\\-2\end{pmatrix}\rangle$$

(c) No, this matrix does not change bases because it is singular.

(d) Yes, this matrix changes bases because it is nonsingular. The calculation of the changed-to basis is as above.

$$D=\langle\begin{pmatrix}1/2\\-1/2\end{pmatrix},\begin{pmatrix}1/2\\1/2\end{pmatrix}\rangle$$

Three.V.1.12 This question has many different solutions. One way to proceed is to make up any basis B for any space, and then compute the appropriate D (necessarily for the same space, of course). Another, easier, way to proceed is to fix the codomain as $\mathbb{R}^3$ and the codomain basis as $\mathcal{E}_3$. This way (recall that the representation of any vector with respect to the standard basis is just the vector itself), we have this.

$$B = \langle \begin{pmatrix} 3 \\ 2 \\ 0 \end{pmatrix}, \begin{pmatrix} 1 \\ -1 \\ 0 \end{pmatrix}, \begin{pmatrix} 4 \\ 1 \\ 4 \end{pmatrix} \rangle \qquad D = \mathcal{E}_3$$

Three.V.1.13 Checking that $B = \langle 2\sin(x)+\cos(x), 3\cos(x)\rangle$ is a basis is routine. Call the natural basis D. To compute the change of basis matrix $\text{Rep}_{B,D}(\text{id})$ we must find $\text{Rep}_D(2\sin(x)+\cos(x))$ and $\text{Rep}_D(3\cos(x))$, that is, we need x_1, y_1, x_2, y_2 such that these equations hold.

$$\begin{aligned} x_1 \cdot \sin(x) + y_1 \cdot \cos(x) &= 2\sin(x) + \cos(x) \\ x_2 \cdot \sin(x) + y_2 \cdot \cos(x) &= 3\cos(x) \end{aligned}$$

Obviously this is the answer.

$$\text{Rep}_{B,D}(\text{id}) = \begin{pmatrix} 2 & 0 \\ 1 & 3 \end{pmatrix}$$

For the change of basis matrix in the other direction we could look for $\text{Rep}_B(\sin(x))$ and $\text{Rep}_B(\cos(x))$ by solving these.

$$\begin{aligned} w_1 \cdot (2\sin(x) + \cos(x)) + z_1 \cdot (3\cos(x)) &= \sin(x) \\ w_2 \cdot (2\sin(x) + \cos(x)) + z_2 \cdot (3\cos(x)) &= \cos(x) \end{aligned}$$

An easier method is to find the inverse of the matrix found above.

$$\text{Rep}_{D,B}(\text{id}) = \begin{pmatrix} 2 & 0 \\ 1 & 3 \end{pmatrix}^{-1} = \frac{1}{6} \cdot \begin{pmatrix} 3 & 0 \\ -1 & 2 \end{pmatrix} = \begin{pmatrix} 1/2 & 0 \\ -1/6 & 1/3 \end{pmatrix}$$

Three.V.1.14 We start by taking the inverse of the matrix, that is, by deciding what is the inverse to the map of interest.

$$\text{Rep}_{D,\mathcal{E}_2}(\text{id})\text{Rep}_{D,\mathcal{E}_2}(\text{id})^{-1} = \frac{1}{-\cos^2(2\theta) - \sin^2(2\theta)} \cdot \begin{pmatrix} -\cos(2\theta) & -\sin(2\theta) \\ -\sin(2\theta) & \cos(2\theta) \end{pmatrix} = \begin{pmatrix} \cos(2\theta) & \sin(2\theta) \\ \sin(2\theta) & -\cos(2\theta) \end{pmatrix}$$

This is more tractable than the representation the other way because this matrix is the concatenation of these two column vectors

$$\text{Rep}_{\mathcal{E}_2}(\vec{\delta}_1) = \begin{pmatrix} \cos(2\theta) \\ \sin(2\theta) \end{pmatrix} \qquad \text{Rep}_{\mathcal{E}_2}(\vec{\delta}_2) = \begin{pmatrix} \sin(2\theta) \\ -\cos(2\theta) \end{pmatrix}$$

and representations with respect to $\mathcal{E}_2$ are transparent.

$$\vec{\delta}_1 = \begin{pmatrix} \cos(2\theta) \\ \sin(2\theta) \end{pmatrix} \qquad \vec{\delta}_2 = \begin{pmatrix} \sin(2\theta) \\ -\cos(2\theta) \end{pmatrix}$$

This pictures the action of the map that transforms D to $\mathcal{E}_2$ (it is, again, the inverse of the map that is the answer to this question). The line lies at an angle θ to the x axis.

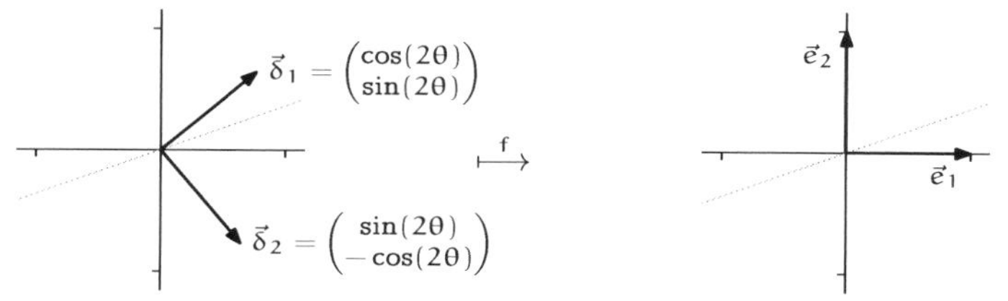

This map reflects vectors over that line. Since reflections are self-inverse, the answer to the question is: the original map reflects about the line through the origin with angle of elevation θ. (Of course, it does this to any basis.)

Three.V.1.15 The appropriately-sized identity matrix.

Three.V.1.16 Each is true if and only if the matrix is nonsingular.

Three.V.1.17 What remains is to show that left multiplication by a reduction matrix represents a change from another basis to $B = \langle \vec{\beta}_1, \ldots, \vec{\beta}_n \rangle$.

Application of a row-multiplication matrix $M_i(k)$ translates a representation with respect to the basis $\langle \vec{\beta}_1, \ldots, k\vec{\beta}_i, \ldots, \vec{\beta}_n \rangle$ to one with respect to B, as here.

$$\vec{v} = c_1 \cdot \vec{\beta}_1 + \cdots + c_i \cdot (k\vec{\beta}_i) + \cdots + c_n \cdot \vec{\beta}_n \mapsto c_1 \cdot \vec{\beta}_1 + \cdots + (kc_i) \cdot \vec{\beta}_i + \cdots + c_n \cdot \vec{\beta}_n = \vec{v}$$

Applying a row-swap matrix $P_{i,j}$ translates a representation with respect to $\langle \vec{\beta}_1, \ldots, \vec{\beta}_j, \ldots, \vec{\beta}_i, \ldots, \vec{\beta}_n \rangle$ to one with respect to $\langle \vec{\beta}_1, \ldots, \vec{\beta}_i, \ldots, \vec{\beta}_j, \ldots, \vec{\beta}_n \rangle$. Finally, applying a row-combination matrix $C_{i,j}(k)$ changes a representation with respect to $\langle \vec{\beta}_1, \ldots, \vec{\beta}_i + k\vec{\beta}_j, \ldots, \vec{\beta}_j, \ldots, \vec{\beta}_n \rangle$ to one with respect to B.

$$\begin{aligned} \vec{v} = c_1 \cdot \vec{\beta}_1 + \cdots + c_i \cdot (\vec{\beta}_i + k\vec{\beta}_j) + \cdots + c_j\vec{\beta}_j + \cdots + c_n \cdot \vec{\beta}_n \\ \mapsto c_1 \cdot \vec{\beta}_1 + \cdots + c_i \cdot \vec{\beta}_i + \cdots + (kc_i + c_j) \cdot \vec{\beta}_j + \cdots + c_n \cdot \vec{\beta}_n = \vec{v} \end{aligned}$$

(As in the part of the proof in the body of this subsection, the various conditions on the row operations, e.g., that the scalar k is nonzero, assure that these are all bases.)

Three.V.1.18 Taking H as a change of basis matrix $H = \mathrm{Rep}_{B,\mathcal{E}_n}(\mathrm{id})$, its columns are

$$\begin{pmatrix} h_{1,i} \\ \vdots \\ h_{n,i} \end{pmatrix} = \mathrm{Rep}_{\mathcal{E}_n}(\mathrm{id}(\vec{\beta}_i)) = \mathrm{Rep}_{\mathcal{E}_n}(\vec{\beta}_i)$$

and, because representations with respect to the standard basis are transparent, we have this.

$$\begin{pmatrix} h_{1,i} \\ \vdots \\ h_{n,i} \end{pmatrix} = \vec{\beta}_i$$

That is, the basis is the one composed of the columns of H.

Three.V.1.19 **(a)** We can change the starting vector representation to the ending one through a sequence of row operations. The proof tells us what how the bases change. We start by swapping the first and second rows of the representation with respect to B to get a representation with respect to a new basis B_1.

$$\mathrm{Rep}_{B_1}(1 - x + 3x^2 - x^3) = \begin{pmatrix} 1 \\ 0 \\ 1 \\ 2 \end{pmatrix}_{B_1} \qquad B_1 = \langle 1 - x, 1 + x, x^2 + x^3, x^2 - x^3 \rangle$$

We next add -2 times the third row of the vector representation to the fourth row.

$$\mathrm{Rep}_{B_3}(1 - x + 3x^2 - x^3) = \begin{pmatrix} 1 \\ 0 \\ 1 \\ 0 \end{pmatrix}_{B_2} \qquad B_2 = \langle 1 - x, 1 + x, 3x^2 - x^3, x^2 - x^3 \rangle$$

(The third element of B_2 is the third element of B_1 minus -2 times the fourth element of B_1.) Now we can finish by doubling the third row.

$$\mathrm{Rep}_{D}(1 - x + 3x^2 - x^3) = \begin{pmatrix} 1 \\ 0 \\ 2 \\ 0 \end{pmatrix}_{D} \qquad D = \langle 1 - x, 1 + x, (3x^2 - x^3)/2, x^2 - x^3 \rangle$$

(b) Here are three different approaches to stating such a result. The first is the assertion: where V is a vector space with basis B and $\vec{v} \in V$ is nonzero, for any nonzero column vector $\vec{z}$ (whose number of components equals the dimension of V) there is a change of basis matrix M such that $M \cdot \mathrm{Rep}_B(\vec{v}) = \vec{z}$. The second possible statement: for any (n-dimensional) vector space V and any nonzero vector $\vec{v} \in V$, where $\vec{z}_1, \vec{z}_2 \in \mathbb{R}^n$ are nonzero, there are bases $B, D \subset V$ such that $\mathrm{Rep}_B(\vec{v}) = \vec{z}_1$ and $\mathrm{Rep}_D(\vec{v}) = \vec{z}_2$. The third is: for any nonzero $\vec{v}$ member of any vector space (of dimension n) and any nonzero column vector (with n components) there is a basis such that $\vec{v}$ is represented with respect to that basis by that column vector.

The first and second statements follow easily from the third. The first follows because the third statement gives a basis D such that $\mathrm{Rep}_D(\vec{v}) = \vec{z}$ and then $\mathrm{Rep}_{B,D}(\mathrm{id})$ is the desired M. The second follows from the third because it is just a doubled application of it.

A way to prove the third is as in the answer to the first part of this question. Here is a sketch. Represent $\vec{v}$ with respect to any basis B with a column vector $\vec{z}_1$. This column vector must have a nonzero component because $\vec{v}$ is a nonzero vector. Use that component in a sequence of row operations to convert $\vec{z}_1$ to $\vec{z}$. (We could fill out this sketch as an induction argument on the dimension of V.)

Three.V.1.20 This is the topic of the next subsection.

Three.V.1.21 A change of basis matrix is nonsingular and thus has rank equal to the number of its columns. Therefore its set of columns is a linearly independent subset of size n in $\mathbb{R}^n$ and it is thus a basis. The answer to the second half is also 'yes'; all implications in the prior sentence reverse (that is, all of the 'if ... then ...' parts of the prior sentence convert to 'if and only if' parts).

Three.V.1.22 In response to the first half of the question, there are infinitely many such matrices. One of them represents with respect to $\mathcal{E}_2$ the transformation of $\mathbb{R}^2$ with this action.

$$\begin{pmatrix}1\\0\end{pmatrix} \mapsto \begin{pmatrix}4\\0\end{pmatrix} \qquad \begin{pmatrix}0\\1\end{pmatrix} \mapsto \begin{pmatrix}0\\-1/3\end{pmatrix}$$

The problem of specifying two distinct input/output pairs is a bit trickier. The fact that matrices have a linear action precludes some possibilities.

(a) Yes, there is such a matrix. These conditions

$$\begin{pmatrix}a&b\\c&d\end{pmatrix}\begin{pmatrix}1\\3\end{pmatrix} = \begin{pmatrix}1\\1\end{pmatrix} \qquad \begin{pmatrix}a&b\\c&d\end{pmatrix}\begin{pmatrix}2\\-1\end{pmatrix} = \begin{pmatrix}-1\\-1\end{pmatrix}$$

can be solved

$$\begin{aligned} a + 3b \qquad\qquad &= 1 \\ c + 3d &= 1 \\ 2a - b \qquad\qquad &= -1 \\ 2c - d &= -1 \end{aligned}$$

to give this matrix.

$$\begin{pmatrix}-2/7&3/7\\-2/7&3/7\end{pmatrix}$$

(b) No, because

$$2 \cdot \begin{pmatrix}1\\3\end{pmatrix} = \begin{pmatrix}2\\6\end{pmatrix} \quad \text{but} \quad 2 \cdot \begin{pmatrix}1\\1\end{pmatrix} \neq \begin{pmatrix}-1\\-1\end{pmatrix}$$

no linear action can produce this effect.

(c) A sufficient condition is that $\{\vec{v}_1, \vec{v}_2\}$ be linearly independent, but that's not a necessary condition. A necessary and sufficient condition is that any linear dependences among the starting vectors appear also among the ending vectors. That is,

$$c_1\vec{v}_1 + c_2\vec{v}_2 = \vec{0} \quad \text{implies} \quad c_1\vec{w}_1 + c_2\vec{w}_2 = \vec{0}.$$

The proof of this condition is routine.

Three.V.2: Changing Map Representations

Three.V.2.10 **(a)** Yes, each has rank two.
(b) Yes, they have the same rank.
(c) No, they have different ranks.

Three.V.2.11 We need only decide what the rank of each is.

(a) $\begin{pmatrix}1&0&0\\0&0&0\end{pmatrix}$ **(b)** $\begin{pmatrix}1&0&0&0\\0&1&0&0\\0&0&1&0\end{pmatrix}$

Three.V.2.12 Recall the diagram and the formula.

$$\begin{array}{ccc}\mathbb{R}^2_{wrt\ B} & \xrightarrow[T]{t} & \mathbb{R}^2_{wrt\ D}\\ \text{id}\downarrow & & \text{id}\downarrow\\ \mathbb{R}^2_{wrt\ \hat{B}} & \xrightarrow[\hat{T}]{t} & \mathbb{R}^2_{wrt\ \hat{D}}\end{array} \qquad \hat{T}=\text{Rep}_{D,\hat{D}}(\text{id})\cdot T\cdot\text{Rep}_{\hat{B},B}(\text{id})$$

(a) These two

$$\begin{pmatrix}1\\1\end{pmatrix}=1\cdot\begin{pmatrix}-1\\0\end{pmatrix}+1\cdot\begin{pmatrix}2\\1\end{pmatrix}\qquad\begin{pmatrix}1\\-1\end{pmatrix}=(-3)\cdot\begin{pmatrix}-1\\0\end{pmatrix}+(-1)\cdot\begin{pmatrix}2\\1\end{pmatrix}$$

show that

$$\text{Rep}_{D,\hat{D}}(\text{id})=\begin{pmatrix}1&-3\\1&-1\end{pmatrix}$$

and similarly these two

$$\begin{pmatrix}0\\1\end{pmatrix}=0\cdot\begin{pmatrix}1\\0\end{pmatrix}+1\cdot\begin{pmatrix}0\\1\end{pmatrix}\qquad\begin{pmatrix}1\\1\end{pmatrix}=1\cdot\begin{pmatrix}1\\0\end{pmatrix}+1\cdot\begin{pmatrix}0\\1\end{pmatrix}$$

give the other nonsingular matrix.

$$\text{Rep}_{\hat{B},B}(\text{id})=\begin{pmatrix}0&1\\1&1\end{pmatrix}$$

Then the answer is this.

$$\hat{T}=\begin{pmatrix}1&-3\\1&-1\end{pmatrix}\begin{pmatrix}1&2\\3&4\end{pmatrix}\begin{pmatrix}0&1\\1&1\end{pmatrix}=\begin{pmatrix}-10&-18\\-2&-4\end{pmatrix}$$

Although not strictly necessary, a check is reassuring. Arbitrarily fixing

$$\vec{v}=\begin{pmatrix}3\\2\end{pmatrix}$$

we have that

$$\text{Rep}_B(\vec{v})=\begin{pmatrix}3\\2\end{pmatrix}_B\qquad\begin{pmatrix}1&2\\3&4\end{pmatrix}_{B,D}\begin{pmatrix}3\\2\end{pmatrix}_B=\begin{pmatrix}7\\17\end{pmatrix}_D$$

and so $t(\vec{v})$ is this.

$$7\cdot\begin{pmatrix}1\\1\end{pmatrix}+17\cdot\begin{pmatrix}1\\-1\end{pmatrix}=\begin{pmatrix}24\\-10\end{pmatrix}$$

Doing the calculation with respect to $\hat{B},\hat{D}$ starts with

$$\text{Rep}_{\hat{B}}(\vec{v})=\begin{pmatrix}-1\\3\end{pmatrix}_{\hat{B}}\qquad\begin{pmatrix}-10&-18\\-2&-4\end{pmatrix}_{\hat{B},\hat{D}}\begin{pmatrix}-1\\3\end{pmatrix}_{\hat{B}}=\begin{pmatrix}-44\\-10\end{pmatrix}_{\hat{D}}$$

and then checks that this is the same result.

$$-44\cdot\begin{pmatrix}-1\\0\end{pmatrix}-10\cdot\begin{pmatrix}2\\1\end{pmatrix}=\begin{pmatrix}24\\-10\end{pmatrix}$$

(b) These two

$$\begin{pmatrix}1\\1\end{pmatrix}=\frac{1}{3}\cdot\begin{pmatrix}1\\2\end{pmatrix}+\frac{1}{3}\cdot\begin{pmatrix}2\\1\end{pmatrix}\qquad\begin{pmatrix}1\\-1\end{pmatrix}=-1\cdot\begin{pmatrix}1\\2\end{pmatrix}+1\cdot\begin{pmatrix}2\\1\end{pmatrix}$$

show that

$$\mathrm{Rep}_{D,\hat{D}}(\mathrm{id})=\begin{pmatrix}1/3&-1\\1/3&1\end{pmatrix}$$

and these two

$$\begin{pmatrix}1\\2\end{pmatrix}=1\cdot\begin{pmatrix}1\\0\end{pmatrix}+2\cdot\begin{pmatrix}0\\1\end{pmatrix}\qquad\begin{pmatrix}1\\0\end{pmatrix}=-1\cdot\begin{pmatrix}1\\0\end{pmatrix}+0\cdot\begin{pmatrix}0\\1\end{pmatrix}$$

show this.

$$\mathrm{Rep}_{\hat{B},B}(\mathrm{id})=\begin{pmatrix}1&1\\2&0\end{pmatrix}$$

With those, the conversion goes in this way.

$$\hat{T}=\begin{pmatrix}1/3&-1\\1/3&1\end{pmatrix}\begin{pmatrix}1&2\\3&4\end{pmatrix}\begin{pmatrix}1&1\\2&0\end{pmatrix}=\begin{pmatrix}-28/3&-8/3\\38/3&10/3\end{pmatrix}$$

As in the prior item, a check provides some confidence that we did this calculation without mistakes. We can for instance, fix the vector

$$\vec{v}=\begin{pmatrix}-1\\2\end{pmatrix}$$

(this is arbitrary, taken from thin air). Now we have

$$\mathrm{Rep}_B(\vec{v})=\begin{pmatrix}-1\\2\end{pmatrix}\qquad\begin{pmatrix}1&2\\3&4\end{pmatrix}_{B,D}\begin{pmatrix}-1\\2\end{pmatrix}_B=\begin{pmatrix}3\\5\end{pmatrix}_D$$

and so $t(\vec{v})$ is this vector.

$$3\cdot\begin{pmatrix}1\\1\end{pmatrix}+5\cdot\begin{pmatrix}1\\-1\end{pmatrix}=\begin{pmatrix}8\\-2\end{pmatrix}$$

With respect to $\hat{B},\hat{D}$ we first calculate

$$\mathrm{Rep}_{\hat{B}}(\vec{v})=\begin{pmatrix}1\\-2\end{pmatrix}\qquad\begin{pmatrix}-28/3&-8/3\\38/3&10/3\end{pmatrix}_{\hat{B},\hat{D}}\begin{pmatrix}1\\-2\end{pmatrix}_{\hat{B}}=\begin{pmatrix}-4\\6\end{pmatrix}_{\hat{D}}$$

and, sure enough, that is the same result for $t(\vec{v})$.

$$-4\cdot\begin{pmatrix}1\\2\end{pmatrix}+6\cdot\begin{pmatrix}2\\1\end{pmatrix}=\begin{pmatrix}8\\-2\end{pmatrix}$$

Three.V.2.13 Where H and $\hat{H}$ are $m\times n$, the matrix P is $m\times m$ while Q is $n\times n$.

Three.V.2.14 Any $n\times n$ matrix is nonsingular if and only if it has rank n, that is, by Theorem 2.6, if and only if it is matrix equivalent to the $n\times n$ matrix whose diagonal is all ones.

Three.V.2.15 If $PAQ=I$ then $QPAQ=Q$, so $QPA=I$, and so $QP=A^{-1}$.

Three.V.2.16 By the definition following Example 2.2, a matrix M is diagonalizable if it represents $M=\mathrm{Rep}_{B,D}(t)$ a transformation with the property that there is some basis $\hat{B}$ such that $\mathrm{Rep}_{\hat{B},\hat{B}}(t)$ is a diagonal matrix—the starting and ending bases must be equal. But Theorem 2.6 says only that there are $\hat{B}$ and $\hat{D}$ such that we can change to a representation $\mathrm{Rep}_{\hat{B},\hat{D}}(t)$ and get a diagonal matrix. We have no reason to suspect that we could pick the two $\hat{B}$ and $\hat{D}$ so that they are equal.

Three.V.2.17 Yes. Row rank equals column rank, so the rank of the transpose equals the rank of the matrix. Same-sized matrices with equal ranks are matrix equivalent.

Three.V.2.18 Only a zero matrix has rank zero.

Three.V.2.19 For reflexivity, to show that any matrix is matrix equivalent to itself, take P and Q to be identity matrices. For symmetry, if $H_1 = PH_2Q$ then $H_2 = P^{-1}H_1Q^{-1}$ (inverses exist because P and Q are nonsingular). Finally, for transitivity, assume that $H_1 = P_2H_2Q_2$ and that $H_2 = P_3H_3Q_3$. Then substitution gives $H_1 = P_2(P_3H_3Q_3)Q_2 = (P_2P_3)H_3(Q_3Q_2)$. A product of nonsingular matrices is nonsingular (we've shown that the product of invertible matrices is invertible; in fact, we've shown how to calculate the inverse) and so H_1 is therefore matrix equivalent to H_3.

Three.V.2.20 By Theorem 2.6, a zero matrix is alone in its class because it is the only $m{\times}n$ of rank zero. No other matrix is alone in its class; any nonzero scalar product of a matrix has the same rank as that matrix.

Three.V.2.21 There are two matrix-equivalence classes of $1{\times}1$ matrices—those of rank zero and those of rank one. The $3{\times}3$ matrices fall into four matrix equivalence classes.

Three.V.2.22 For $m{\times}n$ matrices there are classes for each possible rank: where k is the minimum of m and n there are classes for the matrices of rank 0, 1, ..., k. That's $k+1$ classes. (Of course, totaling over all sizes of matrices we get infinitely many classes.)

Three.V.2.23 They are closed under nonzero scalar multiplication, since a nonzero scalar multiple of a matrix has the same rank as does the matrix. They are not closed under addition, for instance, $H+(-H)$ has rank zero.

Three.V.2.24 **(a)** We have

$$\mathrm{Rep}_{B,\mathcal{E}_2}(\mathrm{id}) = \begin{pmatrix} 1 & -1 \\ 2 & -1 \end{pmatrix} \qquad \mathrm{Rep}_{\mathcal{E}_2,B}(\mathrm{id}) = \mathrm{Rep}_{B,\mathcal{E}_2}(\mathrm{id})^{-1} = \begin{pmatrix} 1 & -1 \\ 2 & -1 \end{pmatrix}^{-1} = \begin{pmatrix} -1 & 1 \\ -2 & 1 \end{pmatrix}$$

and thus the answer is this.

$$\mathrm{Rep}_{B,B}(t) = \begin{pmatrix} 1 & -1 \\ 2 & -1 \end{pmatrix} \begin{pmatrix} 1 & 1 \\ 3 & -1 \end{pmatrix} \begin{pmatrix} -1 & 1 \\ -2 & 1 \end{pmatrix} = \begin{pmatrix} -2 & 0 \\ -5 & 2 \end{pmatrix}$$

As a quick check, we can take a vector at random

$$\vec{v} = \begin{pmatrix} 4 \\ 5 \end{pmatrix}$$

giving

$$\mathrm{Rep}_{\mathcal{E}_2}(\vec{v}) = \begin{pmatrix} 4 \\ 5 \end{pmatrix} \qquad \begin{pmatrix} 1 & 1 \\ 3 & -1 \end{pmatrix} \begin{pmatrix} 4 \\ 5 \end{pmatrix} = \begin{pmatrix} 9 \\ 7 \end{pmatrix} = t(\vec{v})$$

while the calculation with respect to B, B

$$\mathrm{Rep}_{B}(\vec{v}) = \begin{pmatrix} 1 \\ -3 \end{pmatrix} \qquad \begin{pmatrix} -2 & 0 \\ -5 & 2 \end{pmatrix}_{B,B} \begin{pmatrix} 1 \\ -3 \end{pmatrix}_{B} = \begin{pmatrix} -2 \\ -11 \end{pmatrix}_{B}$$

yields the same result.

$$-2 \cdot \begin{pmatrix} 1 \\ 2 \end{pmatrix} - 11 \cdot \begin{pmatrix} -1 \\ -1 \end{pmatrix} = \begin{pmatrix} 9 \\ 7 \end{pmatrix}$$

(b) We have

$$\begin{array}{ccc} \mathbb{R}^2_{wrt\ \mathcal{E}_2} & \xrightarrow[T]{t} & \mathbb{R}^2_{wrt\ \mathcal{E}_2} \\ \mathrm{id}\downarrow & & \mathrm{id}\downarrow \\ \mathbb{R}^2_{wrt\ B} & \xrightarrow[\hat{T}]{t} & \mathbb{R}^2_{wrt\ B} \end{array} \qquad \mathrm{Rep}_{B,B}(t) = \mathrm{Rep}_{\mathcal{E}_2,B}(\mathrm{id}) \cdot T \cdot \mathrm{Rep}_{B,\mathcal{E}_2}(\mathrm{id})$$

and, as in the first item of this question

$$\mathrm{Rep}_{B,\mathcal{E}_2}(\mathrm{id}) = \left(\vec{\beta}_1 \mid \cdots \mid \vec{\beta}_n\right) \qquad \mathrm{Rep}_{\mathcal{E}_2,B}(\mathrm{id}) = \mathrm{Rep}_{B,\mathcal{E}_2}(\mathrm{id})^{-1}$$

so, writing Q for the matrix whose columns are the basis vectors, we have that $\mathrm{Rep}_{B,B}(t) = Q^{-1}TQ$.

Three.V.2.25 **(a)** The adapted form of the arrow diagram is this.

$$\begin{array}{ccc} V_{\mathit{wrt}\ B_1} & \xrightarrow[H]{h} & W_{\mathit{wrt}\ D} \\ \text{id}\big\downarrow Q & & \text{id}\big\downarrow P \\ V_{\mathit{wrt}\ B_2} & \xrightarrow[\hat{H}]{h} & W_{\mathit{wrt}\ D} \end{array}$$

Since there is no need to change bases in W (or we can say that the change of basis matrix P is the identity), we have $\text{Rep}_{B_2,D}(h) = \text{Rep}_{B_1,D}(h) \cdot Q$ where $Q = \text{Rep}_{B_2,B_1}(\text{id})$.

(b) Here, this is the arrow diagram.

$$\begin{array}{ccc} V_{\mathit{wrt}\ B} & \xrightarrow[H]{h} & W_{\mathit{wrt}\ D_1} \\ \text{id}\big\downarrow Q & & \text{id}\big\downarrow P \\ V_{\mathit{wrt}\ B} & \xrightarrow[\hat{H}]{h} & W_{\mathit{wrt}\ D_2} \end{array}$$

We have that $\text{Rep}_{B,D_2}(h) = P \cdot \text{Rep}_{B,D_1}(h)$ where $P = \text{Rep}_{D_1,D_2}(\text{id})$.

Three.V.2.26 **(a)** Here is the arrow diagram, and a version of that diagram for inverse functions.

$$\begin{array}{ccc} V_{\mathit{wrt}\ B} & \xrightarrow[H]{h} & W_{\mathit{wrt}\ D} \\ \text{id}\big\downarrow Q & & \text{id}\big\downarrow P \\ V_{\mathit{wrt}\ \hat{B}} & \xrightarrow[\hat{H}]{h} & W_{\mathit{wrt}\ \hat{D}} \end{array} \qquad \begin{array}{ccc} V_{\mathit{wrt}\ B} & \xleftarrow[H^{-1}]{h^{-1}} & W_{\mathit{wrt}\ D} \\ \text{id}\big\downarrow Q & & \text{id}\big\downarrow P \\ V_{\mathit{wrt}\ \hat{B}} & \xleftarrow[\hat{H}^{-1}]{h^{-1}} & W_{\mathit{wrt}\ \hat{D}} \end{array}$$

Yes, the inverses of the matrices represent the inverses of the maps. That is, we can move from the lower right to the lower left by moving up, then left, then down. In other words, where $\hat{H} = PHQ$ (and P, Q invertible) and $H, \hat{H}$ are invertible then $\hat{H}^{-1} = Q^{-1}H^{-1}P^{-1}$.

(b) Yes; this is the prior part repeated in different terms.

(c) No, we need another assumption: if H represents h with respect to the same starting as ending bases B, B, for some B then H^2 represents $h \circ h$. As a specific example, these two matrices are both rank one and so they are matrix equivalent

$$\begin{pmatrix} 1 & 0 \\ 0 & 0 \end{pmatrix} \quad \begin{pmatrix} 0 & 0 \\ 1 & 0 \end{pmatrix}$$

but the squares are not matrix equivalent—the square of the first has rank one while the square of the second has rank zero.

(d) No. These two are not matrix equivalent but have matrix equivalent squares.

$$\begin{pmatrix} 0 & 0 \\ 0 & 0 \end{pmatrix} \quad \begin{pmatrix} 0 & 0 \\ 1 & 0 \end{pmatrix}$$

Three.V.2.27 **(a)** The arrow diagram suggests the definition.

$$\begin{array}{ccc} V_{\mathit{wrt}\ B_1} & \xrightarrow[T]{t} & V_{\mathit{wrt}\ B_1} \\ \text{id}\big\downarrow & & \text{id}\big\downarrow \\ V_{\mathit{wrt}\ B_2} & \xrightarrow[\hat{T}]{t} & V_{\mathit{wrt}\ B_2} \end{array}$$

Call matrices $T, \hat{T}$ *similar* if there is a nonsingular matrix P such that $\hat{T} = P^{-1}TP$.

(b) Take P^{-1} to be P and take P to be Q.

(c) *This is as in Exercise 19.* Reflexivity is obvious: $T = I^{-1}TI$. Symmetry is also easy: $\hat{T} = P^{-1}TP$ implies that $T = P\hat{T}P^{-1}$ (multiply the first equation from the right by P^{-1} and from the left by P). For transitivity, assume that $T_1 = {P_2}^{-1}T_2P_2$ and that $T_2 = {P_3}^{-1}T_3P_3$. Then $T_1 = {P_2}^{-1}({P_3}^{-1}T_3P_3)P_2 = ({P_2}^{-1}{P_3}^{-1})T_3(P_3P_2)$ and we are finished on noting that P_3P_2 is an invertible matrix with inverse ${P_2}^{-1}{P_3}^{-1}$.

(d) Assume that $\hat{T} = P^{-1}TP$. For the squares: $\hat{T}^2 = (P^{-1}TP)(P^{-1}TP) = P^{-1}T(PP^{-1})TP = P^{-1}T^2P$. Higher powers follow by induction.

(e) These two are matrix equivalent but their squares are not matrix equivalent.

$$\begin{pmatrix} 1 & 0 \\ 0 & 0 \end{pmatrix} \qquad \begin{pmatrix} 0 & 0 \\ 1 & 0 \end{pmatrix}$$

By the prior item, matrix similarity and matrix equivalence are thus different.

Projection

Three.VI.1: Orthogonal Projection Into a Line

Three.VI.1.7 Each is a straightforward application of the formula from Definition 1.1.

(a) $$\frac{\begin{pmatrix} 2 \\ 1 \end{pmatrix} \cdot \begin{pmatrix} 3 \\ -2 \end{pmatrix}}{\begin{pmatrix} 3 \\ -2 \end{pmatrix} \cdot \begin{pmatrix} 3 \\ -2 \end{pmatrix}} \cdot \begin{pmatrix} 3 \\ -2 \end{pmatrix} = \frac{4}{13} \cdot \begin{pmatrix} 3 \\ -2 \end{pmatrix} = \begin{pmatrix} 12/13 \\ -8/13 \end{pmatrix}$$

(b) $$\frac{\begin{pmatrix} 2 \\ 1 \end{pmatrix} \cdot \begin{pmatrix} 3 \\ 0 \end{pmatrix}}{\begin{pmatrix} 3 \\ 0 \end{pmatrix} \cdot \begin{pmatrix} 3 \\ 0 \end{pmatrix}} \cdot \begin{pmatrix} 3 \\ 0 \end{pmatrix} = \frac{2}{3} \cdot \begin{pmatrix} 3 \\ 0 \end{pmatrix} = \begin{pmatrix} 2 \\ 0 \end{pmatrix}$$

(c) $$\frac{\begin{pmatrix} 1 \\ 1 \\ 4 \end{pmatrix} \cdot \begin{pmatrix} 1 \\ 2 \\ -1 \end{pmatrix}}{\begin{pmatrix} 1 \\ 2 \\ -1 \end{pmatrix} \cdot \begin{pmatrix} 1 \\ 2 \\ -1 \end{pmatrix}} \cdot \begin{pmatrix} 1 \\ 2 \\ -1 \end{pmatrix} = \frac{-1}{6} \cdot \begin{pmatrix} 1 \\ 2 \\ -1 \end{pmatrix} = \begin{pmatrix} -1/6 \\ -1/3 \\ 1/6 \end{pmatrix}$$

(d) $$\frac{\begin{pmatrix} 1 \\ 1 \\ 4 \end{pmatrix} \cdot \begin{pmatrix} 3 \\ 3 \\ 12 \end{pmatrix}}{\begin{pmatrix} 3 \\ 3 \\ 12 \end{pmatrix} \cdot \begin{pmatrix} 3 \\ 3 \\ 12 \end{pmatrix}} \cdot \begin{pmatrix} 3 \\ 3 \\ 12 \end{pmatrix} = \frac{1}{3} \cdot \begin{pmatrix} 3 \\ 3 \\ 12 \end{pmatrix} = \begin{pmatrix} 1 \\ 1 \\ 4 \end{pmatrix}$$

Three.VI.1.8 **(a)** $$\frac{\begin{pmatrix} 2 \\ -1 \\ 4 \end{pmatrix} \cdot \begin{pmatrix} -3 \\ 1 \\ -3 \end{pmatrix}}{\begin{pmatrix} -3 \\ 1 \\ -3 \end{pmatrix} \cdot \begin{pmatrix} -3 \\ 1 \\ -3 \end{pmatrix}} \cdot \begin{pmatrix} -3 \\ 1 \\ -3 \end{pmatrix} = \frac{-19}{19} \cdot \begin{pmatrix} -3 \\ 1 \\ -3 \end{pmatrix} = \begin{pmatrix} 3 \\ -1 \\ 3 \end{pmatrix}$$

(b) Writing the line as $\{c \cdot \begin{pmatrix} 1 \\ 3 \end{pmatrix} \mid c \in \mathbb{R}\}$ gives this projection.

$$\frac{\begin{pmatrix} -1 \\ -1 \end{pmatrix} \cdot \begin{pmatrix} 1 \\ 3 \end{pmatrix}}{\begin{pmatrix} 1 \\ 3 \end{pmatrix} \cdot \begin{pmatrix} 1 \\ 3 \end{pmatrix}} \cdot \begin{pmatrix} 1 \\ 3 \end{pmatrix} = \frac{-4}{10} \cdot \begin{pmatrix} 1 \\ 3 \end{pmatrix} = \begin{pmatrix} -2/5 \\ -6/5 \end{pmatrix}$$

Three.VI.1.9 $$\frac{\begin{pmatrix}1\\2\\1\\3\end{pmatrix}\cdot\begin{pmatrix}-1\\1\\-1\\1\end{pmatrix}}{\begin{pmatrix}-1\\1\\-1\\1\end{pmatrix}\cdot\begin{pmatrix}-1\\1\\-1\\1\end{pmatrix}}\cdot\begin{pmatrix}-1\\1\\-1\\1\end{pmatrix}=\frac{3}{4}\cdot\begin{pmatrix}-1\\1\\-1\\1\end{pmatrix}=\begin{pmatrix}-3/4\\3/4\\-3/4\\3/4\end{pmatrix}$$

Three.VI.1.10 **(a)** $$\frac{\begin{pmatrix}1\\2\end{pmatrix}\cdot\begin{pmatrix}3\\1\end{pmatrix}}{\begin{pmatrix}3\\1\end{pmatrix}\cdot\begin{pmatrix}3\\1\end{pmatrix}}\cdot\begin{pmatrix}3\\1\end{pmatrix}=\frac{1}{2}\cdot\begin{pmatrix}3\\1\end{pmatrix}=\begin{pmatrix}3/2\\1/2\end{pmatrix}$$ **(b)** $$\frac{\begin{pmatrix}0\\4\end{pmatrix}\cdot\begin{pmatrix}3\\1\end{pmatrix}}{\begin{pmatrix}3\\1\end{pmatrix}\cdot\begin{pmatrix}3\\1\end{pmatrix}}\cdot\begin{pmatrix}3\\1\end{pmatrix}=\frac{2}{5}\cdot\begin{pmatrix}3\\1\end{pmatrix}=\begin{pmatrix}6/5\\2/5\end{pmatrix}$$

In general the projection is this.

$$\frac{\begin{pmatrix}x_1\\x_2\end{pmatrix}\cdot\begin{pmatrix}3\\1\end{pmatrix}}{\begin{pmatrix}3\\1\end{pmatrix}\cdot\begin{pmatrix}3\\1\end{pmatrix}}\cdot\begin{pmatrix}3\\1\end{pmatrix}=\frac{3x_1+x_2}{10}\cdot\begin{pmatrix}3\\1\end{pmatrix}=\begin{pmatrix}(9x_1+3x_2)/10\\(3x_1+x_2)/10\end{pmatrix}$$

The appropriate matrix is this.

$$\begin{pmatrix}9/10&3/10\\3/10&1/10\end{pmatrix}$$

Three.VI.1.11 Suppose that $\vec{v}_1$ and $\vec{v}_2$ are nonzero and orthogonal. Consider the linear relationship $c_1\vec{v}_1+c_2\vec{v}_2=\vec{0}$. Take the dot product of both sides of the equation with $\vec{v}_1$ to get that

$$\vec{v}_1\cdot(c_1\vec{v}_1+c_2\vec{v}_2)=c_1\cdot(\vec{v}_1\cdot\vec{v}_1)+c_2\cdot(\vec{v}_1\cdot\vec{v}_2)=c_1\cdot(\vec{v}_1\cdot\vec{v}_1)+c_2\cdot 0=c_1\cdot(\vec{v}_1\cdot\vec{v}_1)$$

is equal to $\vec{v}_1\cdot\vec{0}=\vec{0}$. With the assumption that $\vec{v}_1$ is nonzero, this gives that c_1 is zero. Showing that c_2 is zero is similar.

Three.VI.1.12 **(a)** If the vector $\vec{v}$ is in the line then the orthogonal projection is $\vec{v}$. To verify this by calculation, note that since $\vec{v}$ is in the line we have that $\vec{v}=c_{\vec{v}}\cdot\vec{s}$ for some scalar $c_{\vec{v}}$.

$$\frac{\vec{v}\cdot\vec{s}}{\vec{s}\cdot\vec{s}}\cdot\vec{s}=\frac{c_{\vec{v}}\cdot\vec{s}\cdot\vec{s}}{\vec{s}\cdot\vec{s}}\cdot\vec{s}=c_{\vec{v}}\cdot\frac{\vec{s}\cdot\vec{s}}{\vec{s}\cdot\vec{s}}\cdot\vec{s}=c_{\vec{v}}\cdot 1\cdot\vec{s}=\vec{v}$$

(*Remark.* If we assume that $\vec{v}$ is nonzero then we can simplify the above by taking $\vec{s}$ to be $\vec{v}$.)

(b) Write $c_{\vec{p}}\vec{s}$ for the projection $\mathrm{proj}_{[\vec{s}]}(\vec{v})$. Note that, by the assumption that $\vec{v}$ is not in the line, both $\vec{v}$ and $\vec{v}-c_{\vec{p}}\vec{s}$ are nonzero. Note also that if $c_{\vec{p}}$ is zero then we are actually considering the one-element set $\{\vec{v}\}$, and with $\vec{v}$ nonzero, this set is necessarily linearly independent. Therefore, we are left considering the case that $c_{\vec{p}}$ is nonzero.

Setting up a linear relationship

$$a_1(\vec{v})+a_2(\vec{v}-c_{\vec{p}}\vec{s})=\vec{0}$$

leads to the equation $(a_1+a_2)\cdot\vec{v}=a_2c_{\vec{p}}\cdot\vec{s}$. Because $\vec{v}$ isn't in the line, the scalars a_1+a_2 and $a_2c_{\vec{p}}$ must both be zero. We handled the $c_{\vec{p}}=0$ case above, so the remaining case is that $a_2=0$, and this gives that $a_1=0$ also. Hence the set is linearly independent.

Three.VI.1.13 If $\vec{s}$ is the zero vector then the expression

$$\mathrm{proj}_{[\vec{s}]}(\vec{v})=\frac{\vec{v}\cdot\vec{s}}{\vec{s}\cdot\vec{s}}\cdot\vec{s}$$

contains a division by zero, and so is undefined. As for the right definition, for the projection to lie in the span of the zero vector, it must be defined to be $\vec{0}$.

Three.VI.1.14 Any vector in $\mathbb{R}^n$ is the projection of some other into a line, provided that the dimension n is greater than one. (Clearly, any vector is the projection of itself into a line containing itself; the question is to produce some vector other than $\vec{v}$ that projects to $\vec{v}$.)

Suppose that $\vec{v} \in \mathbb{R}^n$ with $n > 1$. If $\vec{v} \neq \vec{0}$ then we consider the line $\ell = \{c\vec{v} \mid c \in \mathbb{R}\}$ and if $\vec{v} = \vec{0}$ we take ℓ to be any (non-degenerate) line at all (actually, we needn't distinguish between these two cases—see the prior exercise). Let $v_1, \ldots, v_n$ be the components of $\vec{v}$; since $n > 1$, there are at least two. If some v_i is zero then the vector $\vec{w} = \vec{e}_i$ is perpendicular to $\vec{v}$. If none of the components is zero then the vector $\vec{w}$ whose components are $v_2, -v_1, 0, \ldots, 0$ is perpendicular to $\vec{v}$. In either case, observe that $\vec{v} + \vec{w}$ does not equal $\vec{v}$, and that $\vec{v}$ is the projection of $\vec{v} + \vec{w}$ into ℓ.

$$\frac{(\vec{v}+\vec{w})\cdot\vec{v}}{\vec{v}\cdot\vec{v}}\cdot\vec{v} = \big(\frac{\vec{v}\cdot\vec{v}}{\vec{v}\cdot\vec{v}} + \frac{\vec{w}\cdot\vec{v}}{\vec{v}\cdot\vec{v}}\big)\cdot\vec{v} = \frac{\vec{v}\cdot\vec{v}}{\vec{v}\cdot\vec{v}}\cdot\vec{v} = \vec{v}$$

We can dispose of the remaining $n = 0$ and $n = 1$ cases. The dimension $n = 0$ case is the trivial vector space, here there is only one vector and so it cannot be expressed as the projection of a different vector. In the dimension $n = 1$ case there is only one (non-degenerate) line, and every vector is in it, hence every vector is the projection only of itself.

Three.VI.1.15 The proof is simply a calculation.

$$\|\frac{\vec{v}\cdot\vec{s}}{\vec{s}\cdot\vec{s}}\cdot\vec{s}\,\| = |\frac{\vec{v}\cdot\vec{s}}{\vec{s}\cdot\vec{s}}|\cdot\|\vec{s}\,\| = \frac{|\vec{v}\cdot\vec{s}\,|}{\|\vec{s}\,\|^2}\cdot\|\vec{s}\,\| = \frac{|\vec{v}\cdot\vec{s}\,|}{\|\vec{s}\,\|}$$

Three.VI.1.16 Because the projection of $\vec{v}$ into the line spanned by $\vec{s}$ is

$$\frac{\vec{v}\cdot\vec{s}}{\vec{s}\cdot\vec{s}}\cdot\vec{s}$$

the distance squared from the point to the line is this (we write a vector dotted with itself $\vec{w}\cdot\vec{w}$ as $\vec{w}^2$).

$$\begin{aligned}
\|\vec{v} - \frac{\vec{v}\cdot\vec{s}}{\vec{s}\cdot\vec{s}}\cdot\vec{s}\,\|^2 &= \vec{v}\cdot\vec{v} - \vec{v}\cdot(\frac{\vec{v}\cdot\vec{s}}{\vec{s}\cdot\vec{s}}\cdot\vec{s}) - (\frac{\vec{v}\cdot\vec{s}}{\vec{s}\cdot\vec{s}}\cdot\vec{s}\,)\cdot\vec{v} + (\frac{\vec{v}\cdot\vec{s}}{\vec{s}\cdot\vec{s}}\cdot\vec{s}\,)^2 \\
&= \vec{v}\cdot\vec{v} - 2\cdot(\frac{\vec{v}\cdot\vec{s}}{\vec{s}\cdot\vec{s}})\cdot\vec{v}\cdot\vec{s} + (\frac{\vec{v}\cdot\vec{s}}{\vec{s}\cdot\vec{s}})\cdot\vec{s}\cdot\vec{s} \\
&= \frac{(\vec{v}\cdot\vec{v})\cdot(\vec{s}\cdot\vec{s}\,) - 2\cdot(\vec{v}\cdot\vec{s}\,)^2 + (\vec{v}\cdot\vec{s}\,)^2}{\vec{s}\cdot\vec{s}} \\
&= \frac{(\vec{v}\cdot\vec{v}\,)(\vec{s}\cdot\vec{s}\,) - (\vec{v}\cdot\vec{s}\,)^2}{\vec{s}\cdot\vec{s}}
\end{aligned}$$

Three.VI.1.17 Because square root is a strictly increasing function, we can minimize $d(c) = (cs_1 - v_1)^2 + (cs_2 - v_2)^2$ instead of the square root of d. The derivative is $dd/dc = 2(cs_1 - v_1)\cdot s_1 + 2(cs_2 - v_2)\cdot s_2$. Setting it equal to zero $2(cs_1 - v_1)\cdot s_1 + 2(cs_2 - v_2)\cdot s_2 = c\cdot(2s_1^2 + 2s_2^2) - (v_1s_1 + v_2s_2) = 0$ gives the only critical point.

$$c = \frac{v_1s_1 + v_2s_2}{{s_1}^2 + {s_2}^2} = \frac{\vec{v}\cdot\vec{s}}{\vec{s}\cdot\vec{s}}$$

Now the second derivative with respect to c

$$\frac{d^2\,d}{dc^2} = 2{s_1}^2 + 2{s_2}^2$$

is strictly positive (as long as neither s_1 nor s_2 is zero, in which case the question is trivial) and so the critical point is a minimum.

The generalization to $\mathbb{R}^n$ is straightforward. Consider $d_n(c) = (cs_1 - v_1)^2 + \cdots + (cs_n - v_n)^2$, take the derivative, etc.

Three.VI.1.18 The Cauchy-Schwartz inequality $|\vec{v}\cdot\vec{s}\,| \leqslant \|\vec{v}\,\|\cdot\|\vec{s}\,\|$ gives that this fraction

$$\|\frac{\vec{v}\cdot\vec{s}}{\vec{s}\cdot\vec{s}}\cdot\vec{s}\,\| = |\frac{\vec{v}\cdot\vec{s}}{\vec{s}\cdot\vec{s}}|\cdot\|\vec{s}\,\| = \frac{|\vec{v}\cdot\vec{s}\,|}{\|\vec{s}\,\|^2}\cdot\|\vec{s}\,\| = \frac{|\vec{v}\cdot\vec{s}\,|}{\|\vec{s}\,\|}$$

when divided by $\|\vec{v}\,\|$ is less than or equal to one. That is, $\|\vec{v}\,\|$ is larger than or equal to the fraction.

Three.VI.1.19 Write $c\vec{s}$ for $\vec{q}$, and calculate: $(\vec{v}\cdot c\vec{s}/c\vec{s}\cdot c\vec{s}\,)\cdot c\vec{s} = (\vec{v}\cdot\vec{s}/\vec{s}\cdot\vec{s}\,)\cdot\vec{s}$.

Three.VI.1.20 **(a)** Fixing

$$\vec{s} = \begin{pmatrix} 1 \\ 1 \end{pmatrix}$$

as the vector whose span is the line, the formula gives this action,

$$\begin{pmatrix} x \\ y \end{pmatrix} \mapsto \frac{\begin{pmatrix} x \\ y \end{pmatrix} \cdot \begin{pmatrix} 1 \\ 1 \end{pmatrix}}{\begin{pmatrix} 1 \\ 1 \end{pmatrix} \cdot \begin{pmatrix} 1 \\ 1 \end{pmatrix}} \cdot \begin{pmatrix} 1 \\ 1 \end{pmatrix} = \frac{x+y}{2} \cdot \begin{pmatrix} 1 \\ 1 \end{pmatrix} = \begin{pmatrix} (x+y)/2 \\ (x+y)/2 \end{pmatrix}$$

which is the effect of this matrix.

$$\begin{pmatrix} 1/2 & 1/2 \\ 1/2 & 1/2 \end{pmatrix}$$

(b) Rotating the entire plane $\pi/4$ radians clockwise brings the $y = x$ line to lie on the x-axis. Now projecting and then rotating back has the desired effect.

Three.VI.1.21 The sequence need not settle down. With

$$\vec{a} = \begin{pmatrix} 1 \\ 0 \end{pmatrix} \qquad \vec{b} = \begin{pmatrix} 1 \\ 1 \end{pmatrix}$$

the projections are these.

$$\vec{v}_1 = \begin{pmatrix} 1/2 \\ 1/2 \end{pmatrix}, \quad \vec{v}_2 = \begin{pmatrix} 1/2 \\ 0 \end{pmatrix}, \quad \vec{v}_3 = \begin{pmatrix} 1/4 \\ 1/4 \end{pmatrix}, \quad \ldots$$

This sequence doesn't repeat.

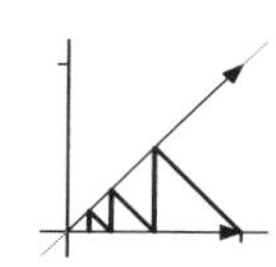

Three.VI.2: Gram-Schmidt Orthogonalization

Three.VI.2.10 **(a)**

$$\vec{\kappa}_1 = \begin{pmatrix} 1 \\ 1 \end{pmatrix}$$

$$\vec{\kappa}_2 = \begin{pmatrix} 2 \\ 1 \end{pmatrix} - \operatorname{proj}_{[\vec{\kappa}_1]}(\begin{pmatrix} 2 \\ 1 \end{pmatrix}) = \begin{pmatrix} 2 \\ 1 \end{pmatrix} - \frac{\begin{pmatrix} 2 \\ 1 \end{pmatrix} \cdot \begin{pmatrix} 1 \\ 1 \end{pmatrix}}{\begin{pmatrix} 1 \\ 1 \end{pmatrix} \cdot \begin{pmatrix} 1 \\ 1 \end{pmatrix}} \cdot \begin{pmatrix} 1 \\ 1 \end{pmatrix} = \begin{pmatrix} 2 \\ 1 \end{pmatrix} - \frac{3}{2} \cdot \begin{pmatrix} 1 \\ 1 \end{pmatrix} = \begin{pmatrix} 1/2 \\ -1/2 \end{pmatrix}$$

(b)

$$\vec{\kappa}_1 = \begin{pmatrix} 0 \\ 1 \end{pmatrix}$$

$$\vec{\kappa}_2 = \begin{pmatrix} -1 \\ 3 \end{pmatrix} - \operatorname{proj}_{[\vec{\kappa}_1]}(\begin{pmatrix} -1 \\ 3 \end{pmatrix}) = \begin{pmatrix} -1 \\ 3 \end{pmatrix} - \frac{\begin{pmatrix} -1 \\ 3 \end{pmatrix} \cdot \begin{pmatrix} 0 \\ 1 \end{pmatrix}}{\begin{pmatrix} 0 \\ 1 \end{pmatrix} \cdot \begin{pmatrix} 0 \\ 1 \end{pmatrix}} \cdot \begin{pmatrix} 0 \\ 1 \end{pmatrix} = \begin{pmatrix} -1 \\ 3 \end{pmatrix} - \frac{3}{1} \cdot \begin{pmatrix} 0 \\ 1 \end{pmatrix} = \begin{pmatrix} -1 \\ 0 \end{pmatrix}$$

(c)

$$\vec{\kappa}_1 = \begin{pmatrix} 0 \\ 1 \end{pmatrix}$$

$$\vec{\kappa}_2 = \begin{pmatrix} -1 \\ 0 \end{pmatrix} - \text{proj}_{[\vec{\kappa}_1]}(\begin{pmatrix} -1 \\ 0 \end{pmatrix}) = \begin{pmatrix} -1 \\ 0 \end{pmatrix} - \frac{\begin{pmatrix} -1 \\ 0 \end{pmatrix} \cdot \begin{pmatrix} 0 \\ 1 \end{pmatrix}}{\begin{pmatrix} 0 \\ 1 \end{pmatrix} \cdot \begin{pmatrix} 0 \\ 1 \end{pmatrix}} \cdot \begin{pmatrix} 0 \\ 1 \end{pmatrix} = \begin{pmatrix} -1 \\ 0 \end{pmatrix} - \frac{0}{1} \cdot \begin{pmatrix} 0 \\ 1 \end{pmatrix} = \begin{pmatrix} -1 \\ 0 \end{pmatrix}$$

The corresponding orthonormal bases for the three parts of this question are these.

$$\langle \begin{pmatrix} 1/\sqrt{2} \\ 1/\sqrt{2} \end{pmatrix}, \begin{pmatrix} \sqrt{2}/2 \\ -\sqrt{2}/2 \end{pmatrix} \rangle \qquad \langle \begin{pmatrix} 0 \\ 1 \end{pmatrix}, \begin{pmatrix} -1 \\ 0 \end{pmatrix} \rangle \qquad \langle \begin{pmatrix} 0 \\ 1 \end{pmatrix}, \begin{pmatrix} -1 \\ 0 \end{pmatrix} \rangle$$

Three.VI.2.11 (a)

$$\vec{\kappa}_1 = \begin{pmatrix} 2 \\ 2 \\ 2 \end{pmatrix}$$

$$\vec{\kappa}_2 = \begin{pmatrix} 1 \\ 0 \\ -1 \end{pmatrix} - \text{proj}_{[\vec{\kappa}_1]}(\begin{pmatrix} 1 \\ 0 \\ -1 \end{pmatrix}) = \begin{pmatrix} 1 \\ 0 \\ -1 \end{pmatrix} - \frac{\begin{pmatrix} 1 \\ 0 \\ -1 \end{pmatrix} \cdot \begin{pmatrix} 2 \\ 2 \\ 2 \end{pmatrix}}{\begin{pmatrix} 2 \\ 2 \\ 2 \end{pmatrix} \cdot \begin{pmatrix} 2 \\ 2 \\ 2 \end{pmatrix}} \cdot \begin{pmatrix} 2 \\ 2 \\ 2 \end{pmatrix} = \begin{pmatrix} 1 \\ 0 \\ -1 \end{pmatrix} - \frac{0}{12} \cdot \begin{pmatrix} 2 \\ 2 \\ 2 \end{pmatrix} = \begin{pmatrix} 1 \\ 0 \\ -1 \end{pmatrix}$$

$$\vec{\kappa}_3 = \begin{pmatrix} 0 \\ 3 \\ 1 \end{pmatrix} - \text{proj}_{[\vec{\kappa}_1]}(\begin{pmatrix} 0 \\ 3 \\ 1 \end{pmatrix}) - \text{proj}_{[\vec{\kappa}_2]}(\begin{pmatrix} 0 \\ 3 \\ 1 \end{pmatrix}) = \begin{pmatrix} 0 \\ 3 \\ 1 \end{pmatrix} - \frac{\begin{pmatrix} 0 \\ 3 \\ 1 \end{pmatrix} \cdot \begin{pmatrix} 2 \\ 2 \\ 2 \end{pmatrix}}{\begin{pmatrix} 2 \\ 2 \\ 2 \end{pmatrix} \cdot \begin{pmatrix} 2 \\ 2 \\ 2 \end{pmatrix}} \cdot \begin{pmatrix} 2 \\ 2 \\ 2 \end{pmatrix} - \frac{\begin{pmatrix} 0 \\ 3 \\ 1 \end{pmatrix} \cdot \begin{pmatrix} 1 \\ 0 \\ -1 \end{pmatrix}}{\begin{pmatrix} 1 \\ 0 \\ -1 \end{pmatrix} \cdot \begin{pmatrix} 1 \\ 0 \\ -1 \end{pmatrix}} \cdot \begin{pmatrix} 1 \\ 0 \\ -1 \end{pmatrix}$$

$$= \begin{pmatrix} 0 \\ 3 \\ 1 \end{pmatrix} - \frac{8}{12} \cdot \begin{pmatrix} 2 \\ 2 \\ 2 \end{pmatrix} - \frac{-1}{2} \cdot \begin{pmatrix} 1 \\ 0 \\ -1 \end{pmatrix} = \begin{pmatrix} -5/6 \\ 5/3 \\ -5/6 \end{pmatrix}$$

(b)

$$\vec{\kappa}_1 = \begin{pmatrix} 1 \\ -1 \\ 0 \end{pmatrix}$$

$$\vec{\kappa}_2 = \begin{pmatrix} 0 \\ 1 \\ 0 \end{pmatrix} - \mathrm{proj}_{[\vec{\kappa}_1]}(\begin{pmatrix} 0 \\ 1 \\ 0 \end{pmatrix}) = \begin{pmatrix} 0 \\ 1 \\ 0 \end{pmatrix} - \frac{\begin{pmatrix} 0 \\ 1 \\ 0 \end{pmatrix} \cdot \begin{pmatrix} 1 \\ -1 \\ 0 \end{pmatrix}}{\begin{pmatrix} 1 \\ -1 \\ 0 \end{pmatrix} \cdot \begin{pmatrix} 1 \\ -1 \\ 0 \end{pmatrix}} \cdot \begin{pmatrix} 1 \\ -1 \\ 0 \end{pmatrix} = \begin{pmatrix} 0 \\ 1 \\ 0 \end{pmatrix} - \frac{-1}{2} \cdot \begin{pmatrix} 1 \\ -1 \\ 0 \end{pmatrix} = \begin{pmatrix} 1/2 \\ 1/2 \\ 0 \end{pmatrix}$$

$$\begin{aligned}
\vec{\kappa}_3 &= \begin{pmatrix} 2 \\ 3 \\ 1 \end{pmatrix} - \mathrm{proj}_{[\vec{\kappa}_1]}(\begin{pmatrix} 2 \\ 3 \\ 1 \end{pmatrix}) - \mathrm{proj}_{[\vec{\kappa}_2]}(\begin{pmatrix} 2 \\ 3 \\ 1 \end{pmatrix}) \\
&= \begin{pmatrix} 2 \\ 3 \\ 1 \end{pmatrix} - \frac{\begin{pmatrix} 2 \\ 3 \\ 1 \end{pmatrix} \cdot \begin{pmatrix} 1 \\ -1 \\ 0 \end{pmatrix}}{\begin{pmatrix} 1 \\ -1 \\ 0 \end{pmatrix} \cdot \begin{pmatrix} 1 \\ -1 \\ 0 \end{pmatrix}} \cdot \begin{pmatrix} 1 \\ -1 \\ 0 \end{pmatrix} - \frac{\begin{pmatrix} 2 \\ 3 \\ 1 \end{pmatrix} \cdot \begin{pmatrix} 1/2 \\ 1/2 \\ 0 \end{pmatrix}}{\begin{pmatrix} 1/2 \\ 1/2 \\ 0 \end{pmatrix} \cdot \begin{pmatrix} 1/2 \\ 1/2 \\ 0 \end{pmatrix}} \cdot \begin{pmatrix} 1/2 \\ 1/2 \\ 0 \end{pmatrix} \\
&= \begin{pmatrix} 2 \\ 3 \\ 1 \end{pmatrix} - \frac{-1}{2} \cdot \begin{pmatrix} 1 \\ -1 \\ 0 \end{pmatrix} - \frac{5/2}{1/2} \cdot \begin{pmatrix} 1/2 \\ 1/2 \\ 0 \end{pmatrix} = \begin{pmatrix} 0 \\ 0 \\ 1 \end{pmatrix}
\end{aligned}$$

The corresponding orthonormal bases for the two parts of this question are these.

$$\langle \begin{pmatrix} 1/\sqrt{3} \\ 1/\sqrt{3} \\ 1/\sqrt{3} \end{pmatrix}, \begin{pmatrix} 1/\sqrt{2} \\ 0 \\ -1/\sqrt{2} \end{pmatrix}, \begin{pmatrix} -1/\sqrt{6} \\ 2/\sqrt{6} \\ -1/\sqrt{6} \end{pmatrix} \rangle \qquad \langle \begin{pmatrix} 1/\sqrt{2} \\ -1/\sqrt{2} \\ 0 \end{pmatrix}, \begin{pmatrix} 1/\sqrt{2} \\ 1/\sqrt{2} \\ 0 \end{pmatrix} \begin{pmatrix} 0 \\ 0 \\ 1 \end{pmatrix} \rangle$$

Three.VI.2.12 We can parametrize the given space can in this way.

$$\{ \begin{pmatrix} x \\ y \\ z \end{pmatrix} \mid x = y - z \} = \{ \begin{pmatrix} 1 \\ 1 \\ 0 \end{pmatrix} \cdot y + \begin{pmatrix} -1 \\ 0 \\ 1 \end{pmatrix} \cdot z \mid y, z \in \mathbb{R} \}$$

So we take the basis

$$\langle \begin{pmatrix} 1 \\ 1 \\ 0 \end{pmatrix}, \begin{pmatrix} -1 \\ 0 \\ 1 \end{pmatrix} \rangle$$

apply the Gram-Schmidt process

$$\vec{\kappa}_1 = \begin{pmatrix} 1 \\ 1 \\ 0 \end{pmatrix}$$

$$\vec{\kappa}_2 = \begin{pmatrix} -1 \\ 0 \\ 1 \end{pmatrix} - \mathrm{proj}_{[\vec{\kappa}_1]}(\begin{pmatrix} -1 \\ 0 \\ 1 \end{pmatrix}) = \begin{pmatrix} -1 \\ 0 \\ 1 \end{pmatrix} - \frac{\begin{pmatrix} -1 \\ 0 \\ 1 \end{pmatrix} \cdot \begin{pmatrix} 1 \\ 1 \\ 0 \end{pmatrix}}{\begin{pmatrix} 1 \\ 1 \\ 0 \end{pmatrix} \cdot \begin{pmatrix} 1 \\ 1 \\ 0 \end{pmatrix}} \cdot \begin{pmatrix} 1 \\ 1 \\ 0 \end{pmatrix} = \begin{pmatrix} -1 \\ 0 \\ 1 \end{pmatrix} - \frac{-1}{2} \cdot \begin{pmatrix} 1 \\ 1 \\ 0 \end{pmatrix} = \begin{pmatrix} -1/2 \\ 1/2 \\ 1 \end{pmatrix}$$

and then normalize.

$$\langle \begin{pmatrix} 1/\sqrt{2} \\ 1/\sqrt{2} \\ 0 \end{pmatrix}, \begin{pmatrix} -1/\sqrt{6} \\ 1/\sqrt{6} \\ 2/\sqrt{6} \end{pmatrix} \rangle$$

Three.VI.2.13 Reducing the linear system

$$\begin{array}{rl} x-y-z+w=0 \\ x \quad +z \quad =0 \end{array} \xrightarrow{-\rho_1+\rho_2} \begin{array}{rl} x-y-\ z+w=0 \\ y+2z-w=0 \end{array}$$

and parametrizing gives this description of the subspace.

$$\{ \begin{pmatrix} -1 \\ -2 \\ 1 \\ 0 \end{pmatrix} \cdot z + \begin{pmatrix} 0 \\ 1 \\ 0 \\ 1 \end{pmatrix} \cdot w \mid z, w \in \mathbb{R} \}$$

So we take the basis,

$$\langle \begin{pmatrix} -1 \\ -2 \\ 1 \\ 0 \end{pmatrix}, \begin{pmatrix} 0 \\ 1 \\ 0 \\ 1 \end{pmatrix} \rangle$$

go through the Gram-Schmidt process

$$\vec{\kappa}_1 = \begin{pmatrix} -1 \\ -2 \\ 1 \\ 0 \end{pmatrix}$$

$$\vec{\kappa}_2 = \begin{pmatrix} 0 \\ 1 \\ 0 \\ 1 \end{pmatrix} - \text{proj}_{[\vec{\kappa}_1]}(\begin{pmatrix} 0 \\ 1 \\ 0 \\ 1 \end{pmatrix}) = \begin{pmatrix} 0 \\ 1 \\ 0 \\ 1 \end{pmatrix} - \frac{\begin{pmatrix} 0 \\ 1 \\ 0 \\ 1 \end{pmatrix} \bullet \begin{pmatrix} -1 \\ -2 \\ 1 \\ 0 \end{pmatrix}}{\begin{pmatrix} -1 \\ -2 \\ 1 \\ 0 \end{pmatrix} \bullet \begin{pmatrix} -1 \\ -2 \\ 1 \\ 0 \end{pmatrix}} \cdot \begin{pmatrix} -1 \\ -2 \\ 1 \\ 0 \end{pmatrix} = \begin{pmatrix} 0 \\ 1 \\ 0 \\ 1 \end{pmatrix} - \frac{-2}{6} \cdot \begin{pmatrix} -1 \\ -2 \\ 1 \\ 0 \end{pmatrix} = \begin{pmatrix} -1/3 \\ 1/3 \\ 1/3 \\ 1 \end{pmatrix}$$

and finish by normalizing.

$$\langle \begin{pmatrix} -1/\sqrt{6} \\ -2/\sqrt{6} \\ 1/\sqrt{6} \\ 0 \end{pmatrix}, \begin{pmatrix} -\sqrt{3}/6 \\ \sqrt{3}/6 \\ \sqrt{3}/6 \\ \sqrt{3}/2 \end{pmatrix} \rangle$$

Three.VI.2.14 A linearly independent subset of $\mathbb{R}^n$ is a basis for its own span. Apply Theorem 2.7.

Remark. Here's why the phrase 'linearly independent' is in the question. Dropping the phrase would require us to worry about two things. The first thing to worry about is that when we do the Gram-Schmidt process on a linearly dependent set then we get some zero vectors. For instance, with

$$S = \{ \begin{pmatrix} 1 \\ 2 \end{pmatrix}, \begin{pmatrix} 3 \\ 6 \end{pmatrix} \}$$

we would get this.

$$\vec{\kappa}_1 = \begin{pmatrix} 1 \\ 2 \end{pmatrix} \qquad \vec{\kappa}_2 = \begin{pmatrix} 3 \\ 6 \end{pmatrix} - \text{proj}_{[\vec{\kappa}_1]}(\begin{pmatrix} 3 \\ 6 \end{pmatrix}) = \begin{pmatrix} 0 \\ 0 \end{pmatrix}$$

This first thing is not so bad because the zero vector is by definition orthogonal to every other vector, so we could accept this situation as yielding an orthogonal set (although it of course can't be normalized), or we just could modify the Gram-Schmidt procedure to throw out any zero vectors. The second thing to worry about if we drop the phrase 'linearly independent' from the question is that the set might be infinite. Of course, any subspace of the finite-dimensional $\mathbb{R}^n$ must also be finite-dimensional so only finitely many of its members are linearly independent, but nonetheless, a "process" that examines the vectors in an infinite set one at a time would at least require some more elaboration in this question. A linearly independent subset of $\mathbb{R}^n$ is automatically finite — in fact, of size n or less — so the 'linearly independent' phrase obviates these concerns.

Three.VI.2.15 If that set is not linearly independent, then we get a zero vector. Otherwise (if our set is linearly independent but does not span the space), we are doing Gram-Schmidt on a set that is a basis for a subspace and so we get an orthogonal basis for a subspace.

Three.VI.2.16 The process leaves the basis unchanged.

Three.VI.2.17 (a) The argument is as in the $i = 3$ case of the proof of Theorem 2.7. The dot product

$$\vec{\kappa}_i \cdot \left(\vec{v} - \mathrm{proj}_{[\vec{\kappa}_1]}(\vec{v}\,) - \cdots - \mathrm{proj}_{[\vec{v}_k]}(\vec{v}\,)\right)$$

can be written as the sum of terms of the form $-\vec{\kappa}_i \cdot \mathrm{proj}_{[\vec{\kappa}_j]}(\vec{v}\,)$ with $j \neq i$, and the term $\vec{\kappa}_i \cdot (\vec{v} - \mathrm{proj}_{[\vec{\kappa}_i]}(\vec{v}\,))$. The first kind of term equals zero because the $\vec{\kappa}$'s are mutually orthogonal. The other term is zero because this projection is orthogonal (that is, the projection definition makes it zero: $\vec{\kappa}_i \cdot (\vec{v} - \mathrm{proj}_{[\vec{\kappa}_i]}(\vec{v}\,)) = \vec{\kappa}_i \cdot \vec{v} - \vec{\kappa}_i \cdot ((\vec{v} \cdot \vec{\kappa}_i)/(\vec{\kappa}_i \cdot \vec{\kappa}_i)) \cdot \vec{\kappa}_i$ equals, after all of the cancellation is done, zero).

(b) The vector $\vec{v}$ is in black and the vector $\mathrm{proj}_{[\vec{\kappa}_1]}(\vec{v}\,) + \mathrm{proj}_{[\vec{v}_2]}(\vec{v}\,) = 1 \cdot \vec{e}_1 + 2 \cdot \vec{e}_2$ is in gray.

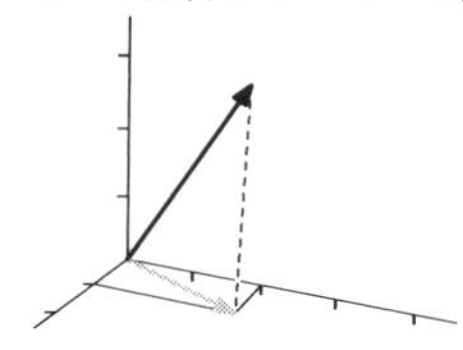

The vector $\vec{v} - (\mathrm{proj}_{[\vec{\kappa}_1]}(\vec{v}\,) + \mathrm{proj}_{[\vec{v}_2]}(\vec{v}\,))$ lies on the dotted line connecting the black vector to the gray one, that is, it is orthogonal to the xy-plane.

(c) We get this diagram by following the hint.

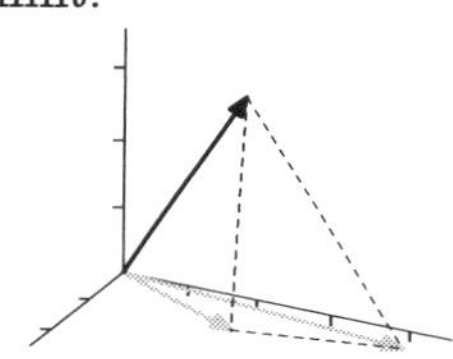

The dashed triangle has a right angle where the gray vector $1 \cdot \vec{e}_1 + 2 \cdot \vec{e}_2$ meets the vertical dashed line $\vec{v} - (1 \cdot \vec{e}_1 + 2 \cdot \vec{e}_2)$; this is what first item of this question proved. The Pythagorean theorem then gives that the hypotenuse — the segment from $\vec{v}$ to any other vector — is longer than the vertical dashed line.

More formally, writing $\mathrm{proj}_{[\vec{\kappa}_1]}(\vec{v}\,) + \cdots + \mathrm{proj}_{[\vec{v}_k]}(\vec{v}\,)$ as $c_1 \cdot \vec{\kappa}_1 + \cdots + c_k \cdot \vec{\kappa}_k$, consider any other vector in the span $d_1 \cdot \vec{\kappa}_1 + \cdots + d_k \cdot \vec{\kappa}_k$. Note that

$$\begin{aligned}\vec{v} - (d_1 \cdot \vec{\kappa}_1 + \cdots + d_k \cdot \vec{\kappa}_k)& \\ &= \left(\vec{v} - (c_1 \cdot \vec{\kappa}_1 + \cdots + c_k \cdot \vec{\kappa}_k)\right) + \left((c_1 \cdot \vec{\kappa}_1 + \cdots + c_k \cdot \vec{\kappa}_k) - (d_1 \cdot \vec{\kappa}_1 + \cdots + d_k \cdot \vec{\kappa}_k)\right)\end{aligned}$$

and that

$$\left(\vec{v} - (c_1 \cdot \vec{\kappa}_1 + \cdots + c_k \cdot \vec{\kappa}_k)\right) \cdot \left((c_1 \cdot \vec{\kappa}_1 + \cdots + c_k \cdot \vec{\kappa}_k) - (d_1 \cdot \vec{\kappa}_1 + \cdots + d_k \cdot \vec{\kappa}_k)\right) = 0$$

(because the first item shows the $\vec{v} - (c_1 \cdot \vec{\kappa}_1 + \cdots + c_k \cdot \vec{\kappa}_k)$ is orthogonal to each $\vec{\kappa}$ and so it is orthogonal to this linear combination of the $\vec{\kappa}$'s). Now apply the Pythagorean Theorem (i.e., the Triangle Inequality).

Three.VI.2.18 One way to proceed is to find a third vector so that the three together make a basis for $\mathbb{R}^3$, e.g.,

$$\vec{\beta}_3 = \begin{pmatrix} 1 \\ 0 \\ 0 \end{pmatrix}$$

(the second vector is not dependent on the third because it has a nonzero second component, and the first is not dependent on the second and third because of its nonzero third component), and then apply the Gram-Schmidt process.

$$\vec{\kappa}_1 = \begin{pmatrix}1\\5\\-1\end{pmatrix}$$

$$\vec{\kappa}_2 = \begin{pmatrix}2\\2\\0\end{pmatrix} - \mathrm{proj}_{[\vec{\kappa}_1]}(\begin{pmatrix}2\\2\\0\end{pmatrix}) = \begin{pmatrix}2\\2\\0\end{pmatrix} - \frac{\begin{pmatrix}2\\2\\0\end{pmatrix}\cdot\begin{pmatrix}1\\5\\-1\end{pmatrix}}{\begin{pmatrix}1\\5\\-1\end{pmatrix}\cdot\begin{pmatrix}1\\5\\-1\end{pmatrix}}\cdot\begin{pmatrix}1\\5\\-1\end{pmatrix}$$

$$= \begin{pmatrix}2\\2\\0\end{pmatrix} - \frac{12}{27}\cdot\begin{pmatrix}1\\5\\-1\end{pmatrix} = \begin{pmatrix}14/9\\-2/9\\4/9\end{pmatrix}$$

$$\vec{\kappa}_3 = \begin{pmatrix}1\\0\\0\end{pmatrix} - \mathrm{proj}_{[\vec{\kappa}_1]}(\begin{pmatrix}1\\0\\0\end{pmatrix}) - \mathrm{proj}_{[\vec{\kappa}_2]}(\begin{pmatrix}1\\0\\0\end{pmatrix})$$

$$= \begin{pmatrix}1\\0\\0\end{pmatrix} - \frac{\begin{pmatrix}1\\0\\0\end{pmatrix}\cdot\begin{pmatrix}1\\5\\-1\end{pmatrix}}{\begin{pmatrix}1\\5\\-1\end{pmatrix}\cdot\begin{pmatrix}1\\5\\-1\end{pmatrix}}\cdot\begin{pmatrix}1\\5\\-1\end{pmatrix} - \frac{\begin{pmatrix}1\\0\\0\end{pmatrix}\cdot\begin{pmatrix}14/9\\-2/9\\4/9\end{pmatrix}}{\begin{pmatrix}14/9\\-2/9\\4/9\end{pmatrix}\cdot\begin{pmatrix}14/9\\-2/9\\4/9\end{pmatrix}}\cdot\begin{pmatrix}14/9\\-2/9\\4/9\end{pmatrix}$$

$$= \begin{pmatrix}1\\0\\0\end{pmatrix} - \frac{1}{27}\cdot\begin{pmatrix}1\\5\\-1\end{pmatrix} - \frac{7}{12}\cdot\begin{pmatrix}14/9\\-2/9\\4/9\end{pmatrix} = \begin{pmatrix}1/18\\-1/18\\-4/18\end{pmatrix}$$

The result $\vec{\kappa}_3$ is orthogonal to both $\vec{\kappa}_1$ and $\vec{\kappa}_2$. It is therefore orthogonal to every vector in the span of the set $\{\vec{\kappa}_1, \vec{\kappa}_2\}$, including the two vectors given in the question.

Three.VI.2.19 **(a)** We can do the representation by eye.

$$\begin{pmatrix}2\\3\end{pmatrix} = 3\cdot\begin{pmatrix}1\\1\end{pmatrix} + (-1)\cdot\begin{pmatrix}1\\0\end{pmatrix} \qquad \mathrm{Rep}_B(\vec{v}) = \begin{pmatrix}3\\-1\end{pmatrix}_B$$

The two projections are also easy.

$$\mathrm{proj}_{[\vec{\beta}_1]}(\begin{pmatrix}2\\3\end{pmatrix}) = \frac{\begin{pmatrix}2\\3\end{pmatrix}\cdot\begin{pmatrix}1\\1\end{pmatrix}}{\begin{pmatrix}1\\1\end{pmatrix}\cdot\begin{pmatrix}1\\1\end{pmatrix}}\cdot\begin{pmatrix}1\\1\end{pmatrix} = \frac{5}{2}\cdot\begin{pmatrix}1\\1\end{pmatrix} \qquad \mathrm{proj}_{[\vec{\beta}_2]}(\begin{pmatrix}2\\3\end{pmatrix}) = \frac{\begin{pmatrix}2\\3\end{pmatrix}\cdot\begin{pmatrix}1\\0\end{pmatrix}}{\begin{pmatrix}1\\0\end{pmatrix}\cdot\begin{pmatrix}1\\0\end{pmatrix}}\cdot\begin{pmatrix}1\\0\end{pmatrix} = \frac{2}{1}\cdot\begin{pmatrix}1\\0\end{pmatrix}$$

(b) As above, we can do the representation by eye

$$\begin{pmatrix}2\\3\end{pmatrix} = (5/2)\cdot\begin{pmatrix}1\\1\end{pmatrix} + (-1/2)\cdot\begin{pmatrix}1\\-1\end{pmatrix}$$

and the two projections are easy.

$$\mathrm{proj}_{[\vec{\beta}_1]}(\begin{pmatrix}2\\3\end{pmatrix}) = \frac{\begin{pmatrix}2\\3\end{pmatrix}\cdot\begin{pmatrix}1\\1\end{pmatrix}}{\begin{pmatrix}1\\1\end{pmatrix}\cdot\begin{pmatrix}1\\1\end{pmatrix}}\cdot\begin{pmatrix}1\\1\end{pmatrix} = \frac{5}{2}\cdot\begin{pmatrix}1\\1\end{pmatrix} \qquad \mathrm{proj}_{[\vec{\beta}_2]}(\begin{pmatrix}2\\3\end{pmatrix}) = \frac{\begin{pmatrix}2\\3\end{pmatrix}\cdot\begin{pmatrix}1\\-1\end{pmatrix}}{\begin{pmatrix}1\\-1\end{pmatrix}\cdot\begin{pmatrix}1\\-1\end{pmatrix}}\cdot\begin{pmatrix}1\\-1\end{pmatrix} = \frac{-1}{2}\cdot\begin{pmatrix}1\\-1\end{pmatrix}$$

Note the recurrence of the $5/2$ and the $-1/2$.

(c) Represent $\vec{v}$ with respect to the basis

$$\mathrm{Rep}_K(\vec{v}) = \begin{pmatrix}r_1\\ \vdots\\ r_k\end{pmatrix}$$

so that $\vec{v} = r_1\vec{\kappa}_1 + \cdots + r_k\vec{\kappa}_k$. To determine r_i, take the dot product of both sides with $\vec{\kappa}_i$.

$$\vec{v}\cdot\vec{\kappa}_i = (r_1\vec{\kappa}_1 + \cdots + r_k\vec{\kappa}_k)\cdot\vec{\kappa}_i = r_1\cdot 0 + \cdots + r_i\cdot(\vec{\kappa}_i\cdot\vec{\kappa}_i) + \cdots + r_k\cdot 0$$

Solving for r_i yields the desired coefficient.

(d) This is a restatement of the prior item.

Three.VI.2.20 First, $\|\vec{v}\,\|^2 = 4^2 + 3^2 + 2^2 + 1^2 = 50$.

(a) $c_1 = 4$ **(b)** $c_1 = 4$, $c_2 = 3$ **(c)** $c_1 = 4$, $c_2 = 3$, $c_3 = 2$, $c_4 = 1$

For the proof, we will do only the $k = 2$ case because the completely general case is messier but no more enlightening. We follow the hint (recall that for any vector $\vec{w}$ we have $\|\vec{w}\,\|^2 = \vec{w}\cdot\vec{w}$).

$$\begin{aligned}
0 &\leqslant \Bigl(\vec{v} - (\frac{\vec{v}\cdot\vec{\kappa}_1}{\vec{\kappa}_1\cdot\vec{\kappa}_1}\cdot\vec{\kappa}_1 + \frac{\vec{v}\cdot\vec{\kappa}_2}{\vec{\kappa}_2\cdot\vec{\kappa}_2}\cdot\vec{\kappa}_2)\Bigr)\cdot\Bigl(\vec{v} - (\frac{\vec{v}\cdot\vec{\kappa}_1}{\vec{\kappa}_1\cdot\vec{\kappa}_1}\cdot\vec{\kappa}_1 + \frac{\vec{v}\cdot\vec{\kappa}_2}{\vec{\kappa}_2\cdot\vec{\kappa}_2}\cdot\vec{\kappa}_2)\Bigr)\\
&= \vec{v}\cdot\vec{v} - 2\cdot\vec{v}\cdot\Bigl(\frac{\vec{v}\cdot\vec{\kappa}_1}{\vec{\kappa}_1\cdot\vec{\kappa}_1}\cdot\vec{\kappa}_1 + \frac{\vec{v}\cdot\vec{\kappa}_2}{\vec{\kappa}_2\cdot\vec{\kappa}_2}\cdot\vec{\kappa}_2\Bigr)\\
&\qquad + \Bigl(\frac{\vec{v}\cdot\vec{\kappa}_1}{\vec{\kappa}_1\cdot\vec{\kappa}_1}\cdot\vec{\kappa}_1 + \frac{\vec{v}\cdot\vec{\kappa}_2}{\vec{\kappa}_2\cdot\vec{\kappa}_2}\cdot\vec{\kappa}_2\Bigr)\cdot\Bigl(\frac{\vec{v}\cdot\vec{\kappa}_1}{\vec{\kappa}_1\cdot\vec{\kappa}_1}\cdot\vec{\kappa}_1 + \frac{\vec{v}\cdot\vec{\kappa}_2}{\vec{\kappa}_2\cdot\vec{\kappa}_2}\cdot\vec{\kappa}_2\Bigr)\\
&= \vec{v}\cdot\vec{v} - 2\cdot\Bigl(\frac{\vec{v}\cdot\vec{\kappa}_1}{\vec{\kappa}_1\cdot\vec{\kappa}_1}\cdot(\vec{v}\cdot\vec{\kappa}_1) + \frac{\vec{v}\cdot\vec{\kappa}_2}{\vec{\kappa}_2\cdot\vec{\kappa}_2}\cdot(\vec{v}\cdot\vec{\kappa}_2)\Bigr) + \Bigl((\frac{\vec{v}\cdot\vec{\kappa}_1}{\vec{\kappa}_1\cdot\vec{\kappa}_1})^2\cdot(\vec{\kappa}_1\cdot\vec{\kappa}_1) + (\frac{\vec{v}\cdot\vec{\kappa}_2}{\vec{\kappa}_2\cdot\vec{\kappa}_2})^2\cdot(\vec{\kappa}_2\cdot\vec{\kappa}_2)\Bigr)
\end{aligned}$$

(The two mixed terms in the third part of the third line are zero because $\vec{\kappa}_1$ and $\vec{\kappa}_2$ are orthogonal.) The result now follows on gathering like terms and on recognizing that $\vec{\kappa}_1\cdot\vec{\kappa}_1 = 1$ and $\vec{\kappa}_2\cdot\vec{\kappa}_2 = 1$ because these vectors are members of an orthonormal set.

Three.VI.2.21 It is true, except for the zero vector. Every vector in $\mathbb{R}^n$ except the zero vector is in a basis, and that basis can be orthogonalized.

Three.VI.2.22 The 3×3 case gives the idea. The set

$$\{\begin{pmatrix}a\\d\\g\end{pmatrix}, \begin{pmatrix}b\\e\\h\end{pmatrix}, \begin{pmatrix}c\\f\\i\end{pmatrix}\}$$

is orthonormal if and only if these nine conditions all hold

$$\begin{array}{ccc}
(a\ \ d\ \ g)\cdot\begin{pmatrix}a\\d\\g\end{pmatrix} = 1 & (a\ \ d\ \ g)\cdot\begin{pmatrix}b\\e\\h\end{pmatrix} = 0 & (a\ \ d\ \ g)\cdot\begin{pmatrix}c\\f\\i\end{pmatrix} = 0\\
(b\ \ e\ \ h)\cdot\begin{pmatrix}a\\d\\g\end{pmatrix} = 0 & (b\ \ e\ \ h)\cdot\begin{pmatrix}b\\e\\h\end{pmatrix} = 1 & (b\ \ e\ \ h)\cdot\begin{pmatrix}c\\f\\i\end{pmatrix} = 0\\
(c\ \ f\ \ i)\cdot\begin{pmatrix}a\\d\\g\end{pmatrix} = 0 & (c\ \ f\ \ i)\cdot\begin{pmatrix}b\\e\\h\end{pmatrix} = 0 & (c\ \ f\ \ i)\cdot\begin{pmatrix}c\\f\\i\end{pmatrix} = 1
\end{array}$$

(the three conditions in the lower left are redundant but nonetheless correct). Those, in turn, hold if and only if

$$\begin{pmatrix} a & d & g \\ b & e & h \\ c & f & i \end{pmatrix}\begin{pmatrix} a & b & c \\ d & e & f \\ g & h & i \end{pmatrix} = \begin{pmatrix} 1 & 0 & 0 \\ 0 & 1 & 0 \\ 0 & 0 & 1 \end{pmatrix}$$

as required.

This is an example, the inverse of this matrix is its transpose.

$$\begin{pmatrix} 1/\sqrt{2} & 1/\sqrt{2} & 0 \\ -1/\sqrt{2} & 1/\sqrt{2} & 0 \\ 0 & 0 & 1 \end{pmatrix}$$

Three.VI.2.23 If the set is empty then the summation on the left side is the linear combination of the empty set of vectors, which by definition adds to the zero vector. In the second sentence, there is not such i, so the 'if ... then ...' implication is vacuously true.

Three.VI.2.24 **(a)** Part of the induction argument proving Theorem 2.7 checks that $\vec{\kappa}_i$ is in the span of $\langle\vec{\beta}_1, \dots, \vec{\beta}_i\rangle$. (The $i = 3$ case in the proof illustrates.) Thus, in the change of basis matrix $\mathrm{Rep}_{K,B}(\mathrm{id})$, the i-th column $\mathrm{Rep}_B(\vec{\kappa}_i)$ has components $i+1$ through k that are zero.

(b) One way to see this is to recall the computational procedure that we use to find the inverse. We write the matrix, write the identity matrix next to it, and then we do Gauss-Jordan reduction. If the matrix starts out upper triangular then the Gauss-Jordan reduction involves only the Jordan half and these steps, when performed on the identity, will result in an upper triangular inverse matrix.

Three.VI.2.25 For the inductive step, we assume that for all j in $[1..i]$, these three conditions are true of each $\vec{\kappa}_j$: (i) each $\vec{\kappa}_j$ is nonzero, (ii) each $\vec{\kappa}_j$ is a linear combination of the vectors $\vec{\beta}_1, \dots, \vec{\beta}_j$, and (iii) each $\vec{\kappa}_j$ is orthogonal to all of the $\vec{\kappa}_m$'s prior to it (that is, with $m < j$). With those inductive hypotheses, consider $\vec{\kappa}_{i+1}$.

$$\begin{aligned} \vec{\kappa}_{i+1} &= \vec{\beta}_{i+1} - \mathrm{proj}_{[\vec{\kappa}_1]}(\beta_{i+1}) - \mathrm{proj}_{[\vec{\kappa}_2]}(\beta_{i+1}) - \cdots - \mathrm{proj}_{[\vec{\kappa}_i]}(\beta_{i+1}) \\ &= \vec{\beta}_{i+1} - \frac{\beta_{i+1} \cdot \vec{\kappa}_1}{\vec{\kappa}_1 \cdot \vec{\kappa}_1} \cdot \vec{\kappa}_1 - \frac{\beta_{i+1} \cdot \vec{\kappa}_2}{\vec{\kappa}_2 \cdot \vec{\kappa}_2} \cdot \vec{\kappa}_2 - \cdots - \frac{\beta_{i+1} \cdot \vec{\kappa}_i}{\vec{\kappa}_i \cdot \vec{\kappa}_i} \cdot \vec{\kappa}_i \end{aligned}$$

By the inductive assumption (ii) we can expand each $\vec{\kappa}_j$ into a linear combination of $\vec{\beta}_1, \dots, \vec{\beta}_j$

$$\begin{aligned} &= \vec{\beta}_{i+1} - \frac{\vec{\beta}_{i+1} \cdot \vec{\kappa}_1}{\vec{\kappa}_1 \cdot \vec{\kappa}_1} \cdot \vec{\beta}_1 \\ &\quad - \frac{\vec{\beta}_{i+1} \cdot \vec{\kappa}_2}{\vec{\kappa}_2 \cdot \vec{\kappa}_2} \cdot \Big(\text{linear combination of } \vec{\beta}_1, \vec{\beta}_2 \Big) - \cdots - \frac{\vec{\beta}_{i+1} \cdot \vec{\kappa}_i}{\vec{\kappa}_i \cdot \vec{\kappa}_i} \cdot \Big(\text{linear combination of } \vec{\beta}_1, \dots, \vec{\beta}_i \Big) \end{aligned}$$

The fractions are scalars so this is a linear combination of linear combinations of $\vec{\beta}_1, \dots, \vec{\beta}_{i+1}$. It is therefore just a linear combination of $\vec{\beta}_1, \dots, \vec{\beta}_{i+1}$. Now, (i) it cannot sum to the zero vector because the equation would then describe a nontrivial linear relationship among the $\vec{\beta}$'s that are given as members of a basis (the relationship is nontrivial because the coefficient of $\vec{\beta}_{i+1}$ is 1). Also, (ii) the equation gives $\vec{\kappa}_{i+1}$ as a combination of $\vec{\beta}_1, \dots, \vec{\beta}_{i+1}$. Finally, for (iii), consider $\vec{\kappa}_j \cdot \vec{\kappa}_{i+1}$; as in the $i = 3$ case, the dot product of $\vec{\kappa}_j$ with $\vec{\kappa}_{i+1} = \vec{\beta}_{i+1} - \mathrm{proj}_{[\vec{\kappa}_1]}(\vec{\beta}_{i+1}) - \cdots - \mathrm{proj}_{[\vec{\kappa}_i]}(\vec{\beta}_{i+1})$ can be rewritten to give two kinds of terms, $\vec{\kappa}_j \cdot \Big(\vec{\beta}_{i+1} - \mathrm{proj}_{[\vec{\kappa}_j]}(\vec{\beta}_{i+1})\Big)$ (which is zero because the projection is orthogonal) and $\vec{\kappa}_j \cdot \mathrm{proj}_{[\vec{\kappa}_m]}(\vec{\beta}_{i+1})$ with $m \neq j$ and $m < i+1$ (which is zero because by the hypothesis (iii) the vectors $\vec{\kappa}_j$ and $\vec{\kappa}_m$ are orthogonal).

Three.VI.3: Projection Into a Subspace

Three.VI.3.10 **(a)** When bases for the subspaces

$$B_M = \langle \begin{pmatrix} 1 \\ -1 \end{pmatrix} \rangle \qquad B_N = \langle \begin{pmatrix} 2 \\ -1 \end{pmatrix} \rangle$$

are concatenated

$$B = B_M ^\frown B_N = \langle \begin{pmatrix} 1 \\ -1 \end{pmatrix}, \begin{pmatrix} 2 \\ -1 \end{pmatrix} \rangle$$

and the given vector is represented

$$\begin{pmatrix} 3 \\ -2 \end{pmatrix} = 1 \cdot \begin{pmatrix} 1 \\ -1 \end{pmatrix} + 1 \cdot \begin{pmatrix} 2 \\ -1 \end{pmatrix}$$

then the answer comes from retaining the M part and dropping the N part.

$$\text{proj}_{M,N}(\begin{pmatrix} 3 \\ -2 \end{pmatrix}) = \begin{pmatrix} 1 \\ -1 \end{pmatrix}$$

(b) When the bases

$$B_M = \langle \begin{pmatrix} 1 \\ 1 \end{pmatrix} \rangle \qquad B_N \langle \begin{pmatrix} 1 \\ -2 \end{pmatrix} \rangle$$

are concatenated, and the vector is represented,

$$\begin{pmatrix} 1 \\ 2 \end{pmatrix} = (4/3) \cdot \begin{pmatrix} 1 \\ 1 \end{pmatrix} - (1/3) \cdot \begin{pmatrix} 1 \\ -2 \end{pmatrix}$$

then retaining only the M part gives this answer.

$$\text{proj}_{M,N}(\begin{pmatrix} 1 \\ 2 \end{pmatrix}) = \begin{pmatrix} 4/3 \\ 4/3 \end{pmatrix}$$

(c) With these bases

$$B_M = \langle \begin{pmatrix} 1 \\ -1 \\ 0 \end{pmatrix}, \begin{pmatrix} 0 \\ 0 \\ 1 \end{pmatrix} \rangle \qquad B_N = \langle \begin{pmatrix} 1 \\ 0 \\ 1 \end{pmatrix} \rangle$$

the representation with respect to the concatenation is this.

$$\begin{pmatrix} 3 \\ 0 \\ 1 \end{pmatrix} = 0 \cdot \begin{pmatrix} 1 \\ -1 \\ 0 \end{pmatrix} - 2 \cdot \begin{pmatrix} 0 \\ 0 \\ 1 \end{pmatrix} + 3 \cdot \begin{pmatrix} 1 \\ 0 \\ 1 \end{pmatrix}$$

and so the projection is this.

$$\text{proj}_{M,N}(\begin{pmatrix} 3 \\ 0 \\ 1 \end{pmatrix}) = \begin{pmatrix} 0 \\ 0 \\ -2 \end{pmatrix}$$

Three.VI.3.11 As in Example 3.5, we can simplify the calculation by just finding the space of vectors perpendicular to all the the vectors in M's basis.

(a) Parametrizing to get

$$M = \{c \cdot \begin{pmatrix} -1 \\ 1 \end{pmatrix} \mid c \in \mathbb{R}\}$$

gives that

$$M^\perp \{ \begin{pmatrix} u \\ v \end{pmatrix} \mid 0 = \begin{pmatrix} u \\ v \end{pmatrix} \cdot \begin{pmatrix} -1 \\ 1 \end{pmatrix} \} = \{ \begin{pmatrix} u \\ v \end{pmatrix} \mid 0 = -u + v \}$$

Parametrizing the one-equation linear system gives this description.

$$M^\perp = \{k \cdot \begin{pmatrix} 1 \\ 1 \end{pmatrix} \mid k \in \mathbb{R}\}$$

(b) As in the answer to the prior part, we can describe M as a span

$$M = \{c \cdot \begin{pmatrix} 3/2 \\ 1 \end{pmatrix} \mid c \in \mathbb{R}\} \qquad B_M = \langle \begin{pmatrix} 3/2 \\ 1 \end{pmatrix} \rangle$$

and then $M^\perp$ is the set of vectors perpendicular to the one vector in this basis.

$$M^\perp = \{\begin{pmatrix} u \\ v \end{pmatrix} \mid (3/2) \cdot u + 1 \cdot v = 0\} = \{k \cdot \begin{pmatrix} -2/3 \\ 1 \end{pmatrix} \mid k \in \mathbb{R}\}$$

(c) Parametrizing the linear requirement in the description of M gives this basis.

$$M = \{c \cdot \begin{pmatrix} 1 \\ 1 \end{pmatrix} \mid c \in \mathbb{R}\} \qquad B_M = \langle \begin{pmatrix} 1 \\ 1 \end{pmatrix} \rangle$$

Now, $M^\perp$ is the set of vectors perpendicular to (the one vector in) B_M.

$$M^\perp = \{\begin{pmatrix} u \\ v \end{pmatrix} \mid u + v = 0\} = \{k \cdot \begin{pmatrix} -1 \\ 1 \end{pmatrix} \mid k \in \mathbb{R}\}$$

(By the way, this answer checks with the first item in this question.)

(d) Every vector in the space is perpendicular to the zero vector so $M^\perp = \mathbb{R}^n$.

(e) The appropriate description and basis for M are routine.

$$M = \{y \cdot \begin{pmatrix} 0 \\ 1 \end{pmatrix} \mid y \in \mathbb{R}\} \qquad B_M = \langle \begin{pmatrix} 0 \\ 1 \end{pmatrix} \rangle$$

Then

$$M^\perp = \{\begin{pmatrix} u \\ v \end{pmatrix} \mid 0 \cdot u + 1 \cdot v = 0\} = \{k \cdot \begin{pmatrix} 1 \\ 0 \end{pmatrix} \mid k \in \mathbb{R}\}$$

and so $(y\text{-axis})^\perp = x\text{-axis}$.

(f) The description of M is easy to find by parametrizing.

$$M = \{c \cdot \begin{pmatrix} 3 \\ 1 \\ 0 \end{pmatrix} + d \cdot \begin{pmatrix} 1 \\ 0 \\ 1 \end{pmatrix} \mid c, d \in \mathbb{R}\} \qquad B_M = \langle \begin{pmatrix} 3 \\ 1 \\ 0 \end{pmatrix}, \begin{pmatrix} 1 \\ 0 \\ 1 \end{pmatrix} \rangle$$

Finding $M^\perp$ here just requires solving a linear system with two equations

$$\begin{array}{rl} 3u + v &= 0 \\ u + w &= 0 \end{array} \xrightarrow{-(1/3)\rho_1 + \rho_2} \begin{array}{rl} 3u + v &= 0 \\ -(1/3)v + w &= 0 \end{array}$$

and parametrizing.

$$M^\perp = \{k \cdot \begin{pmatrix} -1 \\ 3 \\ 1 \end{pmatrix} \mid k \in \mathbb{R}\}$$

(g) Here, M is one-dimensional

$$M = \{c \cdot \begin{pmatrix} 0 \\ -1 \\ 1 \end{pmatrix} \mid c \in \mathbb{R}\} \qquad B_M = \langle \begin{pmatrix} 0 \\ -1 \\ 1 \end{pmatrix} \rangle$$

and as a result, $M^\perp$ is two-dimensional.

$$M^\perp = \{\begin{pmatrix} u \\ v \\ w \end{pmatrix} \mid 0 \cdot u - 1 \cdot v + 1 \cdot w = 0\} = \{j \cdot \begin{pmatrix} 1 \\ 0 \\ 0 \end{pmatrix} + k \cdot \begin{pmatrix} 0 \\ 1 \\ 1 \end{pmatrix} \mid j, k \in \mathbb{R}\}$$

Three.VI.3.12 **(a)** Parametrizing the equation leads to this basis for P.

$$B_P = \langle \begin{pmatrix} 1 \\ 0 \\ 3 \end{pmatrix}, \begin{pmatrix} 0 \\ 1 \\ 2 \end{pmatrix} \rangle$$

(b) Because $\mathbb{R}^3$ is three-dimensional and P is two-dimensional, the complement $P^\perp$ must be a line. Anyway, the calculation as in Example 3.5

$$P^\perp = \{\begin{pmatrix} x \\ y \\ z \end{pmatrix} \mid \begin{pmatrix} 1 & 0 & 3 \\ 0 & 1 & 2 \end{pmatrix}\begin{pmatrix} x \\ y \\ z \end{pmatrix} = \begin{pmatrix} 0 \\ 0 \end{pmatrix}\}$$

gives this basis for $P^\perp$.

$$B_{P^\perp} = \langle \begin{pmatrix} 3 \\ 2 \\ -1 \end{pmatrix} \rangle$$

(c) $\begin{pmatrix} 1 \\ 1 \\ 2 \end{pmatrix} = (5/14)\cdot\begin{pmatrix} 1 \\ 0 \\ 3 \end{pmatrix} + (8/14)\cdot\begin{pmatrix} 0 \\ 1 \\ 2 \end{pmatrix} + (3/14)\cdot\begin{pmatrix} 3 \\ 2 \\ -1 \end{pmatrix}$

(d) $\text{proj}_P(\begin{pmatrix} 1 \\ 1 \\ 2 \end{pmatrix}) = \begin{pmatrix} 5/14 \\ 8/14 \\ 31/14 \end{pmatrix}$

(e) The matrix of the projection

$$\begin{pmatrix} 1 & 0 \\ 0 & 1 \\ 3 & 2 \end{pmatrix}(\begin{pmatrix} 1 & 0 & 3 \\ 0 & 1 & 2 \end{pmatrix}\begin{pmatrix} 1 & 0 \\ 0 & 1 \\ 3 & 2 \end{pmatrix})^{-1}\begin{pmatrix} 1 & 0 & 3 \\ 0 & 1 & 2 \end{pmatrix} = \begin{pmatrix} 1 & 0 \\ 0 & 1 \\ 3 & 2 \end{pmatrix}\begin{pmatrix} 10 & 6 \\ 6 & 5 \end{pmatrix}^{-1}\begin{pmatrix} 1 & 0 & 3 \\ 0 & 1 & 2 \end{pmatrix}$$
$$= \frac{1}{14}\begin{pmatrix} 5 & -6 & 3 \\ -6 & 10 & 2 \\ 3 & 2 & 13 \end{pmatrix}$$

when applied to the vector, yields the expected result.

$$\frac{1}{14}\begin{pmatrix} 5 & -6 & 3 \\ -6 & 10 & 2 \\ 3 & 2 & 13 \end{pmatrix}\begin{pmatrix} 1 \\ 1 \\ 2 \end{pmatrix} = \begin{pmatrix} 5/14 \\ 8/14 \\ 31/14 \end{pmatrix}$$

Three.VI.3.13 **(a)** Parametrizing gives this.

$$M = \{c\cdot\begin{pmatrix} -1 \\ 1 \end{pmatrix} \mid c \in \mathbb{R}\}$$

For the first way, we take the vector spanning the line M to be

$$\vec{s} = \begin{pmatrix} -1 \\ 1 \end{pmatrix}$$

and the Definition 1.1 formula gives this.

$$\text{proj}_{[\vec{s}]}(\begin{pmatrix} 1 \\ -3 \end{pmatrix}) = \frac{\begin{pmatrix} 1 \\ -3 \end{pmatrix}\cdot\begin{pmatrix} -1 \\ 1 \end{pmatrix}}{\begin{pmatrix} -1 \\ 1 \end{pmatrix}\cdot\begin{pmatrix} -1 \\ 1 \end{pmatrix}}\cdot\begin{pmatrix} -1 \\ 1 \end{pmatrix} = \frac{-4}{2}\cdot\begin{pmatrix} -1 \\ 1 \end{pmatrix} = \begin{pmatrix} 2 \\ -2 \end{pmatrix}$$

For the second way, we fix

$$B_M = \langle \begin{pmatrix} -1 \\ 1 \end{pmatrix} \rangle$$

and so (as in Example 3.5 and 3.6, we can just find the vectors perpendicular to all of the members of the basis)

$$M^\perp = \{\begin{pmatrix} u \\ v \end{pmatrix} \mid -1\cdot u + 1\cdot v = 0\} = \{k\cdot\begin{pmatrix} 1 \\ 1 \end{pmatrix} \mid k \in \mathbb{R}\} \qquad B_{M^\perp} = \langle \begin{pmatrix} 1 \\ 1 \end{pmatrix} \rangle$$

and representing the vector with respect to the concatenation gives this.

$$\begin{pmatrix}1\\-3\end{pmatrix}=-2\cdot\begin{pmatrix}-1\\1\end{pmatrix}-1\cdot\begin{pmatrix}1\\1\end{pmatrix}$$

Keeping the M part yields the answer.

$$\mathrm{proj}_{M,M^{\perp}}(\begin{pmatrix}1\\-3\end{pmatrix})=\begin{pmatrix}2\\-2\end{pmatrix}$$

The third part is also a simple calculation (there is a 1×1 matrix in the middle, and the inverse of it is also 1×1)

$$\begin{aligned}A\left(A^{\mathsf{T}}A\right)^{-1}A^{\mathsf{T}}&=\begin{pmatrix}-1\\1\end{pmatrix}\left(\begin{pmatrix}-1&1\end{pmatrix}\begin{pmatrix}-1\\1\end{pmatrix}\right)^{-1}\begin{pmatrix}-1&1\end{pmatrix}=\begin{pmatrix}-1\\1\end{pmatrix}\begin{pmatrix}2\end{pmatrix}^{-1}\begin{pmatrix}-1&1\end{pmatrix}\\&=\begin{pmatrix}-1\\1\end{pmatrix}\begin{pmatrix}1/2\end{pmatrix}\begin{pmatrix}-1&1\end{pmatrix}=\begin{pmatrix}-1\\1\end{pmatrix}\begin{pmatrix}-1/2&1/2\end{pmatrix}=\begin{pmatrix}1/2&-1/2\\-1/2&1/2\end{pmatrix}\end{aligned}$$

which of course gives the same answer.

$$\mathrm{proj}_{M}(\begin{pmatrix}1\\-3\end{pmatrix})=\begin{pmatrix}1/2&-1/2\\-1/2&1/2\end{pmatrix}\begin{pmatrix}1\\-3\end{pmatrix}=\begin{pmatrix}2\\-2\end{pmatrix}$$

(b) Parametrization gives this.

$$M=\{c\cdot\begin{pmatrix}-1\\0\\1\end{pmatrix}\mid c\in\mathbb{R}\}$$

With that, the formula for the first way gives this.

$$\frac{\begin{pmatrix}0\\1\\2\end{pmatrix}\cdot\begin{pmatrix}-1\\0\\1\end{pmatrix}}{\begin{pmatrix}-1\\0\\1\end{pmatrix}\cdot\begin{pmatrix}-1\\0\\1\end{pmatrix}}\cdot\begin{pmatrix}-1\\0\\1\end{pmatrix}=\frac{2}{2}\cdot\begin{pmatrix}-1\\0\\1\end{pmatrix}=\begin{pmatrix}-1\\0\\1\end{pmatrix}$$

To proceed by the second method we find $M^{\perp}$,

$$M^{\perp}=\{\begin{pmatrix}u\\v\\w\end{pmatrix}\mid -u+w=0\}=\{j\cdot\begin{pmatrix}1\\0\\1\end{pmatrix}+k\cdot\begin{pmatrix}0\\1\\0\end{pmatrix}\mid j,k\in\mathbb{R}\}$$

find the representation of the given vector with respect to the concatenation of the bases B_M and $B_{M^{\perp}}$

$$\begin{pmatrix}0\\1\\2\end{pmatrix}=1\cdot\begin{pmatrix}-1\\0\\1\end{pmatrix}+1\cdot\begin{pmatrix}1\\0\\1\end{pmatrix}+1\cdot\begin{pmatrix}0\\1\\0\end{pmatrix}$$

and retain only the M part.

$$\mathrm{proj}_{M}(\begin{pmatrix}0\\1\\2\end{pmatrix})=1\cdot\begin{pmatrix}-1\\0\\1\end{pmatrix}=\begin{pmatrix}-1\\0\\1\end{pmatrix}$$

Finally, for the third method, the matrix calculation

$$\begin{aligned}A\left(A^{\mathsf{T}}A\right)^{-1}A^{\mathsf{T}}&=\begin{pmatrix}-1\\0\\1\end{pmatrix}\left(\begin{pmatrix}-1&0&1\end{pmatrix}\begin{pmatrix}-1\\0\\1\end{pmatrix}\right)^{-1}\begin{pmatrix}-1&0&1\end{pmatrix}=\begin{pmatrix}-1\\0\\1\end{pmatrix}\begin{pmatrix}2\end{pmatrix}^{-1}\begin{pmatrix}-1&0&1\end{pmatrix}\\&=\begin{pmatrix}-1\\0\\1\end{pmatrix}\begin{pmatrix}1/2\end{pmatrix}\begin{pmatrix}-1&0&1\end{pmatrix}=\begin{pmatrix}-1\\0\\1\end{pmatrix}\begin{pmatrix}-1/2&0&1/2\end{pmatrix}=\begin{pmatrix}1/2&0&-1/2\\0&0&0\\-1/2&0&1/2\end{pmatrix}\end{aligned}$$

followed by matrix-vector multiplication

$$\text{proj}_M(\begin{pmatrix}0\\1\\2\end{pmatrix})\begin{pmatrix}1/2&0&-1/2\\0&0&0\\-1/2&0&1/2\end{pmatrix}\begin{pmatrix}0\\1\\2\end{pmatrix}=\begin{pmatrix}-1\\0\\1\end{pmatrix}$$

gives the answer.

Three.VI.3.14 No, a decomposition of vectors $\vec{v}=\vec{m}+\vec{n}$ into $\vec{m}\in M$ and $\vec{n}\in N$ does not depend on the bases chosen for the subspaces, as we showed in the Direct Sum subsection.

Three.VI.3.15 The orthogonal projection of a vector into a subspace is a member of that subspace. Since a trivial subspace has only one member, $\vec{0}$, the projection of any vector must equal $\vec{0}$.

Three.VI.3.16 The projection into M along N of a $\vec{v}\in M$ is $\vec{v}$. Decomposing $\vec{v}=\vec{m}+\vec{n}$ gives $\vec{m}=\vec{v}$ and $\vec{n}=\vec{0}$, and dropping the N part but retaining the M part results in a projection of $\vec{m}=\vec{v}$.

Three.VI.3.17 The proof of Lemma 3.7 shows that each vector $\vec{v}\in\mathbb{R}^n$ is the sum of its orthogonal projections onto the lines spanned by the basis vectors.

$$\vec{v}=\text{proj}_{[\vec{\kappa}_1]}(\vec{v})+\cdots+\text{proj}_{[\vec{\kappa}_n]}(\vec{v})=\frac{\vec{v}\cdot\vec{\kappa}_1}{\vec{\kappa}_1\cdot\vec{\kappa}_1}\cdot\vec{\kappa}_1+\cdots+\frac{\vec{v}\cdot\vec{\kappa}_n}{\vec{\kappa}_n\cdot\vec{\kappa}_n}\cdot\vec{\kappa}_n$$

Since the basis is orthonormal, the bottom of each fraction has $\vec{\kappa}_i\cdot\vec{\kappa}_i=1$.

Three.VI.3.18 If $V=M\oplus N$ then every vector decomposes uniquely as $\vec{v}=\vec{m}+\vec{n}$. For all $\vec{v}$ the map p gives $p(\vec{v})=\vec{m}$ if and only if $\vec{v}-p(\vec{v})=\vec{n}$, as required.

Three.VI.3.19 Let $\vec{v}$ be perpendicular to every $\vec{w}\in S$. Then $\vec{v}\cdot(c_1\vec{w}_1+\cdots+c_n\vec{w}_n)=\vec{v}\cdot(c_1\vec{w}_1)+\cdots+\vec{v}\cdot(c_n\cdot\vec{w}_n)=c_1(\vec{v}\cdot\vec{w}_1)+\cdots+c_n(\vec{v}\cdot\vec{w}_n)=c_1\cdot 0+\cdots+c_n\cdot 0=0$.

Three.VI.3.20 True; the only vector orthogonal to itself is the zero vector.

Three.VI.3.21 This is immediate from the statement in Lemma 3.7 that the space is the direct sum of the two.

Three.VI.3.22 The two must be equal, even only under the seemingly weaker condition that they yield the same result on all orthogonal projections. Consider the subspace M spanned by the set $\{\vec{v}_1,\vec{v}_2\}$. Since each is in M, the orthogonal projection of $\vec{v}_1$ into M is $\vec{v}_1$ and the orthogonal projection of $\vec{v}_2$ into M is $\vec{v}_2$. For their projections into M to be equal, they must be equal.

Three.VI.3.23 (a) We will show that the sets are mutually inclusive, $M\subseteq(M^\perp)^\perp$ and $(M^\perp)^\perp\subseteq M$. For the first, if $\vec{m}\in M$ then by the definition of the perp operation, $\vec{m}$ is perpendicular to every $\vec{v}\in M^\perp$, and therefore (again by the definition of the perp operation) $\vec{m}\in(M^\perp)^\perp$. For the other direction, consider $\vec{v}\in(M^\perp)^\perp$. Lemma 3.7's proof shows that $\mathbb{R}^n=M\oplus M^\perp$ and that we can give an orthogonal basis for the space $\langle\vec{\kappa}_1,\ldots,\vec{\kappa}_k,\vec{\kappa}_{k+1},\ldots,\vec{\kappa}_n\rangle$ such that the first half $\langle\vec{\kappa}_1,\ldots,\vec{\kappa}_k\rangle$ is a basis for M and the second half is a basis for $M^\perp$. The proof also checks that each vector in the space is the sum of its orthogonal projections onto the lines spanned by these basis vectors.

$$\vec{v}=\text{proj}_{[\vec{\kappa}_1]}(\vec{v})+\cdots+\text{proj}_{[\vec{\kappa}_n]}(\vec{v})$$

Because $\vec{v}\in(M^\perp)^\perp$, it is perpendicular to every vector in $M^\perp$, and so the projections in the second half are all zero. Thus $\vec{v}=\text{proj}_{[\vec{\kappa}_1]}(\vec{v})+\cdots+\text{proj}_{[\vec{\kappa}_k]}(\vec{v})$, which is a linear combination of vectors from M, and so $\vec{v}\in M$. (*Remark.* Here is a slicker way to do the second half: write the space both as $M\oplus M^\perp$ and as $M^\perp\oplus(M^\perp)^\perp$. Because the first half showed that $M\subseteq(M^\perp)^\perp$ and the prior sentence shows that the dimension of the two subspaces M and $(M^\perp)^\perp$ are equal, we can conclude that M equals $(M^\perp)^\perp$.)

(b) Because $M\subseteq N$, any $\vec{v}$ that is perpendicular to every vector in N is also perpendicular to every vector in M. But that sentence simply says that $N^\perp\subseteq M^\perp$.

(c) We will again show that the sets are equal by mutual inclusion. The first direction is easy; any $\vec{v}$ perpendicular to every vector in $M+N=\{\vec{m}+\vec{n}\mid\vec{m}\in M,\ \vec{n}\in N\}$ is perpendicular to every vector of the form $\vec{m}+\vec{0}$ (that is, every vector in M) and every vector of the form $\vec{0}+\vec{n}$ (every vector in N), and so $(M+N)^\perp\subseteq M^\perp\cap N^\perp$. The second direction is also routine; any vector $\vec{v}\in M^\perp\cap N^\perp$ is perpendicular to any vector of the form $c\vec{m}+d\vec{n}$ because $\vec{v}\cdot(c\vec{m}+d\vec{n})=c\cdot(\vec{v}\cdot\vec{m})+d\cdot(\vec{v}\cdot\vec{n})=c\cdot 0+d\cdot 0=0$.

Three.VI.3.24 **(a)** The representation of

$$\begin{pmatrix} v_1 \\ v_2 \\ v_3 \end{pmatrix} \stackrel{f}{\longmapsto} 1v_1 + 2v_2 + 3v_3$$

is this.

$$\mathrm{Rep}_{\mathcal{E}_3,\mathcal{E}_1}(f) = \begin{pmatrix} 1 & 2 & 3 \end{pmatrix}$$

By the definition of f

$$\mathscr{N}(f) = \{ \begin{pmatrix} v_1 \\ v_2 \\ v_3 \end{pmatrix} \mid 1v_1 + 2v_2 + 3v_3 = 0 \} = \{ \begin{pmatrix} v_1 \\ v_2 \\ v_3 \end{pmatrix} \mid \begin{pmatrix} 1 \\ 2 \\ 3 \end{pmatrix} \cdot \begin{pmatrix} v_1 \\ v_2 \\ v_3 \end{pmatrix} = 0 \}$$

and this second description exactly says this.

$$\mathscr{N}(f)^{\perp} = [\{ \begin{pmatrix} 1 \\ 2 \\ 3 \end{pmatrix} \}]$$

(b) The generalization is that for any $f\colon \mathbb{R}^n \to \mathbb{R}$ there is a vector $\vec{h}$ so that

$$\begin{pmatrix} v_1 \\ \vdots \\ v_n \end{pmatrix} \stackrel{f}{\longmapsto} h_1v_1 + \cdots + h_nv_n$$

and $\vec{h} \in \mathscr{N}(f)^{\perp}$. We can prove this by, as in the prior item, representing f with respect to the standard bases and taking $\vec{h}$ to be the column vector gotten by transposing the one row of that matrix representation.

(c) Of course,

$$\mathrm{Rep}_{\mathcal{E}_3,\mathcal{E}_2}(f) = \begin{pmatrix} 1 & 2 & 3 \\ 4 & 5 & 6 \end{pmatrix}$$

and so the null space is this set.

$$\mathscr{N}(f)\{ \begin{pmatrix} v_1 \\ v_2 \\ v_3 \end{pmatrix} \mid \begin{pmatrix} 1 & 2 & 3 \\ 4 & 5 & 6 \end{pmatrix} \begin{pmatrix} v_1 \\ v_2 \\ v_3 \end{pmatrix} = \begin{pmatrix} 0 \\ 0 \end{pmatrix} \}$$

That description makes clear that

$$\begin{pmatrix} 1 \\ 2 \\ 3 \end{pmatrix}, \begin{pmatrix} 4 \\ 5 \\ 6 \end{pmatrix} \in \mathscr{N}(f)^{\perp}$$

and since $\mathscr{N}(f)^{\perp}$ is a subspace of $\mathbb{R}^n$, the span of the two vectors is a subspace of the perp of the null space. To see that this containment is an equality, take

$$M = [\{ \begin{pmatrix} 1 \\ 2 \\ 3 \end{pmatrix} \}] \qquad N = [\{ \begin{pmatrix} 4 \\ 5 \\ 6 \end{pmatrix} \}]$$

in the third item of Exercise 23, as suggested in the hint.

(d) As above, generalizing from the specific case is easy: for any $f\colon \mathbb{R}^n \to \mathbb{R}^m$ the matrix H representing the map with respect to the standard bases describes the action

$$\begin{pmatrix} v_1 \\ \vdots \\ v_n \end{pmatrix} \stackrel{f}{\longmapsto} \begin{pmatrix} h_{1,1}v_1 + h_{1,2}v_2 + \cdots + h_{1,n}v_n \\ \vdots \\ h_{m,1}v_1 + h_{m,2}v_2 + \cdots + h_{m,n}v_n \end{pmatrix}$$

and the description of the null space gives that on transposing the m rows of H

$$\vec{h}_1 = \begin{pmatrix} h_{1,1} \\ h_{1,2} \\ \vdots \\ h_{1,n} \end{pmatrix}, \ldots \vec{h}_m = \begin{pmatrix} h_{m,1} \\ h_{m,2} \\ \vdots \\ h_{m,n} \end{pmatrix}$$

we have $\mathscr{N}(f)^{\perp} = [\{\vec{h}_1, \ldots, \vec{h}_m\}]$. ([Strang 93] describes this space as the transpose of the row space of H.)

Three.VI.3.25 **(a)** First note that if a vector $\vec{v}$ is already in the line then the orthogonal projection gives $\vec{v}$ itself. One way to verify this is to apply the formula for projection into the line spanned by a vector $\vec{s}$, namely $(\vec{v} \cdot \vec{s}/\vec{s} \cdot \vec{s}) \cdot \vec{s}$. Taking the line as $\{k \cdot \vec{v} \mid k \in \mathbb{R}\}$ (the $\vec{v} = \vec{0}$ case is separate but easy) gives $(\vec{v} \cdot \vec{v}/\vec{v} \cdot \vec{v}) \cdot \vec{v}$, which simplifies to $\vec{v}$, as required.

Now, that answers the question because after once projecting into the line, the result $\text{proj}_{\ell}(\vec{v})$ is in that line. The prior paragraph says that projecting into the same line again will have no effect.

(b) The argument here is similar to the one in the prior item. With $V = M \oplus N$, the projection of $\vec{v} = \vec{m} + \vec{n}$ is $\text{proj}_{M,N}(\vec{v}) = \vec{m}$. Now repeating the projection will give $\text{proj}_{M,N}(\vec{m}) = \vec{m}$, as required, because the decomposition of a member of M into the sum of a member of M and a member of N is $\vec{m} = \vec{m} + \vec{0}$. Thus, projecting twice into M along N has the same effect as projecting once.

(c) As suggested by the prior items, the condition gives that t leaves vectors in the range space unchanged, and hints that we should take $\vec{\beta}_1, \ldots, \vec{\beta}_r$ to be basis vectors for the range, that is, that we should take the range space of t for M (so that $\dim(M) = r$). As for the complement, we write N for the null space of t and we will show that $V = M \oplus N$.

To show this, we can show that their intersection is trivial $M \cap N = \{\vec{0}\}$ and that they sum to the entire space $M + N = V$. For the first, if a vector $\vec{m}$ is in the range space then there is a $\vec{v} \in V$ with $t(\vec{v}) = \vec{m}$, and the condition on t gives that $t(\vec{m}) = (t \circ t)(\vec{v}) = t(\vec{v}) = \vec{m}$, while if that same vector is also in the null space then $t(\vec{m}) = \vec{0}$ and so the intersection of the range space and null space is trivial. For the second, to write an arbitrary $\vec{v}$ as the sum of a vector from the range space and a vector from the null space, the fact that the condition $t(\vec{v}) = t(t(\vec{v}))$ can be rewritten as $t(\vec{v} - t(\vec{v})) = \vec{0}$ suggests taking $\vec{v} = t(\vec{v}) + (\vec{v} - t(\vec{v}))$.

To finish we taking a basis $B = \langle \vec{\beta}_1, \ldots, \vec{\beta}_n \rangle$ for V where $\langle \vec{\beta}_1, \ldots, \vec{\beta}_r \rangle$ is a basis for the range space M and $\langle \vec{\beta}_{r+1}, \ldots, \vec{\beta}_n \rangle$ is a basis for the null space N.

(d) Every projection (as defined in this exercise) is a projection into its range space and along its null space.

(e) This also follows immediately from the third item.

Three.VI.3.26 For any matrix M we have that $(M^{-1})^{T} = (M^{T})^{-1}$, and for any two matrices M, N we have that $MN^{T} = N^{T}M^{T}$ (provided, of course, that the inverse and product are defined). Applying these two gives that the matrix equals its transpose.

$$\begin{aligned}(A(A^{T}A)^{-1}A^{T})^{T} &= (A^{T^{T}})(((A^{T}A)^{-1})^{T})(A^{T}) \\ &= (A^{T^{T}})(((A^{T}A)^{T})^{-1})(A^{T}) = A(A^{T}A^{T^{T}})^{-1}A^{T} = A(A^{T}A)^{-1}A^{T}\end{aligned}$$

Topic: Line of Best Fit

Data on the progression of the world's records (taken from the Runner's World web site) is below.

Progression of Men's Mile Record

time	name	date
4:52.0	Cadet Marshall (GBR)	02Sep52
4:45.0	Thomas Finch (GBR)	03Nov58
4:40.0	Gerald Surman (GBR)	24Nov59
4:33.0	George Farran (IRL)	23May62
4:29 3/5	Walter Chinnery (GBR)	10Mar68
4:28 4/5	William Gibbs (GBR)	03Apr68
4:28 3/5	Charles Gunton (GBR)	31Mar73
4:26.0	Walter Slade (GBR)	30May74
4:24 1/2	Walter Slade (GBR)	19Jun75
4:23 1/5	Walter George (GBR)	16Aug80
4:19 2/5	Walter George (GBR)	03Jun82
4:18 2/5	Walter George (GBR)	21Jun84
4:17 4/5	Thomas Conneff (USA)	26Aug93
4:17.0	Fred Bacon (GBR)	06Jul95
4:15 3/5	Thomas Conneff (USA)	28Aug95
4:15 2/5	John Paul Jones (USA)	27May11
4:14.4	John Paul Jones (USA)	31May13
4:12.6	Norman Taber (USA)	16Jul15
4:10.4	Paavo Nurmi (FIN)	23Aug23
4:09 1/5	Jules Ladoumegue (FRA)	04Oct31
4:07.6	Jack Lovelock (NZL)	15Jul33
4:06.8	Glenn Cunningham (USA)	16Jun34
4:06.4	Sydney Wooderson (GBR)	28Aug37
4:06.2	Gunder Hagg (SWE)	01Jul42
4:04.6	Gunder Hagg (SWE)	04Sep42
4:02.6	Arne Andersson (SWE)	01Jul43
4:01.6	Arne Andersson (SWE)	18Jul44
4:01.4	Gunder Hagg (SWE)	17Jul45
3:59.4	Roger Bannister (GBR)	06May54
3:58.0	John Landy (AUS)	21Jun54
3:57.2	Derek Ibbotson (GBR)	19Jul57
3:54.5	Herb Elliott (AUS)	06Aug58
3:54.4	Peter Snell (NZL)	27Jan62
3:54.1	Peter Snell (NZL)	17Nov64
3:53.6	Michel Jazy (FRA)	09Jun65
3:51.3	Jim Ryun (USA)	17Jul66
3:51.1	Jim Ryun (USA)	23Jun67
3:51.0	Filbert Bayi (TAN)	17May75
3:49.4	John Walker (NZL)	12Aug75
3:49.0	Sebastian Coe (GBR)	17Jul79
3:48.8	Steve Ovett (GBR)	01Jul80
3:48.53	Sebastian Coe (GBR)	19Aug81
3:48.40	Steve Ovett (GBR)	26Aug81
3:47.33	Sebastian Coe (GBR)	28Aug81
3:46.32	Steve Cram (GBR)	27Jul85
3:44.39	Noureddine Morceli (ALG)	05Sep93
3:43.13	Hicham el Guerrouj (MOR)	07Jul99

Progression of Men's 1500 Meter Record

time	name	date
4:09.0	John Bray (USA)	30May00
4:06.2	Charles Bennett (GBR)	15Jul00
4:05.4	James Lightbody (USA)	03Sep04
3:59.8	Harold Wilson (GBR)	30May08
3:59.2	Abel Kiviat (USA)	26May12
3:56.8	Abel Kiviat (USA)	02Jun12
3:55.8	Abel Kiviat (USA)	08Jun12
3:55.0	Norman Taber (USA)	16Jul15
3:54.7	John Zander (SWE)	05Aug17
3:53.0	Paavo Nurmi (FIN)	23Aug23
3:52.6	Paavo Nurmi (FIN)	19Jun24
3:51.0	Otto Peltzer (GER)	11Sep26
3:49.2	Jules Ladoumegue (FRA)	05Oct30
3:49.0	Luigi Beccali (ITA)	17Sep33
3:48.8	William Bonthron (USA)	30Jun34
3:47.8	Jack Lovelock (NZL)	06Aug36
3:47.6	Gunder Hagg (SWE)	10Aug41
3:45.8	Gunder Hagg (SWE)	17Jul42
3:45.0	Arne Andersson (SWE)	17Aug43
3:43.0	Gunder Hagg (SWE)	07Jul44
3:42.8	Wes Santee (USA)	04Jun54
3:41.8	John Landy (AUS)	21Jun54
3:40.8	Sandor Iharos (HUN)	28Jul55
3:40.6	Istvan Rozsavolgyi (HUN)	03Aug56
3:40.2	Olavi Salsola (FIN)	11Jul57
3:38.1	Stanislav Jungwirth (CZE)	12Jul57
3:36.0	Herb Elliott (AUS)	28Aug58
3:35.6	Herb Elliott (AUS)	06Sep60
3:33.1	Jim Ryun (USA)	08Jul67
3:32.2	Filbert Bayi (TAN)	02Feb74
3:32.1	Sebastian Coe (GBR)	15Aug79
3:31.36	Steve Ovett (GBR)	27Aug80
3:31.24	Sydney Maree (usa)	28Aug83
3:30.77	Steve Ovett (GBR)	04Sep83
3:29.67	Steve Cram (GBR)	16Jul85
3:29.46	Said Aouita (MOR)	23Aug85
3:28.86	Noureddine Morceli (ALG)	06Sep92
3:27.37	Noureddine Morceli (ALG)	12Jul95
3:26.00	Hicham el Guerrouj (MOR)	14Jul98

Progression of Women's Mile Record

time	name	date
6:13.2	Elizabeth Atkinson (GBR)	24Jun21
5:27.5	Ruth Christmas (GBR)	20Aug32
5:24.0	Gladys Lunn (GBR)	01Jun36
5:23.0	Gladys Lunn (GBR)	18Jul36
5:20.8	Gladys Lunn (GBR)	08May37
5:17.0	Gladys Lunn (GBR)	07Aug37
5:15.3	Evelyne Forster (GBR)	22Jul39
5:11.0	Anne Oliver (GBR)	14Jun52
5:09.8	Enid Harding (GBR)	04Jul53
5:08.0	Anne Oliver (GBR)	12Sep53
5:02.6	Diane Leather (GBR)	30Sep53
5:00.3	Edith Treybal (ROM)	01Nov53
5:00.2	Diane Leather (GBR)	26May54
4:59.6	Diane Leather (GBR)	29May54
4:50.8	Diane Leather (GBR)	24May55
4:45.0	Diane Leather (GBR)	21Sep55
4:41.4	Marise Chamberlain (NZL)	08Dec62
4:39.2	Anne Smith (GBR)	13May67
4:37.0	Anne Smith (GBR)	03Jun67
4:36.8	Maria Gommers (HOL)	14Jun69
4:35.3	Ellen Tittel (FRG)	20Aug71
4:34.9	Glenda Reiser (CAN)	07Jul73
4:29.5	Paola Pigni-Cacchi (ITA)	08Aug73
4:23.8	Natalia Marasescu (ROM)	21May77
4:22.1	Natalia Marasescu (ROM)	27Jan79
4:21.7	Mary Decker (USA)	26Jan80
4:20.89	Lyudmila Veselkova (SOV)	12Sep81
4:18.08	Mary Decker-Tabb (USA)	09Jul82
4:17.44	Maricica Puica (ROM)	16Sep82
4:15.8	Natalya Artyomova (SOV)	05Aug84
4:16.71	Mary Decker-Slaney (USA)	21Aug85
4:15.61	Paula Ivan (ROM)	10Jul89
4:12.56	Svetlana Masterkova (RUS)	14Aug96

1 As with the first example discussed above, we are trying to find a best m to "solve" this system.

$$\begin{aligned} 8m &= 4 \\ 16m &= 9 \\ 24m &= 13 \\ 32m &= 17 \\ 40m &= 20 \end{aligned}$$

Projecting into the linear subspace gives this

$$\frac{\begin{pmatrix} 4 \\ 9 \\ 13 \\ 17 \\ 20 \end{pmatrix} \cdot \begin{pmatrix} 8 \\ 16 \\ 24 \\ 32 \\ 40 \end{pmatrix}}{\begin{pmatrix} 8 \\ 16 \\ 24 \\ 32 \\ 40 \end{pmatrix} \cdot \begin{pmatrix} 8 \\ 16 \\ 24 \\ 32 \\ 40 \end{pmatrix}} \cdot \begin{pmatrix} 8 \\ 16 \\ 24 \\ 32 \\ 40 \end{pmatrix} = \frac{1832}{3520} \cdot \begin{pmatrix} 8 \\ 16 \\ 24 \\ 32 \\ 40 \end{pmatrix}$$

so the slope of the line of best fit is approximately 0.52.

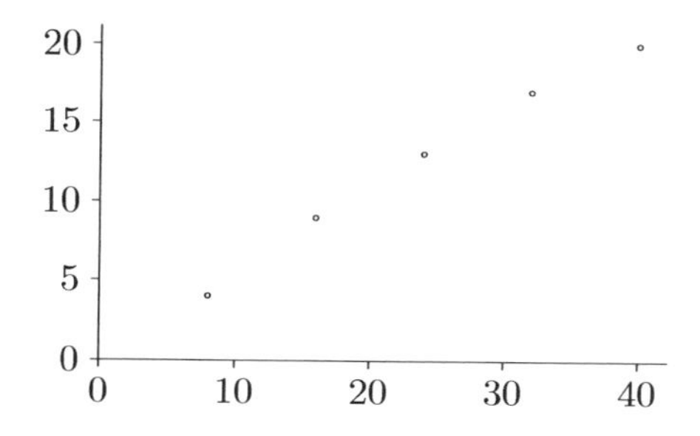

2 With this input

$$A = \begin{pmatrix} 1 & 1852.71 \\ 1 & 1858.88 \\ \vdots & \vdots \\ 1 & 1985.54 \\ 1 & 1993.71 \end{pmatrix} \qquad b = \begin{pmatrix} 292.0 \\ 285.0 \\ \vdots \\ 226.32 \\ 224.39 \end{pmatrix}$$

(the dates have been rounded to months, e.g., for a September record, the decimal $.71 \approx (8.5/12)$ was used), Maple responded with an intercept of $b = 994.8276974$ and a slope of $m = -0.3871993827$.

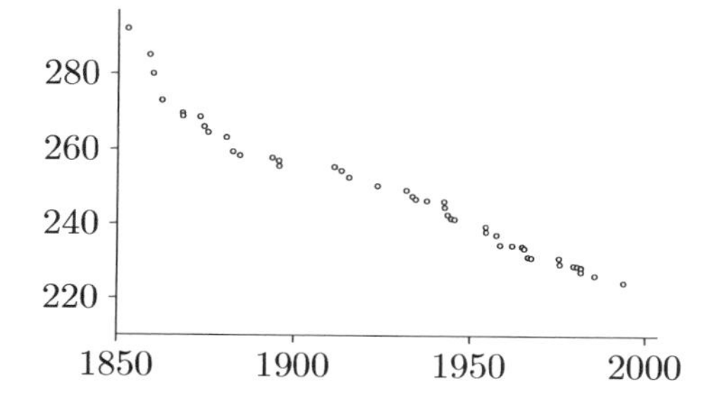

3 With this input (the years are zeroed at 1900)

$$A := \begin{pmatrix} 1 & .38 \\ 1 & .54 \\ \vdots & \vdots \\ 1 & 92.71 \\ 1 & 95.54 \end{pmatrix} \qquad b = \begin{pmatrix} 249.0 \\ 246.2 \\ \vdots \\ 208.86 \\ 207.37 \end{pmatrix}$$

(the dates have been rounded to months, e.g., for a September record, the decimal $.71 \approx (8.5/12)$ was used), Maple gives an intercept of $b = 243.1590327$ and a slope of $m = -0.401647703$. The slope given in the body of this Topic for the men's mile is quite close to this.

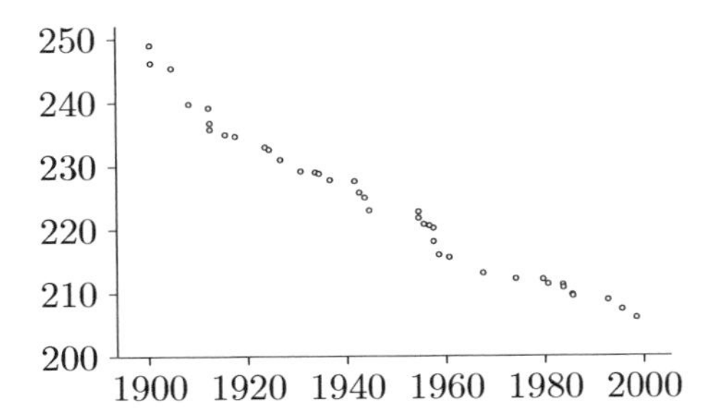

4 With this input (the years are zeroed at 1900)

$$A = \begin{pmatrix} 1 & 21.46 \\ 1 & 32.63 \\ \vdots & \vdots \\ 1 & 89.54 \\ 1 & 96.63 \end{pmatrix} \qquad b = \begin{pmatrix} 373.2 \\ 327.5 \\ \vdots \\ 255.61 \\ 252.56 \end{pmatrix}$$

(the dates have been rounded to months, e.g., for a September record, the decimal $.71 \approx (8.5/12)$ was used), MAPLE gave an intercept of $b = 378.7114894$ and a slope of $m = -1.445753225$.

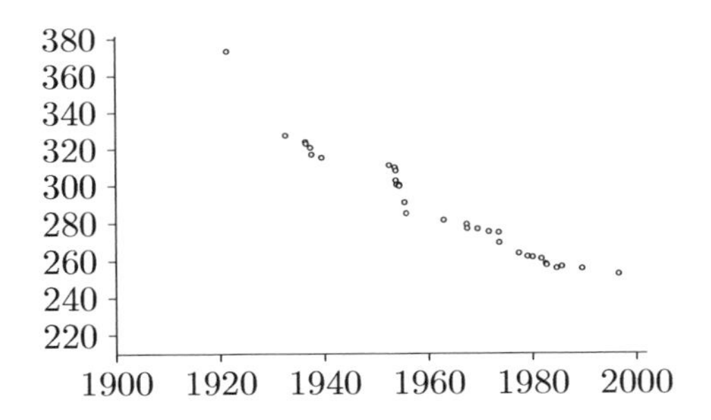

5 These are the equations of the lines for men's and women's mile (the vertical intercept term of the equation for the women's mile has been adjusted from the answer above, to zero it at the year 0, because that's how the men's mile equation was done).

$$y = 994.8276974 - 0.3871993827x$$
$$y = 3125.6426 - 1.445753225x$$

Obviously the lines cross. A computer program is the easiest way to do the arithmetic: MuPAD gives $x = 2012.949004$ and $y = 215.4150856$ (215 seconds is 3 minutes and 35 seconds). *Remark.* Of course all of this projection is highly dubious — for one thing, the equation for the women is influenced by the quite slow early times — but it is nonetheless fun.

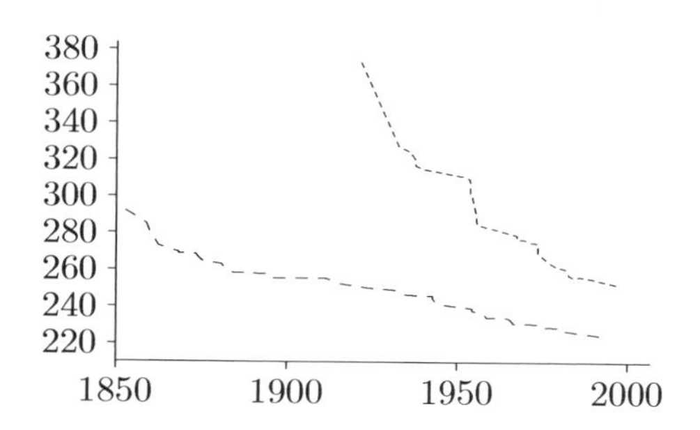

6 *Sage* gives the line of best fit as toll $= -0.05$dist $+ 5.63$. But the graph shows that the equation has no predictive value.

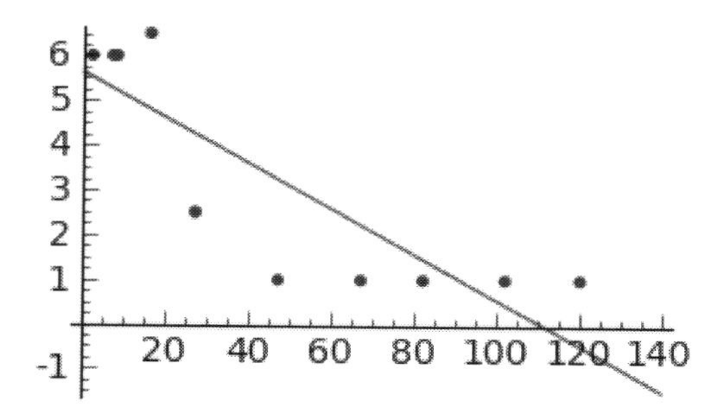

Apparently a better model is that (with only one intermediate exception) crossings in the city cost roughly the same as each other, and crossings upstate cost the same as each other.

7 **(a)** A computer algebra system like MAPLE or MuPAD will give an intercept of $b = 4259/1398 \approx 3.239628$ and a slope of $m = -71/2796 \approx -0.025393419$ Plugging $x = 31$ into the equation yields a predicted number of O-ring failures of $y = 2.45$ (rounded to two places). Plugging in $y = 4$ and solving gives a temperature of $x = -29.94°$F.

(b) On the basis of this information

$$A = \begin{pmatrix} 1 & 53 \\ 1 & 75 \\ \vdots & \\ 1 & 80 \\ 1 & 81 \end{pmatrix} \qquad b = \begin{pmatrix} 3 \\ 2 \\ \vdots \\ 0 \\ 0 \end{pmatrix}$$

MAPLE gives the intercept $b = 187/40 = 4.675$ and the slope $m = -73/1200 \approx -0.060833$. Here, plugging $x = 31$ into the equation predicts $y = 2.79$ O-ring failures (rounded to two places). Plugging in $y = 4$ failures gives a temperature of $x = 11°$F.

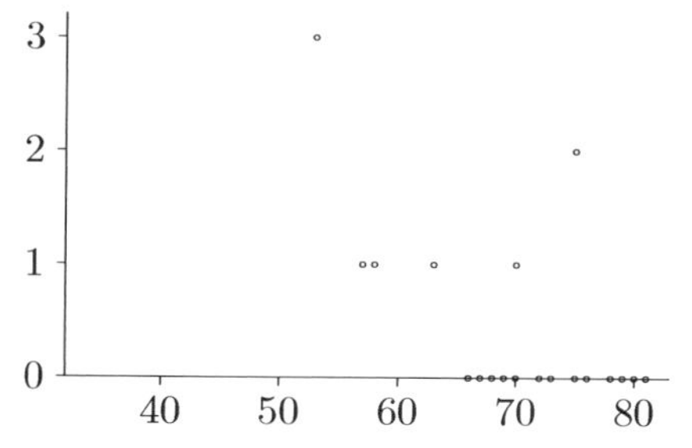

8 **(a)** The plot is nonlinear.

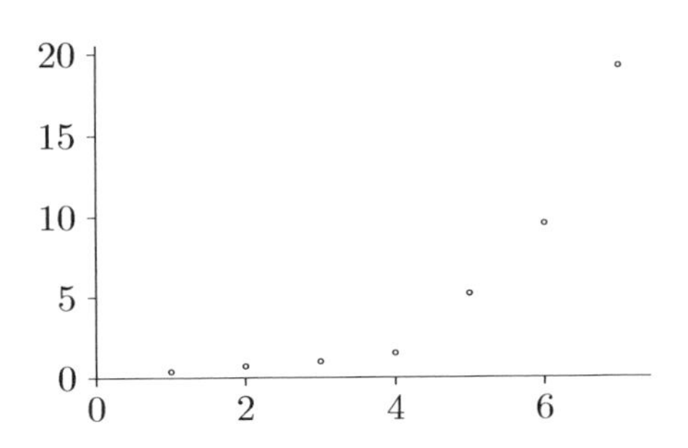

(b) Here is the plot.

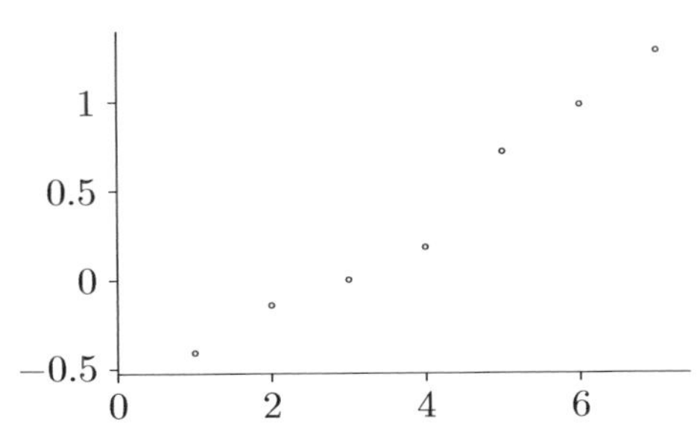

There is perhaps a jog up between planet 4 and planet 5.

(c) This plot seems even more linear.

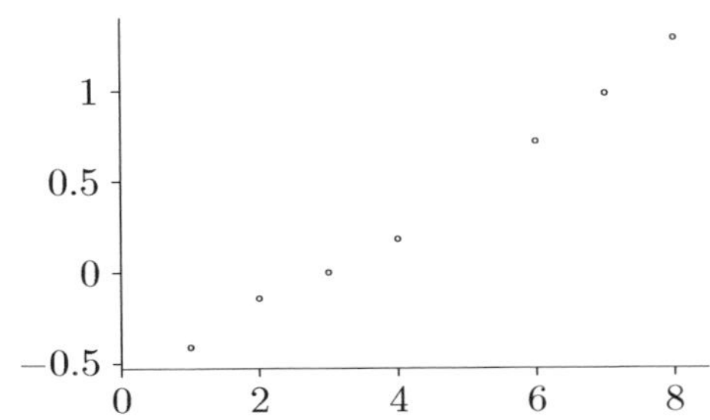

(d) With this input

$$A = \begin{pmatrix} 1 & 1 \\ 1 & 2 \\ 1 & 3 \\ 1 & 4 \\ 1 & 6 \\ 1 & 7 \\ 1 & 8 \end{pmatrix} \qquad b = \begin{pmatrix} -0.40893539 \\ -0.1426675 \\ 0 \\ 0.18184359 \\ 0.71600334 \\ 0.97954837 \\ 1.2833012 \end{pmatrix}$$

MuPAD gives that the intercept is $b = -0.6780677466$ and the slope is $m = 0.2372763818$.

(e) Plugging $x = 9$ into the equation $y = -0.6780677466 + 0.2372763818x$ from the prior item gives that the log of the distance is 1.4574197, so the expected distance is 28.669472. The actual distance is about 30.003.

(f) Plugging $x = 10$ into the same equation gives that the log of the distance is 1.6946961, so the expected distance is 49.510362. The actual distance is about 39.503.

Topic: Geometry of Linear Maps

1 **(a)** To represent H, recall that rotation counterclockwise by θ radians is represented with respect to the standard basis in this way.

$$\mathrm{Rep}_{\mathcal{E}_2,\mathcal{E}_2}(h) = \begin{pmatrix} \cos\theta & -\sin\theta \\ \sin\theta & \cos\theta \end{pmatrix}$$

A clockwise angle is the negative of a counterclockwise one.

$$\mathrm{Rep}_{\mathcal{E}_2,\mathcal{E}_2}(h) = \begin{pmatrix} \cos(-\pi/4) & -\sin(-\pi/4) \\ \sin(-\pi/4) & \cos(-\pi/4) \end{pmatrix} = \begin{pmatrix} \sqrt{2}/2 & \sqrt{2}/2 \\ -\sqrt{2}/2 & \sqrt{2}/2 \end{pmatrix}$$

This Gauss-Jordan reduction

$$\xrightarrow{\rho_1+\rho_2} \begin{pmatrix} \sqrt{2}/2 & \sqrt{2}/2 \\ 0 & \sqrt{2} \end{pmatrix} \xrightarrow[(1/\sqrt{2})\rho_2]{(2/\sqrt{2})\rho_1} \begin{pmatrix} 1 & 1 \\ 0 & 1 \end{pmatrix} \xrightarrow{-\rho_2+\rho_1} \begin{pmatrix} 1 & 0 \\ 0 & 1 \end{pmatrix}$$

produces the identity matrix so there is no need for column-swapping operations to end with a partial-identity.

(b) In matrix multiplication the reduction is

$$\begin{pmatrix} 1 & -1 \\ 0 & 1 \end{pmatrix} \begin{pmatrix} 2/\sqrt{2} & 0 \\ 0 & 1/\sqrt{2} \end{pmatrix} \begin{pmatrix} 1 & 0 \\ 1 & 1 \end{pmatrix} H = I$$

(note that composition of the Gaussian operations is from right to left).

(c) Taking inverses

$$H = \underbrace{\begin{pmatrix} 1 & 0 \\ -1 & 1 \end{pmatrix} \begin{pmatrix} \sqrt{2}/2 & 0 \\ 0 & \sqrt{2} \end{pmatrix} \begin{pmatrix} 1 & 1 \\ 0 & 1 \end{pmatrix}}_{P} I$$

gives the desired factorization of H (here, the partial identity is I, and Q is trivial, that is, it is also an identity matrix).

(d) Reading the composition from right to left (and ignoring the identity matrices as trivial) gives that H has the same effect as first performing this skew

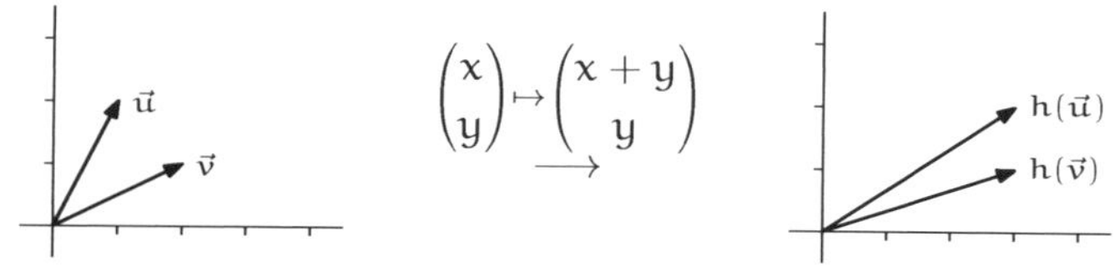

followed by a dilation that multiplies all first components by $\sqrt{2}/2$ (this is a "shrink" in that $\sqrt{2}/2 \approx 0.707$) and all second components by $\sqrt{2}$, followed by another skew.

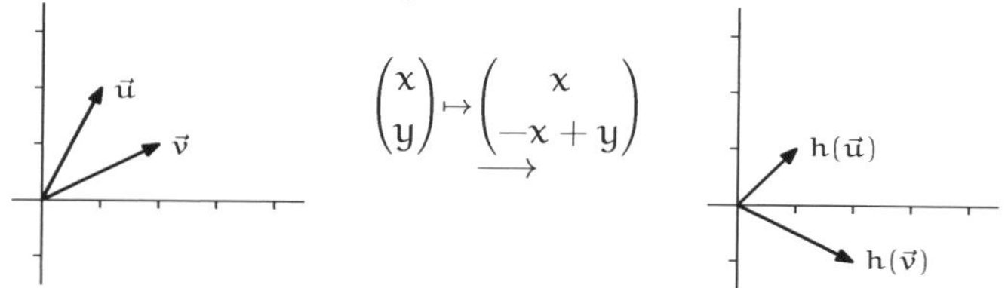

For instance, the effect of H on the unit vector whose angle with the x-axis is $\pi/3$ is this.

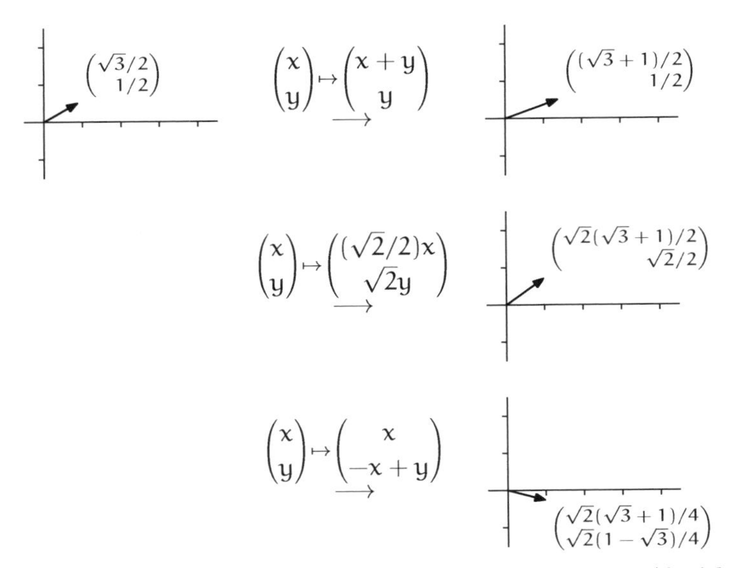

Verifying that the resulting vector has unit length and forms an angle of $-\pi/6$ with the x-axis is routine.

2 We will first represent the map with a matrix H, perform the row operations and, if needed, column operations to reduce it to a partial-identity matrix. We will then translate that into a factorization $H = PBQ$. Substituting into the general matrix

$$\mathrm{Rep}_{\mathcal{E}_2,\mathcal{E}_2}(r_\theta)\begin{pmatrix}\cos\theta & -\sin\theta\\ \sin\theta & \cos\theta\end{pmatrix}$$

gives this representation.

$$\mathrm{Rep}_{\mathcal{E}_2,\mathcal{E}_2}(r_{2\pi/3})\begin{pmatrix}-1/2 & -\sqrt{3}/2\\ \sqrt{3}/2 & -1/2\end{pmatrix}$$

Gauss's Method is routine.

$$\xrightarrow{\sqrt{3}\rho_1+\rho_2}\begin{pmatrix}-1/2 & -\sqrt{3}/2\\ 0 & -2\end{pmatrix}\xrightarrow[(-1/2)\rho_2]{-2\rho_1}\begin{pmatrix}1 & \sqrt{3}\\ 0 & 1\end{pmatrix}\xrightarrow{-\sqrt{3}\rho_2+\rho_1}\begin{pmatrix}1 & 0\\ 0 & 1\end{pmatrix}$$

That translates to a matrix equation in this way.

$$\begin{pmatrix}1 & -\sqrt{3}\\ 0 & 1\end{pmatrix}\begin{pmatrix}-2 & 0\\ 0 & -1/2\end{pmatrix}\begin{pmatrix}1 & 0\\ \sqrt{3} & 1\end{pmatrix}\begin{pmatrix}-1/2 & -\sqrt{3}/2\\ \sqrt{3}/2 & -1/2\end{pmatrix}=I$$

Taking inverses to solve for H yields this factorization.

$$\begin{pmatrix}-1/2 & -\sqrt{3}/2\\ \sqrt{3}/2 & -1/2\end{pmatrix}=\begin{pmatrix}1 & 0\\ -\sqrt{3} & 1\end{pmatrix}\begin{pmatrix}-1/2 & 0\\ 0 & -2\end{pmatrix}\begin{pmatrix}1 & \sqrt{3}\\ 0 & 1\end{pmatrix}I$$

3 This Gaussian reduction

$$\xrightarrow[-\rho_1+\rho_3]{-3\rho_1+\rho_2}\begin{pmatrix}1 & 2 & 1\\ 0 & 0 & -3\\ 0 & 0 & 1\end{pmatrix}\xrightarrow{(1/3)\rho_2+\rho_3}\begin{pmatrix}1 & 2 & 1\\ 0 & 0 & -3\\ 0 & 0 & 0\end{pmatrix}\xrightarrow{(-1/3)\rho_2}\begin{pmatrix}1 & 2 & 1\\ 0 & 0 & 1\\ 0 & 0 & 0\end{pmatrix}\xrightarrow{-\rho_2+\rho_1}\begin{pmatrix}1 & 2 & 0\\ 0 & 0 & 1\\ 0 & 0 & 0\end{pmatrix}$$

gives the reduced echelon form of the matrix. Now the two column operations of taking -2 times the first column and adding it to the second, and then of swapping columns two and three produce this partial identity.

$$B=\begin{pmatrix}1 & 0 & 0\\ 0 & 1 & 0\\ 0 & 0 & 0\end{pmatrix}$$

All of that translates into matrix terms as: where

$$P=\begin{pmatrix}1 & -1 & 0\\ 0 & 1 & 0\\ 0 & 0 & 1\end{pmatrix}\begin{pmatrix}1 & 0 & 0\\ 0 & -1/3 & 0\\ 0 & 0 & 1\end{pmatrix}\begin{pmatrix}1 & 0 & 0\\ 0 & 1 & 0\\ 0 & 1/3 & 1\end{pmatrix}\begin{pmatrix}1 & 0 & 0\\ 0 & 1 & 0\\ -1 & 0 & 1\end{pmatrix}\begin{pmatrix}1 & 0 & 0\\ -3 & 1 & 0\\ 0 & 0 & 1\end{pmatrix}$$

and

$$Q = \begin{pmatrix} 1 & -2 & 0 \\ 0 & 1 & 0 \\ 0 & 0 & 1 \end{pmatrix} \begin{pmatrix} 0 & 1 & 0 \\ 1 & 0 & 0 \\ 0 & 0 & 1 \end{pmatrix}$$

the given matrix factors as PBQ.

4 Represent it with respect to the standard bases $\mathcal{E}_1, \mathcal{E}_1$, then the only entry in the resulting 1×1 matrix is the scalar k.

5 We can show this by induction on the number of components in the vector. In the $n = 1$ base case the only permutation is the trivial one, and the map

$$\begin{pmatrix} x_1 \end{pmatrix} \mapsto \begin{pmatrix} x_1 \end{pmatrix}$$

is expressible as a composition of swaps—as zero swaps. For the inductive step we assume that the map induced by any permutation of fewer than n numbers can be expressed with swaps only, and we consider the map induced by a permutation p of n numbers.

$$\begin{pmatrix} x_1 \\ x_2 \\ \vdots \\ x_n \end{pmatrix} \mapsto \begin{pmatrix} x_{p(1)} \\ x_{p(2)} \\ \vdots \\ x_{p(n)} \end{pmatrix}$$

Consider the number i such that $p(i) = n$. The map

$$\begin{pmatrix} x_1 \\ x_2 \\ \vdots \\ x_i \\ \vdots \\ x_n \end{pmatrix} \stackrel{\hat{p}}{\longmapsto} \begin{pmatrix} x_{p(1)} \\ x_{p(2)} \\ \vdots \\ x_{p(n)} \\ \vdots \\ x_n \end{pmatrix}$$

will, when followed by the swap of the i-th and n-th components, give the map p. Now, the inductive hypothesis gives that $\hat{p}$ is achievable as a composition of swaps.

6 **(a)** A line is a subset of $\mathbb{R}^n$ of the form $\{\vec{v} = \vec{u} + t \cdot \vec{w} \mid t \in \mathbb{R}\}$. The image of a point on that line is $h(\vec{v}) = h(\vec{u} + t \cdot \vec{w}) = h(\vec{u}) + t \cdot h(\vec{w})$, and the set of such vectors, as t ranges over the reals, is a line (albeit, degenerate if $h(\vec{w}) = \vec{0}$).

(b) This is an obvious extension of the prior argument.

(c) If the point B is between the points A and C then the line from A to C has B in it. That is, there is a $t \in (0\,..\,1)$ such that $\vec{b} = \vec{a} + t \cdot (\vec{c} - \vec{a})$ (where B is the endpoint of $\vec{b}$, etc.). Now, as in the argument of the first item, linearity shows that $h(\vec{b}) = h(\vec{a}) + t \cdot h(\vec{c} - \vec{a})$.

7 The two are inverse. For instance, for a fixed $x \in \mathbb{R}$, if $f'(x) = k$ (with $k \neq 0$) then $(f^{-1})'(x) = 1/k$.

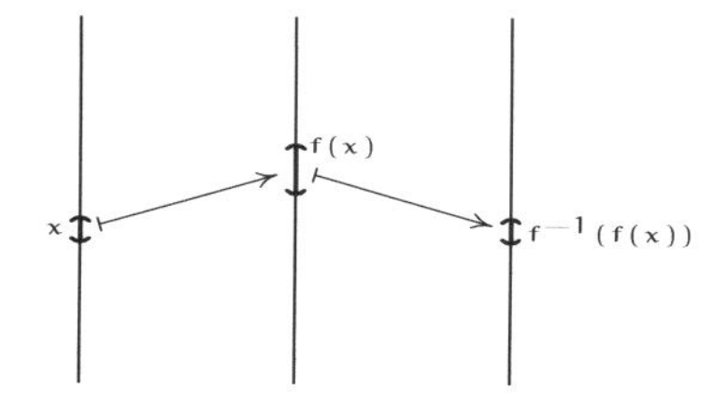

Topic: Magic Squares

1 **(a)** The sum of the entries of M is the sum of the sums of the three rows.

(b) The constraints on entries of M involving the center entry make this system.

$$\begin{aligned} m_{2,1}+m_{2,2}+m_{2,3}&=s\\ m_{1,2}+m_{2,2}+m_{3,2}&=s\\ m_{1,1}+m_{2,2}+m_{3,3}&=s\\ m_{1,3}+m_{2,2}+m_{3,1}&=s \end{aligned}$$

Adding those four equations counts each matrix entry once and only once, except that we count the center entry four times. Thus the left side sums to $3s+3m_{2,2}$ while the right sums to $4s$. So $3m_{2,2}=s$.

(c) The second row adds to s so $m_{2,1}+m_{2,2}+m_{2,3}=3m_{2,2}$, giving that $(1/2)\cdot(m_{2,1}+m_{2,3})=m_{2,2}$. The same goes for the column and the diagonals.

(d) By the prior exercise either both $m_{2,1}$ and $m_{2,3}$ are equal to $m_{2,2}$ or else one is greater while one is smaller. Thus $m_{2,2}$ is the median of the set $\{m_{2,1},m_{2,2},m_{2,3}\}$. The same reasoning applied to the second column shows that Thus $m_{2,2}$ is the median of the set $\{m_{1,2},m_{2,1},m_{2,2},m_{2,3},m_{3,2}\}$. Extending to the two diagonals shows it is the median of the set of all entries.

2 For any k we have this.

$$\left(\begin{array}{cccc|c} 1&1&0&0&s\\ 0&0&1&1&s\\ 1&0&1&0&s\\ 0&1&0&1&s\\ 1&0&0&1&s\\ 0&1&1&0&s \end{array}\right) \xrightarrow[-\rho_1+\rho_5]{-\rho_1+\rho_3} \left(\begin{array}{cccc|c} 1&1&0&0&s\\ 0&0&1&1&s\\ 0&-1&1&0&0\\ 0&1&0&1&s\\ 0&-1&0&1&0\\ 0&1&1&0&s \end{array}\right) \xrightarrow{-\rho_2\leftrightarrow\rho_6} \left(\begin{array}{cccc|c} 1&1&0&0&s\\ 0&1&1&0&s\\ 0&-1&1&0&0\\ 0&1&0&1&s\\ 0&-1&0&1&0\\ 0&0&1&1&s \end{array}\right) \xrightarrow[\substack{-\rho_2+\rho_4\\ \rho_2+\rho_5}]{-\rho_2+\rho_3} \left(\begin{array}{cccc|c} 1&1&0&0&s\\ 0&1&1&0&s\\ 0&0&2&0&s\\ 0&1&-1&1&0\\ 0&0&1&1&s\\ 0&0&1&1&s \end{array}\right)$$

The unique solution is $a=b=c=d=s/2$.

3 By the prior exercise the only member is $Z_{2\times 2}$.

4 **(a)** Where $M,N\in\mathcal{M}_{n\times n}$ we have $\operatorname{Tr}(cM+dN)=(cm_{1,1}+dn_{1,1})+\cdots+(cm_{n,n}+dn_{n,n})=(cm_{1,1}+\cdots+cm_{n,n})+(dn_{1,1}+\cdots+dn_{n,n})=c\cdot\operatorname{Tr}(M)+d\cdot\operatorname{Tr}(N)$ where all numbers are real, so the trace preserves linear combinations. The argument for Tr^* is similar.

(b) It preserves linear combinations: where all numbers are real, $\theta(cM+dN)=(\operatorname{Tr}(cM+dN),\operatorname{Tr}^*(cM+dN))=(c\cdot\operatorname{Tr}(M)+d\cdot\operatorname{Tr}(N),c\cdot\operatorname{Tr}^*(M)+d\cdot\operatorname{Tr}^*(N))=c\cdot\theta(M)+d\cdot\theta(N)$.

(c) Where $h_1,\ldots,h_n\colon V\to W$ are linear then so is $g\colon V\to W^n$ given by $g(\vec{v})=(h_1(\vec{v}),\ldots,h_n(\vec{v}))$. The proof just follows the proof of the prior item.

5 **(a)** The sum of two semimagic squares is semimagic, as is a scalar multiple of a semimagic square.

(b) As with the prior item, a linear combination of two semimagic squares with magic number zero is also such a matrix.

6 **(a)** Consider the matrix $C\in\mathcal{H}_n$ that has all entries zero except that the four corners are $c_{1,1}=c_{n,n}=1$ and $c_{1,n}=c_{n,1}=-1$. Also consider the matrix $D\in\mathcal{H}_n$ with all entries zero except that $d_{1,1}=d_{2,2}=1$ and $d_{1,2}=d_{2,1}=-1$. We have

$$\theta(C)=\begin{pmatrix}2\\-2\end{pmatrix}\qquad \theta(D)=\begin{cases}\binom{2}{-1} & \text{if } n=3\\ \binom{2}{0} & \text{if } n>3\end{cases}$$

and so the image of θ includes a basis for $\mathbb{R}^2$ and thus θ is onto. With that, because for any linear map the dimension of the domain equals its rank plus its nullity we conclude that $\dim(\mathcal{H}_n)=2+\dim(\mathscr{M}_{n,0})$, as desired.

(b) We claim that $\phi\colon\mathscr{H}_{n,0}\to\mathcal{M}_{(n-1)\times(n-1)}$. is one-to-one and onto.

To show that it is one-to-one we will show that the only member of $\mathscr{H}_{n,0}$ mapped to the zero matrix $Z_{(n-1)\times(n-1)}$ is the zero matrix $Z_{n\times n}$. Suppose that $M\in\mathscr{H}_{n\times n}$ and $\phi(M)=Z_{(n-1)\times(n-1)}$. On all but the final row and column ϕ is the identity so the entries in M in all but the final row and column are zero: $m_{i,j} = 0$ for $i, j \in \{1 \ldots n-1\}$. The first row of M adds to zero and hence the final entry in that row $m_{1,n}$ is zero. Similarly the final entry in each row $i \in \{1 \ldots n-1\}$ and column $j \in \{1 \ldots n-1\}$ is zero. Then, the final column adds to zero so $m_{n,n} = 0$. Therefore M is the zero matrix $Z_{n\times n}$ and the restriction of ϕ is one-to-one.

(c) Consider a member $\hat{M}$ of the codomain $\mathcal{M}_{(n-1)\times(n-1)}$. We will produce a matrix M from the domain $\mathscr{H}_{n,0}$ that maps to it. The function ϕ is the identity on all but the final row and column of M so for $i, j \in \{1 \ldots n-1\}$ the entries are $m_{i,j} = \hat{m}_{i,j}$.

$$M = \begin{pmatrix} \hat{m}_{1,1} & \hat{m}_{1,2} & \ldots & \hat{m}_{1,n-1} & m_{1,n} \\ \vdots & \vdots & & & \\ \hat{m}_{n-1,1} & \hat{m}_{n-1,2} & \ldots & \hat{m}_{n-1,n-1} & m_{n-1,n} \\ m_{n,1} & m_{n,2} & \ldots & m_{n,n-1} & m_{n,n} \end{pmatrix}$$

The first row of M must add to zero so we take $m_{1,n}$ to be $-(\hat{m}_{1,1} + \cdots + \hat{m}_{1,n-1})$. In the same way we get the final entries $m_{i,n} = -(\hat{m}_{i,1} + \cdots + \hat{m}_{i,n-1})$ in all the rows but the bottom $i \in \{1 \ldots n-1\}$, and the final entries $m_{n,j} = -(\hat{m}_{1,j} + \cdots + \hat{m}_{n-1,j})$ in all the columns but the last $j \in \{1 \ldots n-1\}$. The entry remaining is the one in the lower right $m_{n,n}$. The final column adds to zero so we set it to $-(m_{1,n} + \cdots + m_{n-1,n})$ but we must check that the final row now also adds to zero. We have $m_{n,n} = -m_{1,n} - \cdots - m_{n-1,n}$ and expanding each of the $m_{i,n}$ as $-\hat{m}_{1,1} - \cdots - \hat{m}_{1,n-1}$ gives that we have defined $m_{n,n}$ to be the sum of all the entries of $\hat{M}$. The sum of the all the entries but the last in the final row is $m_{1,n} + m_{2,n} + \cdots + m_{n-1,n}$ and expanding each $m_{n,j} = -\hat{m}_{1,j} - \cdots - \hat{m}_{n-1,j}$ verifies that the sum of the final row is zero. Thus M is semimagic with magic number zero and so ϕ is onto.

(d) Theorem Two.II.2.15 says that for any linear map the dimension of the domain equals its rank plus its nullity. Because $\phi\colon \mathscr{H}_n \to \mathcal{M}_{(n-1)\times(n-1)}$ is one-to-one its nullity is zero. Because it is onto its rank is $\dim(\mathcal{M}_{(n-1)\times(n-1)}) = (n-1)^2$. Thus the domain of ϕ, the subspace $\mathscr{H}_{n,0}$ of semimagic squares with magic number zero, has dimension $(n-1)^2$.

(e) We have that $\dim \mathscr{M}_n = \dim \mathscr{M}_{n,0} + 1 = (\dim \mathscr{H}_n - 2) + 1 = (n-1)^2 - 1 = n^2 - n$ when $n \geqslant 3$.

Topic: Markov Chains

1 (a) With this file coin.m

```
# Octave function for Markov coin game.  p is chance of going down.
function w = coin(p,v)
  q = 1-p;
  A=[1,p,0,0,0,0;
     0,0,p,0,0,0;
     0,q,0,p,0,0;
     0,0,q,0,p,0;
     0,0,0,q,0,0;
     0,0,0,0,q,1];
  w = A * v;
endfunction
```

This Octave session produced the output given here.

```
octave:1> v0=[0;0;0;1;0;0]
v0 =
  0
  0
```

```
  0
  1
  0
  0
octave:2> p=.5
p = 0.50000
octave:3> v1=coin(p,v0)
v1 =
  0.00000
  0.00000
  0.50000
  0.00000
  0.50000
  0.00000
octave:4> v2=coin(p,v1)
v2 =
  0.00000
  0.25000
  0.00000
  0.50000
  0.00000
  0.25000
```

This continued for too many steps to list here.

```
octave:26> v24=coin(p,v23)
v24 =
  0.39600
  0.00276
  0.00000
  0.00447
  0.00000
  0.59676
```

(b) Using these formulas

$$p_1(n+1) = 0.5 \cdot p_2(n) \qquad p_2(n+1) = 0.5 \cdot p_1(n) + 0.5 \cdot p_3(n)$$
$$p_3(n+1) = 0.5 \cdot p_2(n) + 0.5 \cdot p_4(n) \qquad p_5(n+1) = 0.5 \cdot p_4(n)$$

and these initial conditions

$$\begin{pmatrix} p_0(0) \\ p_1(0) \\ p_2(0) \\ p_3(0) \\ p_4(0) \\ p_5(0) \end{pmatrix} = \begin{pmatrix} 0 \\ 0 \\ 0 \\ 1 \\ 0 \\ 0 \end{pmatrix}$$

we will prove by induction that when n is odd then $p_1(n) = p_3(n) = 0$ and when n is even then $p_2(n) = p_4(n) = 0$. Note first that this is true in the $n = 0$ base case by the initial conditions. For the inductive step, suppose that it is true in the $n = 0$, $n = 1$, ..., $n = k$ cases and consider the $n = k + 1$ case. If $k + 1$ is odd then the two

$$p_1(k+1) = 0.5 \cdot p_2(k) = 0.5 \cdot 0 = 0$$
$$p_3(k+1) = 0.5 \cdot p_2(k) + 0.5 \cdot p_4(k) = 0.5 \cdot 0 + 0.5 \cdot 0 = 0$$

follow from the inductive hypothesis that $p_2(k) = p_4(k) = 0$ since k is even. The case where $k + 1$ is even is similar.

(c) We can use, say, $n = 100$. This Octave session

```
octave:1> B=[1,.5,0,0,0,0;
>             0,0,.5,0,0,0;
>             0,.5,0,.5,0,0;
>             0,0,.5,0,.5,0;
>             0,0,0,.5,0,0;
>             0,0,0,0,.5,1];
octave:2> B100=B**100
B100 =
  1.00000  0.80000  0.60000  0.40000  0.20000  0.00000
  0.00000  0.00000  0.00000  0.00000  0.00000  0.00000
  0.00000  0.00000  0.00000  0.00000  0.00000  0.00000
  0.00000  0.00000  0.00000  0.00000  0.00000  0.00000
  0.00000  0.00000  0.00000  0.00000  0.00000  0.00000
  0.00000  0.20000  0.40000  0.60000  0.80000  1.00000
octave:3> B100*[0;1;0;0;0;0]
octave:4> B100*[0;1;0;0;0;0]
octave:5> B100*[0;0;0;1;0;0]
octave:6> B100*[0;1;0;0;0;0]
```

yields these outputs.

starting with:	\$1	\$2	\$3	\$4
$s_0(100)$	0.80000	0.60000	0.40000	0.20000
$s_1(100)$	0.00000	0.00000	0.00000	0.00000
$s_2(100)$	0.00000	0.00000	0.00000	0.00000
$s_3(100)$	0.00000	0.00000	0.00000	0.00000
$s_4(100)$	0.00000	0.00000	0.00000	0.00000
$s_5(100)$	0.20000	0.40000	0.60000	0.80000

2 **(a)** From these equations

$$\begin{aligned}
(1/6)s_1(n) + 0s_2(n) + 0s_3(n) + 0s_4(n) + 0s_5(n) + 0s_6(n) &= s_1(n+1)\\
(1/6)s_1(n) + (2/6)s_2(n) + 0s_3(n) + 0s_4(n) + 0s_5(n) + 0s_6(n) &= s_2(n+1)\\
(1/6)s_1(n) + (1/6)s_2(n) + (3/6)s_3(n) + 0s_4(n) + 0s_5(n) + 0s_6(n) &= s_3(n+1)\\
(1/6)s_1(n) + (1/6)s_2(n) + (1/6)s_3(n) + (4/6)s_4(n) + 0s_5(n) + 0s_6(n) &= s_4(n+1)\\
(1/6)s_1(n) + (1/6)s_2(n) + (1/6)s_3(n) + (1/6)s_4(n) + (5/6)s_5(n) + 0s_6(n) &= s_5(n+1)\\
(1/6)s_1(n) + (1/6)s_2(n) + (1/6)s_3(n) + (1/6)s_4(n) + (1/6)s_5(n) + (6/6)s_6(n) &= s_6(n+1)
\end{aligned}$$

We get this transition matrix.

$$\begin{pmatrix}
1/6 & 0 & 0 & 0 & 0 & 0\\
1/6 & 2/6 & 0 & 0 & 0 & 0\\
1/6 & 1/6 & 3/6 & 0 & 0 & 0\\
1/6 & 1/6 & 1/6 & 4/6 & 0 & 0\\
1/6 & 1/6 & 1/6 & 1/6 & 5/6 & 0\\
1/6 & 1/6 & 1/6 & 1/6 & 1/6 & 6/6
\end{pmatrix}$$

(b) This is the Octave session, with outputs edited out and condensed into the table at the end.

```
octave:1>    F=[1/6,  0,    0,    0,    0,    0;
>       1/6,  2/6, 0,    0,    0,    0;
>       1/6,  1/6, 3/6, 0,    0,    0;
>       1/6,  1/6, 1/6, 4/6, 0,    0;
>       1/6,  1/6, 1/6, 1/6, 5/6, 0;
>       1/6,  1/6, 1/6, 1/6, 1/6, 6/6];
octave:2> v0=[1;0;0;0;0;0]
octave:3> v1=F*v0
octave:4> v2=F*v1
octave:5> v3=F*v2
```

```
octave:6> v4=F*v3
octave:7> v5=F*v4
```

These are the results.

	1	2	3	4	5
1	0.16667	0.027778	0.0046296	0.00077160	0.00012860
0	0.16667	0.083333	0.0324074	0.01157407	0.00398663
0	0.16667	0.138889	0.0879630	0.05015432	0.02713477
0	0.16667	0.194444	0.1712963	0.13503086	0.10043724
0	0.16667	0.250000	0.2824074	0.28472222	0.27019033
0	0.16667	0.305556	0.4212963	0.51774691	0.59812243

3 **(a)** It does seem reasonable that, while the firm's present location should strongly influence where it is next time (for instance, whether it stays), any locations in the prior stages should have little influence. That is, while a company may move or stay because of where it is, it is unlikely to move or stay because of where it was.

(b) This is the Octave session, slightly edited, with the outputs put together in a table at the end.

```
octave:1> M=[.787,0,0,.111,.102;
>            0,.966,.034,0,0;
>            0,.063,.937,0,0;
>            0,0,.074,.612,.314;
>            .021,.009,.005,.010,.954]
M =
  0.78700  0.00000  0.00000  0.11100  0.10200
  0.00000  0.96600  0.03400  0.00000  0.00000
  0.00000  0.06300  0.93700  0.00000  0.00000
  0.00000  0.00000  0.07400  0.61200  0.31400
  0.02100  0.00900  0.00500  0.01000  0.95400
octave:2> v0=[.025;.025;.025;.025;.900]
octave:3> v1=M*v0
octave:4> v2=M*v1
octave:5> v3=M*v2
octave:6> v4=M*v3
```

This table summarizes.

$\vec{p}_0$	$\vec{p}_1$	$\vec{p}_2$	$\vec{p}_3$	$\vec{p}_4$
0.025000	0.114250	0.210879	0.300739	0.377920
0.025000	0.025000	0.025000	0.025000	0.025000
0.025000	0.025000	0.025000	0.025000	0.025000
0.025000	0.299750	0.455251	0.539804	0.582550
0.900000	0.859725	0.825924	0.797263	0.772652

(c) This is a continuation of the Octave session from the prior item.

```
octave:7> p0=[.0000;.6522;.3478;.0000;.0000]
octave:8> p1=M*p0
octave:9> p2=M*p1
octave:10> p3=M*p2
octave:11> p4=M*p3
```

This summarizes the output.

$\vec{p}_0$	$\vec{p}_1$	$\vec{p}_2$	$\vec{p}_3$	$\vec{p}_4$
0.00000	0.00000	0.0036329	0.0094301	0.016485
0.65220	0.64185	0.6325047	0.6240656	0.616445
0.34780	0.36698	0.3842942	0.3999315	0.414052
0.00000	0.02574	0.0452966	0.0609094	0.073960
0.00000	0.00761	0.0151277	0.0225751	0.029960

(d) This is more of the same Octave session.

```
octave:12> M50=M**50
M50 =
  0.03992  0.33666  0.20318  0.02198  0.37332
  0.00000  0.65162  0.34838  0.00000  0.00000
```

```
  0.00000  0.64553  0.35447  0.00000  0.00000
  0.03384  0.38235  0.22511  0.01864  0.31652
  0.04003  0.33316  0.20029  0.02204  0.37437
octave:13> p50=M50*p0
p50 =
  0.29024
  0.54615
  0.54430
  0.32766
  0.28695
octave:14> p51=M*p50
p51 =
  0.29406
  0.54609
  0.54442
  0.33091
  0.29076
```

This is close to a steady state.

4 **(a)** This is the relevant system of equations.

$$\begin{array}{rcl}
(1-2p)\cdot s_U(n) + p\cdot t_A(n) + p\cdot t_B(n) & = & s_U(n+1) \\
p\cdot s_U(n) + (1-2p)\cdot t_A(n) & = & t_A(n+1) \\
p\cdot s_U(n) + (1-2p)\cdot t_B(n) & = & t_B(n+1) \\
p\cdot t_A(n) + s_A(n) & = & s_A(n+1) \\
p\cdot t_B(n) + s_B(n) & = & s_B(n+1)
\end{array}$$

Thus we have this.

$$\begin{pmatrix} 1-2p & p & p & 0 & 0 \\ p & 1-2p & 0 & 0 & 0 \\ p & 0 & 1-2p & 0 & 0 \\ 0 & p & 0 & 1 & 0 \\ 0 & 0 & p & 0 & 1 \end{pmatrix} \begin{pmatrix} s_U(n) \\ t_A(n) \\ t_B(n) \\ s_A(n) \\ s_B(n) \end{pmatrix} = \begin{pmatrix} s_U(n+1) \\ t_A(n+1) \\ t_B(n+1) \\ s_A(n+1) \\ s_B(n+1) \end{pmatrix}$$

(b) This is the Octave code, with the output removed.

```
octave:1> T=[.5,.25,.25,0,0;
>         .25,.5,0,0,0;
>         .25,0,.5,0,0;
>         0,.25,0,1,0;
>         0,0,.25,0,1]
T =
  0.50000  0.25000  0.25000  0.00000  0.00000
  0.25000  0.50000  0.00000  0.00000  0.00000
  0.25000  0.00000  0.50000  0.00000  0.00000
  0.00000  0.25000  0.00000  1.00000  0.00000
  0.00000  0.00000  0.25000  0.00000  1.00000
octave:2> p0=[1;0;0;0;0]
octave:3> p1=T*p0
octave:4> p2=T*p1
octave:5> p3=T*p2
octave:6> p4=T*p3
octave:7> p5=T*p4
```

Here is the output. The probability of ending at s_A is about 0.23.

	$\vec{p}_0$	$\vec{p}_1$	$\vec{p}_2$	$\vec{p}_3$	$\vec{p}_4$	$\vec{p}_5$
s_U	1	0.50000	0.375000	0.31250	0.26562	0.22656
t_A	0	0.25000	0.250000	0.21875	0.18750	0.16016
t_B	0	0.25000	0.250000	0.21875	0.18750	0.16016
s_A	0	0.00000	0.062500	0.12500	0.17969	0.22656
s_B	0	0.00000	0.062500	0.12500	0.17969	0.22656

(c) With this file as `learn.m`

```
# Octave script file for learning model.
function w = learn(p)
  T = [1-2*p,p,    p,    0, 0;
       p,    1-2*p,0,    0, 0;
```

```
      p,    0,    1-2*p,0, 0;
      0,    p,    0,    1, 0;
      0,    0,    p,    0, 1];
   T5 = T**5;
   p5 = T5*[1;0;0;0;0];
   w = p5(4);
endfunction
```

issuing the command `octave:1> learn(.20)` yields `ans = 0.17664`.

(d) This Octave session

```
octave:1> x=(.01:.01:.50)';
octave:2> y=(.01:.01:.50)';
octave:3> for i=.01:.01:.50
>           y(100*i)=learn(i);
>         endfor
octave:4> z=[x, y];
octave:5> gplot z
```

yields this plot. There is no threshold value — no probability above which the curve rises sharply.

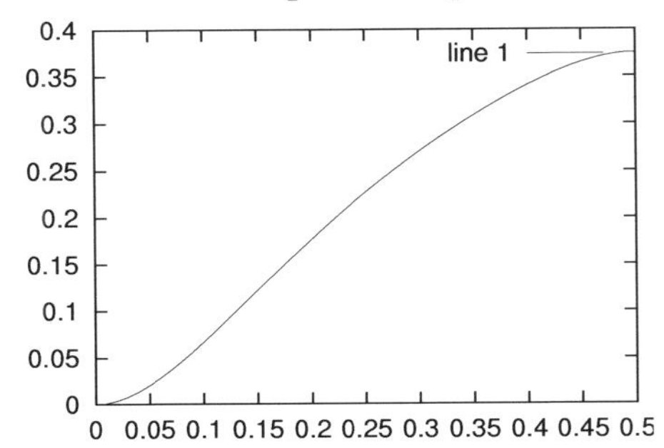

5 **(a)** From these equations

$$\begin{aligned} 0.90 \cdot p_T(n) + 0.01 \cdot p_C(n) &= p_T(n+1) \\ 0.10 \cdot p_T(n) + 0.99 \cdot p_C(n) &= p_C(n+1) \end{aligned}$$

we get this matrix.

$$\begin{pmatrix} 0.90 & 0.01 \\ 0.10 & 0.99 \end{pmatrix} \begin{pmatrix} p_T(n) \\ p_C(n) \end{pmatrix} = \begin{pmatrix} p_T(n+1) \\ p_C(n+1) \end{pmatrix}$$

(b) This is the result from Octave.

$n=0$	1	2	3	4	5
0.30000	0.27700	0.25653	0.23831	0.22210	0.20767
0.70000	0.72300	0.74347	0.76169	0.77790	0.79233

6	7	8	9	10
0.19482	0.18339	0.17322	0.16417	0.15611
0.80518	0.81661	0.82678	0.83583	0.84389

(c) This is the $s_T = 0.2$ result.

$n=0$	1	2	3	4	5
0.20000	0.18800	0.17732	0.16781	0.15936	0.15183
0.80000	0.81200	0.82268	0.83219	0.84064	0.84817

6	7	8	9	10
0.14513	0.13916	0.13385	0.12913	0.12493
0.85487	0.86084	0.86615	0.87087	0.87507

(d) Although the probability vectors start 0.1 apart, they end only 0.032 apart. So they are alike.

6 These are the $p = .55$ vectors,

	n = 0	n = 1	n = 2	n = 3	n = 4	n = 5	n = 6	n = 7
0-0	1	0	0	0	0	0	0	0
1-0	0	0.55000	0	0	0	0	0	0
0-1	0	0.45000	0	0	0	0	0	0
2-0	0	0	0.30250	0	0	0	0	0
1-1	0	0	0.49500	0	0	0	0	0
0-2	0	0	0.20250	0	0	0	0	0
3-0	0	0	0	0.16638	0	0	0	0
2-1	0	0	0	0.40837	0	0	0	0
1-2	0	0	0	0.33412	0	0	0	0
0-3	0	0	0	0.09112	0	0	0	0
4-0	0	0	0	0	0.09151	0.09151	0.09151	0.09151
3-1	0	0	0	0	0.29948	0	0	0
2-2	0	0	0	0	0.36754	0	0	0
1-3	0	0	0	0	0.20047	0	0	0
0-4	0	0	0	0	0.04101	0.04101	0.04101	0.04101
4-1	0	0	0	0	0	0.16471	0.16471	0.16471
3-2	0	0	0	0	0	0.33691	0	0
2-3	0	0	0	0	0	0.27565	0	0
1-4	0	0	0	0	0	0.09021	0.09021	0.09021
4-2	0	0	0	0	0	0	0.18530	0.18530
3-3	0	0	0	0	0	0	0.30322	0
2-4	0	0	0	0	0	0	0.12404	0.12404
4-3	0	0	0	0	0	0	0	0.16677
3-4	0	0	0	0	0	0	0	0.13645

and these are the $p = .60$ vectors.

	n = 0	n = 1	n = 2	n = 3	n = 4	n = 5	n = 6	n = 7
0-0	1	0	0	0	0	0	0	0
1-0	0	0.60000	0	0	0	0	0	0
0-1	0	0.40000	0	0	0	0	0	0
2-0	0	0	0.36000	0	0	0	0	0
1-1	0	0	0.48000	0	0	0	0	0
0-2	0	0	0.16000	0	0	0	0	0
3-0	0	0	0	0.21600	0	0	0	0
2-1	0	0	0	0.43200	0	0	0	0
1-2	0	0	0	0.28800	0	0	0	0
0-3	0	0	0	0.06400	0	0	0	0
4-0	0	0	0	0	0.12960	0.12960	0.12960	0.12960
3-1	0	0	0	0	0.34560	0	0	0
2-2	0	0	0	0	0.34560	0	0	0
1-3	0	0	0	0	0.15360	0	0	0
0-4	0	0	0	0	0.02560	0.02560	0.02560	0.02560
4-1	0	0	0	0	0	0.20736	0.20736	0.20736
3-2	0	0	0	0	0	0.34560	0	0
2-3	0	0	0	0	0	0.23040	0	0
1-4	0	0	0	0	0	0.06144	0.06144	0.06144
4-2	0	0	0	0	0	0	0.20736	0.20736
3-3	0	0	0	0	0	0	0.27648	0
2-4	0	0	0	0	0	0	0.09216	0.09216
4-3	0	0	0	0	0	0	0	0.16589
3-4	0	0	0	0	0	0	0	0.11059

(a) We can adapt the script from the end of this Topic.

```
# Octave script file to compute chance of World Series outcomes.
function w = markov(p,v)
```

```
  q = 1-p;
  A=[0,0,0,0,0,0, 0,0,0,0,0,0, 0,0,0,0,0,0, 0,0,0,0,0,0;  # 0-0
     p,0,0,0,0,0, 0,0,0,0,0,0, 0,0,0,0,0,0, 0,0,0,0,0,0;  # 1-0
     q,0,0,0,0,0, 0,0,0,0,0,0, 0,0,0,0,0,0, 0,0,0,0,0,0;  # 0-1_
     0,p,0,0,0,0, 0,0,0,0,0,0, 0,0,0,0,0,0, 0,0,0,0,0,0;  # 2-0
     0,q,p,0,0,0, 0,0,0,0,0,0, 0,0,0,0,0,0, 0,0,0,0,0,0;  # 1-1
     0,0,q,0,0,0, 0,0,0,0,0,0, 0,0,0,0,0,0, 0,0,0,0,0,0;  # 0-2__
     0,0,0,p,0,0, 0,0,0,0,0,0, 0,0,0,0,0,0, 0,0,0,0,0,0;  # 3-0
     0,0,0,q,p,0, 0,0,0,0,0,0, 0,0,0,0,0,0, 0,0,0,0,0,0;  # 2-1
     0,0,0,0,q,p, 0,0,0,0,0,0, 0,0,0,0,0,0, 0,0,0,0,0,0;  # 1-2_
     0,0,0,0,0,q, 0,0,0,0,0,0, 0,0,0,0,0,0, 0,0,0,0,0,0;  # 0-3
     0,0,0,0,0,0, p,0,0,0,1,0, 0,0,0,0,0,0, 0,0,0,0,0,0;  # 4-0
     0,0,0,0,0,0, q,p,0,0,0,0, 0,0,0,0,0,0, 0,0,0,0,0,0;  # 3-1__
     0,0,0,0,0,0, 0,q,p,0,0,0, 0,0,0,0,0,0, 0,0,0,0,0,0;  # 2-2
     0,0,0,0,0,0, 0,0,q,p,0,0, 0,0,0,0,0,0, 0,0,0,0,0,0;  # 1-3
     0,0,0,0,0,0, 0,0,0,q,0,0, 0,0,1,0,0,0, 0,0,0,0,0,0;  # 0-4_
     0,0,0,0,0,0, 0,0,0,0,0,p, 0,0,0,1,0,0, 0,0,0,0,0,0;  # 4-1
     0,0,0,0,0,0, 0,0,0,0,0,q, p,0,0,0,0,0, 0,0,0,0,0,0;  # 3-2
     0,0,0,0,0,0, 0,0,0,0,0,0, q,p,0,0,0,0, 0,0,0,0,0,0;  # 2-3__
     0,0,0,0,0,0, 0,0,0,0,0,0, 0,q,0,0,0,0, 1,0,0,0,0,0;  # 1-4
     0,0,0,0,0,0, 0,0,0,0,0,0, 0,0,0,0,p,0, 0,1,0,0,0,0;  # 4-2
     0,0,0,0,0,0, 0,0,0,0,0,0, 0,0,0,0,q,p, 0,0,0,0,0,0;  # 3-3_
     0,0,0,0,0,0, 0,0,0,0,0,0, 0,0,0,0,0,q, 0,0,0,1,0,0;  # 2-4
     0,0,0,0,0,0, 0,0,0,0,0,0, 0,0,0,0,0,0, 0,0,p,0,1,0;  # 4-3
     0,0,0,0,0,0, 0,0,0,0,0,0, 0,0,0,0,0,0, 0,0,q,0,0,1]; # 3-4
  v7 = (A**7) * v;
  w = v7(11)+v7(16)+v7(20)+v7(23)
endfunction
```

Using this script, we get that when the American League has a $p = 0.55$ probability of winning each game then their probability of winning the first-to-win-four series is 0.60829. When their probability of winning any one game is $p = 0.6$ then their probability of winning the series is 0.71021.

(b) This Octave session

```
octave:1> v0=[1;0;0;0;0;0;0;0;0;0;0;0;0;0;0;0;0;0;0;0;0;0;0;0];
octave:2> x=(.01:.01:.99)';
octave:3> y=(.01:.01:.99)';
octave:4> for i=.01:.01:.99
>           y(100*i)=markov(i,v0);
>         endfor
octave:5> z=[x, y];
octave:6> gplot z
```

yields this graph. By eye we judge that if $p > 0.7$ then the team is close to assured of the series.

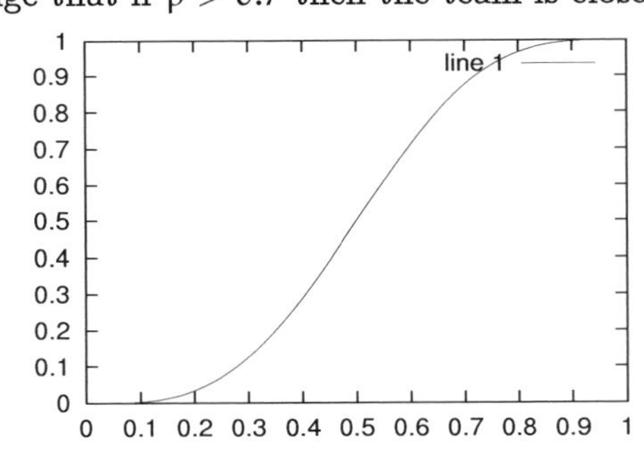

7 **(a)** They must satisfy this condition because the total probability of a state transition (including back to the same state) is 100%.

(b) See the answer to the third item.

(c) We will do the 2×2 case; bigger-sized cases are just notational problems. This product

$$\begin{pmatrix} a_{1,1} & a_{1,2} \\ a_{2,1} & a_{2,2} \end{pmatrix} \begin{pmatrix} b_{1,1} & b_{1,2} \\ b_{2,1} & b_{2,2} \end{pmatrix} = \begin{pmatrix} a_{1,1}b_{1,1} + a_{1,2}b_{2,1} & a_{1,1}b_{1,2} + a_{1,2}b_{2,2} \\ a_{2,1}b_{1,1} + a_{2,2}b_{2,1} & a_{2,1}b_{1,2} + a_{2,2}b_{2,2} \end{pmatrix}$$

has these two column sums

$$(a_{1,1}b_{1,1}+a_{1,2}b_{2,1})+(a_{2,1}b_{1,1}+a_{2,2}b_{2,1}) = (a_{1,1}+a_{2,1})\cdot b_{1,1}+(a_{1,2}+a_{2,2})\cdot b_{2,1} = 1\cdot b_{1,1}+1\cdot b_{2,1} = 1$$

and

$$(a_{1,1}b_{1,2}+a_{1,2}b_{2,2})+(a_{2,1}b_{1,2}+a_{2,2}b_{2,2}) = (a_{1,1}+a_{2,1})\cdot b_{1,2}+(a_{1,2}+a_{2,2})\cdot b_{2,2} = 1\cdot b_{1,2}+1\cdot b_{2,2} = 1$$

as required.

Topic: Orthonormal Matrices

1 (a) Yes.

(b) No, the columns do not have length one.

(c) Yes.

2 Some of these are nonlinear, because they involve a nontrivial translation.

(a) $\begin{pmatrix} x \\ y \end{pmatrix} \mapsto \begin{pmatrix} x\cdot\cos(\pi/6) - y\cdot\sin(\pi/6) \\ x\cdot\sin(\pi/6) + y\cdot\cos(\pi/6) \end{pmatrix} + \begin{pmatrix} 0 \\ 1 \end{pmatrix} = \begin{pmatrix} x\cdot(\sqrt{3}/2) - y\cdot(1/2) + 0 \\ x\cdot(1/2) + y\cdot\cos(\sqrt{3}/2) + 1 \end{pmatrix}$

(b) The line $y = 2x$ makes an angle of $\arctan(2/1)$ with the x-axis. Thus $\sin\theta = 2/\sqrt{5}$ and $\cos\theta = 1/\sqrt{5}$.

$$\begin{pmatrix} x \\ y \end{pmatrix} \mapsto \begin{pmatrix} x\cdot(1/\sqrt{5}) - y\cdot(2/\sqrt{5}) \\ x\cdot(2/\sqrt{5}) + y\cdot(1/\sqrt{5}) \end{pmatrix}$$

(c) $\begin{pmatrix} x \\ y \end{pmatrix} \mapsto \begin{pmatrix} x\cdot(1/\sqrt{5}) - y\cdot(-2/\sqrt{5}) \\ x\cdot(-2/\sqrt{5}) + y\cdot(1/\sqrt{5}) \end{pmatrix} + \begin{pmatrix} 1 \\ 1 \end{pmatrix} = \begin{pmatrix} x/\sqrt{5} + 2y/\sqrt{5} + 1 \\ -2x/\sqrt{5} + y/\sqrt{5} + 1 \end{pmatrix}$

3 (a) Let f be distance-preserving and consider f^{-1}. Any two points in the codomain can be written as $f(P_1)$ and $f(P_2)$. Because f is distance-preserving, the distance from $f(P_1)$ to $f(P_2)$ equals the distance from P_1 to P_2. But this is exactly what is required for f^{-1} to be distance-preserving.

(b) Any plane figure F is congruent to itself via the identity map $\mathrm{id}\colon \mathbb{R}^2 \to \mathbb{R}^2$, which is obviously distance-preserving. If F_1 is congruent to F_2 (via some f) then F_2 is congruent to F_1 via f^{-1}, which is distance-preserving by the prior item. Finally, if F_1 is congruent to F_2 (via some f) and F_2 is congruent to F_3 (via some g) then F_1 is congruent to F_3 via $g \circ f$, which is easily checked to be distance-preserving.

4 The first two components of each are $ax + cy + e$ and $bx + dy + f$.

5 (a) The Pythagorean Theorem gives that three points are colinear if and only if (for some ordering of them into P_1, P_2, and P_3), $\mathrm{dist}(P_1, P_2) + \mathrm{dist}(P_2, P_3) = \mathrm{dist}(P_1, P_3)$. Of course, where f is distance-preserving, this holds if and only if $\mathrm{dist}(f(P_1), f(P_2)) + \mathrm{dist}(f(P_2), f(P_3)) = \mathrm{dist}(f(P_1), f(P_3))$, which, again by Pythagoras, is true if and only if $f(P_1)$, $f(P_2)$, and $f(P_3)$ are colinear.

The argument for betweeness is similar (above, P_2 is between P_1 and P_3).

If the figure F is a triangle then it is the union of three line segments P_1P_2, P_2P_3, and P_1P_3. The prior two paragraphs together show that the property of being a line segment is invariant. So $f(F)$ is the union of three line segments, and so is a triangle.

A circle C centered at P and of radius r is the set of all points Q such that $\mathrm{dist}(P, Q) = r$. Applying the distance-preserving map f gives that the image $f(C)$ is the set of all $f(Q)$ subject to the condition that $\mathrm{dist}(P, Q) = r$. Since $\mathrm{dist}(P, Q) = \mathrm{dist}(f(P), f(Q))$, the set $f(C)$ is also a circle, with center $f(P)$ and radius r.

(b) Here are two that are easy to verify: (i) the property of being a right triangle, and (ii) the property of two lines being parallel.

(c) One that was mentioned in the section is the 'sense' of a figure. A triangle whose vertices read clockwise as P_1, P_2, P_3 may, under a distance-preserving map, be sent to a triangle read P_1, P_2, P_3 counterclockwise.

Chapter Four

Chapter Four: Determinants

Definition

Four.I.1: Exploration

Four.I.1.1 **(a)** 4 **(b)** 3 **(c)** -12

Four.I.1.2 **(a)** 6 **(b)** 21 **(c)** 27

Four.I.1.3 For the first, apply the formula in this section, note that any term with a d, g, or h is zero, and simplify. Lower-triangular matrices work the same way.

Four.I.1.4 **(a)** Nonsingular, the determinant is -1.
(b) Nonsingular, the determinant is -1.
(c) Singular, the determinant is 0.

Four.I.1.5 **(a)** Nonsingular, the determinant is 3.
(b) Singular, the determinant is 0.
(c) Singular, the determinant is 0.

Four.I.1.6 **(a)** $\det(B) = \det(A)$ via $-2\rho_1 + \rho_2$
(b) $\det(B) = -\det(A)$ via $\rho_2 \leftrightarrow \rho_3$
(c) $\det(B) = (1/2) \cdot \det(A)$ via $(1/2)\rho_2$

Four.I.1.7 Using the formula for the determinant of a 3×3 matrix we expand the left side

$$1 \cdot b \cdot c^2 + 1 \cdot c \cdot a^2 + 1 \cdot a \cdot b^2 - b^2 \cdot c \cdot 1 - c^2 \cdot a \cdot 1 - a^2 \cdot b \cdot 1$$

and by distributing we expand the right side.

$$(bc - ba - ac + a^2) \cdot (c - b) = c^2b - b^2c - bac + b^2a - ac^2 + acb + a^2c - a^2b$$

Now we can just check that the two are equal. (*Remark.* This is the 3×3 case of *Vandermonde's determinant* which arises in applications).

Four.I.1.8 This equation

$$0 = \det(\begin{pmatrix} 12-x & 4 \\ 8 & 8-x \end{pmatrix}) = 64 - 20x + x^2 = (x-16)(x-4)$$

has roots $x = 16$ and $x = 4$.

Four.I.1.9 We first reduce the matrix to echelon form. To begin, assume that $a \neq 0$ and that $ae - bd \neq 0$.

$$\xrightarrow{(1/a)\rho_1} \begin{pmatrix} 1 & b/a & c/a \\ d & e & f \\ g & h & i \end{pmatrix} \xrightarrow[-g\rho_1+\rho_3]{-d\rho_1+\rho_2} \begin{pmatrix} 1 & b/a & c/a \\ 0 & (ae-bd)/a & (af-cd)/a \\ 0 & (ah-bg)/a & (ai-cg)/a \end{pmatrix}$$

$$\xrightarrow{(a/(ae-bd))\rho_2} \begin{pmatrix} 1 & b/a & c/a \\ 0 & 1 & (af-cd)/(ae-bd) \\ 0 & (ah-bg)/a & (ai-cg)/a \end{pmatrix}$$

This step finishes the calculation.

$$\xrightarrow{((ah-bg)/a)\rho_2+\rho_3}\begin{pmatrix}1 & b/a & c/a \\ 0 & 1 & (af-cd)/(ae-bd) \\ 0 & 0 & (aei+bgf+cdh-hfa-idb-gec)/(ae-bd)\end{pmatrix}$$

Now assuming that $a \neq 0$ and $ae-bd \neq 0$, the original matrix is nonsingular if and only if the $3,3$ entry above is nonzero. That is, under the assumptions, the original matrix is nonsingular if and only if $aei+bgf+cdh-hfa-idb-gec \neq 0$, as required.

We finish by running down what happens if the assumptions that were taken for convenience in the prior paragraph do not hold. First, if $a \neq 0$ but $ae-bd=0$ then we can swap

$$\begin{pmatrix}1 & b/a & c/a \\ 0 & 0 & (af-cd)/a \\ 0 & (ah-bg)/a & (ai-cg)/a\end{pmatrix}\xrightarrow{\rho_2\leftrightarrow\rho_3}\begin{pmatrix}1 & b/a & c/a \\ 0 & (ah-bg)/a & (ai-cg)/a \\ 0 & 0 & (af-cd)/a\end{pmatrix}$$

and conclude that the matrix is nonsingular if and only if either $ah-bg=0$ or $af-cd=0$. The condition '$ah-bg=0$ or $af-cd=0$' is equivalent to the condition '$(ah-bg)(af-cd)=0$'. Multiplying out and using the case assumption that $ae-bd=0$ to substitute ae for bd gives this.

$$0 = ahaf-ahcd-bgaf+bgcd = ahaf-ahcd-bgaf+aegc = a(haf-hcd-bgf+egc)$$

Since $a \neq 0$, we have that the matrix is nonsingular if and only if $haf-hcd-bgf+egc=0$. Therefore, in this $a \neq 0$ and $ae-bd=0$ case, the matrix is nonsingular when $haf-hcd-bgf+egc-i(ae-bd)=0$.

The remaining cases are routine. Do the $a=0$ but $d \neq 0$ case and the $a=0$ and $d=0$ but $g \neq 0$ case by first swapping rows and then going on as above. The $a=0$, $d=0$, and $g=0$ case is easy—that matrix is singular since the columns form a linearly dependent set, and the determinant comes out to be zero.

Four.I.1.10 Figuring the determinant and doing some algebra gives this.

$$\begin{aligned}0 &= y_1x+x_2y+x_1y_2-y_2x-x_1y-x_2y_1 \\ (x_2-x_1)\cdot y &= (y_2-y_1)\cdot x+x_2y_1-x_1y_2 \\ y &= \frac{y_2-y_1}{x_2-x_1}\cdot x+\frac{x_2y_1-x_1y_2}{x_2-x_1}\end{aligned}$$

Note that this is the equation of a line (in particular, in contains the familiar expression for the slope), and note that (x_1,y_1) and (x_2,y_2) satisfy it.

Four.I.1.11 **(a)** The comparison with the formula given in the preamble to this section is easy.

(b) While it holds for 2×2 matrices

$$\left(\begin{array}{cc|c}h_{1,1} & h_{1,2} & h_{1,1} \\ h_{2,1} & h_{2,2} & h_{2,1}\end{array}\right)\begin{aligned}&= h_{1,1}h_{2,2}+h_{1,2}h_{2,1} \\ &\quad -h_{2,1}h_{1,2}-h_{2,2}h_{1,1} \\ &= h_{1,1}h_{2,2}-h_{1,2}h_{2,1}\end{aligned}$$

it does not hold for 4×4 matrices. An example is that this matrix is singular because the second and third rows are equal

$$\begin{pmatrix}1 & 0 & 0 & 1 \\ 0 & 1 & 1 & 0 \\ 0 & 1 & 1 & 0 \\ -1 & 0 & 0 & 1\end{pmatrix}$$

but following the scheme of the mnemonic does not give zero.

$$\left(\begin{array}{cccc|ccc}1 & 0 & 0 & 1 & 1 & 0 & 0 \\ 0 & 1 & 1 & 0 & 0 & 1 & 1 \\ 0 & 1 & 1 & 0 & 0 & 1 & 1 \\ -1 & 0 & 0 & 1 & -1 & 0 & 0\end{array}\right)\begin{aligned}&= 1+0+0+0 \\ &\quad -(-1)-0-0-0\end{aligned}$$

Four.I.1.12 The determinant is $(x_2y_3-x_3y_2)\vec{e}_1+(x_3y_1-x_1y_3)\vec{e}_2+(x_1y_2-x_2y_1)\vec{e}_3$. To check perpendicularity, we check that the dot product with the first vector is zero

$$\begin{pmatrix} x_1 \\ x_2 \\ x_3 \end{pmatrix} \cdot \begin{pmatrix} x_2y_3-x_3y_2 \\ x_3y_1-x_1y_3 \\ x_1y_2-x_2y_1 \end{pmatrix} = x_1x_2y_3-x_1x_3y_2+x_2x_3y_1-x_1x_2y_3+x_1x_3y_2-x_2x_3y_1=0$$

and the dot product with the second vector is also zero.

$$\begin{pmatrix} y_1 \\ y_2 \\ y_3 \end{pmatrix} \cdot \begin{pmatrix} x_2y_3-x_3y_2 \\ x_3y_1-x_1y_3 \\ x_1y_2-x_2y_1 \end{pmatrix} = x_2y_1y_3-x_3y_1y_2+x_3y_1y_2-x_1y_2y_3+x_1y_2y_3-x_2y_1y_3=0$$

Four.I.1.13 **(a)** Plug and chug: the determinant of the product is this

$$\begin{aligned} \det(\begin{pmatrix} a & b \\ c & d \end{pmatrix}\begin{pmatrix} w & x \\ y & z \end{pmatrix}) &= \det(\begin{pmatrix} aw+by & ax+bz \\ cw+dy & cx+dz \end{pmatrix}) \\ &= acwx+adwz+bcxy+bdyz \\ &\quad -acwx-bcwz-adxy-bdyz \end{aligned}$$

while the product of the determinants is this.

$$\det(\begin{pmatrix} a & b \\ c & d \end{pmatrix})\cdot\det(\begin{pmatrix} w & x \\ y & z \end{pmatrix}) = (ad-bc)\cdot(wz-xy)$$

Verification that they are equal is easy.

(b) Use the prior item.

That similar matrices have the same determinant is immediate from the above two: $\det(PTP^{-1}) = \det(P)\cdot\det(T)\cdot\det(P^{-1})$.

Four.I.1.14 One way is to count these areas

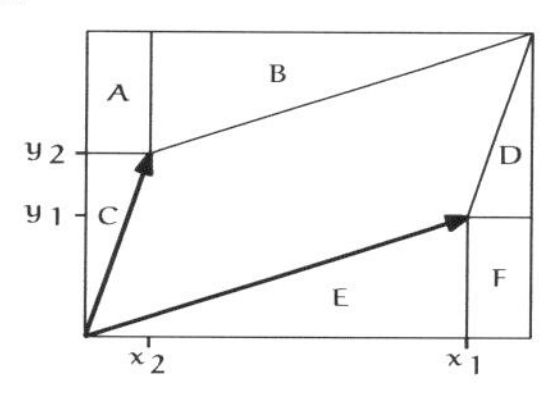

by taking the area of the entire rectangle and subtracting the area of A the upper-left rectangle, B the upper-middle triangle, D the upper-right triangle, C the lower-left triangle, E the lower-middle triangle, and F the lower-right rectangle $(x_1+x_2)(y_1+y_2)-x_2y_1-(1/2)x_1y_1-(1/2)x_2y_2-(1/2)x_2y_2-(1/2)x_1y_1-x_2y_1$. Simplification gives the determinant formula.

This determinant is the negative of the one above; the formula distinguishes whether the second column is counterclockwise from the first.

Four.I.1.15 The computation for 2×2 matrices, using the formula quoted in the preamble, is easy. It does also hold for 3×3 matrices; the computation is routine.

Four.I.1.16 No. Recall that constants come out one row at a time.

$$\det(\begin{pmatrix} 2 & 4 \\ 2 & 6 \end{pmatrix}) = 2\cdot\det(\begin{pmatrix} 1 & 2 \\ 2 & 6 \end{pmatrix}) = 2\cdot2\cdot\det(\begin{pmatrix} 1 & 2 \\ 1 & 3 \end{pmatrix})$$

This contradicts linearity (here we didn't need S, i.e., we can take S to be the matrix of zeros).

Four.I.1.17 Bring out the c's one row at a time.

Four.I.1.18 There are no real numbers θ that make the matrix singular because the determinant of the matrix $\cos^2\theta+\sin^2\theta$ is never 0, it equals 1 for all θ. Geometrically, with respect to the standard basis, this matrix represents a rotation of the plane through an angle of θ. Each such map is one-to-one — for one thing, it is invertible.

Four.I.1.19 *This is how the answer was given in the cited source.* Let P be the sum of the three positive terms of the determinant and $-N$ the sum of the three negative terms. The maximum value of P is

$$9\cdot 8\cdot 7+6\cdot 5\cdot 4+3\cdot 2\cdot 1=630.$$

The minimum value of N consistent with P is

$$9\cdot 6\cdot 1+8\cdot 5\cdot 2+7\cdot 4\cdot 3=218.$$

Any change in P would result in lowering that sum by more than 4. Therefore 412 the maximum value for the determinant and one form for the determinant is

$$\begin{vmatrix}9&4&2\\3&8&6\\5&1&7\end{vmatrix}.$$

Four.I.2: Properties of Determinants

Four.I.2.7 **(a)** $\begin{vmatrix}3&1&2\\3&1&0\\0&1&4\end{vmatrix}=\begin{vmatrix}3&1&2\\0&0&-2\\0&1&4\end{vmatrix}=-\begin{vmatrix}3&1&2\\0&1&4\\0&0&-2\end{vmatrix}=6$

(b) $\begin{vmatrix}1&0&0&1\\2&1&1&0\\-1&0&1&0\\1&1&1&0\end{vmatrix}=\begin{vmatrix}1&0&0&1\\0&1&1&-2\\0&0&1&1\\0&1&1&-1\end{vmatrix}=\begin{vmatrix}1&0&0&1\\0&1&1&-2\\0&0&1&1\\0&0&0&1\end{vmatrix}=1$

Four.I.2.8 **(a)** $\begin{vmatrix}2&-1\\-1&-1\end{vmatrix}=\begin{vmatrix}2&-1\\0&-3/2\end{vmatrix}=-3$; **(b)** $\begin{vmatrix}1&1&0\\3&0&2\\5&2&2\end{vmatrix}=\begin{vmatrix}1&1&0\\0&-3&2\\0&-3&2\end{vmatrix}=\begin{vmatrix}1&1&0\\0&-3&2\\0&0&0\end{vmatrix}=0$

Four.I.2.9 When is the determinant not zero?

$$\begin{vmatrix}1&0&1&-1\\0&1&-2&0\\1&0&k&0\\0&0&1&-1\end{vmatrix}=\begin{vmatrix}1&0&1&-1\\0&1&-2&0\\0&0&k-1&1\\0&0&1&-1\end{vmatrix}$$

Obviously, $k=1$ gives nonsingularity and hence a nonzero determinant. If $k\neq 1$ then we get echelon form with a $(-1/k-1)\rho_3+\rho_4$ combination.

$$=\begin{vmatrix}1&0&1&-1\\0&1&-2&0\\0&0&k-1&1\\0&0&0&-1-(1/k-1)\end{vmatrix}$$

Multiplying down the diagonal gives $(k-1)(-1-(1/k-1))=-(k-1)-1=-k$. Thus the matrix has a nonzero determinant, and so the system has a unique solution, if and only if $k\neq 0$.

Four.I.2.10 **(a)** Property (2) of the definition of determinants applies via the swap $\rho_1\leftrightarrow\rho_3$.

$$\begin{vmatrix}h_{3,1}&h_{3,2}&h_{3,3}\\h_{2,1}&h_{2,2}&h_{2,3}\\h_{1,1}&h_{1,2}&h_{1,3}\end{vmatrix}=-\begin{vmatrix}h_{1,1}&h_{1,2}&h_{1,3}\\h_{2,1}&h_{2,2}&h_{2,3}\\h_{3,1}&h_{3,2}&h_{3,3}\end{vmatrix}$$

(b) Property (3) applies.

$$\begin{vmatrix}-h_{1,1}&-h_{1,2}&-h_{1,3}\\-2h_{2,1}&-2h_{2,2}&-2h_{2,3}\\-3h_{3,1}&-3h_{3,2}&-3h_{3,3}\end{vmatrix}=(-1)\cdot(-2)\cdot(-3)\cdot\begin{vmatrix}h_{1,1}&h_{1,2}&h_{1,3}\\h_{2,1}&h_{2,2}&h_{2,3}\\h_{3,1}&h_{3,2}&h_{3,3}\end{vmatrix}=(-6)\cdot\begin{vmatrix}h_{1,1}&h_{1,2}&h_{1,3}\\h_{2,1}&h_{2,2}&h_{2,3}\\h_{3,1}&h_{3,2}&h_{3,3}\end{vmatrix}$$

(c)

$$\begin{vmatrix} h_{1,1}+h_{3,1} & h_{1,2}+h_{3,2} & h_{1,3}+h_{3,3} \\ h_{2,1} & h_{2,2} & h_{2,3} \\ 5h_{3,1} & 5h_{3,2} & 5h_{3,3} \end{vmatrix} = 5\cdot\begin{vmatrix} h_{1,1}+h_{3,1} & h_{1,2}+h_{3,2} & h_{1,3}+h_{3,3} \\ h_{2,1} & h_{2,2} & h_{2,3} \\ h_{3,1} & h_{3,2} & h_{3,3} \end{vmatrix}$$
$$= 5\cdot\begin{vmatrix} h_{1,1} & h_{1,2} & h_{1,3} \\ h_{2,1} & h_{2,2} & h_{2,3} \\ h_{3,1} & h_{3,2} & h_{3,3} \end{vmatrix}$$

Four.I.2.11 A diagonal matrix is in echelon form, so the determinant is the product down the diagonal.

Four.I.2.12 It is the trivial subspace.

Four.I.2.13 Adding the second row to the first gives a matrix whose first row is $x+y+z$ times its third row.

Four.I.2.14 (a) $\begin{pmatrix}1\end{pmatrix}$, $\begin{pmatrix}1 & -1 \\ -1 & 1\end{pmatrix}$, $\begin{pmatrix}1 & -1 & 1 \\ -1 & 1 & -1 \\ 1 & -1 & 1\end{pmatrix}$

(b) The determinant in the 1×1 case is 1. In every other case the second row is the negative of the first, and so matrix is singular and the determinant is zero.

Four.I.2.15 (a) $\begin{pmatrix}2\end{pmatrix}$, $\begin{pmatrix}2 & 3 \\ 3 & 4\end{pmatrix}$, $\begin{pmatrix}2 & 3 & 4 \\ 3 & 4 & 5 \\ 4 & 5 & 6\end{pmatrix}$

(b) The 1×1 and 2×2 cases yield these.

$$\begin{vmatrix}2\end{vmatrix} = 2 \qquad \begin{vmatrix}2 & 3 \\ 3 & 4\end{vmatrix} = -1$$

And $n\times n$ matrices with $n \geqslant 3$ are singular, e.g.,

$$\begin{vmatrix}2 & 3 & 4 \\ 3 & 4 & 5 \\ 4 & 5 & 6\end{vmatrix} = 0$$

because twice the second row minus the first row equals the third row. Checking this is routine.

Four.I.2.16 This one

$$A = B = \begin{pmatrix}1 & 2 \\ 3 & 4\end{pmatrix}$$

is easy to check.

$$|A+B| = \begin{vmatrix}2 & 4 \\ 6 & 8\end{vmatrix} = -8 \qquad |A|+|B| = -2-2 = -4$$

By the way, this also gives an example where scalar multiplication is not preserved $|2\cdot A| \neq 2\cdot|A|$.

Four.I.2.17 No, we cannot replace it. Remark 2.2 shows that the four conditions after the replacement would conflict — no function satisfies all four.

Four.I.2.18 A upper-triangular matrix is in echelon form.

A lower-triangular matrix is either singular or nonsingular. If it is singular then it has a zero on its diagonal and so its determinant (namely, zero) is indeed the product down its diagonal. If it is nonsingular then it has no zeroes on its diagonal, and we can reduce it by Gauss's Method to echelon form without changing the diagonal.

Four.I.2.19 (a) The properties in the definition of determinant show that $|M_i(k)| = k$, $|P_{i,j}| = -1$, and $|C_{i,j}(k)| = 1$.

(b) The three cases are easy to check by recalling the action of left multiplication by each type of matrix.

(c) If TS is invertible $(TS)M = I$ then the associative property of matrix multiplication $T(SM) = I$ shows that T is invertible. So if T is not invertible then neither is TS.

(d) If T is singular then apply the prior answer: $|TS| = 0$ and $|T|\cdot|S| = 0\cdot|S| = 0$. If T is not singular then we can write it as a product of elementary matrices $|TS| = |E_r \cdots E_1 S| = |E_r| \cdots |E_1|\cdot|S| = |E_r \cdots E_1||S| = |T||S|$.
(e) $1 = |I| = |T \cdot T^{-1}| = |T||T^{-1}|$

Four.I.2.20 (a) We must show that if

$$T \xrightarrow{k\rho_i+\rho_j} \hat{T}$$

then $d(T) = |TS|/|S| = |\hat{T}S|/|S| = d(\hat{T})$. We will be done if we show that combining rows first and then multiplying to get $\hat{T}S$ gives the same result as multiplying first to get TS and then combining (because the determinant $|TS|$ is unaffected by the combination so we'll then have $|\hat{T}S| = |TS|$, and hence $d(\hat{T}) = d(T)$). That argument runs: after adding k times row i of TS to row j of TS, the j,p entry is $(kt_{i,1} + t_{j,1})s_{1,p} + \cdots + (kt_{i,r} + t_{j,r})s_{r,p}$, which is the j,p entry of $\hat{T}S$.
(b) We need only show that swapping $T \xrightarrow{\rho_i \leftrightarrow \rho_j} \hat{T}$ and then multiplying to get $\hat{T}S$ gives the same result as multiplying T by S and then swapping (because, as the determinant $|TS|$ changes sign on the row swap, we'll then have $|\hat{T}S| = -|TS|$, and so $d(\hat{T}) = -d(T)$). That argument runs just like the prior one.
(c) Not surprisingly by now, we need only show that multiplying a row by a scalar $T \xrightarrow{k\rho_i} \hat{T}$ and then computing $\hat{T}S$ gives the same result as first computing TS and then multiplying the row by k (as the determinant $|TS|$ is rescaled by k the multiplication, we'll have $|\hat{T}S| = k|TS|$, so $d(\hat{T}) = k\,d(T)$). The argument runs just as above.
(d) Clear.
(e) Because we've shown that $d(T)$ is a determinant and that determinant functions (if they exist) are unique, we have that so $|T| = d(T) = |TS|/|S|$.

Four.I.2.21 We will first argue that a rank r matrix has a $r\times r$ submatrix with nonzero determinant. A rank r matrix has a linearly independent set of r rows. A matrix made from those rows will have row rank r and thus has column rank r. Conclusion: from those r rows we can extract a linearly independent set of r columns, and so the original matrix has a $r\times r$ submatrix of rank r.

We finish by showing that if r is the largest such integer then the rank of the matrix is r. We need only show, by the maximality of r, that if a matrix has a $k\times k$ submatrix of nonzero determinant then the rank of the matrix is at least k. Consider such a $k\times k$ submatrix. Its rows are parts of the rows of the original matrix, clearly the set of whole rows is linearly independent. Thus the row rank of the original matrix is at least k, and the row rank of a matrix equals its rank.

Four.I.2.22 A matrix with only rational entries reduces with Gauss's Method to an echelon form matrix using only rational arithmetic. Thus the entries on the diagonal must be rationals, and so the product down the diagonal is rational.

Four.I.2.23 *This is how the answer was given in the cited source.* The value $(1-a^4)^3$ of the determinant is independent of the values B, C, D. Hence operation (e) does not change the value of the determinant but merely changes its appearance. Thus the element of likeness in (a), (b), (c), (d), and (e) is only that the appearance of the principle entity is changed. The same element appears in (f) changing the name-label of a rose, (g) writing a decimal integer in the scale of 12, (h) gilding the lily, (i) whitewashing a politician, and (j) granting an honorary degree.

Four.I.3: The Permutation Expansion

Four.I.3.15 **(a)** This matrix is singular.

$$\begin{vmatrix} 1 & 2 & 3 \\ 4 & 5 & 6 \\ 7 & 8 & 9 \end{vmatrix} = (1)(5)(9)\,|P_{\phi_1}| + (1)(6)(8)\,|P_{\phi_2}| + (2)(4)(9)\,|P_{\phi_3}| + (2)(6)(7)\,|P_{\phi_4}| + (3)(4)(8)\,|P_{\phi_5}| + (7)(5)(3)\,|P_{\phi_6}| = 0$$

(b) This matrix is nonsingular.

$$\begin{vmatrix} 2 & 2 & 1 \\ 3 & -1 & 0 \\ -2 & 0 & 5 \end{vmatrix} = (2)(-1)(5)\,|P_{\phi_1}| + (2)(0)(0)\,|P_{\phi_2}| + (2)(3)(5)\,|P_{\phi_3}| + (2)(0)(-2)\,|P_{\phi_4}| + (1)(3)(0)\,|P_{\phi_5}| + (-2)(-1)(1)\,|P_{\phi_6}| = -42$$

Four.I.3.16 **(a)** Gauss's Method gives this

$$\begin{vmatrix} 2 & 1 \\ 3 & 1 \end{vmatrix} = \begin{vmatrix} 2 & 1 \\ 0 & -1/2 \end{vmatrix} = -1$$

and permutation expansion gives this.

$$\begin{vmatrix} 2 & 1 \\ 3 & 1 \end{vmatrix} = \begin{vmatrix} 2 & 0 \\ 0 & 1 \end{vmatrix} + \begin{vmatrix} 0 & 1 \\ 3 & 0 \end{vmatrix} = (2)(1)\begin{vmatrix} 1 & 0 \\ 0 & 1 \end{vmatrix} + (1)(3)\begin{vmatrix} 0 & 1 \\ 1 & 0 \end{vmatrix} = -1$$

(b) Gauss's Method gives this

$$\begin{vmatrix} 0 & 1 & 4 \\ 0 & 2 & 3 \\ 1 & 5 & 1 \end{vmatrix} = -\begin{vmatrix} 1 & 5 & 1 \\ 0 & 2 & 3 \\ 0 & 1 & 4 \end{vmatrix} = -\begin{vmatrix} 1 & 5 & 1 \\ 0 & 2 & 3 \\ 0 & 0 & 5/2 \end{vmatrix} = -5$$

and the permutation expansion gives this.

$$\begin{vmatrix} 0 & 1 & 4 \\ 0 & 2 & 3 \\ 1 & 5 & 1 \end{vmatrix} = (0)(2)(1)\,|P_{\phi_1}| + (0)(3)(5)\,|P_{\phi_2}| + (1)(0)(1)\,|P_{\phi_3}| + (1)(3)(1)\,|P_{\phi_4}| + (4)(0)(5)\,|P_{\phi_5}| + (1)(2)(0)\,|P_{\phi_6}| = -5$$

Four.I.3.17 Following Example 3.6 gives this.

$$\begin{aligned} \begin{vmatrix} t_{1,1} & t_{1,2} & t_{1,3} \\ t_{2,1} & t_{2,2} & t_{2,3} \\ t_{3,1} & t_{3,2} & t_{3,3} \end{vmatrix} &= t_{1,1}t_{2,2}t_{3,3}\,|P_{\phi_1}| + t_{1,1}t_{2,3}t_{3,2}\,|P_{\phi_2}| \\ &\quad + t_{1,2}t_{2,1}t_{3,3}\,|P_{\phi_3}| + t_{1,2}t_{2,3}t_{3,1}\,|P_{\phi_4}| \\ &\quad + t_{1,3}t_{2,1}t_{3,2}\,|P_{\phi_5}| + t_{1,3}t_{2,2}t_{3,1}\,|P_{\phi_6}| \\ &= t_{1,1}t_{2,2}t_{3,3}(+1) + t_{1,1}t_{2,3}t_{3,2}(-1) \\ &\quad + t_{1,2}t_{2,1}t_{3,3}(-1) + t_{1,2}t_{2,3}t_{3,1}(+1) \\ &\quad + t_{1,3}t_{2,1}t_{3,2}(+1) + t_{1,3}t_{2,2}t_{3,1}(-1) \end{aligned}$$

Four.I.3.18 This is all of the permutations where $\phi(1) = 1$

$$\phi_1 = \langle 1,2,3,4\rangle \quad \phi_2 = \langle 1,2,4,3\rangle \quad \phi_3 = \langle 1,3,2,4\rangle$$
$$\phi_4 = \langle 1,3,4,2\rangle \quad \phi_5 = \langle 1,4,2,3\rangle \quad \phi_6 = \langle 1,4,3,2\rangle$$

the ones where $\phi(1) = 1$

$$\phi_7 = \langle 2,1,3,4\rangle \quad \phi_8 = \langle 2,1,4,3\rangle \quad \phi_9 = \langle 2,3,1,4\rangle$$
$$\phi_{10} = \langle 2,3,4,1\rangle \quad \phi_{11} = \langle 2,4,1,3\rangle \quad \phi_{12} = \langle 2,4,3,1\rangle$$

the ones where $\phi(1) = 3$

$$\phi_{13} = \langle 3,1,2,4\rangle \quad \phi_{14} = \langle 3,1,4,2\rangle \quad \phi_{15} = \langle 3,2,1,4\rangle$$
$$\phi_{16} = \langle 3,2,4,1\rangle \quad \phi_{17} = \langle 3,4,1,2\rangle \quad \phi_{18} = \langle 3,4,2,1\rangle$$

and the ones where $\phi(1) = 4$.

$$\phi_{19} = \langle 4,1,2,3\rangle \quad \phi_{20} = \langle 4,1,3,2\rangle \quad \phi_{21} = \langle 4,2,1,3\rangle$$
$$\phi_{22} = \langle 4,2,3,1\rangle \quad \phi_{23} = \langle 4,3,1,2\rangle \quad \phi_{24} = \langle 4,3,2,1\rangle$$

Four.I.3.19 Each of these is easy to check.

(a)

permutation	ϕ_1	ϕ_2
inverse	ϕ_1	ϕ_2

(b)

permutation	ϕ_1	ϕ_2	ϕ_3	ϕ_4	ϕ_5	ϕ_6
inverse	ϕ_1	ϕ_2	ϕ_3	ϕ_5	ϕ_4	ϕ_6

Four.I.3.20 For the 'if' half, the first condition of Definition 3.2 follows from taking $k_1 = k_2 = 1$ and the second condition follows from taking $k_2 = 0$.

The 'only if' half also routine. From $f(\vec{\rho}_1,\ldots,k_1\vec{v}_1 + k_2\vec{v}_2,\ldots,\vec{\rho}_n)$ the first condition of Definition 3.2 gives $= f(\vec{\rho}_1,\ldots,k_1\vec{v}_1,\ldots,\vec{\rho}_n) + f(\vec{\rho}_1,\ldots,k_2\vec{v}_2,\ldots,\vec{\rho}_n)$ and the second condition, applied twice, gives the result.

Four.I.3.21 They would all double.

Four.I.3.22 For the second statement, given a matrix, transpose it, swap rows, and transpose back. The result is swapped columns, and the determinant changes by a factor of -1. The third statement is similar: given a matrix, transpose it, apply multilinearity to what are now rows, and then transpose back the resulting matrices.

Four.I.3.23 An $n\times n$ matrix with a nonzero determinant has rank n so its columns form a basis for $\mathbb{R}^n$.

Four.I.3.24 False.

$$\begin{vmatrix} 1 & -1 \\ 1 & 1 \end{vmatrix} = 2$$

Four.I.3.25 (a) For the column index of the entry in the first row there are five choices. Then, for the column index of the entry in the second row there are four choices. Continuing, we get $5\cdot 4\cdot 3\cdot 2\cdot 1 = 120$. (See also the next question.)

(b) Once we choose the second column in the first row, we can choose the other entries in $4\cdot 3\cdot 2\cdot 1 = 24$ ways.

Four.I.3.26 $n\cdot(n-1)\cdots 2\cdot 1 = n!$

Four.I.3.27 [Schmidt] We will show that $PP^{\mathsf{T}} = I$; the $P^{\mathsf{T}}P = I$ argument is similar. The i,j entry of PP^{T} is the sum of terms of the form $p_{i,k}q_{k,j}$ where the entries of P^{T} are denoted with q's, that is, $q_{k,j} = p_{j,k}$. Thus the i,j entry of PP^{T} is the sum $\sum_{k=1}^{n} p_{i,k}p_{j,k}$. But $p_{i,k}$ is usually 0, and so $P_{i,k}P_{j,k}$ is usually 0. The only time $P_{i,k}$ is nonzero is when it is 1, but then there are no other $i' \neq i$ such that $P_{i',k}$ is nonzero (i is the only row with a 1 in column k). In other words,

$$\sum_{k=1}^{n} p_{i,k}p_{j,k} = \begin{cases} 1 & i=j \\ 0 & \text{otherwise} \end{cases}$$

and this is exactly the formula for the entries of the identity matrix.

Four.I.3.28 In $|A| = |A^{\mathsf{T}}| = |-A| = (-1)^n|A|$ the exponent n must be even.

Four.I.3.29 Showing that no placement of three zeros suffices is routine. Four zeroes does suffice; put them all in the same row or column.

Four.I.3.30 The $n = 3$ case shows what to do. The row combination operations of $-x_1\rho_2+\rho_3$ and $-x_1\rho_1+\rho_2$ give this.

$$\begin{vmatrix} 1 & 1 & 1 \\ x_1 & x_2 & x_3 \\ x_1^2 & x_2^2 & x_3^2 \end{vmatrix} = \begin{vmatrix} 1 & 1 & 1 \\ x_1 & x_2 & x_3 \\ 0 & (-x_1+x_2)x_2 & (-x_1+x_3)x_3 \end{vmatrix} = \begin{vmatrix} 1 & 1 & 1 \\ 0 & -x_1+x_2 & -x_1+x_3 \\ 0 & (-x_1+x_2)x_2 & (-x_1+x_3)x_3 \end{vmatrix}$$

Then the row combination operation of $x_2\rho_2 + \rho_3$ gives the desired result.

$$= \begin{vmatrix} 1 & 1 & 1 \\ 0 & -x_1 + x_2 & -x_1 + x_3 \\ 0 & 0 & (-x_1 + x_3)(-x_2 + x_3) \end{vmatrix} = (x_2 - x_1)(x_3 - x_1)(x_3 - x_2)$$

Four.I.3.31 Let T be $n \times n$, let J be $p \times p$, and let K be $q \times q$. Apply the permutation expansion formula

$$|T| = \sum_{\text{permutations } \phi} t_{1,\phi(1)} t_{2,\phi(2)} \dots t_{n,\phi(n)} |P_\phi|$$

Because the upper right of T is all zeroes, if a ϕ has at least one of $p + 1, \dots, n$ among its first p column numbers $\phi(1), \dots, \phi(p)$ then the term arising from ϕ is 0 (e.g., if $\phi(1) = n$ then $t_{1,\phi(1)} t_{2,\phi(2)} \dots t_{n,\phi(n)}$ is 0). So the above formula reduces to a sum over all permutations with two halves: first rearrange $1, \dots, p$ and after that comes a permutation of $p + 1, \dots, p + q$. To see this gives $|J| \cdot |K|$, distribute.

$$\Big[\sum_{\substack{\text{perms } \phi_1 \\ \text{of } 1,\dots,p}} t_{1,\phi_1(1)} \cdots t_{p,\phi_1(p)} |P_{\phi_1}| \Big] \cdot \Big[\sum_{\substack{\text{perms } \phi_2 \\ \text{of } p+1,\dots,p+q}} t_{p+1,\phi_2(p+1)} \cdots t_{p+q,\phi_2(p+q)} |P_{\phi_2}| \Big]$$

Four.I.3.32 The $n = 3$ case shows what happens.

$$|T - rI| = \begin{vmatrix} t_{1,1} - x & t_{1,2} & t_{1,3} \\ t_{2,1} & t_{2,2} - x & t_{2,3} \\ t_{3,1} & t_{3,2} & t_{3,3} - x \end{vmatrix}$$

Each term in the permutation expansion has three factors drawn from entries in the matrix (e.g., $(t_{1,1} - x)(t_{2,2} - x)(t_{3,3} - x)$ and $(t_{1,1} - x)(t_{2,3})(t_{3,2})$), and so the determinant is expressible as a polynomial in x of degree 3. Such a polynomial has at most 3 roots.

In general, the permutation expansion shows that the determinant is a sum of terms, each with n factors, giving a polynomial of degree n. A polynomial of degree n has at most n roots.

Four.I.3.33 *This is how the answer was given in the cited source.* When two rows of a determinant are interchanged, the sign of the determinant is changed. When the rows of a three-by-three determinant are permuted, 3 positive and 3 negative determinants equal in absolute value are obtained. Hence the $9!$ determinants fall into $9!/6$ groups, each of which sums to zero.

Four.I.3.34 *This is how the answer was given in the cited source.* When the elements of any column are subtracted from the elements of each of the other two, the elements in two of the columns of the derived determinant are proportional, so the determinant vanishes. That is,

$$\begin{vmatrix} 2 & 1 & x-4 \\ 4 & 2 & x-3 \\ 6 & 3 & x-10 \end{vmatrix} = \begin{vmatrix} 1 & x-3 & -1 \\ 2 & x-1 & -2 \\ 3 & x-7 & -3 \end{vmatrix} = \begin{vmatrix} x-2 & -1 & -2 \\ x+1 & -2 & -4 \\ x-4 & -3 & -6 \end{vmatrix} = 0.$$

Four.I.3.35 *This is how the answer was given in the cited source.* Let

$$\begin{matrix} a & b & c \\ d & e & f \\ g & h & i \end{matrix}$$

have magic sum $N = S/3$. Then

$$\begin{aligned} N &= (a + e + i) + (d + e + f) + (g + e + c) \\ &\quad - (a + d + g) - (c + f + i) = 3e \end{aligned}$$

and $S = 9e$. Hence, adding rows and columns,

$$D = \begin{vmatrix} a & b & c \\ d & e & f \\ g & h & i \end{vmatrix} = \begin{vmatrix} a & b & c \\ d & e & f \\ 3e & 3e & 3e \end{vmatrix} = \begin{vmatrix} a & b & 3e \\ d & e & 3e \\ 3e & 3e & 9e \end{vmatrix} = \begin{vmatrix} a & b & e \\ d & e & e \\ 1 & 1 & 1 \end{vmatrix} S.$$

Four.I.3.36 *This is how the answer was given in the cited source.* Denote by D_n the determinant in question and by $a_{i,j}$ the element in the i-th row and j-th column. Then from the law of formation of the elements we have

$$a_{i,j} = a_{i,j-1} + a_{i-1,j}, \qquad a_{1,j} = a_{i,1} = 1.$$

Subtract each row of D_n from the row following it, beginning the process with the last pair of rows. After the $n-1$ subtractions the above equality shows that the element $a_{i,j}$ is replaced by the element $a_{i,j-1}$, and all the elements in the first column, except $a_{1,1} = 1$, become zeroes. Now subtract each column from the one following it, beginning with the last pair. After this process the element $a_{i,j-1}$ is replaced by $a_{i-1,j-1}$, as shown in the above relation. The result of the two operations is to replace $a_{i,j}$ by $a_{i-1,j-1}$, and to reduce each element in the first row and in the first column to zero. Hence $D_n = D_{n+i}$ and consequently

$$D_n = D_{n-1} = D_{n-2} = \cdots = D_2 = 1.$$

Four.I.4: Determinants Exist

Four.I.4.9 This is the permutation expansion of the determinant of a 2×2 matrix

$$\begin{vmatrix} a & b \\ c & d \end{vmatrix} = ad \cdot \begin{vmatrix} 1 & 0 \\ 0 & 1 \end{vmatrix} + bc \cdot \begin{vmatrix} 0 & 1 \\ 1 & 0 \end{vmatrix}$$

and the permutation expansion of the determinant of its transpose.

$$\begin{vmatrix} a & c \\ b & d \end{vmatrix} = ad \cdot \begin{vmatrix} 1 & 0 \\ 0 & 1 \end{vmatrix} + cb \cdot \begin{vmatrix} 0 & 1 \\ 1 & 0 \end{vmatrix}$$

As with the 3×3 expansions described in the subsection, the permutation matrices from corresponding terms are transposes (although this is disguised by the fact that each is self-transpose).

Four.I.4.10 Each of these is easy to check.

(a)

permutation	ϕ_1	ϕ_2
inverse	ϕ_1	ϕ_2

(b)

permutation	ϕ_1	ϕ_2	ϕ_3	ϕ_4	ϕ_5	ϕ_6
inverse	ϕ_1	ϕ_2	ϕ_3	ϕ_5	ϕ_4	ϕ_6

Four.I.4.11 (a) $\text{sgn}(\phi_1) = +1$, $\text{sgn}(\phi_2) = -1$
(b) $\text{sgn}(\phi_1) = +1$, $\text{sgn}(\phi_2) = -1$, $\text{sgn}(\phi_3) = -1$, $\text{sgn}(\phi_4) = +1$, $\text{sgn}(\phi_5) = +1$, $\text{sgn}(\phi_6) = -1$

Four.I.4.12 To get a nonzero term in the permutation expansion we must use the $1,2$ entry and the $4,3$ entry. Having fixed on those two we must also use the $2,1$ entry and the $3,4$ entry. The signum of $\langle 2,1,4,3\rangle$ is $+1$ because from

$$\begin{pmatrix} 0 & 1 & 0 & 0 \\ 1 & 0 & 0 & 0 \\ 0 & 0 & 0 & 1 \\ 0 & 0 & 1 & 0 \end{pmatrix}$$

the two row swaps $\rho_1 \leftrightarrow \rho_2$ and $\rho_3 \leftrightarrow \rho_4$ will produce the identity matrix.

Four.I.4.13 The pattern is this.

i	1	2	3	4	5	6	...
$\text{sgn}(\phi_i)$	$+1$	-1	-1	$+1$	$+1$	-1	...

So to find the signum of $\phi_{n!}$, we subtract one $n!-1$ and look at the remainder on division by four. If the remainder is 1 or 2 then the signum is -1, otherwise it is $+1$. For $n>4$, the number $n!$ is divisible by four, so $n!-1$ leaves a remainder of -1 on division by four (more properly said, a remainder or 3), and so the signum is $+1$. The $n=1$ case has a signum of $+1$, the $n=2$ case has a signum of -1 and the $n=3$ case has a signum of -1.

Four.I.4.14 **(a)** We can view permutations as one-one and onto maps $\phi\colon\{1,\ldots,n\}\to\{1,\ldots,n\}$. Any one-one and onto map has an inverse.

(b) If it always takes an odd number of swaps to get from P_ϕ to the identity, then it always takes an odd number of swaps to get from the identity to P_ϕ (any swap is reversible).

(c) This is the first question again.

Four.I.4.15 If $\phi(i)=j$ then $\phi^{-1}(j)=i$. The result now follows on the observation that P_ϕ has a 1 in entry i,j if and only if $\phi(i)=j$, and $P_{\phi^{-1}}$ has a 1 in entry j,i if and only if $\phi^{-1}(j)=i$,

Four.I.4.16 This does not say that m is the least number of swaps to produce an identity, nor does it say that m is the most. It instead says that there is a way to swap to the identity in exactly m steps.

Let ι_j be the first row that is inverted with respect to a prior row and let ι_k be the first row giving that inversion. We have this interval of rows.

$$\begin{pmatrix} \vdots \\ \iota_k \\ \iota_{r_1} \\ \vdots \\ \iota_{r_s} \\ \iota_j \\ \vdots \end{pmatrix} \qquad j<k<r_1<\cdots<r_s$$

Swap.

$$\begin{pmatrix} \vdots \\ \iota_j \\ \iota_{r_1} \\ \vdots \\ \iota_{r_s} \\ \iota_k \\ \vdots \end{pmatrix}$$

The second matrix has one fewer inversion because there is one fewer inversion in the interval (s vs. $s+1$) and inversions involving rows outside the interval are not affected.

Proceed in this way, at each step reducing the number of inversions by one with each row swap. When no inversions remain the result is the identity.

The contrast with Corollary 4.4 is that the statement of this exercise is a 'there exists' statement: there exists a way to swap to the identity in exactly m steps. But the corollary is a 'for all' statement: for all ways to swap to the identity, the parity (evenness or oddness) is the same.

Four.I.4.17 **(a)** First, $g(\phi_1)$ is the product of the single factor $2-1$ and so $g(\phi_1)=1$. Second, $g(\phi_2)$ is the product of the single factor $1-2$ and so $g(\phi_2)=-1$.

(b)

permutation ϕ	ϕ_1	ϕ_2	ϕ_3	ϕ_4	ϕ_5	ϕ_6
$g(\phi)$	2	-2	-2	2	2	-2

(c) It is a product of nonzero terms.

(d) Note that $\phi(j)-\phi(i)$ is negative if and only if $\iota_{\phi(j)}$ and $\iota_{\phi(i)}$ are in an inversion of their usual order.

Geometry of Determinants

Four.II.1: Determinants as Size Functions

Four.II.1.8 For each, find the determinant and take the absolute value.
(a) 7 (b) 0 (c) 58

Four.II.1.9 Solving

$$c_1\begin{pmatrix}3\\3\\1\end{pmatrix}+c_2\begin{pmatrix}2\\6\\1\end{pmatrix}+c_3\begin{pmatrix}1\\0\\5\end{pmatrix}=\begin{pmatrix}4\\1\\2\end{pmatrix}$$

gives the unique solution $c_3 = 11/57$, $c_2 = -40/57$ and $c_1 = 99/57$. Because $c_1 > 1$, the vector is not in the box.

Four.II.1.10 Move the parallelepiped to start at the origin, so that it becomes the box formed by

$$\langle\begin{pmatrix}3\\0\end{pmatrix},\begin{pmatrix}2\\1\end{pmatrix}\rangle$$

and now the absolute value of this determinant is easily computed as 3.

$$\begin{vmatrix}3&2\\0&1\end{vmatrix}=3$$

Four.II.1.11 (a) 3 (b) 9 (c) 1/9

Four.II.1.12 Express each transformation with respect to the standard bases and find the determinant.

(a) 6 (b) −1 (c) −5

Four.II.1.13 The starting area is 6 and the matrix changes sizes by −14. Thus the area of the image is 84.

Four.II.1.14 By a factor of 21/2.

Four.II.1.15 For a box we take a sequence of vectors (as described in the remark, the order of the vectors matters), while for a span we take a set of vectors. Also, for a box subset of $\mathbb{R}^n$ there must be n vectors; of course for a span there can be any number of vectors. Finally, for a box the coefficients $t_1, \ldots, t_n$ are in the interval $[0..1]$, while for a span the coefficients are free to range over all of $\mathbb{R}$.

Four.II.1.16 We have drawn that picture to mislead. The picture on the left is not the box formed by two vectors. If we slide it to the origin then it becomes the box formed by this sequence.

$$\langle\begin{pmatrix}0\\1\end{pmatrix},\begin{pmatrix}2\\0\end{pmatrix}\rangle$$

Then the image under the action of the matrix is the box formed by this sequence.

$$\langle\begin{pmatrix}1\\1\end{pmatrix},\begin{pmatrix}4\\0\end{pmatrix}\rangle$$

which has an area of 4.

Four.II.1.17 Yes to both. For instance, the first is $|TS| = |T|\cdot|S| = |S|\cdot|T| = |ST|$.

Four.II.1.18 (a) If it is defined then it is $(3^2)\cdot(2)\cdot(2^{-2})\cdot(3)$.
(b) $|6A^3+5A^2+2A| = |A|\cdot|6A^2+5A+2I|$.

Four.II.1.19 $\begin{vmatrix}\cos\theta&-\sin\theta\\\sin\theta&\cos\theta\end{vmatrix}=1$

Four.II.1.20 No, for instance the determinant of

$$T=\begin{pmatrix}2&0\\0&1/2\end{pmatrix}$$

is 1 so it preserves areas, but the vector $T\vec{e}_1$ has length 2.

Four.II.1.21 It is zero.

Four.II.1.22 Two of the three sides of the triangle are formed by these vectors.

$$\begin{pmatrix}2\\2\\2\end{pmatrix}-\begin{pmatrix}1\\2\\1\end{pmatrix}=\begin{pmatrix}1\\0\\1\end{pmatrix}\qquad\begin{pmatrix}3\\-1\\4\end{pmatrix}-\begin{pmatrix}1\\2\\1\end{pmatrix}=\begin{pmatrix}2\\-3\\3\end{pmatrix}$$

One way to find the area of this triangle is to produce a length-one vector orthogonal to these two. From these two relations

$$\begin{pmatrix}1\\0\\1\end{pmatrix}\cdot\begin{pmatrix}x\\y\\z\end{pmatrix}=\begin{pmatrix}0\\0\\0\end{pmatrix}\qquad\begin{pmatrix}2\\-3\\3\end{pmatrix}\cdot\begin{pmatrix}x\\y\\z\end{pmatrix}=\begin{pmatrix}0\\0\\0\end{pmatrix}$$

we get a system

$$\begin{array}{rl}x \qquad + z=0\\ 2x-3y+3z=0\end{array}\xrightarrow{-2\rho_1+\rho_2}\begin{array}{rl}x \qquad + z=0\\ -3y+z=0\end{array}$$

with this solution set.

$$\{\begin{pmatrix}-1\\1/3\\1\end{pmatrix}z \mid z\in\mathbb{R}\},$$

A solution of length one is this.

$$\frac{1}{\sqrt{19/9}}\begin{pmatrix}-1\\1/3\\1\end{pmatrix}$$

Thus the area of the triangle is the absolute value of this determinant.

$$\begin{vmatrix}1&2&-3/\sqrt{19}\\0&-3&1/\sqrt{19}\\1&3&3/\sqrt{19}\end{vmatrix}=-12/\sqrt{19}$$

Four.II.1.23 **(a)** Because the image of a linearly dependent set is linearly dependent, if the vectors forming S make a linearly dependent set, so that $|S|=0$, then the vectors forming $t(S)$ make a linearly dependent set, so that $|TS|=0$, and in this case the equation holds.

(b) We must check that if $T\xrightarrow{k\rho_i+\rho_j}\hat{T}$ then $d(T)=|TS|/|S|=|\hat{T}S|/|S|=d(\hat{T})$. We can do this by checking that combining rows first and then multiplying to get $\hat{T}S$ gives the same result as multiplying first to get TS and then combining (because the determinant $|TS|$ is unaffected by the combining rows so we'll then have that $|\hat{T}S|=|TS|$ and hence that $d(\hat{T})=d(T)$). This check runs: after adding k times row i of TS to row j of TS, the j,p entry is $(kt_{i,1}+t_{j,1})s_{1,p}+\cdots+(kt_{i,r}+t_{j,r})s_{r,p}$, which is the j,p entry of $\hat{T}S$.

(c) For the second property, we need only check that swapping $T\xrightarrow{\rho_i\leftrightarrow\rho_j}\hat{T}$ and then multiplying to get $\hat{T}S$ gives the same result as multiplying T by S first and then swapping (because, as the determinant $|TS|$ changes sign on the row swap, we'll then have $|\hat{T}S|=-|TS|$, and so $d(\hat{T})=-d(T)$). This check runs just like the one for the first property.

For the third property, we need only show that performing $T\xrightarrow{k\rho_i}\hat{T}$ and then computing $\hat{T}S$ gives the same result as first computing TS and then performing the scalar multiplication (as the determinant $|TS|$ is rescaled by k, we'll have $|\hat{T}S|=k|TS|$ and so $d(\hat{T})=k\,d(T)$). Here too, the argument runs just as above.

The fourth property, that if T is I then the result is 1, is obvious.

(d) Determinant functions are unique, so $|TS|/|S|=d(T)=|T|$, and so $|TS|=|T||S|$.

Four.II.1.24 Any permutation matrix has the property that the transpose of the matrix is its inverse.

For the implication, we know that $|A^{\mathsf{T}}|=|A|$. Then $1=|A\cdot A^{-1}|=|A\cdot A^{\mathsf{T}}|=|A|\cdot|A^{\mathsf{T}}|=|A|^2$.

The converse does not hold; here is an example.

$$\begin{pmatrix} 3 & 1 \\ 2 & 1 \end{pmatrix}$$

Four.II.1.25 Where the sides of the box are c times longer, the box has c^3 times as many cubic units of volume.

Four.II.1.26 If $H = P^{-1}GP$ then $|H| = |P^{-1}||G||P| = |P^{-1}||P||G| = |P^{-1}P||G| = |G|$.

Four.II.1.27 (a) The new basis is the old basis rotated by $\pi/4$.

(b) $\langle\begin{pmatrix} -1 \\ 0 \end{pmatrix},\begin{pmatrix} 0 \\ -1 \end{pmatrix}\rangle$, $\langle\begin{pmatrix} 0 \\ -1 \end{pmatrix},\begin{pmatrix} 1 \\ 0 \end{pmatrix}\rangle$

(c) In each case the determinant is $+1$ (we say that these bases have *positive orientation*).

(d) Because only one sign can change at a time, the only other cycle possible is

$$\cdots \longrightarrow \begin{pmatrix} + \\ + \end{pmatrix} \longrightarrow \begin{pmatrix} + \\ - \end{pmatrix} \longrightarrow \begin{pmatrix} - \\ - \end{pmatrix} \longrightarrow \begin{pmatrix} - \\ + \end{pmatrix} \longrightarrow \cdots.$$

Here each associated determinant is -1 (we say that such bases have a *negative orientation*).

(e) There is one positively oriented basis $\langle(1)\rangle$ and one negatively oriented basis $\langle(-1)\rangle$.

(f) There are 48 bases (6 half-axis choices are possible for the first unit vector, 4 for the second, and 2 for the last). Half are positively oriented like the standard basis on the left below, and half are negatively oriented like the one on the right

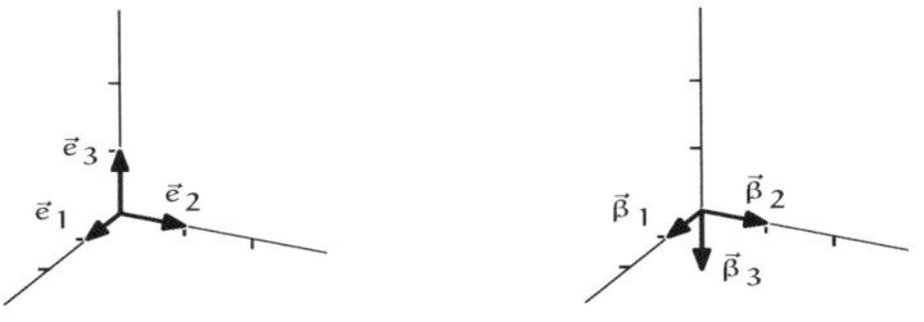

In $\mathbb{R}^3$ positive orientation is sometimes called 'right hand orientation' because if a person places their right hand with their fingers curling from $\vec{e}_1$ to $\vec{e}_2$ then the thumb will point with $\vec{e}_3$.

Four.II.1.28 We will compare $\det(\vec{s}_1,\ldots,\vec{s}_n)$ with $\det(t(\vec{s}_1),\ldots,t(\vec{s}_n))$ to show that the second differs from the first by a factor of $|T|$. We represent the $\vec{s}$'s with respect to the standard bases

$$\text{Rep}_{\mathcal{E}_n}(\vec{s}_i) = \begin{pmatrix} s_{1,i} \\ s_{2,i} \\ \vdots \\ s_{n,i} \end{pmatrix}$$

and then we represent the map application with matrix-vector multiplication

$$\begin{aligned}
\text{Rep}_{\mathcal{E}_n}(\,t(\vec{s}_i)\,) &= \begin{pmatrix} t_{1,1} & t_{1,2} & \ldots & t_{1,n} \\ t_{2,1} & t_{2,2} & \ldots & t_{2,n} \\ & \vdots & & \\ t_{n,1} & t_{n,2} & \ldots & t_{n,n} \end{pmatrix}\begin{pmatrix} s_{1,j} \\ s_{2,j} \\ \vdots \\ s_{n,j} \end{pmatrix} \\
&= s_{1,j}\begin{pmatrix} t_{1,1} \\ t_{2,1} \\ \vdots \\ t_{n,1} \end{pmatrix} + s_{2,j}\begin{pmatrix} t_{1,2} \\ t_{2,2} \\ \vdots \\ t_{n,2} \end{pmatrix} + \cdots + s_{n,j}\begin{pmatrix} t_{1,n} \\ t_{2,n} \\ \vdots \\ t_{n,n} \end{pmatrix} \\
&= s_{1,j}\vec{t}_1 + s_{2,j}\vec{t}_2 + \cdots + s_{n,j}\vec{t}_n
\end{aligned}$$

where $\vec{t}_i$ is column i of T. Then $\det(t(\vec{s}_1),\,\ldots,\,t(\vec{s}_n))$ equals $\det(s_{1,1}\vec{t}_1+s_{2,1}\vec{t}_2+\ldots+s_{n,1}\vec{t}_n,\,\ldots,\,s_{1,n}\vec{t}_1+s_{2,n}\vec{t}_2+\ldots+s_{n,n}\vec{t}_n)$.

As in the derivation of the permutation expansion formula, we apply multilinearity, first splitting along the sum in the first argument

$$\det(s_{1,1}\vec{t}_1,\,\ldots,\,s_{1,n}\vec{t}_1 + s_{2,n}\vec{t}_2 + \cdots + s_{n,n}\vec{t}_n) + \cdots + \det(s_{n,1}\vec{t}_n,\,\ldots,\,s_{1,n}\vec{t}_1 + s_{2,n}\vec{t}_2 + \cdots + s_{n,n}\vec{t}_n)$$

and then splitting each of those n summands along the sums in the second arguments, etc. We end with, as in the derivation of the permutation expansion, n^n summand determinants, each of the form $\det(s_{i_1,1}\vec{t}_{i_1}, s_{i_2,2}\vec{t}_{i_2}, \ldots, s_{i_n,n}\vec{t}_{i_n})$. Factor out each of the $s_{i,j}$'s $= s_{i_1,1}s_{i_2,2}\ldots s_{i_n,n}\cdot\det(\vec{t}_{i_1}, \vec{t}_{i_2}, \ldots, \vec{t}_{i_n})$.

As in the permutation expansion derivation, whenever two of the indices in $i_1, \ldots, i_n$ are equal then the determinant has two equal arguments, and evaluates to 0. So we need only consider the cases where $i_1, \ldots, i_n$ form a permutation of the numbers $1, \ldots, n$. We thus have

$$\det(t(\vec{s}_1), \ldots, t(\vec{s}_n)) = \sum_{\text{permutations } \phi} s_{\phi(1),1}\ldots s_{\phi(n),n}\det(\vec{t}_{\phi(1)}, \ldots, \vec{t}_{\phi(n)}).$$

Swap the columns in $\det(\vec{t}_{\phi(1)}, \ldots, \vec{t}_{\phi(n)})$ to get the matrix T back, which changes the sign by a factor of $\operatorname{sgn}\phi$, and then factor out the determinant of T.

$$= \sum_{\phi} s_{\phi(1),1}\ldots s_{\phi(n),n}\det(\vec{t}_1, \ldots, \vec{t}_n)\cdot\operatorname{sgn}\phi = \det(T)\sum_{\phi} s_{\phi(1),1}\ldots s_{\phi(n),n}\cdot\operatorname{sgn}\phi.$$

As in the proof that the determinant of a matrix equals the determinant of its transpose, we commute the s's to list them by ascending row number instead of by ascending column number (and we substitute $\operatorname{sgn}(\phi^{-1})$ for $\operatorname{sgn}(\phi)$).

$$= \det(T)\sum_{\phi} s_{1,\phi^{-1}(1)}\ldots s_{n,\phi^{-1}(n)}\cdot\operatorname{sgn}\phi^{-1} = \det(T)\det(\vec{s}_1, \vec{s}_2, \ldots, \vec{s}_n)$$

Four.II.1.29 **(a)** An algebraic check is easy.

$$0 = xy_2 + x_2y_3 + x_3y - x_3y_2 - xy_3 - x_2y = x\cdot(y_2 - y_3) + y\cdot(x_3 - x_2) + x_2y_3 - x_3y_2$$

simplifies to the familiar form

$$y = x\cdot(x_3 - x_2)/(y_3 - y_2) + (x_2y_3 - x_3y_2)/(y_3 - y_2)$$

(the $y_3 - y_2 = 0$ case is easily handled).

For geometric insight, this picture shows that the box formed by the three vectors. Note that all three vectors end in the $z = 1$ plane. Below the two vectors on the right is the line through (x_2, y_2) and (x_3, y_3).

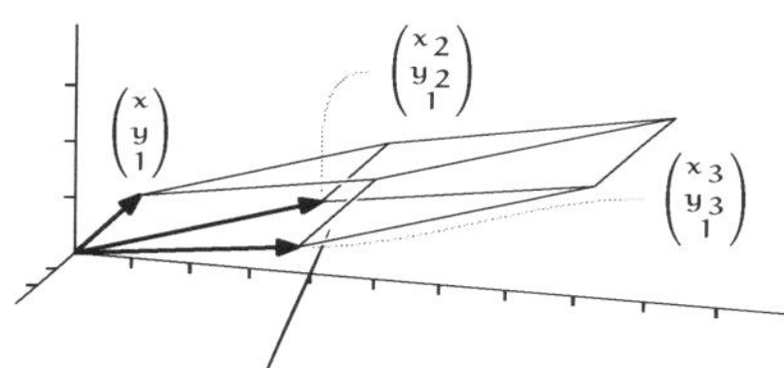

The box will have a nonzero volume unless the triangle formed by the ends of the three is degenerate. That only happens (assuming that $(x_2, y_3) \neq (x_3, y_3)$) if (x, y) lies on the line through the other two.

(b) *This is how the answer was given in the cited source.* We find the altitude through (x_1, y_1) of a triangle with vertices (x_1, y_1) (x_2, y_2) and (x_3, y_3) in the usual way from the normal form of the above:

$$\frac{1}{\sqrt{(x_2 - x_3)^2 + (y_2 - y_3)^2}}\begin{vmatrix} x_1 & x_2 & x_3 \\ y_1 & y_2 & y_3 \\ 1 & 1 & 1 \end{vmatrix}.$$

Another step shows the area of the triangle to be

$$\frac{1}{2}\begin{vmatrix} x_1 & x_2 & x_3 \\ y_1 & y_2 & y_3 \\ 1 & 1 & 1 \end{vmatrix}.$$

This exposition reveals the *modus operandi* more clearly than the usual proof of showing a collection of terms to be identical with the determinant.

(c) *This is how the answer was given in the cited source.* Let

$$D = \begin{vmatrix} x_1 & x_2 & x_3 \\ y_1 & y_2 & y_3 \\ 1 & 1 & 1 \end{vmatrix}$$

then the area of the triangle is $(1/2)|D|$. Now if the coordinates are all integers, then D is an integer.

Laplace's Expansion

Four.III.1: Laplace's Expansion Formula

Four.III.1.13 (a) $(-1)^{2+3}\begin{vmatrix}1&0\\0&2\end{vmatrix}=-2$ (b) $(-1)^{3+2}\begin{vmatrix}1&2\\-1&3\end{vmatrix}=-5$ (c) $(-1)^{4}\begin{vmatrix}-1&1\\0&2\end{vmatrix}=-2$

Four.III.1.14 (a) $3\cdot(+1)\begin{vmatrix}2&2\\3&0\end{vmatrix}+0\cdot(-1)\begin{vmatrix}1&2\\-1&0\end{vmatrix}+1\cdot(+1)\begin{vmatrix}1&2\\-1&3\end{vmatrix}=-13$

(b) $1\cdot(-1)\begin{vmatrix}0&1\\3&0\end{vmatrix}+2\cdot(+1)\begin{vmatrix}3&1\\-1&0\end{vmatrix}+2\cdot(-1)\begin{vmatrix}3&0\\-1&3\end{vmatrix}=-13$

(c) $1\cdot(+1)\begin{vmatrix}1&2\\-1&3\end{vmatrix}+2\cdot(-1)\begin{vmatrix}3&0\\-1&3\end{vmatrix}+0\cdot(+1)\begin{vmatrix}3&0\\1&2\end{vmatrix}=-13$

Four.III.1.15

$$\text{adj}(T)=\begin{pmatrix}T_{1,1}&T_{2,1}&T_{3,1}\\T_{1,2}&T_{2,2}&T_{3,2}\\T_{1,3}&T_{2,3}&T_{3,3}\end{pmatrix}=\begin{pmatrix}+\begin{vmatrix}5&6\\8&9\end{vmatrix}&-\begin{vmatrix}2&3\\8&9\end{vmatrix}&+\begin{vmatrix}2&3\\5&6\end{vmatrix}\\-\begin{vmatrix}4&6\\7&9\end{vmatrix}&+\begin{vmatrix}1&3\\7&9\end{vmatrix}&-\begin{vmatrix}1&3\\4&6\end{vmatrix}\\+\begin{vmatrix}4&5\\7&8\end{vmatrix}&-\begin{vmatrix}1&2\\7&8\end{vmatrix}&+\begin{vmatrix}1&2\\4&5\end{vmatrix}\end{pmatrix}=\begin{pmatrix}-3&6&-3\\6&-12&6\\-3&6&-3\end{pmatrix}$$

Four.III.1.16 (a)

$$\begin{pmatrix}T_{1,1}&T_{2,1}&T_{3,1}\\T_{1,2}&T_{2,2}&T_{3,2}\\T_{1,3}&T_{2,3}&T_{3,3}\end{pmatrix}=\begin{pmatrix}\begin{vmatrix}0&2\\0&1\end{vmatrix}&-\begin{vmatrix}1&4\\0&1\end{vmatrix}&\begin{vmatrix}1&4\\0&2\end{vmatrix}\\-\begin{vmatrix}-1&2\\1&1\end{vmatrix}&\begin{vmatrix}2&4\\1&1\end{vmatrix}&-\begin{vmatrix}2&4\\-1&2\end{vmatrix}\\\begin{vmatrix}-1&0\\1&0\end{vmatrix}&-\begin{vmatrix}2&1\\1&0\end{vmatrix}&\begin{vmatrix}2&1\\-1&0\end{vmatrix}\end{pmatrix}=\begin{pmatrix}0&-1&2\\3&-2&-8\\0&1&1\end{pmatrix}$$

(b) The minors are 1×1: $\begin{pmatrix}T_{1,1}&T_{2,1}\\T_{1,2}&T_{2,2}\end{pmatrix}=\begin{pmatrix}\begin{vmatrix}4\end{vmatrix}&-\begin{vmatrix}-1\end{vmatrix}\\-\begin{vmatrix}2\end{vmatrix}&\begin{vmatrix}3\end{vmatrix}\end{pmatrix}=\begin{pmatrix}4&1\\-2&3\end{pmatrix}$.

(c) $\begin{pmatrix}0&-1\\-5&1\end{pmatrix}$

(d)

$$\begin{pmatrix}T_{1,1}&T_{2,1}&T_{3,1}\\T_{1,2}&T_{2,2}&T_{3,2}\\T_{1,3}&T_{2,3}&T_{3,3}\end{pmatrix}=\begin{pmatrix}\begin{vmatrix}0&3\\8&9\end{vmatrix}&-\begin{vmatrix}4&3\\8&9\end{vmatrix}&\begin{vmatrix}4&3\\0&3\end{vmatrix}\\-\begin{vmatrix}-1&3\\1&9\end{vmatrix}&\begin{vmatrix}1&3\\1&9\end{vmatrix}&-\begin{vmatrix}1&3\\-1&3\end{vmatrix}\\\begin{vmatrix}-1&0\\1&8\end{vmatrix}&-\begin{vmatrix}1&4\\1&8\end{vmatrix}&\begin{vmatrix}1&4\\-1&0\end{vmatrix}\end{pmatrix}=\begin{pmatrix}-24&-12&12\\12&6&-6\\-8&-4&4\end{pmatrix}$$

Four.III.1.17 (a) $(1/3)\cdot\begin{pmatrix}0&-1&2\\3&-2&-8\\0&1&1\end{pmatrix}=\begin{pmatrix}0&-1/3&2/3\\1&-2/3&-8/3\\0&1/3&1/3\end{pmatrix}$

(b) $(1/14)\cdot\begin{pmatrix}4&1\\-2&3\end{pmatrix}=\begin{pmatrix}2/7&1/14\\-1/7&3/14\end{pmatrix}$

(c) $(1/-5)\cdot\begin{pmatrix}0&-1\\-5&1\end{pmatrix}=\begin{pmatrix}0&1/5\\1&-1/5\end{pmatrix}$

(d) The matrix has a zero determinant, and so has no inverse.

Four.III.1.18
$$\begin{pmatrix} T_{1,1} & T_{2,1} & T_{3,1} & T_{4,1} \\ T_{1,2} & T_{2,2} & T_{3,2} & T_{4,2} \\ T_{1,3} & T_{2,3} & T_{3,3} & T_{4,3} \\ T_{1,4} & T_{2,4} & T_{3,4} & T_{4,4} \end{pmatrix} = \begin{pmatrix} 4 & -3 & 2 & -1 \\ -3 & 6 & -4 & 2 \\ 2 & -4 & 6 & -3 \\ -1 & 2 & -3 & 4 \end{pmatrix}$$

Four.III.1.19 The determinant
$$\begin{vmatrix} a & b \\ c & d \end{vmatrix}$$
expanded on the first row gives $a\cdot(+1)|d| + b\cdot(-1)|c| = ad - bc$ (note the two 1×1 minors).

Four.III.1.20 The determinant of
$$\begin{pmatrix} a & b & c \\ d & e & f \\ g & h & i \end{pmatrix}$$
is this.
$$a\cdot\begin{vmatrix} e & f \\ h & i \end{vmatrix} - b\cdot\begin{vmatrix} d & f \\ g & i \end{vmatrix} + c\cdot\begin{vmatrix} d & e \\ g & h \end{vmatrix} = a(ei - fh) - b(di - fg) + c(dh - eg)$$

Four.III.1.21 **(a)**
$$\begin{pmatrix} T_{1,1} & T_{2,1} \\ T_{1,2} & T_{2,2} \end{pmatrix} = \begin{pmatrix} |t_{2,2}| & -|t_{1,2}| \\ -|t_{2,1}| & |t_{1,1}| \end{pmatrix} = \begin{pmatrix} t_{2,2} & -t_{1,2} \\ -t_{2,1} & t_{1,1} \end{pmatrix}$$

(b) $(1/t_{1,1}t_{2,2} - t_{1,2}t_{2,1})\cdot\begin{pmatrix} t_{2,2} & -t_{1,2} \\ -t_{2,1} & t_{1,1} \end{pmatrix}$

Four.III.1.22 No. Here is a determinant whose value
$$\begin{vmatrix} 1 & 0 & 0 \\ 0 & 1 & 0 \\ 0 & 0 & 1 \end{vmatrix} = 1$$
doesn't equal the result of expanding down the diagonal.
$$1\cdot(+1)\begin{vmatrix} 1 & 0 \\ 0 & 1 \end{vmatrix} + 1\cdot(+1)\begin{vmatrix} 1 & 0 \\ 0 & 1 \end{vmatrix} + 1\cdot(+1)\begin{vmatrix} 1 & 0 \\ 0 & 1 \end{vmatrix} = 3$$

Four.III.1.23 Consider this diagonal matrix.
$$D = \begin{pmatrix} d_1 & 0 & 0 & \ldots & \\ 0 & d_2 & 0 & & \\ 0 & 0 & d_3 & & \\ & & & \ddots & \\ & & & & d_n \end{pmatrix}$$
If $i \neq j$ then the i,j minor is an $(n-1)\times(n-1)$ matrix with only $n-2$ nonzero entries, because we have deleted both d_i and d_j. Thus, at least one row or column of the minor is all zeroes, and so the cofactor $D_{i,j}$ is zero. If $i = j$ then the minor is the diagonal matrix with entries $d_1, \ldots, d_{i-1}, d_{i+1}, \ldots, d_n$. Its determinant is obviously $(-1)^{i+j} = (-1)^{2i} = 1$ times the product of those.
$$\mathrm{adj}(D) = \begin{pmatrix} d_2\cdots d_n & 0 & & 0 \\ 0 & d_1 d_3\cdots d_n & & 0 \\ & & \ddots & \\ & & & d_1\cdots d_{n-1} \end{pmatrix}$$

By the way, Theorem 1.9 provides a slicker way to derive this conclusion.

Four.III.1.24 Just note that if $S = T^{\mathsf{T}}$ then the cofactor $S_{j,i}$ equals the cofactor $T_{i,j}$ because $(-1)^{j+i} = (-1)^{i+j}$ and because the minors are the transposes of each other (and the determinant of a transpose equals the determinant of the matrix).

Four.III.1.25 It is false; here is an example.

$$T=\begin{pmatrix}1&2&3\\4&5&6\\7&8&9\end{pmatrix}\qquad \operatorname{adj}(T)=\begin{pmatrix}-3&6&-3\\6&-12&6\\-3&6&-3\end{pmatrix}\qquad \operatorname{adj}(\operatorname{adj}(T))=\begin{pmatrix}0&0&0\\0&0&0\\0&0&0\end{pmatrix}$$

Four.III.1.26 **(a)** An example

$$M=\begin{pmatrix}1&2&3\\0&4&5\\0&0&6\end{pmatrix}$$

suggests the right answer.

$$\operatorname{adj}(M)=\begin{pmatrix}M_{1,1}&M_{2,1}&M_{3,1}\\M_{1,2}&M_{2,2}&M_{3,2}\\M_{1,3}&M_{2,3}&M_{3,3}\end{pmatrix}=\begin{pmatrix}\begin{vmatrix}4&5\\0&6\end{vmatrix}&-\begin{vmatrix}2&3\\0&6\end{vmatrix}&\begin{vmatrix}2&3\\4&5\end{vmatrix}\\-\begin{vmatrix}0&5\\0&6\end{vmatrix}&\begin{vmatrix}1&3\\0&6\end{vmatrix}&-\begin{vmatrix}1&3\\0&5\end{vmatrix}\\\begin{vmatrix}0&4\\0&0\end{vmatrix}&-\begin{vmatrix}1&2\\0&0\end{vmatrix}&\begin{vmatrix}1&2\\0&4\end{vmatrix}\end{pmatrix}=\begin{pmatrix}24&-12&-2\\0&6&-5\\0&0&4\end{pmatrix}$$

The result is indeed upper triangular.

This check is detailed but not hard. The entries in the upper triangle of the adjoint are $M_{a,b}$ where $a>b$. We need to verify that the cofactor $M_{a,b}$ is zero if $a>b$. With $a>b$, row a and column b of M,

$$\begin{pmatrix}m_{1,1}&\ldots&m_{1,b}&&\\m_{2,1}&\ldots&m_{2,b}&&\\\vdots&&\vdots&&\\m_{a,1}&\ldots&m_{a,b}&\ldots&m_{a,n}\\&&\vdots&&\\&&m_{n,b}&&\end{pmatrix}$$

when deleted, leave an upper triangular minor, because entry i,j of the minor is either entry i,j of M (this happens if $a>i$ and $b>j$; in this case $i<j$ implies that the entry is zero) or it is entry $i,j+1$ of M (this happens if $i<a$ and $j>b$; in this case, $i<j$ implies that $i<j+1$, which implies that the entry is zero), or it is entry $i+1,j+1$ of M (this last case happens when $i>a$ and $j>b$; obviously here $i<j$ implies that $i+1<j+1$ and so the entry is zero). Thus the determinant of the minor is the product down the diagonal. Observe that the $a-1,a$ entry of M is the $a-1,a-1$ entry of the minor (it doesn't get deleted because the relation $a>b$ is strict). But this entry is zero because M is upper triangular and $a-1<a$. Therefore the cofactor is zero, and the adjoint is upper triangular. (The lower triangular case is similar.)

(b) This is immediate from the prior part, by Corollary 1.11.

Four.III.1.27 We will show that each determinant can be expanded along row i. The argument for column j is similar.

Each term in the permutation expansion contains one and only one entry from each row. As in Example 1.1, factor out each row i entry to get $|T|=t_{i,1}\cdot\hat{T}_{i,1}+\cdots+t_{i,n}\cdot\hat{T}_{i,n}$, where each $\hat{T}_{i,j}$ is a sum of terms not containing any elements of row i. We will show that $\hat{T}_{i,j}$ is the i,j cofactor.

Consider the $i,j=n,n$ case first:

$$t_{n,n}\cdot\hat{T}_{n,n}=t_{n,n}\cdot\sum_{\phi}t_{1,\phi(1)}t_{2,\phi(2)}\ldots t_{n-1,\phi(n-1)}\operatorname{sgn}(\phi)$$

where the sum is over all n-permutations ϕ such that $\phi(n)=n$. To show that $\hat{T}_{i,j}$ is the minor $T_{i,j}$, we need only show that if ϕ is an n-permutation such that $\phi(n)=n$ and σ is an $n-1$-permutation with

$\sigma(1)=\phi(1)$, ..., $\sigma(n-1)=\phi(n-1)$ then $\operatorname{sgn}(\sigma)=\operatorname{sgn}(\phi)$. But that's true because ϕ and σ have the same number of inversions.

Back to the general i,j case. Swap adjacent rows until the i-th is last and swap adjacent columns until the j-th is last. Observe that the determinant of the i,j-th minor is not affected by these adjacent swaps because inversions are preserved (since the minor has the i-th row and j-th column omitted). On the other hand, the sign of $|T|$ and $\hat{T}_{i,j}$ changes $n-i$ plus $n-j$ times. Thus $\hat{T}_{i,j}=(-1)^{n-i+n-j}|T_{i,j}|=(-1)^{i+j}|T_{i,j}|$.

Four.III.1.28 This is obvious for the 1×1 base case.

For the inductive case, assume that the determinant of a matrix equals the determinant of its transpose for all 1×1, ..., $(n-1)\times(n-1)$ matrices. Expanding on row i gives $|T|=t_{i,1}T_{i,1}+\dots+t_{i,n}T_{i,n}$ and expanding on column i gives $|T^{\mathsf{T}}|=t_{1,i}(T^{\mathsf{T}})_{1,i}+\dots+t_{n,i}(T^{\mathsf{T}})_{n,i}$ Since $(-1)^{i+j}=(-1)^{j+i}$ the signs are the same in the two summations. Since the j,i minor of T^{T} is the transpose of the i,j minor of T, the inductive hypothesis gives $|(T^{\mathsf{T}})_{i,j}|=|T_{i,j}|$.

Four.III.1.29 *This is how the answer was given in the cited source.* Denoting the above determinant by D_n, it is seen that $D_2=1$, $D_3=2$. It remains to show that $D_n=D_{n-1}+D_{n-2}$, $n\geqslant 4$. In D_n subtract the $(n-3)$-th column from the $(n-1)$-th, the $(n-4)$-th from the $(n-2)$-th, ..., the first from the third, obtaining

$$F_n=\begin{vmatrix}1&-1&0&0&0&0&\dots\\1&1&-1&0&0&0&\dots\\0&1&1&-1&0&0&\dots\\0&0&1&1&-1&0&\dots\\.&.&.&.&.&.&\dots\end{vmatrix}.$$

By expanding this determinant with reference to the first row, there results the desired relation.

Topic: Cramer's Rule

1 (a)

$$x=\frac{\begin{vmatrix}4&-1\\-7&2\end{vmatrix}}{\begin{vmatrix}1&-1\\-1&2\end{vmatrix}}=\frac{1}{1}=1\qquad y=\frac{\begin{vmatrix}1&4\\-1&-7\end{vmatrix}}{\begin{vmatrix}1&-1\\-1&2\end{vmatrix}}=\frac{-3}{1}=-3$$

(b) $x=2$, $y=2$

2 $z=1$

3 Determinants are unchanged by combinations, including column combinations, so $\det(B_i)=\det(\vec{a}_1,\dots,x_1\vec{a}_1+\dots+x_i\vec{a}_i+\dots+x_n\vec{a}_n,\dots,\vec{a}_n)$ is equal to $\det(\vec{a}_1,\dots,x_i\vec{a}_i,\dots,\vec{a}_n)$ (use the operation of taking $-x_1$ times the first column and adding it to the i-th column, etc.). That is equal to $x_i\cdot\det(\vec{a}_1,\dots,\vec{a}_i,\dots,\vec{a}_n)=x_i\cdot\det(A)$, as required.

4 (a) Here is the case of a 2×2 system with $i=2$.

$$\begin{aligned}a_{1,1}x_1+a_{1,2}x_2&=b_1\\a_{2,1}x_1+a_{2,2}x_2&=b_2\end{aligned}\quad\Longleftrightarrow\quad\begin{pmatrix}a_{1,1}&a_{1,2}\\a_{2,1}&a_{2,2}\end{pmatrix}\begin{pmatrix}1&x_1\\0&x_2\end{pmatrix}=\begin{pmatrix}a_{1,1}&b_1\\a_{2,1}&b_2\end{pmatrix}$$

(b) The determinant function is multiplicative $\det(B_i)=\det(AX_i)=\det(A)\cdot\det(X_i)$. The Laplace expansion shows that $\det(X_i)=x_i$, and solving for x_i gives Cramer's Rule.

5 Because the determinant of A is nonzero, Cramer's Rule applies and shows that $x_i=|B_i|/1$. Since B_i is a matrix of integers, its determinant is an integer.

6 The solution of

$$\begin{aligned} ax + by &= e \\ cx + dy &= f \end{aligned}$$

is

$$x = \frac{ed - fb}{ad - bc} \qquad y = \frac{af - ec}{ad - bc}$$

provided of course that the denominators are not zero.

7 Of course, singular systems have $|A|$ equal to zero, but we can characterize the infinitely many solutions case is by the fact that all of the $|B_i|$ are zero as well.

8 We can consider the two nonsingular cases together with this system

$$\begin{aligned} x_1 + 2x_2 &= 6 \\ x_1 + 2x_2 &= c \end{aligned}$$

where $c = 6$ of course yields infinitely many solutions, and any other value for c yields no solutions. The corresponding vector equation

$$x_1 \cdot \begin{pmatrix} 1 \\ 1 \end{pmatrix} + x_2 \cdot \begin{pmatrix} 2 \\ 2 \end{pmatrix} = \begin{pmatrix} 6 \\ c \end{pmatrix}$$

gives a picture of two overlapping vectors. Both lie on the line $y = x$. In the $c = 6$ case the vector on the right side also lies on the line $y = x$ but in any other case it does not.

Topic: Speed of Calculating Determinants

1 **(a)** Under Octave, rank(rand(5)) finds the rank of a 5×5 matrix whose entries are (uniformly distributed) in the interval $[0..1)$. This loop which runs the test 5000 times

```
octave:1> for i=1:5000
> if rank(rand(5))<5 printf("That's one."); endif
> endfor
```

produces (after a few seconds) returns the prompt, with no output.

The Octave script

```
function elapsed_time = detspeed (size)
  a=rand(size);
  tic();
  for i=1:10
     det(a);
  endfor
  elapsed_time=toc();
endfunction
```

lead to this session (obviously, your times will vary).

```
octave:1> detspeed(5)
ans = 0.019505
octave:2> detspeed(15)
ans = 0.0054691
octave:3> detspeed(25)
ans = 0.0097431
octave:4> detspeed(35)
ans = 0.017398
```

(b) Here is the data (rounded a bit), and the graph.

matrix rows	15	25	35	45	55	65	75	85	95
time per ten	0.0034	0.0098	0.0675	0.0285	0.0443	0.0663	0.1428	0.2282	0.1686

(This data is from an average of twenty runs of the above script, because of the possibility that the randomly chosen matrix happens to take an unusually long or short time. Even so, the timing cannot be relied on too heavily; this is just an experiment.)

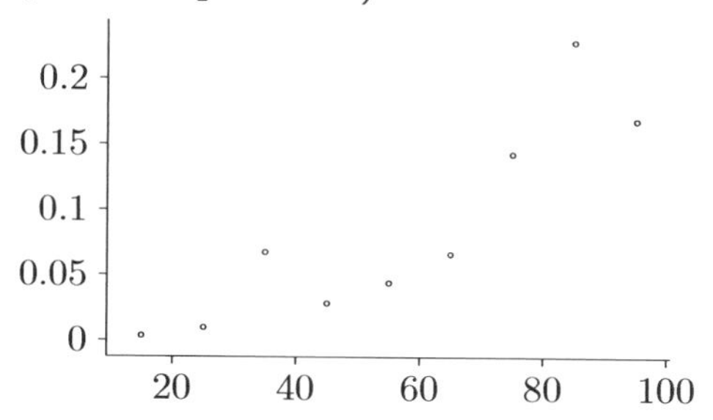

2 The number of operations depends on exactly how we do the operations.

(a) The determinant is -11. To row reduce takes a single row combination with two multiplications ($-5/2$ times 2 plus 5, and $-5/2$ times 1 plus -3) and the product down the diagonal takes one more multiplication. The permutation expansion takes two multiplications (2 times -3 and 5 times 1).

(b) The determinant is -39. Counting the operations is routine.

(c) The determinant is 4.

3 One way to get started is to compare these under Octave: `det(rand(10));`, versus `det(hilb(10));`, versus `det(eye(10));`, versus `det(zeroes(10));`. You can time them as in `tic(); det(rand(10)); toc()`.

4 Yes, because the J is in the innermost loop.

Topic: Chiò's Method

1 **(a)** Chiò's matrix is

$$C = \begin{pmatrix} -3 & -6 \\ -6 & -12 \end{pmatrix}$$

and its determinant is 0 **(b)** Start with

$$C_3 = \begin{pmatrix} 2 & 8 & 0 \\ 1 & -2 & 2 \\ 4 & 2 & 2 \end{pmatrix}$$

and then the next step

$$C_2 = \begin{pmatrix} -12 & 4 \\ -28 & 4 \end{pmatrix}$$

with determinant $\det(C_2) = 64$. The determinant of the original matrix is thus $64/(2^2 \cdot 2^1) = 8$

2 The same construction as was used for the 3×3 case above shows that in place of $a_{1,1}$ we can select any nonzero entry $a_{i,j}$. Entry $c_{p,q}$ of Chiò's matrix is the value of this determinant

$$\begin{vmatrix} a_{1,1} & a_{1,q+1} \\ a_{p+1,1} & a_{p+1,q+1} \end{vmatrix}$$

where $p + 1 \neq i$ and $q + 1 \neq j$.

3 Sarrus's formula uses 12 multiplications and 5 additions (including the subtractions). Chiò's formula uses two multiplications and an addition (which is actually a subtraction) for each of the four 2×2 determinants,

and another two multiplications and an addition for the 2×2 Chio's determinant, as well as a final division by $a_{1,1}$. That's eleven multiplication/divisions and five addition/subtractions. So Chiò is the winner.

4 Consider an $n\times n$ matrix.

$$A=\begin{pmatrix} a_{1,1} & a_{1,2} & \cdots & a_{1,n-1} & a_{1,n} \\ a_{2,1} & a_{2,2} & \cdots & a_{2,n-1} & a_{2,n} \\ & & \vdots & & \\ a_{n-1,1} & a_{n-1,2} & \cdots & a_{n-1,n-1} & a_{n-1,n} \\ a_{n,1} & a_{n,2} & \cdots & a_{n,n-1} & a_{n,n} \end{pmatrix}$$

Rescale every row but the first by $a_{1,1}$.

$$\xrightarrow[\substack{a_{1,1}\rho_3 \\ \vdots \\ a_{1,1}\rho_n}]{a_{1,1}\rho_2}\begin{pmatrix} a_{1,1} & a_{1,2} & \cdots & a_{1,n-1} & a_{1,n} \\ a_{2,1}a_{1,1} & a_{2,2}a_{1,1} & \cdots & a_{2,n-1}a_{1,1} & a_{2,n}a_{1,1} \\ & & \vdots & & \\ a_{n-1,1}a_{1,1} & a_{n-1,2}a_{1,1} & \cdots & a_{n-1,n-1}a_{1,1} & a_{n-1,n}a_{1,1} \\ a_{n,1}a_{1,1} & a_{n,2}a_{1,1} & \cdots & a_{n,n-1}a_{1,1} & a_{n,n}a_{1,1} \end{pmatrix}$$

That rescales the determinant by a factor of $a_{1,1}^{n-1}$.

Next perform the row operation $-a_{i,1}\rho_1+\rho_i$ on each row $i>1$. These row operations don't change the determinant.

$$\xrightarrow[\substack{-a_{3,1}\rho_1+\rho_3 \\ \vdots \\ -a_{n,1}\rho_1+\rho_n}]{-a_{2,1}\rho_1+\rho_2}\begin{pmatrix} a_{1,1} & a_{1,2} & \cdots & a_{1,n-1} & a_{1,n} \\ 0 & a_{2,2}a_{1,1}-a_{2,1}a_{1,2} & \cdots & a_{2,n-1}a_{n,n}-a_{2,n-1}a_{1,n-1} & a_{2,n}a_{n,n}-a_{2,n}a_{1,n} \\ \vdots & & & & \\ 0 & a_{n,2}a_{1,1}-a_{n,1}a_{1,2} & \cdots & a_{n,n-1}a_{1,1}-a_{n,1}a_{1,n-1} & a_{n,n}a_{1,1}-a_{n,1}a_{1,n} \end{pmatrix}$$

The determinant of this matrix is $a_{1,1}^{n-1}$ times the determinant of A.

Denote by C the $1,1$ minor of the matrix, that is, the submatrix consisting of the first $n-1$ rows and columns. The Laplace expansion down the final column of the above matrix gives that its determinant is $(-1)^{1+1}a_{1,1}\det(C)$.

If $a_{1,1}\neq 0$ then setting the two equal and canceling gives $\det(A)=\det(C)/a_{n,n}^{n-2}$.

Topic: Projective Geometry

1 From the dot product

$$0=\begin{pmatrix}1\\0\\0\end{pmatrix}\cdot\begin{pmatrix}L_1 & L_2 & L_3\end{pmatrix}=L_1$$

we get that the equation is $L_1=0$.

2 **(a)** This determinant

$$0=\begin{vmatrix}1&4&x\\2&5&y\\3&6&z\end{vmatrix}=-3x+6y-3z$$

shows that the line is $L=\begin{pmatrix}-3 & 6 & -3\end{pmatrix}$.

(b) $\begin{pmatrix}-3\\6\\-3\end{pmatrix}$

3 The line incident on

$$u=\begin{pmatrix}u_1\\u_2\\u_3\end{pmatrix}\qquad v=\begin{pmatrix}v_1\\v_2\\v_3\end{pmatrix}$$

comes from this determinant equation.

$$0=\begin{vmatrix}u_1&v_1&x\\u_2&v_2&y\\u_3&v_3&z\end{vmatrix}=(u_2v_3-u_3v_2)\cdot x+(u_3v_1-u_1v_3)\cdot y+(u_1v_2-u_2v_1)\cdot z$$

The equation for the point incident on two lines is the same.

4 If p_1, p_2, p_3, and q_1, q_2, q_3 are two triples of homogeneous coordinates for p then the two column vectors are in proportion, that is, lie on the same line through the origin. Similarly, the two row vectors are in proportion.

$$k\cdot\begin{pmatrix}p_1\\p_2\\p_3\end{pmatrix}=\begin{pmatrix}q_1\\q_2\\q_3\end{pmatrix}\qquad m\cdot\begin{pmatrix}L_1&L_2&L_3\end{pmatrix}=\begin{pmatrix}M_1&M_2&M_3\end{pmatrix}$$

Then multiplying gives the answer $(km)\cdot(p_1L_1+p_2L_2+p_3L_3)=q_1M_1+q_2M_2+q_3M_3=0$.

5 The picture of the solar eclipse—unless the image plane is exactly perpendicular to the line from the sun through the pinhole—shows the circle of the sun projecting to an image that is an ellipse. (Another example is that in many pictures in this Topic, we've shown the circle that is the sphere's equator as an ellipse, that is, a viewer of the drawing sees a circle as an ellipse.)

The solar eclipse picture also shows the converse. If we picture the projection as going from left to right through the pinhole then the ellipse I projects through P to a circle S.

6 A spot on the unit sphere

$$\begin{pmatrix}p_1\\p_2\\p_3\end{pmatrix}$$

is non-equatorial if and only if $p_3\neq 0$. In that case it corresponds to this point on the $z=1$ plane

$$\begin{pmatrix}p_1/p_3\\p_2/p_3\\1\end{pmatrix}$$

since that is intersection of the line containing the vector and the plane.

7 **(a)** Other pictures are possible, but this is one.

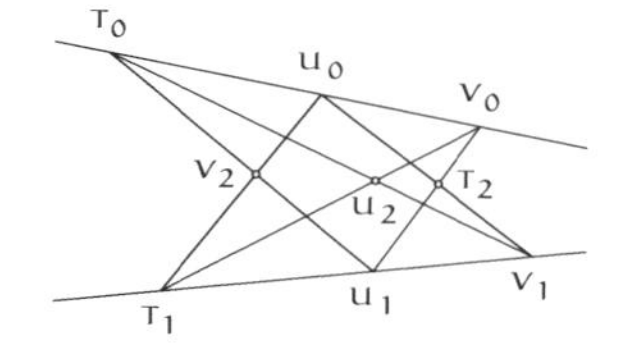

The intersections $T_0U_1\cap T_1U_0=V_2$, $T_0V_1\cap T_1V_0=U_2$, and $U_0V_1\cap U_1V_0=T_2$ are labeled so that on each line is a T, a U, and a V.

(b) The lemma used in Desargue's Theorem gives a basis B with respect to which the points have these homogeneous coordinate vectors.

$$\operatorname{Rep}_B(\vec{t}_0)=\begin{pmatrix}1\\0\\0\end{pmatrix}\quad\operatorname{Rep}_B(\vec{t}_1)=\begin{pmatrix}0\\1\\0\end{pmatrix}\quad\operatorname{Rep}_B(\vec{t}_2)=\begin{pmatrix}0\\0\\1\end{pmatrix}\quad\operatorname{Rep}_B(\vec{v}_0)=\begin{pmatrix}1\\1\\1\end{pmatrix}$$

(c) First, any U_0 on T_0V_0

$$\mathrm{Rep}_B(\vec{u}_0)=a\begin{pmatrix}1\\0\\0\end{pmatrix}+b\begin{pmatrix}1\\1\\1\end{pmatrix}=\begin{pmatrix}a+b\\b\\b\end{pmatrix}$$

has homogeneous coordinate vectors of this form

$$\begin{pmatrix}u_0\\1\\1\end{pmatrix}$$

(u_0 is a parameter; it depends on where on the T_0V_0 line the point U_0 is, but any point on that line has a homogeneous coordinate vector of this form for some $u_0\in\mathbb{R}$). Similarly, U_2 is on T_1V_0

$$\mathrm{Rep}_B(\vec{u}_2)=c\begin{pmatrix}0\\1\\0\end{pmatrix}+d\begin{pmatrix}1\\1\\1\end{pmatrix}=\begin{pmatrix}d\\c+d\\d\end{pmatrix}$$

and so has this homogeneous coordinate vector.

$$\begin{pmatrix}1\\u_2\\1\end{pmatrix}$$

Also similarly, U_1 is incident on T_2V_0

$$\mathrm{Rep}_B(\vec{u}_1)=e\begin{pmatrix}0\\0\\1\end{pmatrix}+f\begin{pmatrix}1\\1\\1\end{pmatrix}=\begin{pmatrix}f\\f\\e+f\end{pmatrix}$$

and has this homogeneous coordinate vector.

$$\begin{pmatrix}1\\1\\u_1\end{pmatrix}$$

(d) Because V_1 is $T_0U_2\cap U_0T_2$ we have this.

$$g\begin{pmatrix}1\\0\\0\end{pmatrix}+h\begin{pmatrix}1\\u_2\\1\end{pmatrix}=i\begin{pmatrix}u_0\\1\\1\end{pmatrix}+j\begin{pmatrix}0\\0\\1\end{pmatrix}\qquad\Longrightarrow\qquad\begin{aligned}g+h&=iu_0\\hu_2&=i\\h&=i+j\end{aligned}$$

Substituting hu_2 for i in the first equation

$$\begin{pmatrix}hu_0u_2\\hu_2\\h\end{pmatrix}$$

shows that V_1 has this two-parameter homogeneous coordinate vector.

$$\begin{pmatrix}u_0u_2\\u_2\\1\end{pmatrix}$$

(e) Since V_2 is the intersection $T_0U_1\cap T_1U_0$

$$k\begin{pmatrix}1\\0\\0\end{pmatrix}+l\begin{pmatrix}1\\1\\u_1\end{pmatrix}=m\begin{pmatrix}0\\1\\0\end{pmatrix}+n\begin{pmatrix}u_0\\1\\1\end{pmatrix}\qquad\Longrightarrow\qquad\begin{aligned}k+l&=nu_0\\l&=m+n\\lu_1&=n\end{aligned}$$

and substituting lu_1 for n in the first equation

$$\begin{pmatrix}lu_0u_1\\l\\lu_1\end{pmatrix}$$

gives that V_2 has this two-parameter homogeneous coordinate vector.

$$\begin{pmatrix} u_0u_1 \\ 1 \\ u_1 \end{pmatrix}$$

(f) Because V_1 is on the T_1U_1 line its homogeneous coordinate vector has the form

$$p\begin{pmatrix} 0 \\ 1 \\ 0 \end{pmatrix} + q\begin{pmatrix} 1 \\ 1 \\ u_1 \end{pmatrix} = \begin{pmatrix} q \\ p+q \\ qu_1 \end{pmatrix} \qquad (*)$$

but a previous part of this question established that V_1's homogeneous coordinate vectors have the form

$$\begin{pmatrix} u_0u_2 \\ u_2 \\ 1 \end{pmatrix}$$

and so this a homogeneous coordinate vector for V_1.

$$\begin{pmatrix} u_0u_1u_2 \\ u_1u_2 \\ u_1 \end{pmatrix} \qquad (**)$$

By $(*)$ and $(**)$, there is a relationship among the three parameters: $u_0u_1u_2 = 1$.

(g) The homogeneous coordinate vector of V_2 can be written in this way.

$$\begin{pmatrix} u_0u_1u_2 \\ u_2 \\ u_1u_2 \end{pmatrix} = \begin{pmatrix} 1 \\ u_2 \\ u_1u_2 \end{pmatrix}$$

Now, the T_2U_2 line consists of the points whose homogeneous coordinates have this form.

$$r\begin{pmatrix} 0 \\ 0 \\ 1 \end{pmatrix} + s\begin{pmatrix} 1 \\ u_2 \\ 1 \end{pmatrix} = \begin{pmatrix} s \\ su_2 \\ r+s \end{pmatrix}$$

Taking $s = 1$ and $r = u_1u_2 - 1$ shows that the homogeneous coordinate vectors of V_2 have this form.

Chapter Five

Chapter Five: Similarity

Complex Vector Spaces

Similarity

Five.II.1: Definition and Examples

Five.II.1.4 One way to proceed is left to right.

$$PSP^{-1} = \begin{pmatrix} 4 & 2 \\ -3 & 2 \end{pmatrix} \begin{pmatrix} 1 & 3 \\ -2 & -6 \end{pmatrix} \begin{pmatrix} 2/14 & -2/14 \\ 3/14 & 4/14 \end{pmatrix} = \begin{pmatrix} 0 & 0 \\ -7 & -21 \end{pmatrix} \begin{pmatrix} 2/14 & -2/14 \\ 3/14 & 4/14 \end{pmatrix} = \begin{pmatrix} 0 & 0 \\ -11/2 & -5 \end{pmatrix}$$

Five.II.1.5 **(a)** Because the matrix (2) is 1×1, the matrices P and P^{-1} are also 1×1 and so where $P=(p)$ the inverse is $P^{-1}=(1/p)$. Thus $P(2)P^{-1}=(p)(2)(1/p)=(2)$.

(b) Yes: recall that we can bring scalar multiples out of a matrix $P(cI)P^{-1}=cPIP^{-1}=cI$. By the way, the zero and identity matrices are the special cases $c=0$ and $c=1$.

(c) No, as this example shows.

$$\begin{pmatrix} 1 & -2 \\ -1 & 1 \end{pmatrix} \begin{pmatrix} -1 & 0 \\ 0 & -3 \end{pmatrix} \begin{pmatrix} -1 & -2 \\ -1 & -1 \end{pmatrix} = \begin{pmatrix} -5 & -4 \\ 2 & 1 \end{pmatrix}$$

Five.II.1.6 Gauss's Method shows that the first matrix represents maps of rank two while the second matrix represents maps of rank three.

Five.II.1.7 **(a)** Because we describe t with the members of B, finding the matrix representation is easy:

$$\text{Rep}_B(t(x^2)) = \begin{pmatrix} 0 \\ 1 \\ 1 \end{pmatrix}_B \qquad \text{Rep}_B(t(x)) = \begin{pmatrix} 1 \\ 0 \\ -1 \end{pmatrix}_B \qquad \text{Rep}_B(t(1)) = \begin{pmatrix} 0 \\ 0 \\ 3 \end{pmatrix}_B$$

gives this.

$$\text{Rep}_{B,B}(t) \begin{pmatrix} 0 & 1 & 0 \\ 1 & 0 & 0 \\ 1 & -1 & 3 \end{pmatrix}$$

(b) We will find $t(1)$, $t(1+x)$, and $t(1+x+x^2$, to find how each is represented with respect to D. We are given that $t(1)=3$, and the other two are easy to see: $t(1+x)=x^2+2$ and $t(1+x+x^2)=x^2+x+3$. By eye, we get the representation of each vector

$$\text{Rep}_D(t(1)) = \begin{pmatrix} 3 \\ 0 \\ 0 \end{pmatrix}_D \qquad \text{Rep}_D(t(1+x)) = \begin{pmatrix} 2 \\ -1 \\ 1 \end{pmatrix}_D \qquad \text{Rep}_D(t(1+x+x^2)) = \begin{pmatrix} 2 \\ 0 \\ 1 \end{pmatrix}_D$$

and thus the representation of the map.

$$\mathrm{Rep}_{D,D}(t) = \begin{pmatrix} 3 & 2 & 2 \\ 0 & -1 & 0 \\ 0 & 1 & 1 \end{pmatrix}$$

(c) The diagram, adapted for this T and S,

$$\begin{array}{ccc} V_{wrt\ D} & \xrightarrow[S]{t} & V_{wrt\ D} \\ \mathrm{id}\big\downarrow P & & \mathrm{id}\big\downarrow P \\ V_{wrt\ B} & \xrightarrow[T]{t} & V_{wrt\ B} \end{array}$$

shows that $P = \mathrm{Rep}_{D,B}(\mathrm{id})$.

$$P = \begin{pmatrix} 0 & 0 & 1 \\ 0 & 1 & 1 \\ 1 & 1 & 1 \end{pmatrix}$$

Five.II.1.8 One possible choice of the bases is

$$B = \langle \begin{pmatrix} 1 \\ 2 \end{pmatrix}, \begin{pmatrix} -1 \\ 1 \end{pmatrix} \rangle \qquad D = \mathcal{E}_2 = \langle \begin{pmatrix} 1 \\ 0 \end{pmatrix}, \begin{pmatrix} 0 \\ 1 \end{pmatrix} \rangle$$

(this B comes from the map description). To find the matrix $T = \mathrm{Rep}_{B,B}(t)$, solve the relations

$$c_1 \begin{pmatrix} 1 \\ 2 \end{pmatrix} + c_2 \begin{pmatrix} -1 \\ 1 \end{pmatrix} = \begin{pmatrix} 3 \\ 0 \end{pmatrix} \qquad \hat{c}_1 \begin{pmatrix} 1 \\ 2 \end{pmatrix} + \hat{c}_2 \begin{pmatrix} -1 \\ 1 \end{pmatrix} = \begin{pmatrix} -1 \\ 2 \end{pmatrix}$$

to get $c_1 = 1$, $c_2 = -2$, $\hat{c}_1 = 1/3$ and $\hat{c}_2 = 4/3$.

$$\mathrm{Rep}_{B,B}(t) = \begin{pmatrix} 1 & 1/3 \\ -2 & 4/3 \end{pmatrix}$$

Finding $\mathrm{Rep}_{D,D}(t)$ involves a bit more computation. We first find $t(\vec{e}_1)$. The relation

$$c_1 \begin{pmatrix} 1 \\ 2 \end{pmatrix} + c_2 \begin{pmatrix} -1 \\ 1 \end{pmatrix} = \begin{pmatrix} 1 \\ 0 \end{pmatrix}$$

gives $c_1 = 1/3$ and $c_2 = -2/3$, and so

$$\mathrm{Rep}_B(\vec{e}_1) = \begin{pmatrix} 1/3 \\ -2/3 \end{pmatrix}_B$$

making

$$\mathrm{Rep}_B(t(\vec{e}_1)) = \begin{pmatrix} 1 & 1/3 \\ -2 & 4/3 \end{pmatrix}_{B,B} \begin{pmatrix} 1/3 \\ -2/3 \end{pmatrix}_B = \begin{pmatrix} 1/9 \\ -14/9 \end{pmatrix}_B$$

and hence t acts on the first basis vector $\vec{e}_1$ in this way.

$$t(\vec{e}_1) = (1/9) \cdot \begin{pmatrix} 1 \\ 2 \end{pmatrix} - (14/9) \cdot \begin{pmatrix} -1 \\ 1 \end{pmatrix} = \begin{pmatrix} 5/3 \\ -4/3 \end{pmatrix}$$

The computation for $t(\vec{e}_2)$ is similar. The relation

$$c_1 \begin{pmatrix} 1 \\ 2 \end{pmatrix} + c_2 \begin{pmatrix} -1 \\ 1 \end{pmatrix} = \begin{pmatrix} 0 \\ 1 \end{pmatrix}$$

gives $c_1 = 1/3$ and $c_2 = 1/3$, so

$$\mathrm{Rep}_B(\vec{e}_1) = \begin{pmatrix} 1/3 \\ 1/3 \end{pmatrix}_B$$

making

$$\mathrm{Rep}_B(t(\vec{e}_1)) = \begin{pmatrix} 1 & 1/3 \\ -2 & 4/3 \end{pmatrix}_{B,B} \begin{pmatrix} 1/3 \\ 1/3 \end{pmatrix}_B = \begin{pmatrix} 4/9 \\ -2/9 \end{pmatrix}_B$$

and hence t acts on the second basis vector $\vec{e}_2$ in this way.

$$t(\vec{e}_2)=(4/9)\cdot\begin{pmatrix}1\\2\end{pmatrix}-(2/9)\cdot\begin{pmatrix}-1\\1\end{pmatrix}=\begin{pmatrix}2/3\\2/3\end{pmatrix}$$

Therefore

$$\mathrm{Rep}_{D,D}(t)=\begin{pmatrix}5/3&2/3\\-4/3&2/3\end{pmatrix}$$

and these are the change of basis matrices.

$$P=\mathrm{Rep}_{B,D}(\mathrm{id})=\begin{pmatrix}1&-1\\2&1\end{pmatrix}\qquad P^{-1}=\left(\mathrm{Rep}_{B,D}(\mathrm{id})\right)^{-1}=\begin{pmatrix}1&-1\\2&1\end{pmatrix}^{-1}=\begin{pmatrix}1/3&1/3\\-2/3&1/3\end{pmatrix}$$

The check of these computations is routine.

$$\begin{pmatrix}1&-1\\2&1\end{pmatrix}\begin{pmatrix}1&1/3\\-2&4/3\end{pmatrix}\begin{pmatrix}1/3&1/3\\-2/3&1/3\end{pmatrix}=\begin{pmatrix}5/3&2/3\\-4/3&2/3\end{pmatrix}$$

Five.II.1.9 The only representation of a zero map is a zero matrix, no matter what the pair of bases $\mathrm{Rep}_{B,D}(z)=Z$, and so in particular for any single basis B we have $\mathrm{Rep}_{B,B}(z)=Z$. The case of the identity is slightly different: the only representation of the identity map, with respect to any B,B, is the identity $\mathrm{Rep}_{B,B}(\mathrm{id})=I$. (*Remark:* of course, we have seen examples where $B\neq D$ and $\mathrm{Rep}_{B,D}(\mathrm{id})\neq I$—in fact, we have seen that any nonsingular matrix is a representation of the identity map with respect to some B,D.)

Five.II.1.10 No. If $A=PBP^{-1}$ then $A^2=(PBP^{-1})(PBP^{-1})=PB^2P^{-1}$.

Five.II.1.11 Matrix similarity is a special case of matrix equivalence (if matrices are similar then they are matrix equivalent) and matrix equivalence preserves nonsingularity.

Five.II.1.12 A matrix is similar to itself; take P to be the identity matrix: $IPI^{-1}=IPI=P$.

If T is similar to S then $T=PSP^{-1}$ and so $P^{-1}TP=S$. Rewrite this as $S=(P^{-1})T(P^{-1})^{-1}$ to conclude that S is similar to T.

If T is similar to S and S is similar to U then $T=PSP^{-1}$ and $S=QUQ^{-1}$. Then $T=PQUQ^{-1}P^{-1}=(PQ)U(PQ)^{-1}$, showing that T is similar to U.

Five.II.1.13 Let f_x and f_y be the reflection maps (sometimes called 'flip's). For any bases B and D, the matrices $\mathrm{Rep}_{B,B}(f_x)$ and $\mathrm{Rep}_{D,D}(f_y)$ are similar. First note that

$$S=\mathrm{Rep}_{\mathcal{E}_2,\mathcal{E}_2}(f_x)=\begin{pmatrix}1&0\\0&-1\end{pmatrix}\qquad T=\mathrm{Rep}_{\mathcal{E}_2,\mathcal{E}_2}(f_y)=\begin{pmatrix}-1&0\\0&1\end{pmatrix}$$

are similar because the second matrix is the representation of f_x with respect to the basis $A=\langle\vec{e}_2,\vec{e}_1\rangle$:

$$\begin{pmatrix}1&0\\0&-1\end{pmatrix}=P\begin{pmatrix}-1&0\\0&1\end{pmatrix}P^{-1}$$

where $P=\mathrm{Rep}_{A,\mathcal{E}_2}(\mathrm{id})$.

$$\begin{array}{ccc}\mathbb{R}^2_{wrt\ A} & \xrightarrow[T]{f_x} & V\mathbb{R}^2_{wrt\ A}\\ \mathrm{id}\big\downarrow P & & \mathrm{id}\big\downarrow P\\ \mathbb{R}^2_{wrt\ \mathcal{E}_2} & \xrightarrow[S]{f_x} & \mathbb{R}^2_{wrt\ \mathcal{E}_2}\end{array}$$

Now the conclusion follows from the transitivity part of Exercise 12.

To finish without relying on that exercise, write $\mathrm{Rep}_{B,B}(f_x)=QTQ^{-1}=Q\mathrm{Rep}_{\mathcal{E}_2,\mathcal{E}_2}(f_x)Q^{-1}$ and $\mathrm{Rep}_{D,D}(f_y)=RSR^{-1}=R\mathrm{Rep}_{\mathcal{E}_2,\mathcal{E}_2}(f_y)R^{-1}$. Using the equation in the first paragraph, the first of these two becomes $\mathrm{Rep}_{B,B}(f_x)=QP\mathrm{Rep}_{\mathcal{E}_2,\mathcal{E}_2}(f_y)P^{-1}Q^{-1}$ and rewriting the second of these two as $R^{-1}\cdot\mathrm{Rep}_{D,D}(f_y)\cdot R=\mathrm{Rep}_{\mathcal{E}_2,\mathcal{E}_2}(f_y)$ and substituting gives the desired relationship

$$\begin{aligned}\mathrm{Rep}_{B,B}(f_x)&=QP\mathrm{Rep}_{\mathcal{E}_2,\mathcal{E}_2}(f_y)P^{-1}Q^{-1}\\&=QPR^{-1}\cdot\mathrm{Rep}_{D,D}(f_y)\cdot RP^{-1}Q^{-1}=(QPR^{-1})\cdot\mathrm{Rep}_{D,D}(f_y)\cdot(QPR^{-1})^{-1}\end{aligned}$$

Thus the matrices $\mathrm{Rep}_{B,B}(f_x)$ and $\mathrm{Rep}_{D,D}(f_y)$ are similar.

Five.II.1.14 We must show that if two matrices are similar then they have the same determinant and the same rank. Both determinant and rank are properties of matrices that are preserved by matrix equivalence. They are therefore preserved by similarity (which is a special case of matrix equivalence: if two matrices are similar then they are matrix equivalent).

To prove the statement without quoting the results about matrix equivalence, note first that rank is a property of the map (it is the dimension of the range space) and since we've shown that the rank of a map is the rank of a representation, it must be the same for all representations. As for determinants, $|PSP^{-1}| = |P| \cdot |S| \cdot |P^{-1}| = |P| \cdot |S| \cdot |P|^{-1} = |S|$.

The converse of the statement does not hold; for instance, there are matrices with the same determinant that are not similar. To check this, consider a nonzero matrix with a determinant of zero. It is not similar to the zero matrix, the zero matrix is similar only to itself, but they have they same determinant. The argument for rank is much the same.

Five.II.1.15 The matrix equivalence class containing all $n \times n$ rank zero matrices contains only a single matrix, the zero matrix. Therefore it has as a subset only one similarity class.

In contrast, the matrix equivalence class of 1×1 matrices of rank one consists of those 1×1 matrices (k) where $k \neq 0$. For any basis B, the representation of multiplication by the scalar k is $\mathrm{Rep}_{B,B}(t_k) = (k)$, so each such matrix is alone in its similarity class. So this is a case where a matrix equivalence class splits into infinitely many similarity classes.

Five.II.1.16 Yes, these are similar

$$\begin{pmatrix} 1 & 0 \\ 0 & 3 \end{pmatrix} \qquad \begin{pmatrix} 3 & 0 \\ 0 & 1 \end{pmatrix}$$

since, where the first matrix is $\mathrm{Rep}_{B,B}(t)$ for $B = \langle \vec{\beta}_1, \vec{\beta}_2 \rangle$, the second matrix is $\mathrm{Rep}_{D,D}(t)$ for $D = \langle \vec{\beta}_2, \vec{\beta}_1 \rangle$.

Five.II.1.17 The k-th powers are similar because, where each matrix represents the map t, the k-th powers represent t^k, the composition of k-many t's. (For instance, if $T = reptB, B$ then $T^2 = \mathrm{Rep}_{B,B}(t \circ t)$.)

Restated more computationally, if $T = PSP^{-1}$ then $T^2 = (PSP^{-1})(PSP^{-1}) = PS^2P^{-1}$. Induction extends that to all powers.

For the $k \leqslant 0$ case, suppose that S is invertible and that $T = PSP^{-1}$. Note that T is invertible: $T^{-1} = (PSP^{-1})^{-1} = PS^{-1}P^{-1}$, and that same equation shows that T^{-1} is similar to S^{-1}. Other negative powers are now given by the first paragraph.

Five.II.1.18 In conceptual terms, both represent $p(t)$ for some transformation t. In computational terms, we have this.

$$\begin{aligned} p(T) &= c_n(PSP^{-1})^n + \cdots + c_1(PSP^{-1}) + c_0 I \\ &= c_n PS^n P^{-1} + \cdots + c_1 PSP^{-1} + c_0 I \\ &= Pc_n S^n P^{-1} + \cdots + Pc_1 SP^{-1} + Pc_0 P^{-1} \\ &= P(c_n S^n + \cdots + c_1 S + c_0)P^{-1} \end{aligned}$$

Five.II.1.19 There are two equivalence classes, (i) the class of rank zero matrices, of which there is one: $\mathscr{C}_1 = \{(0)\}$, and (2) the class of rank one matrices, of which there are infinitely many: $\mathscr{C}_2 = \{(k) \mid k \neq 0\}$.

Each 1×1 matrix is alone in its similarity class. That's because any transformation of a one-dimensional space is multiplication by a scalar $t_k \colon V \to V$ given by $\vec{v} \mapsto k \cdot \vec{v}$. Thus, for any basis $B = \langle \vec{\beta} \rangle$, the matrix representing a transformation t_k with respect to B, B is $(\mathrm{Rep}_B(t_k(\vec{\beta}))) = (k)$.

So, contained in the matrix equivalence class $\mathscr{C}_1$ is (obviously) the single similarity class consisting of the matrix (0). And, contained in the matrix equivalence class $\mathscr{C}_2$ are the infinitely many, one-member-each, similarity classes consisting of (k) for $k \neq 0$.

Five.II.1.20 No. Here is an example that has two pairs, each of two similar matrices:

$$\begin{pmatrix} 1 & -1 \\ 1 & 2 \end{pmatrix} \begin{pmatrix} 1 & 0 \\ 0 & 3 \end{pmatrix} \begin{pmatrix} 2/3 & 1/3 \\ -1/3 & 1/3 \end{pmatrix} = \begin{pmatrix} 5/3 & -2/3 \\ -4/3 & 7/3 \end{pmatrix}$$

and

$$\begin{pmatrix}1&-2\\-1&1\end{pmatrix}\begin{pmatrix}-1&0\\0&-3\end{pmatrix}\begin{pmatrix}-1&-2\\-1&-1\end{pmatrix}=\begin{pmatrix}-5&-4\\2&1\end{pmatrix}$$

(this example is not entirely arbitrary because the center matrices on the two left sides add to the zero matrix). Note that the sums of these similar matrices are not similar

$$\begin{pmatrix}1&0\\0&3\end{pmatrix}+\begin{pmatrix}-1&0\\0&-3\end{pmatrix}=\begin{pmatrix}0&0\\0&0\end{pmatrix}\qquad\begin{pmatrix}5/3&-2/3\\-4/3&7/3\end{pmatrix}+\begin{pmatrix}-5&-4\\2&1\end{pmatrix}\neq\begin{pmatrix}0&0\\0&0\end{pmatrix}$$

since the zero matrix is similar only to itself.

Five.II.1.21 If $N = P(T-\lambda I)P^{-1}$ then $N = PTP^{-1} - P(\lambda I)P^{-1}$. The diagonal matrix λI commutes with anything, so $P(\lambda I)P^{-1} = PP^{-1}(\lambda I) = \lambda I$. Thus $N = PTP^{-1} - \lambda I$ and consequently $N + \lambda I = PTP^{-1}$. (So not only are they similar, in fact they are similar via the same P.)

Five.II.2: Diagonalizability

Five.II.2.6 Because we chose the basis vectors arbitrarily, many different answers are possible. However, here is one way to go; to diagonalize

$$T=\begin{pmatrix}4&-2\\1&1\end{pmatrix}$$

take it as the representation of a transformation with respect to the standard basis $T = \text{Rep}_{\mathcal{E}_2,\mathcal{E}_2}(t)$ and look for $B = \langle\vec{\beta}_1, \vec{\beta}_2\rangle$ such that

$$\text{Rep}_{B,B}(t)=\begin{pmatrix}\lambda_1&0\\0&\lambda_2\end{pmatrix}$$

that is, such that $t(\vec{\beta}_1) = \lambda_1$ and $t(\vec{\beta}_2) = \lambda_2$.

$$\begin{pmatrix}4&-2\\1&1\end{pmatrix}\vec{\beta}_1=\lambda_1\cdot\vec{\beta}_1\qquad\begin{pmatrix}4&-2\\1&1\end{pmatrix}\vec{\beta}_2=\lambda_2\cdot\vec{\beta}_2$$

We are looking for scalars x such that this equation

$$\begin{pmatrix}4&-2\\1&1\end{pmatrix}\begin{pmatrix}b_1\\b_2\end{pmatrix}=x\cdot\begin{pmatrix}b_1\\b_2\end{pmatrix}$$

has solutions b_1 and b_2, which are not both zero. Rewrite that as a linear system

$$\begin{aligned}(4-x)\cdot b_1 + \quad -2\cdot b_2&=0\\ 1\cdot b_1+(1-x)\cdot b_2&=0\end{aligned}$$

If $x = 4$ then the first equation gives that $b_2 = 0$, and then the second equation gives that $b_1 = 0$. We have disallowed the case where both b's are zero so we can assume that $x \neq 4$.

$$\xrightarrow{(-1/(4-x))\rho_1+\rho_2}\begin{aligned}(4-x)\cdot b_1 + \qquad\qquad -2\cdot b_2&=0\\ ((x^2-5x+6)/(4-x))\cdot b_2&=0\end{aligned}$$

Consider the bottom equation. If $b_2 = 0$ then the first equation gives $b_1 = 0$ or $x = 4$. The $b_1 = b_2 = 0$ case is not allowed. The other possibility for the bottom equation is that the numerator of the fraction $x^2-5x+6 = (x-2)(x-3)$ is zero. The $x = 2$ case gives a first equation of $2b_1 - 2b_2 = 0$, and so associated with $x = 2$ we have vectors whose first and second components are equal:

$$\vec{\beta}_1=\begin{pmatrix}1\\1\end{pmatrix}\qquad(\text{so }\begin{pmatrix}4&-2\\1&1\end{pmatrix}\begin{pmatrix}1\\1\end{pmatrix}=2\cdot\begin{pmatrix}1\\1\end{pmatrix},\text{ and }\lambda_1=2).$$

If $x = 3$ then the first equation is $b_1 - 2b_2 = 0$ and so the associated vectors are those whose first component is twice their second:

$$\vec{\beta}_2 = \begin{pmatrix} 2 \\ 1 \end{pmatrix} \qquad (\text{so } \begin{pmatrix} 4 & -2 \\ 1 & 1 \end{pmatrix}\begin{pmatrix} 2 \\ 1 \end{pmatrix} = 3 \cdot \begin{pmatrix} 2 \\ 1 \end{pmatrix}, \text{ and so } \lambda_2 = 3).$$

This picture

$$\begin{array}{ccc} \mathbb{R}^2_{wrt\ \mathcal{E}_2} & \xrightarrow[T]{t} & \mathbb{R}^2_{wrt\ \mathcal{E}_2} \\ \text{id}\big\downarrow & & \text{id}\big\downarrow \\ \mathbb{R}^2_{wrt\ B} & \xrightarrow[D]{t} & \mathbb{R}^2_{wrt\ B} \end{array}$$

shows how to get the diagonalization.

$$\begin{pmatrix} 2 & 0 \\ 0 & 3 \end{pmatrix} = \begin{pmatrix} 1 & 2 \\ 1 & 1 \end{pmatrix}^{-1}\begin{pmatrix} 4 & -2 \\ 1 & 1 \end{pmatrix}\begin{pmatrix} 1 & 2 \\ 1 & 1 \end{pmatrix}$$

Comment. This equation matches the $T = PSP^{-1}$ definition under this renaming.

$$T = \begin{pmatrix} 2 & 0 \\ 0 & 3 \end{pmatrix} \quad P = \begin{pmatrix} 1 & 2 \\ 1 & 1 \end{pmatrix}^{-1} \quad P^{-1} = \begin{pmatrix} 1 & 2 \\ 1 & 1 \end{pmatrix} \quad S = \begin{pmatrix} 4 & -2 \\ 1 & 1 \end{pmatrix}$$

Five.II.2.7 **(a)** Setting up

$$\begin{pmatrix} -2 & 1 \\ 0 & 2 \end{pmatrix}\begin{pmatrix} b_1 \\ b_2 \end{pmatrix} = x \cdot \begin{pmatrix} b_1 \\ b_2 \end{pmatrix} \qquad \Longrightarrow \qquad \begin{aligned} (-2 - x) \cdot b_1 + \quad b_2 &= 0 \\ (2 - x) \cdot b_2 &= 0 \end{aligned}$$

gives the two possibilities that $b_2 = 0$ and $x = 2$. Following the $b_2 = 0$ possibility leads to the first equation $(-2 - x)b_1 = 0$ with the two cases that $b_1 = 0$ and that $x = -2$. Thus, under this first possibility, we find $x = -2$ and the associated vectors whose second component is zero, and whose first component is free.

$$\begin{pmatrix} -2 & 1 \\ 0 & 2 \end{pmatrix}\begin{pmatrix} b_1 \\ 0 \end{pmatrix} = -2 \cdot \begin{pmatrix} b_1 \\ 0 \end{pmatrix} \qquad \vec{\beta}_1 = \begin{pmatrix} 1 \\ 0 \end{pmatrix}$$

Following the other possibility leads to a first equation of $-4b_1 + b_2 = 0$ and so the vectors associated with this solution have a second component that is four times their first component.

$$\begin{pmatrix} -2 & 1 \\ 0 & 2 \end{pmatrix}\begin{pmatrix} b_1 \\ 4b_1 \end{pmatrix} = 2 \cdot \begin{pmatrix} b_1 \\ 4b_1 \end{pmatrix} \qquad \vec{\beta}_2 = \begin{pmatrix} 1 \\ 4 \end{pmatrix}$$

The diagonalization is this.

$$\begin{pmatrix} 1 & 1 \\ 0 & 4 \end{pmatrix}\begin{pmatrix} -2 & 1 \\ 0 & 2 \end{pmatrix}\begin{pmatrix} 1 & 1 \\ 0 & 4 \end{pmatrix}^{-1} = \begin{pmatrix} -2 & 0 \\ 0 & 2 \end{pmatrix}$$

(b) The calculations are like those in the prior part.

$$\begin{pmatrix} 5 & 4 \\ 0 & 1 \end{pmatrix}\begin{pmatrix} b_1 \\ b_2 \end{pmatrix} = x \cdot \begin{pmatrix} b_1 \\ b_2 \end{pmatrix} \qquad \Longrightarrow \qquad \begin{aligned} (5 - x) \cdot b_1 + \quad 4 \cdot b_2 &= 0 \\ (1 - x) \cdot b_2 &= 0 \end{aligned}$$

The bottom equation gives the two possibilities that $b_2 = 0$ and $x = 1$. Following the $b_2 = 0$ possibility, and discarding the case where both b_2 and b_1 are zero, gives that $x = 5$, associated with vectors whose second component is zero and whose first component is free.

$$\vec{\beta}_1 = \begin{pmatrix} 1 \\ 0 \end{pmatrix}$$

The $x = 1$ possibility gives a first equation of $4b_1 + 4b_2 = 0$ and so the associated vectors have a second component that is the negative of their first component.

$$\vec{\beta}_1 = \begin{pmatrix} 1 \\ -1 \end{pmatrix}$$

We thus have this diagonalization.

$$\begin{pmatrix}1&1\\0&-1\end{pmatrix}\begin{pmatrix}5&4\\0&1\end{pmatrix}\begin{pmatrix}1&1\\0&-1\end{pmatrix}^{-1}=\begin{pmatrix}5&0\\0&1\end{pmatrix}$$

Five.II.2.8 For any integer p,

$$\begin{pmatrix}d_1&0&\\0&\ddots&\\&&d_n\end{pmatrix}^p=\begin{pmatrix}d_1^p&0&\\0&\ddots&\\&&d_n^p\end{pmatrix}.$$

Five.II.2.9 These two are not similar

$$\begin{pmatrix}0&0\\0&0\end{pmatrix}\qquad\begin{pmatrix}1&0\\0&1\end{pmatrix}$$

because each is alone in its similarity class.

For the second half, these

$$\begin{pmatrix}2&0\\0&3\end{pmatrix}\qquad\begin{pmatrix}3&0\\0&2\end{pmatrix}$$

are similar via the matrix that changes bases from $\langle\vec{\beta}_1,\vec{\beta}_2\rangle$ to $\langle\vec{\beta}_2,\vec{\beta}_1\rangle$. (*Question.* Are two diagonal matrices similar if and only if their diagonal entries are permutations of each others?)

Five.II.2.10 Contrast these two.

$$\begin{pmatrix}2&0\\0&1\end{pmatrix}\qquad\begin{pmatrix}2&0\\0&0\end{pmatrix}$$

The first is nonsingular, the second is singular.

Five.II.2.11 To check that the inverse of a diagonal matrix is the diagonal matrix of the inverses, just multiply.

$$\begin{pmatrix}a_{1,1}&0&&\\0&a_{2,2}&&\\&&\ddots&\\&&&a_{n,n}\end{pmatrix}\begin{pmatrix}1/a_{1,1}&0&&\\0&1/a_{2,2}&&\\&&\ddots&\\&&&1/a_{n,n}\end{pmatrix}$$

(Showing that it is a left inverse is just as easy.)

If a diagonal entry is zero then the diagonal matrix is singular; it has a zero determinant.

Five.II.2.12 (a) The check is easy.

$$\begin{pmatrix}1&1\\0&-1\end{pmatrix}\begin{pmatrix}3&2\\0&1\end{pmatrix}=\begin{pmatrix}3&3\\0&-1\end{pmatrix}\qquad\begin{pmatrix}3&3\\0&-1\end{pmatrix}\begin{pmatrix}1&1\\0&-1\end{pmatrix}^{-1}=\begin{pmatrix}3&0\\0&1\end{pmatrix}$$

(b) It is a coincidence, in the sense that if $T=PSP^{-1}$ then T need not equal $P^{-1}SP$. Even in the case of a diagonal matrix D, the condition that $D=PTP^{-1}$ does not imply that D equals $P^{-1}TP$. The matrices from Example 2.2 show this.

$$\begin{pmatrix}1&2\\1&1\end{pmatrix}\begin{pmatrix}4&-2\\1&1\end{pmatrix}=\begin{pmatrix}6&0\\5&-1\end{pmatrix}\qquad\begin{pmatrix}6&0\\5&-1\end{pmatrix}\begin{pmatrix}1&2\\1&1\end{pmatrix}^{-1}=\begin{pmatrix}-6&12\\-6&11\end{pmatrix}$$

Five.II.2.13 The columns of the matrix are the vectors associated with the x's. The exact choice, and the order of the choice was arbitrary. We could, for instance, get a different matrix by swapping the two columns.

Five.II.2.14 Diagonalizing and then taking powers of the diagonal matrix shows that

$$\begin{pmatrix}-3&1\\-4&2\end{pmatrix}^k=\frac{1}{3}\begin{pmatrix}-1&1\\-4&4\end{pmatrix}+\left(\frac{-2}{3}\right)^k\begin{pmatrix}4&-1\\4&-1\end{pmatrix}.$$

Five.II.2.15 (a) $\begin{pmatrix}1&1\\0&-1\end{pmatrix}^{-1}\begin{pmatrix}1&1\\0&0\end{pmatrix}\begin{pmatrix}1&1\\0&-1\end{pmatrix}=\begin{pmatrix}1&0\\0&0\end{pmatrix}$

(b) $\begin{pmatrix}1&1\\0&-1\end{pmatrix}^{-1}\begin{pmatrix}0&1\\1&0\end{pmatrix}\begin{pmatrix}1&1\\0&-1\end{pmatrix}=\begin{pmatrix}1&0\\0&-1\end{pmatrix}$

Five.II.2.16 Yes, ct is diagonalizable by the final theorem of this subsection.

No, $t+s$ need not be diagonalizable. Intuitively, the problem arises when the two maps diagonalize with respect to different bases (that is, when they are not *simultaneously diagonalizable*). Specifically, these two are diagonalizable but their sum is not:

$$\begin{pmatrix}1&1\\0&0\end{pmatrix}\qquad\begin{pmatrix}-1&0\\0&0\end{pmatrix}$$

(the second is already diagonal; for the first, see Exercise 15). The sum is not diagonalizable because its square is the zero matrix.

The same intuition suggests that $t\circ s$ is not be diagonalizable. These two are diagonalizable but their product is not:

$$\begin{pmatrix}1&0\\0&0\end{pmatrix}\qquad\begin{pmatrix}0&1\\1&0\end{pmatrix}$$

(for the second, see Exercise 15).

Five.II.2.17 If

$$P\begin{pmatrix}1&c\\0&1\end{pmatrix}P^{-1}=\begin{pmatrix}a&0\\0&b\end{pmatrix}$$

then

$$P\begin{pmatrix}1&c\\0&1\end{pmatrix}=\begin{pmatrix}a&0\\0&b\end{pmatrix}P$$

so

$$\begin{pmatrix}p&q\\r&s\end{pmatrix}\begin{pmatrix}1&c\\0&1\end{pmatrix}=\begin{pmatrix}a&0\\0&b\end{pmatrix}\begin{pmatrix}p&q\\r&s\end{pmatrix}$$

$$\begin{pmatrix}p&cp+q\\r&cr+s\end{pmatrix}=\begin{pmatrix}ap&aq\\br&bs\end{pmatrix}$$

The $1,1$ entries show that $a=1$ and the $1,2$ entries then show that $pc=0$. Since $c\neq 0$ this means that $p=0$. The $2,1$ entries show that $b=1$ and the $2,2$ entries then show that $rc=0$. Since $c\neq 0$ this means that $r=0$. But if both p and r are 0 then P is not invertible.

Five.II.2.18 (a) Using the formula for the inverse of a 2×2 matrix gives this.

$$\begin{pmatrix}a&b\\c&d\end{pmatrix}\begin{pmatrix}1&2\\2&1\end{pmatrix}\cdot\frac{1}{ad-bc}\cdot\begin{pmatrix}d&-b\\-c&a\end{pmatrix}=\frac{1}{ad-bc}\begin{pmatrix}ad+2bd-2ac-bc&-ab-2b^2+2a^2+ab\\cd+2d^2-2c^2-cd&-bc-2bd+2ac+ad\end{pmatrix}$$

Now pick scalars $a,\ldots,d$ so that $ad-bc\neq 0$ and $2d^2-2c^2=0$ and $2a^2-2b^2=0$. For example, these will do.

$$\begin{pmatrix}1&1\\1&-1\end{pmatrix}\begin{pmatrix}1&2\\2&1\end{pmatrix}\cdot\frac{1}{-2}\cdot\begin{pmatrix}-1&-1\\-1&1\end{pmatrix}=\frac{1}{-2}\begin{pmatrix}-6&0\\0&2\end{pmatrix}$$

(b) As above,

$$\begin{pmatrix}a&b\\c&d\end{pmatrix}\begin{pmatrix}x&y\\y&z\end{pmatrix}\cdot\frac{1}{ad-bc}\cdot\begin{pmatrix}d&-b\\-c&a\end{pmatrix}=\frac{1}{ad-bc}\begin{pmatrix}adx+bdy-acy-bcz&-abx-b^2y+a^2y+abz\\cdx+d^2y-c^2y-cdz&-bcx-bdy+acy+adz\end{pmatrix}$$

we are looking for scalars $a,\ldots,d$ so that $ad-bc\neq 0$ and $-abx-b^2y+a^2y+abz=0$ and $cdx+d^2y-c^2y-cdz=0$, no matter what values x, y, and z have.

For starters, we assume that $y\neq 0$, else the given matrix is already diagonal. We shall use that assumption because if we (arbitrarily) let $a=1$ then we get

$$-bx-b^2y+y+bz=0$$
$$(-y)b^2+(z-x)b+y=0$$

and the quadratic formula gives

$$b = \frac{-(z-x) \pm \sqrt{(z-x)^2 - 4(-y)(y)}}{-2y} \qquad y \neq 0$$

(note that if x, y, and z are real then these two b's are real as the discriminant is positive). By the same token, if we (arbitrarily) let $c = 1$ then

$$\begin{aligned} dx + d^2y - y - dz &= 0 \\ (y)d^2 + (x-z)d - y &= 0 \end{aligned}$$

and we get here

$$d = \frac{-(x-z) \pm \sqrt{(x-z)^2 - 4(y)(-y)}}{2y} \qquad y \neq 0$$

(as above, if $x, y, z \in \mathbb{R}$ then this discriminant is positive so a symmetric, real, 2×2 matrix is similar to a real diagonal matrix).

For a check we try $x = 1$, $y = 2$, $z = 1$.

$$b = \frac{0 \pm \sqrt{0+16}}{-4} = \mp 1 \qquad d = \frac{0 \pm \sqrt{0+16}}{4} = \pm 1$$

Note that not all four choices $(b, d) = (+1, +1), \ldots, (-1, -1)$ satisfy $ad - bc \neq 0$.

Five.II.3: Eigenvalues and Eigenvectors

Five.II.3.22 **(a)** This

$$0 = \begin{vmatrix} 10-x & -9 \\ 4 & -2-x \end{vmatrix} = (10-x)(-2-x) - (-36)$$

simplifies to the characteristic equation $x^2 - 8x + 16 = 0$. Because the equation factors into $(x-4)^2$ there is only one eigenvalue $\lambda_1 = 4$.

(b) $0 = (1-x)(3-x) - 8 = x^2 - 4x - 5$; $\lambda_1 = 5$, $\lambda_2 = -1$

(c) $x^2 - 21 = 0$; $\lambda_1 = \sqrt{21}$, $\lambda_2 = -\sqrt{21}$

(d) $x^2 = 0$; $\lambda_1 = 0$

(e) $x^2 - 2x + 1 = 0$; $\lambda_1 = 1$

Five.II.3.23 **(a)** The characteristic equation is $(3-x)(-1-x) = 0$. Its roots, the eigenvalues, are $\lambda_1 = 3$ and $\lambda_2 = -1$. For the eigenvectors we consider this equation.

$$\begin{pmatrix} 3-x & 0 \\ 8 & -1-x \end{pmatrix} \begin{pmatrix} b_1 \\ b_2 \end{pmatrix} = \begin{pmatrix} 0 \\ 0 \end{pmatrix}$$

For the eigenvector associated with $\lambda_1 = 3$, we consider the resulting linear system.

$$\begin{aligned} 0 \cdot b_1 + 0 \cdot b_2 &= 0 \\ 8 \cdot b_1 + -4 \cdot b_2 &= 0 \end{aligned}$$

The eigenspace is the set of vectors whose second component is twice the first component.

$$\{\begin{pmatrix} b_2/2 \\ b_2 \end{pmatrix} \mid b_2 \in \mathbb{C}\} \qquad \begin{pmatrix} 3 & 0 \\ 8 & -1 \end{pmatrix} \begin{pmatrix} b_2/2 \\ b_2 \end{pmatrix} = 3 \cdot \begin{pmatrix} b_2/2 \\ b_2 \end{pmatrix}$$

(Here, the parameter is b_2 only because that is the variable that is free in the above system.) Hence, this is an eigenvector associated with the eigenvalue 3.

$$\begin{pmatrix} 1 \\ 2 \end{pmatrix}$$

Finding an eigenvector associated with $\lambda_2 = -1$ is similar. This system

$$\begin{aligned} 4\cdot b_1 + 0\cdot b_2 &= 0 \\ 8\cdot b_1 + 0\cdot b_2 &= 0 \end{aligned}$$

leads to the set of vectors whose first component is zero.

$$\{\begin{pmatrix} 0 \\ b_2 \end{pmatrix} \mid b_2 \in \mathbb{C}\} \qquad \begin{pmatrix} 3 & 0 \\ 8 & -1 \end{pmatrix}\begin{pmatrix} 0 \\ b_2 \end{pmatrix} = -1\cdot\begin{pmatrix} 0 \\ b_2 \end{pmatrix}$$

And so this is an eigenvector associated with λ_2.

$$\begin{pmatrix} 0 \\ 1 \end{pmatrix}$$

(b) The characteristic equation is

$$0 = \begin{vmatrix} 3-x & 2 \\ -1 & -x \end{vmatrix} = x^2 - 3x + 2 = (x-2)(x-1)$$

and so the eigenvalues are $\lambda_1 = 2$ and $\lambda_2 = 1$. To find eigenvectors, consider this system.

$$\begin{aligned} (3-x)\cdot b_1 + 2\cdot b_2 &= 0 \\ -1\cdot b_1 - x\cdot b_2 &= 0 \end{aligned}$$

For $\lambda_1 = 2$ we get

$$\begin{aligned} 1\cdot b_1 + 2\cdot b_2 &= 0 \\ -1\cdot b_1 - 2\cdot b_2 &= 0 \end{aligned}$$

leading to this eigenspace and eigenvector.

$$\{\begin{pmatrix} -2b_2 \\ b_2 \end{pmatrix} \mid b_2 \in \mathbb{C}\} \qquad \begin{pmatrix} -2 \\ 1 \end{pmatrix}$$

For $\lambda_2 = 1$ the system is

$$\begin{aligned} 2\cdot b_1 + 2\cdot b_2 &= 0 \\ -1\cdot b_1 - 1\cdot b_2 &= 0 \end{aligned}$$

leading to this.

$$\{\begin{pmatrix} -b_2 \\ b_2 \end{pmatrix} \mid b_2 \in \mathbb{C}\} \qquad \begin{pmatrix} -1 \\ 1 \end{pmatrix}$$

Five.II.3.24 The characteristic equation

$$0 = \begin{vmatrix} -2-x & -1 \\ 5 & 2-x \end{vmatrix} = x^2 + 1$$

has the complex roots $\lambda_1 = i$ and $\lambda_2 = -i$. This system

$$\begin{aligned} (-2-x)\cdot b_1 - \quad 1\cdot b_2 &= 0 \\ 5\cdot b_1 \quad (2-x)\cdot b_2 &= 0 \end{aligned}$$

For $\lambda_1 = i$ Gauss's Method gives this reduction.

$$\begin{aligned} (-2-i)\cdot b_1 - \quad 1\cdot b_2 &= 0 \\ 5\cdot b_1 - (2-i)\cdot b_2 &= 0 \end{aligned} \quad \xrightarrow{(-5/(-2-i))\rho_1+\rho_2} \quad \begin{aligned} (-2-i)\cdot b_1 - 1\cdot b_2 &= 0 \\ 0 &= 0 \end{aligned}$$

(For the calculation in the lower right get a common denominator

$$\frac{5}{-2-i} - (2-i) = \frac{5}{-2-i} - \frac{-2-i}{-2-i}\cdot(2-i) = \frac{5-(-5)}{-2-i}$$

to see that it gives a $0 = 0$ equation.) These are the resulting eigenspace and eigenvector.

$$\{\begin{pmatrix} (1/(-2-i))b_2 \\ b_2 \end{pmatrix} \mid b_2 \in \mathbb{C}\} \qquad \begin{pmatrix} 1/(-2-i) \\ 1 \end{pmatrix}$$

For $\lambda_2 = -i$ the system

$$\begin{aligned} (-2+i)\cdot b_1 - \quad 1\cdot b_2 &= 0 \\ 5\cdot b_1 - (2+i)\cdot b_2 &= 0 \end{aligned} \quad \xrightarrow{(-5/(-2+i))\rho_1+\rho_2} \quad \begin{aligned} (-2+i)\cdot b_1 - 1\cdot b_2 &= 0 \\ 0 &= 0 \end{aligned}$$

leads to this.

$$\{\begin{pmatrix} (1/(-2+i))b_2 \\ b_2 \end{pmatrix} \mid b_2 \in \mathbb{C}\} \qquad \begin{pmatrix} 1/(-2+i) \\ 1 \end{pmatrix}$$

Five.II.3.25 The characteristic equation is

$$0 = \begin{vmatrix} 1-x & 1 & 1 \\ 0 & -x & 1 \\ 0 & 0 & 1-x \end{vmatrix} = (1-x)^2(-x)$$

and so the eigenvalues are $\lambda_1 = 1$ (this is a repeated root of the equation) and $\lambda_2 = 0$. For the rest, consider this system.

$$\begin{aligned} (1-x)\cdot b_1 + \quad b_2 + \quad b_3 &= 0 \\ -x\cdot b_2 + \quad b_3 &= 0 \\ (1-x)\cdot b_3 &= 0 \end{aligned}$$

When $x = \lambda_1 = 1$ then the solution set is this eigenspace.

$$\{\begin{pmatrix} b_1 \\ 0 \\ 0 \end{pmatrix} \mid b_1 \in \mathbb{C}\}$$

When $x = \lambda_2 = 0$ then the solution set is this eigenspace.

$$\{\begin{pmatrix} -b_2 \\ b_2 \\ 0 \end{pmatrix} \mid b_2 \in \mathbb{C}\}$$

So these are eigenvectors associated with $\lambda_1 = 1$ and $\lambda_2 = 0$.

$$\begin{pmatrix} 1 \\ 0 \\ 0 \end{pmatrix} \qquad \begin{pmatrix} -1 \\ 1 \\ 0 \end{pmatrix}$$

Five.II.3.26 (a) The characteristic equation is

$$0 = \begin{vmatrix} 3-x & -2 & 0 \\ -2 & 3-x & 0 \\ 0 & 0 & 5-x \end{vmatrix} = x^3 - 11x^2 + 35x - 25 = (x-1)(x-5)^2$$

and so the eigenvalues are $\lambda_1 = 1$ and also the repeated eigenvalue $\lambda_2 = 5$. To find eigenvectors, consider this system.

$$\begin{aligned} (3-x)\cdot b_1 - \quad 2\cdot b_2 \quad &= 0 \\ -2\cdot b_1 + (3-x)\cdot b_2 \quad &= 0 \\ (5-x)\cdot b_3 &= 0 \end{aligned}$$

For $\lambda_1 = 1$ we get

$$\begin{aligned} 2\cdot b_1 - 2\cdot b_2 \quad &= 0 \\ -2\cdot b_1 + 2\cdot b_2 \quad &= 0 \\ 4\cdot b_3 &= 0 \end{aligned}$$

leading to this eigenspace and eigenvector.

$$\{\begin{pmatrix} b_2 \\ b_2 \\ 0 \end{pmatrix} \mid b_2 \in \mathbb{C}\} \qquad \begin{pmatrix} 1 \\ 1 \\ 0 \end{pmatrix}$$

For $\lambda_2 = 5$ the system is

$$\begin{aligned} -2\cdot b_1 - 2\cdot b_2 \quad &= 0 \\ -2\cdot b_1 - 2\cdot b_2 \quad &= 0 \\ 0\cdot b_3 &= 0 \end{aligned}$$

leading to this.

$$\{\begin{pmatrix} -b_2 \\ b_2 \\ 0 \end{pmatrix} + \begin{pmatrix} 0 \\ 0 \\ b_3 \end{pmatrix} \mid b_2, b_3 \in \mathbb{C}\} \qquad \begin{pmatrix} -1 \\ 1 \\ 0 \end{pmatrix}, \begin{pmatrix} 0 \\ 0 \\ 1 \end{pmatrix}$$

(b) The characteristic equation is

$$0=\begin{vmatrix}-x & 1 & 0\\ 0 & -x & 1\\ 4 & -17 & 8-x\end{vmatrix}=-x^3+8x^2-17x+4=-1\cdot(x-4)(x^2-4x+1)$$

and the eigenvalues are $\lambda_1=4$ and (by using the quadratic equation) $\lambda_2=2+\sqrt{3}$ and $\lambda_3=2-\sqrt{3}$. To find eigenvectors, consider this system.

$$\begin{aligned}-x\cdot b_1+\quad b_2\qquad\qquad &=0\\ -x\cdot b_2+\qquad b_3&=0\\ 4\cdot b_1-17\cdot b_2+(8-x)\cdot b_3&=0\end{aligned}$$

Substituting $x=\lambda_1=4$ gives the system

$$\begin{aligned}-4\cdot b_1+\quad b_2\qquad\qquad&=0\\ -4\cdot b_2+\quad b_3&=0\\ 4\cdot b_1-17\cdot b_2+4\cdot b_3&=0\end{aligned}\;\xrightarrow{\rho_1+\rho_3}\;\begin{aligned}-4\cdot b_1+\quad b_2\qquad\qquad&=0\\ -4\cdot b_2+\quad b_3&=0\\ -16\cdot b_2+4\cdot b_3&=0\end{aligned}\;\xrightarrow{-4\rho_2+\rho_3}\;\begin{aligned}-4\cdot b_1+\quad b_2\qquad&=0\\ -4\cdot b_2+b_3&=0\\ 0&=0\end{aligned}$$

leading to this eigenspace and eigenvector.

$$V_4=\{\begin{pmatrix}(1/16)\cdot b_3\\ (1/4)\cdot b_3\\ b_3\end{pmatrix}\bigm| b_2\in\mathbb{C}\}\qquad\begin{pmatrix}1\\4\\16\end{pmatrix}$$

Substituting $x=\lambda_2=2+\sqrt{3}$ gives the system

$$\begin{aligned}(-2-\sqrt{3})\cdot b_1+\qquad b_2\qquad\qquad&=0\\ (-2-\sqrt{3})\cdot b_2+\qquad b_3&=0\\ 4\cdot b_1-\qquad 17\cdot b_2+(6-\sqrt{3})\cdot b_3&=0\end{aligned}$$

$$\xrightarrow{(-4/(-2-\sqrt{3}))\rho_1+\rho_3}\;\begin{aligned}(-2-\sqrt{3})\cdot b_1+\qquad b_2\qquad\qquad&=0\\ (-2-\sqrt{3})\cdot b_2+\qquad b_3&=0\\ +(-9-4\sqrt{3})\cdot b_2+(6-\sqrt{3})\cdot b_3&=0\end{aligned}$$

(the middle coefficient in the third equation equals the number $(-4/(-2-\sqrt{3}))-17$; find a common denominator of $-2-\sqrt{3}$ and then rationalize the denominator by multiplying the top and bottom of the fraction by $-2+\sqrt{3}$)

$$\xrightarrow{((9+4\sqrt{3})/(-2-\sqrt{3}))\rho_2+\rho_3}\;\begin{aligned}(-2-\sqrt{3})\cdot b_1+\qquad b_2\qquad&=0\\ (-2-\sqrt{3})\cdot b_2+b_3&=0\\ 0&=0\end{aligned}$$

which leads to this eigenspace and eigenvector.

$$V_{2+\sqrt{3}}=\{\begin{pmatrix}(1/(2+\sqrt{3})^2)\cdot b_3\\ (1/(2+\sqrt{3}))\cdot b_3\\ b_3\end{pmatrix}\bigm| b_3\in\mathbb{C}\}\qquad\begin{pmatrix}(1/(2+\sqrt{3})^2)\\ (1/(2+\sqrt{3}))\\ 1\end{pmatrix}$$

Finally, substituting $x=\lambda_3=2-\sqrt{3}$ gives the system

$$\begin{aligned}(-2+\sqrt{3})\cdot b_1+\qquad b_2\qquad\qquad&=0\\ (-2+\sqrt{3})\cdot b_2+\qquad b_3&=0\\ 4\cdot b_1-\qquad 17\cdot b_2+(6+\sqrt{3})\cdot b_3&=0\end{aligned}$$

$$\xrightarrow{(-4/(-2+\sqrt{3}))\rho_1+\rho_3}\;\begin{aligned}(-2+\sqrt{3})\cdot b_1+\qquad b_2\qquad\qquad&=0\\ (-2+\sqrt{3})\cdot b_2+\qquad b_3&=0\\ (-9+4\sqrt{3})\cdot b_2+(6+\sqrt{3})\cdot b_3&=0\end{aligned}$$

$$\xrightarrow{((9-4\sqrt{3})/(-2+\sqrt{3}))\rho_2+\rho_3}\;\begin{aligned}(-2+\sqrt{3})\cdot b_1+\qquad b_2\qquad&=0\\ (-2+\sqrt{3})\cdot b_2+b_3&=0\\ 0&=0\end{aligned}$$

which gives this eigenspace and eigenvector.

$$V_{2-\sqrt{3}} = \{ \begin{pmatrix} (1/(2+\sqrt{3})^2)\cdot b_3 \\ (1/(2-\sqrt{3}))\cdot b_3 \\ b_3 \end{pmatrix} \mid b_3 \in \mathbb{C}\} \qquad \begin{pmatrix} (1/(-2+\sqrt{3})^2) \\ (1/(-2+\sqrt{3})) \\ 1 \end{pmatrix}$$

Five.II.3.27 With respect to the natural basis $B = \langle 1, x, x^2 \rangle$ the matrix representation is this.

$$\text{Rep}_{B,B}(t) = \begin{pmatrix} 5 & 6 & 2 \\ 0 & -1 & -8 \\ 1 & 0 & -2 \end{pmatrix}$$

Thus the characteristic equation

$$0 = \begin{pmatrix} 5-x & 6 & 2 \\ 0 & -1-x & -8 \\ 1 & 0 & -2-x \end{pmatrix} = (5-x)(-1-x)(-2-x) - 48 - 2\cdot(-1-x)$$

is $0 = -x^3 + 2x^2 + 15x - 36 = -1\cdot(x+4)(x-3)^2$. To find the associated eigenvectors, consider this system.

$$\begin{aligned} (5-x)\cdot b_1 + \quad 6\cdot b_2 + \quad 2\cdot b_3 &= 0 \\ (-1-x)\cdot b_2 - \quad 8\cdot b_3 &= 0 \\ b_1 \quad + (-2-x)\cdot b_3 &= 0 \end{aligned}$$

Plugging in $x = \lambda_1 = -4$ gives

$$\begin{aligned} 9b_1 + 6\cdot b_2 + 2\cdot b_3 &= 0 \\ 3\cdot b_2 - 8\cdot b_3 &= 0 \\ b_1 \quad + 2\cdot b_3 &= 0 \end{aligned} \xrightarrow{-(1/9)\rho_1+\rho_3} \xrightarrow{(2/9)\rho_2+\rho_3} \begin{aligned} 9b_1 + 6\cdot b_2 + 2\cdot b_3 &= 0 \\ 3\cdot b_2 - 8\cdot b_3 &= 0 \end{aligned}$$

The eigenspace and eigenvector are this.

$$V_{-4} = \{ \begin{pmatrix} (14/9)\cdot b_3 \\ (-8/3)\cdot b_3 \\ b_3 \end{pmatrix} \mid b_3 \in \mathbb{C}\} \qquad \begin{pmatrix} 14/9 \\ -8/3 \\ 1 \end{pmatrix}$$

Similarly, plugging in $x = \lambda_2 = 3$ gives

$$\begin{aligned} 2b_1 + \quad 6\cdot b_2 + 2\cdot b_3 &= 0 \\ -4\cdot b_2 - 8\cdot b_3 &= 0 \\ b_1 \quad - 5\cdot b_3 &= 0 \end{aligned} \xrightarrow{-(1/2)\rho_1+\rho_3} \xrightarrow{-(3/4)\rho_2+\rho_3} \begin{aligned} 2b_1 + \quad 6\cdot b_2 + 2\cdot b_3 &= 0 \\ -4\cdot b_2 - 8\cdot b_3 &= 0 \end{aligned}$$

with this eigenspace and eigenvector.

$$V_3 = \{ \begin{pmatrix} 5\cdot b_3 \\ -2\cdot b_3 \\ b_3 \end{pmatrix} \mid b_3 \in \mathbb{C}\} \qquad \begin{pmatrix} 5 \\ -2 \\ 1 \end{pmatrix}$$

Five.II.3.28 $\lambda = 1, \begin{pmatrix} 0 & 0 \\ 0 & 1 \end{pmatrix}$ and $\begin{pmatrix} 2 & 3 \\ 1 & 0 \end{pmatrix}, \lambda = -2, \begin{pmatrix} -1 & 0 \\ 1 & 0 \end{pmatrix}, \lambda = -1, \begin{pmatrix} -2 & 1 \\ 1 & 0 \end{pmatrix}$

Five.II.3.29 Fix the natural basis $B = \langle 1, x, x^2, x^3 \rangle$. The map's action is $1 \mapsto 0$, $x \mapsto 1$, $x^2 \mapsto 2x$, and $x^3 \mapsto 3x^2$ and its representation is easy to compute.

$$T = \text{Rep}_{B,B}(d/dx) = \begin{pmatrix} 0 & 1 & 0 & 0 \\ 0 & 0 & 2 & 0 \\ 0 & 0 & 0 & 3 \\ 0 & 0 & 0 & 0 \end{pmatrix}_{B,B}$$

We find the eigenvalues with this computation.

$$0 = |T - xI| = \begin{vmatrix} -x & 1 & 0 & 0 \\ 0 & -x & 2 & 0 \\ 0 & 0 & -x & 3 \\ 0 & 0 & 0 & -x \end{vmatrix} = x^4$$

Thus the map has the single eigenvalue $\lambda=0$. To find the associated eigenvectors, we solve

$$\begin{pmatrix}0&1&0&0\\0&0&2&0\\0&0&0&3\\0&0&0&0\end{pmatrix}_{B,B}\begin{pmatrix}b_1\\b_2\\b_3\\b_4\end{pmatrix}_B=0\cdot\begin{pmatrix}b_1\\b_2\\b_3\\b_4\end{pmatrix}_B \quad\Longrightarrow\quad b_2=0,\ b_3=0,\ b_4=0$$

to get this eigenspace.

$$\{\begin{pmatrix}b_1\\0\\0\\0\end{pmatrix}_B \bigm| b_1\in\mathbb{C}\}=\{b_1+0\cdot x+0\cdot x^2+0\cdot x^3 \bigm| b_1\in\mathbb{C}\}=\{b_1 \bigm| b_1\in\mathbb{C}\}$$

Five.II.3.30 The determinant of the triangular matrix $T-xI$ is the product down the diagonal, and so it factors into the product of the terms $t_{i,i}-x$.

Five.II.3.31 Just expand the determinant of $T-xI$.

$$\begin{vmatrix}a-x&c\\b&d-x\end{vmatrix}=(a-x)(d-x)-bc=x^2+(-a-d)\cdot x+(ad-bc)$$

Five.II.3.32 Any two representations of that transformation are similar, and similar matrices have the same characteristic polynomial.

Five.II.3.33 It is not true. All of the eigenvalues of this matrix are 0.

$$\begin{pmatrix}0&1\\0&0\end{pmatrix}$$

Five.II.3.34 **(a)** Use $\lambda=1$ and the identity map.

(b) Yes, use the transformation that multiplies all vectors by the scalar λ.

Five.II.3.35 If $t(\vec{v})=\lambda\cdot\vec{v}$ then $\vec{v}\mapsto\vec{0}$ under the map $t-\lambda\cdot\mathrm{id}$.

Five.II.3.36 The characteristic equation

$$0=\begin{vmatrix}a-x&b\\c&d-x\end{vmatrix}=(a-x)(d-x)-bc$$

simplifies to $x^2+(-a-d)\cdot x+(ad-bc)$. Checking that the values $x=a+b$ and $x=a-c$ satisfy the equation (under the $a+b=c+d$ condition) is routine.

Five.II.3.37 Consider an eigenspace V_λ. Any $\vec{w}\in V_\lambda$ is the image $\vec{w}=\lambda\cdot\vec{v}$ of some $\vec{v}\in V_\lambda$ (namely, $\vec{v}=(1/\lambda)\cdot\vec{w}$). Thus, on V_λ (which is a nontrivial subspace) the action of t^{-1} is $t^{-1}(\vec{w})=\vec{v}=(1/\lambda)\cdot\vec{w}$, and so $1/\lambda$ is an eigenvalue of t^{-1}.

Five.II.3.38 **(a)** We have $(cT+dI)\vec{v}=cT\vec{v}+dI\vec{v}=c\lambda\vec{v}+d\vec{v}=(c\lambda+d)\cdot\vec{v}$.

(b) Suppose that $S=PTP^{-1}$ is diagonal. Then $P(cT+dI)P^{-1}=P(cT)P^{-1}+P(dI)P^{-1}=cPTP^{-1}+dI=cS+dI$ is also diagonal.

Five.II.3.39 The scalar λ is an eigenvalue if and only if the transformation $t-\lambda\,\mathrm{id}$ is singular. A transformation is singular if and only if it is not an isomorphism (that is, a transformation is an isomorphism if and only if it is nonsingular).

Five.II.3.40 **(a)** Where the eigenvalue λ is associated with the eigenvector $\vec{x}$ then $A^k\vec{x}=A\cdots A\vec{x}=A^{k-1}\lambda\vec{x}=\lambda A^{k-1}\vec{x}=\cdots=\lambda^k\vec{x}$. (The full details require induction on k.)

(b) The eigenvector associated with λ might not be an eigenvector associated with μ.

Five.II.3.41 No. These are two same-sized, equal rank, matrices with different eigenvalues.

$$\begin{pmatrix}1&0\\0&1\end{pmatrix}\qquad\begin{pmatrix}1&0\\0&2\end{pmatrix}$$

Five.II.3.42 The characteristic polynomial has an odd power and so has at least one real root.

Five.II.3.43 The characteristic polynomial $x^3 - 5x^2 + 6x$ has distinct roots $\lambda_1 = 0$, $\lambda_2 = -2$, and $\lambda_3 = -3$. Thus the matrix can be diagonalized into this form.

$$\begin{pmatrix} 0 & 0 & 0 \\ 0 & -2 & 0 \\ 0 & 0 & -3 \end{pmatrix}$$

Five.II.3.44 We must show that it is one-to-one and onto, and that it respects the operations of matrix addition and scalar multiplication.

To show that it is one-to-one, suppose that $t_P(T) = t_P(S)$, that is, suppose that $PTP^{-1} = PSP^{-1}$, and note that multiplying both sides on the left by P^{-1} and on the right by P gives that $T = S$. To show that it is onto, consider $S \in \mathcal{M}_{n \times n}$ and observe that $S = t_P(P^{-1}SP)$.

The map t_P preserves matrix addition since $t_P(T+S) = P(T+S)P^{-1} = (PT+PS)P^{-1} = PTP^{-1}+PSP^{-1} = t_P(T+S)$ follows from properties of matrix multiplication and addition that we have seen. Scalar multiplication is similar: $t_P(cT) = P(c \cdot T)P^{-1} = c \cdot (PTP^{-1}) = c \cdot t_P(T)$.

Five.II.3.45 *This is how the answer was given in the cited source.* If the argument of the characteristic function of A is set equal to c, adding the first $(n-1)$ rows (columns) to the nth row (column) yields a determinant whose nth row (column) is zero. Thus c is a characteristic root of A.

Nilpotence

Five.III.1: Self-Composition

Five.III.1.9 For the zero transformation, no matter what the space, the chain of range spaces is $V \supset \{\vec{0}\} = \{\vec{0}\} = \cdots$ and the chain of null spaces is $\{\vec{0}\} \subset V = V = \cdots$. For the identity transformation the chains are $V = V = V = \cdots$ and $\{\vec{0}\} = \{\vec{0}\} = \cdots$.

Five.III.1.10 **(a)** Iterating t_0 twice $a + bx + cx^2 \mapsto b + cx^2 \mapsto cx^2$ gives

$$a + bx + cx^2 \stackrel{t_0^2}{\longmapsto} cx^2$$

and any higher power is the same map. Thus, while $\mathscr{R}(t_0)$ is the space of quadratic polynomials with no linear term $\{p + rx^2 \mid p, r \in \mathbb{C}\}$, and $\mathscr{R}(t_0^2)$ is the space of purely-quadratic polynomials $\{rx^2 \mid r \in \mathbb{C}\}$, this is where the chain stabilizes $\mathscr{R}_\infty(t_0) = \{rx^2 \mid n \in \mathbb{C}\}$. As for null spaces, $\mathscr{N}(t_0)$ is the space of purely-linear quadratic polynomials $\{qx \mid q \in \mathbb{C}\}$, and $\mathscr{N}(t_0^2)$ is the space of quadratic polynomials with no x^2 term $\{p + qx \mid p, q \in \mathbb{C}\}$, and this is the end $\mathscr{N}_\infty(t_0) = \mathscr{N}(t_0^2)$.

(b) The second power

$$\begin{pmatrix} a \\ b \end{pmatrix} \stackrel{t_1}{\longmapsto} \begin{pmatrix} 0 \\ a \end{pmatrix} \stackrel{t_1}{\longmapsto} \begin{pmatrix} 0 \\ 0 \end{pmatrix}$$

is the zero map. Consequently, the chain of range spaces

$$\mathbb{R}^2 \supset \{\begin{pmatrix} 0 \\ p \end{pmatrix} \mid p \in \mathbb{C}\} \supset \{\vec{0}\} = \cdots$$

and the chain of null spaces

$$\{\vec{0}\} \subset \{\begin{pmatrix} q \\ 0 \end{pmatrix} \mid q \in \mathbb{C}\} \subset \mathbb{R}^2 = \cdots$$

each has length two. The generalized range space is the trivial subspace and the generalized null space is the entire space.

(c) Iterates of this map cycle around

$$a+bx+cx^2 \stackrel{t_2}{\longmapsto} b+cx+ax^2 \stackrel{t_2}{\longmapsto} c+ax+bx^2 \stackrel{t_2}{\longmapsto} a+bx+cx^2 \ \cdots$$

and the chains of range spaces and null spaces are trivial.

$$\mathcal{P}_2 = \mathcal{P}_2 = \cdots \qquad \{\vec{0}\} = \{\vec{0}\} = \cdots$$

Thus, obviously, generalized spaces are $\mathscr{R}_\infty(t_2) = \mathcal{P}_2$ and $\mathscr{N}_\infty(t_2) = \{\vec{0}\}$.

(d) We have

$$\begin{pmatrix} a \\ b \\ c \end{pmatrix} \mapsto \begin{pmatrix} a \\ a \\ b \end{pmatrix} \mapsto \begin{pmatrix} a \\ a \\ a \end{pmatrix} \mapsto \begin{pmatrix} a \\ a \\ a \end{pmatrix} \mapsto \cdots$$

and so the chain of range spaces

$$\mathbb{R}^3 \supset \{\begin{pmatrix} p \\ p \\ r \end{pmatrix} \mid p, r \in \mathbb{C}\} \supset \{\begin{pmatrix} p \\ p \\ p \end{pmatrix} \mid p \in \mathbb{C}\} = \cdots$$

and the chain of null spaces

$$\{\vec{0}\} \subset \{\begin{pmatrix} 0 \\ 0 \\ r \end{pmatrix} \mid r \in \mathbb{C}\} \subset \{\begin{pmatrix} 0 \\ q \\ r \end{pmatrix} \mid q, r \in \mathbb{C}\} = \cdots$$

each has length two. The generalized spaces are the final ones shown above in each chain.

Five.III.1.11 Each maps $x \mapsto t(t(t(x)))$.

Five.III.1.12 Recall that if W is a subspace of V then we can enlarge any basis B_W for W to make a basis B_V for V. From this the first sentence is immediate. The second sentence is also not hard: W is the span of B_W and if W is a proper subspace then V is not the span of B_W, and so B_V must have at least one vector more than does B_W.

Five.III.1.13 It is both 'if' and 'only if'. A linear map is nonsingular if and only if it preserves dimension, that is, if the dimension of its range equals the dimension of its domain. With a transformation $t\colon V \to V$ that means that the map is nonsingular if and only if it is onto: $\mathscr{R}(t) = V$ (and thus $\mathscr{R}(t^2) = V$, etc).

Five.III.1.14 The null spaces form chains because because if $\vec{v} \in \mathscr{N}(t^j)$ then $t^j(\vec{v}) = \vec{0}$ and $t^{j+1}(\vec{v}) = t(\,t^j(\vec{v})\,) = t(\vec{0}) = \vec{0}$ and so $\vec{v} \in \mathscr{N}(t^{j+1})$.

Now, the "further" property for null spaces follows from that fact that it holds for range spaces, along with the prior exercise. Because the dimension of $\mathscr{R}(t^j)$ plus the dimension of $\mathscr{N}(t^j)$ equals the dimension n of the starting space V, when the dimensions of the range spaces stop decreasing, so do the dimensions of the null spaces. The prior exercise shows that from this point k on, the containments in the chain are not proper — the null spaces are equal.

Five.III.1.15 (Many examples are correct but here is one.) An example is the shift operator on triples of reals $(x, y, z) \mapsto (0, x, y)$. The null space is all triples that start with two zeros. The map stabilizes after three iterations.

Five.III.1.16 The differentiation operator $d/dx\colon \mathcal{P}_1 \to \mathcal{P}_1$ has the same range space as null space. For an example of where they are disjoint — except for the zero vector — consider an identity map, or any nonsingular map.

Five.III.2: Strings

Five.III.2.19 Three. It is at least three because $\ell^2(\,(1,1,1)\,) = (0,0,1) \neq \vec{0}$. It is at most three because $(x,y,z) \mapsto (0,x,y) \mapsto (0,0,x) \mapsto (0,0,0)$.

Five.III.2.20 **(a)** The domain has dimension four. The map's action is that any vector in the space $c_1 \cdot \vec{\beta}_1 + c_2 \cdot \vec{\beta}_2 + c_3 \cdot \vec{\beta}_3 + c_4 \cdot \vec{\beta}_4$ goes to $c_1 \cdot \vec{\beta}_2 + c_2 \cdot \vec{0} + c_3 \cdot \vec{\beta}_4 + c_4 \cdot \vec{0} = c_1 \cdot \vec{\beta}_3 + c_3 \cdot \vec{\beta}_4$. The first application of the map sends two basis vectors $\vec{\beta}_2$ and $\vec{\beta}_4$ to zero, and therefore the null space has dimension two and the range space has dimension two. With a second application, all four basis vectors go to zero and so the null space of the second power has dimension four while the range space of the second power has dimension zero. Thus the index of nilpotency is two. This is the canonical form.

$$\begin{pmatrix} 0 & 0 & 0 & 0 \\ 1 & 0 & 0 & 0 \\ 0 & 0 & 0 & 0 \\ 0 & 0 & 1 & 0 \end{pmatrix}$$

(b) The dimension of the domain of this map is six. For the first power the dimension of the null space is four and the dimension of the range space is two. For the second power the dimension of the null space is five and the dimension of the range space is one. Then the third iteration results in a null space of dimension six and a range space of dimension zero. The index of nilpotency is three, and this is the canonical form.

$$\begin{pmatrix} 0 & 0 & 0 & 0 & 0 & 0 \\ 1 & 0 & 0 & 0 & 0 & 0 \\ 0 & 1 & 0 & 0 & 0 & 0 \\ 0 & 0 & 0 & 0 & 0 & 0 \\ 0 & 0 & 0 & 0 & 0 & 0 \\ 0 & 0 & 0 & 0 & 0 & 0 \end{pmatrix}$$

(c) The dimension of the domain is three, and the index of nilpotency is three. The first power's null space has dimension one and its range space has dimension two. The second power's null space has dimension two and its range space has dimension one. Finally, the third power's null space has dimension three and its range space has dimension zero. Here is the canonical form matrix.

$$\begin{pmatrix} 0 & 0 & 0 \\ 1 & 0 & 0 \\ 0 & 1 & 0 \end{pmatrix}$$

Five.III.2.21 By Lemma 1.4 the nullity has grown as large as possible by the n-th iteration where n is the dimension of the domain. Thus, for the 2×2 matrices, we need only check whether the square is the zero matrix. For the 3×3 matrices, we need only check the cube.

(a) Yes, this matrix is nilpotent because its square is the zero matrix.

(b) No, the square is not the zero matrix.

$$\begin{pmatrix} 3 & 1 \\ 1 & 3 \end{pmatrix}^2 = \begin{pmatrix} 10 & 6 \\ 6 & 10 \end{pmatrix}$$

(c) Yes, the cube is the zero matrix. In fact, the square is zero.

(d) No, the third power is not the zero matrix.

$$\begin{pmatrix} 1 & 1 & 4 \\ 3 & 0 & -1 \\ 5 & 2 & 7 \end{pmatrix}^3 = \begin{pmatrix} 206 & 86 & 304 \\ 26 & 8 & 26 \\ 438 & 180 & 634 \end{pmatrix}$$

(e) Yes, the cube of this matrix is the zero matrix.

Another way to see that the second and fourth matrices are not nilpotent is to note that they are nonsingular.

Five.III.2.22 The table of calculations

p	N^p	$\mathscr{N}(N^p)$
1	$\begin{pmatrix} 0 & 1 & 1 & 0 & 1 \\ 0 & 0 & 1 & 1 & 1 \\ 0 & 0 & 0 & 0 & 0 \\ 0 & 0 & 0 & 0 & 0 \\ 0 & 0 & 0 & 0 & 0 \end{pmatrix}$	$\{\begin{pmatrix} r \\ u \\ -u-v \\ u \\ v \end{pmatrix} \mid r, u, v \in \mathbb{C}\}$
2	$\begin{pmatrix} 0 & 0 & 1 & 1 & 1 \\ 0 & 0 & 0 & 0 & 0 \\ 0 & 0 & 0 & 0 & 0 \\ 0 & 0 & 0 & 0 & 0 \\ 0 & 0 & 0 & 0 & 0 \end{pmatrix}$	$\{\begin{pmatrix} r \\ s \\ -u-v \\ u \\ v \end{pmatrix} \mid r, s, u, v \in \mathbb{C}\}$
2	*–zero matrix–*	$\mathbb{C}^5$

gives these requirements of the string basis: three basis vectors map directly to zero, one more basis vector maps to zero by a second application, and the final basis vector maps to zero by a third application. Thus, the string basis has this form.

$$\begin{array}{l} \vec{\beta}_1 \mapsto \vec{\beta}_2 \mapsto \vec{\beta}_3 \mapsto \vec{0} \\ \vec{\beta}_4 \mapsto \vec{0} \\ \vec{\beta}_5 \mapsto \vec{0} \end{array}$$

From that the canonical form is immediate.

$$\begin{pmatrix} 0 & 0 & 0 & 0 & 0 \\ 1 & 0 & 0 & 0 & 0 \\ 0 & 1 & 0 & 0 & 0 \\ 0 & 0 & 0 & 0 & 0 \\ 0 & 0 & 0 & 0 & 0 \end{pmatrix}$$

Five.III.2.23 **(a)** The canonical form has a 3×3 block and a 2×2 block

$$\left(\begin{array}{ccc|cc} 0 & 0 & 0 & 0 & 0 \\ 1 & 0 & 0 & 0 & 0 \\ 0 & 1 & 0 & 0 & 0 \\ \hline 0 & 0 & 0 & 0 & 0 \\ 0 & 0 & 0 & 1 & 0 \end{array}\right)$$

corresponding to the length three string and the length two string in the basis.

(b) Assume that N is the representation of the underlying map with respect to the standard basis. Let B be the basis to which we will change. By the similarity diagram

$$\begin{array}{ccc} \mathbb{C}^2_{wrt\ \mathcal{E}_2} & \xrightarrow[N]{n} & \mathbb{C}^2_{wrt\ \mathcal{E}_2} \\ \text{id}\big\downarrow P & & \text{id}\big\downarrow P \\ \mathbb{C}^2_{wrt\ B} & \xrightarrow{n} & \mathbb{C}^2_{wrt\ B} \end{array}$$

we have that the canonical form matrix is PNP^{-1} where

$$P^{-1} = \text{Rep}_{B,\mathcal{E}_5}(\text{id}) = \begin{pmatrix} 1 & 0 & 0 & 0 & 0 \\ 0 & 1 & 0 & 1 & 0 \\ 1 & 0 & 1 & 0 & 0 \\ 0 & 0 & 1 & 1 & 1 \\ 0 & 0 & 0 & 0 & 1 \end{pmatrix}$$

and P is the inverse of that.

$$P=\mathrm{Rep}_{\mathcal{E}_5,B}(\mathrm{id})=(P^{-1})^{-1}=\begin{pmatrix}1&0&0&0&0\\-1&1&1&-1&1\\-1&0&1&0&0\\1&0&-1&1&-1\\0&0&0&0&1\end{pmatrix}$$

(c) The calculation to check this is routine.

Five.III.2.24 **(a)** The calculation

p	N^p	$\mathscr{N}(N^p)$
1	$\begin{pmatrix}1/2&-1/2\\1/2&-1/2\end{pmatrix}$	$\{\begin{pmatrix}u\\u\end{pmatrix}\mid u\in\mathbb{C}\}$
2	*–zero matrix–*	$\mathbb{C}^2$

shows that any map represented by the matrix must act on the string basis in this way

$$\vec{\beta}_1\mapsto\vec{\beta}_2\mapsto\vec{0}$$

because the null space after one application has dimension one and exactly one basis vector, $\vec{\beta}_2$, maps to zero. Therefore, this representation with respect to $\langle\vec{\beta}_1,\vec{\beta}_2\rangle$ is the canonical form.

$$\begin{pmatrix}0&0\\1&0\end{pmatrix}$$

(b) The calculation here is similar to the prior one.

p	N^p	$\mathscr{N}(N^p)$
1	$\begin{pmatrix}0&0&0\\0&-1&1\\0&-1&1\end{pmatrix}$	$\{\begin{pmatrix}u\\v\\v\end{pmatrix}\mid u,v\in\mathbb{C}\}$
2	*–zero matrix–*	$\mathbb{C}^3$

The table shows that the string basis is of the form

$$\begin{array}{l}\vec{\beta}_1\mapsto\vec{\beta}_2\mapsto\vec{0}\\\vec{\beta}_3\mapsto\vec{0}\end{array}$$

because the null space after one application of the map has dimension two — $\vec{\beta}_2$ and $\vec{\beta}_3$ are both sent to zero — and one more iteration results in the additional vector going to zero.

(c) The calculation

p	N^p	$\mathscr{N}(N^p)$
1	$\begin{pmatrix}-1&1&-1\\1&0&1\\1&-1&1\end{pmatrix}$	$\{\begin{pmatrix}u\\0\\-u\end{pmatrix}\mid u\in\mathbb{C}\}$
2	$\begin{pmatrix}1&0&1\\0&0&0\\-1&0&-1\end{pmatrix}$	$\{\begin{pmatrix}u\\v\\-u\end{pmatrix}\mid u,v\in\mathbb{C}\}$
3	*–zero matrix–*	$\mathbb{C}^3$

shows that any map represented by this basis must act on a string basis in this way.

$$\vec{\beta}_1\mapsto\vec{\beta}_2\mapsto\vec{\beta}_3\mapsto\vec{0}$$

Therefore, this is the canonical form.

$$\begin{pmatrix}0&0&0\\1&0&0\\0&1&0\end{pmatrix}$$

Five.III.2.25 A couple of examples

$$\begin{pmatrix}0&0\\1&0\end{pmatrix}\begin{pmatrix}a&b\\c&d\end{pmatrix}=\begin{pmatrix}0&0\\a&b\end{pmatrix}\qquad\begin{pmatrix}0&0&0\\1&0&0\\0&1&0\end{pmatrix}\begin{pmatrix}a&b&c\\d&e&f\\g&h&i\end{pmatrix}=\begin{pmatrix}0&0&0\\a&b&c\\d&e&f\end{pmatrix}$$

suggest that left multiplication by a block of subdiagonal ones shifts the rows of a matrix downward. Distinct blocks

$$\begin{pmatrix}0&0&0&0\\1&0&0&0\\0&0&0&0\\0&0&1&0\end{pmatrix}\begin{pmatrix}a&b&c&d\\e&f&g&h\\i&j&k&l\\m&n&o&p\end{pmatrix}=\begin{pmatrix}0&0&0&0\\a&b&c&d\\0&0&0&0\\i&j&k&l\end{pmatrix}$$

act to shift down distinct parts of the matrix.

Right multiplication does an analogous thing to columns. See Exercise 19.

Five.III.2.26 Yes. Generalize the last sentence in Example 2.10. As to the index, that same last sentence shows that the index of the new matrix is less than or equal to the index of $\hat{N}$, and reversing the roles of the two matrices gives inequality in the other direction.

Another answer to this question is to show that a matrix is nilpotent if and only if any associated map is nilpotent, and with the same index. Then, because similar matrices represent the same map, the conclusion follows. This is Exercise 32 below.

Five.III.2.27 Observe that a canonical form nilpotent matrix has only zero eigenvalues; e.g., the determinant of this lower-triangular matrix

$$\begin{pmatrix}-x&0&0\\1&-x&0\\0&1&-x\end{pmatrix}$$

is $(-x)^3$, the only root of which is zero. But similar matrices have the same eigenvalues and every nilpotent matrix is similar to one in canonical form.

Another way to see this is to observe that a nilpotent matrix sends all vectors to zero after some number of iterations, but that conflicts with an action on an eigenspace $\vec{v}\mapsto\lambda\vec{v}$ unless λ is zero.

Five.III.2.28 No, by Lemma 1.4 for a map on a two-dimensional space, the nullity has grown as large as possible by the second iteration.

Five.III.2.29 The index of nilpotency of a transformation can be zero only when the vector starting the string must be $\vec{0}$, that is, only when V is a trivial space.

Five.III.2.30 **(a)** Any member $\vec{w}$ of the span is a linear combination $\vec{w}=c_0\cdot\vec{v}+c_1\cdot t(\vec{v})+\cdots+c_{k-1}\cdot t^{k-1}(\vec{v})$. But then, by the linearity of the map, $t(\vec{w})=c_0\cdot t(\vec{v})+c_1\cdot t^2(\vec{v})+\cdots+c_{k-2}\cdot t^{k-1}(\vec{v})+c_{k-1}\cdot\vec{0}$ is also in the span.

(b) The operation in the prior item, when iterated k times, will result in a linear combination of zeros.

(c) If $\vec{v}=\vec{0}$ then the set is empty and so is linearly independent by definition. Otherwise write $c_1\vec{v}+\cdots+c_{k-1}t^{k-1}(\vec{v})=\vec{0}$ and apply t^{k-1} to both sides. The right side gives $\vec{0}$ while the left side gives $c_1t^{k-1}(\vec{v})$; conclude that $c_1=0$. Continue in this way by applying t^{k-2} to both sides, etc.

(d) Of course, t acts on the span by acting on this basis as a single, k-long, t-string.

$$\begin{pmatrix}0&0&0&0&\ldots&0&0\\1&0&0&0&\ldots&0&0\\0&1&0&0&\ldots&0&0\\0&0&1&0&&0&0\\&&&&\ddots&&\\0&0&0&0&&1&0\end{pmatrix}$$

Five.III.2.31 We must check that $B \cup \hat{C} \cup \{\vec{v}_1, \ldots, \vec{v}_j\}$ is linearly independent where B is a t-string basis for $\mathscr{R}(t)$, where $\hat{C}$ is a basis for $\mathscr{N}(t)$, and where $t(\vec{v}_1) = \vec{\beta}_1, \ldots, t(\vec{v}_i = \vec{\beta}_i$. Write

$$\vec{0} = c_{1,-1}\vec{v}_1 + c_{1,0}\vec{\beta}_1 + c_{1,1}t(\vec{\beta}_1) + \cdots + c_{1,h_1}t^{h_1}(\vec{\beta}_1) + c_{2,-1}\vec{v}_2 + \cdots + c_{j,h_i}t^{h_i}(\vec{\beta}_i)$$

and apply t.

$$\vec{0} = c_{1,-1}\vec{\beta}_1 + c_{1,0}t(\vec{\beta}_1) + \cdots + c_{1,h_1-1}t^{h_1}(\vec{\vec{\beta}}_1) + c_{1,h_1}\vec{0} + c_{2,-1}\vec{\beta}_2 + \cdots + c_{i,h_i-1}t^{h_i}(\vec{\beta}_i) + c_{i,h_i}\vec{0}$$

Conclude that the coefficients $c_{1,-1}, \ldots, c_{1,h_i-1}, c_{2,-1}, \ldots, c_{i,h_i-1}$ are all zero as $B \cup \hat{C}$ is a basis. Substitute back into the first displayed equation to conclude that the remaining coefficients are zero also.

Five.III.2.32 For any basis B, a transformation n is nilpotent if and only if $N = \text{Rep}_{B,B}(n)$ is a nilpotent matrix. This is because only the zero matrix represents the zero map and so n^j is the zero map if and only if N^j is the zero matrix.

Five.III.2.33 It can be of any size greater than or equal to one. To have a transformation that is nilpotent of index four, whose cube has range space of dimension k, take a vector space, a basis for that space, and a transformation that acts on that basis in this way.

$$\begin{array}{ccccccccc} \vec{\beta}_1 & \mapsto & \vec{\beta}_2 & \mapsto & \vec{\beta}_3 & \mapsto & \vec{\beta}_4 & \mapsto & \vec{0} \\ \vec{\beta}_5 & \mapsto & \vec{\beta}_6 & \mapsto & \vec{\beta}_7 & \mapsto & \vec{\beta}_8 & \mapsto & \vec{0} \\ & \vdots & & & & & & & \\ \vec{\beta}_{4k-3} & \mapsto & \vec{\beta}_{4k-2} & \mapsto & \vec{\beta}_{4k-1} & \mapsto & \vec{\beta}_{4k} & \mapsto & \vec{0} \\ & \vdots & & & & & & & \end{array}$$

–possibly other, shorter, strings–

So the dimension of the range space of T^3 can be as large as desired. The smallest that it can be is one—there must be at least one string or else the map's index of nilpotency would not be four.

Five.III.2.34 These two have only zero for eigenvalues

$$\begin{pmatrix} 0 & 0 \\ 0 & 0 \end{pmatrix} \qquad \begin{pmatrix} 0 & 0 \\ 1 & 0 \end{pmatrix}$$

but are not similar (they have different canonical representatives, namely, themselves).

Five.III.2.35 It is onto by Lemma 1.4. It need not be the identity: consider this map $t\colon \mathbb{R}^2 \to \mathbb{R}^2$.

$$\begin{pmatrix} x \\ y \end{pmatrix} \overset{t}{\longmapsto} \begin{pmatrix} y \\ x \end{pmatrix}$$

For that map $\mathscr{R}_\infty(t) = \mathbb{R}^2$, and t is not the identity.

Five.III.2.36 A simple reordering of the string basis will do. For instance, a map that is associated with this string basis

$$\vec{\beta}_1 \mapsto \vec{\beta}_2 \mapsto \vec{0}$$

is represented with respect to $B = \langle \vec{\beta}_1, \vec{\beta}_2 \rangle$ by this matrix

$$\begin{pmatrix} 0 & 0 \\ 1 & 0 \end{pmatrix}$$

but is represented with respect to $B = \langle \vec{\beta}_2, \vec{\beta}_1 \rangle$ in this way.

$$\begin{pmatrix} 0 & 1 \\ 0 & 0 \end{pmatrix}$$

Five.III.2.37 Let $t\colon V \to V$ be the transformation. If $\text{rank}(t) = \text{nullity}(t)$ then the equation $\text{rank}(t) + \text{nullity}(t) = \dim(V)$ shows that $\dim(V)$ is even.

Five.III.2.38 For the matrices to be nilpotent they must be square. For them to commute they must be the same size. Thus their product and sum are defined.

Call the matrices A and B. To see that AB is nilpotent, multiply $(AB)^2 = ABAB = AABB = A^2B^2$, and $(AB)^3 = A^3B^3$, etc., and, as A is nilpotent, that product is eventually zero.

The sum is similar; use the Binomial Theorem.

Five.III.2.39 Some experimentation gives the idea for the proof. Expansion of the second power

$$t_S^2(T) = S(ST - TS) - (ST - TS)S = S^2 - 2STS + TS^2$$

the third power

$$\begin{aligned} t_S^3(T) &= S(S^2 - 2STS + TS^2) - (S^2 - 2STS + TS^2)S \\ &= S^3T - 3S^2TS + 3STS^2 - TS^3 \end{aligned}$$

and the fourth power

$$\begin{aligned} t_S^4(T) &= S(S^3T - 3S^2TS + 3STS^2 - TS^3) - (S^3T - 3S^2TS + 3STS^2 - TS^3)S \\ &= S^4T - 4S^3TS + 6S^2TS^2 - 4STS^3 + TS^4 \end{aligned}$$

suggest that the expansions follow the Binomial Theorem. Verifying this by induction on the power of t_S is routine. This answers the question because, where the index of nilpotency of S is k, in the expansion of t_S^{2k}

$$t_S^{2k}(T) = \sum_{0 \leqslant i \leqslant 2k} (-1)^i \binom{2k}{i} S^i T S^{2k-i}$$

for any i at least one of the S^i and S^{2k-i} has a power higher than k, and so the term gives the zero matrix.

Five.III.2.40 Use the geometric series: $I - N^{k+1} = (I - N)(N^k + N^{k-1} + \cdots + I)$. If N^{k+1} is the zero matrix then we have a right inverse for $I - N$. It is also a left inverse.

This statement is not 'only if' since

$$\begin{pmatrix} 1 & 0 \\ 0 & 1 \end{pmatrix} - \begin{pmatrix} -1 & 0 \\ 0 & -1 \end{pmatrix}$$

is invertible.

Jordan Form

Five.IV.1: Polynomials of Maps and Matrices

Five.IV.1.13 For each, the minimal polynomial must have a leading coefficient of 1 and Theorem 1.8, the Cayley-Hamilton Theorem, says that the minimal polynomial must contain the same linear factors as the characteristic polynomial, although possibly of lower degree but not of zero degree.

(a) The possibilities are $m_1(x) = x - 3$, $m_2(x) = (x-3)^2$, $m_3(x) = (x-3)^3$, and $m_4(x) = (x-3)^4$. Note that the 8 has been dropped because a minimal polynomial must have a leading coefficient of one. The first is a degree one polynomial, the second is degree two, the third is degree three, and the fourth is degree four.

(b) The possibilities are $m_1(x) = (x+1)(x-4)$, $m_2(x) = (x+1)^2(x-4)$, and $m_3(x) = (x+1)^3(x-4)$. The first is a quadratic polynomial, that is, it has degree two. The second has degree three, and the third has degree four.

(c) We have $m_1(x) = (x-2)(x-5)$, $m_2(x) = (x-2)^2(x-5)$, $m_3(x) = (x-2)(x-5)^2$, and $m_4(x) = (x-2)^2(x-5)^2$. They are polynomials of degree two, three, three, and four.

(d) The possibilities are $m_1(x) = (x+3)(x-1)(x-2)$, $m_2(x) = (x+3)^2(x-1)(x-2)$, $m_3(x) = (x+3)(x-1)(x-2)^2$, and $m_4(x) = (x+3)^2(x-1)(x-2)^2$. The degree of m_1 is three, the degree of m_2 is four, the degree of m_3 is four, and the degree of m_4 is five.

Five.IV.1.14 In each case we will use the method of Example 1.12.

(a) Because T is triangular, $T - xI$ is also triangular

$$T - xI = \begin{pmatrix} 3-x & 0 & 0 \\ 1 & 3-x & 0 \\ 0 & 0 & 4-x \end{pmatrix}$$

the characteristic polynomial is easy $c(x) = |T - xI| = (3-x)^2(4-x) = -1 \cdot (x-3)^2(x-4)$. There are only two possibilities for the minimal polynomial, $m_1(x) = (x-3)(x-4)$ and $m_2(x) = (x-3)^2(x-4)$. (Note that the characteristic polynomial has a negative sign but the minimal polynomial does not since it must have a leading coefficient of one). Because $m_1(T)$ is not the zero matrix

$$(T-3I)(T-4I) = \begin{pmatrix} 0 & 0 & 0 \\ 1 & 0 & 0 \\ 0 & 0 & 1 \end{pmatrix} \begin{pmatrix} -1 & 0 & 0 \\ 1 & -1 & 0 \\ 0 & 0 & 0 \end{pmatrix} = \begin{pmatrix} 0 & 0 & 0 \\ -1 & 0 & 0 \\ 0 & 0 & 0 \end{pmatrix}$$

the minimal polynomial is $m(x) = m_2(x)$.

$$(T-3I)^2(T-4I) = (T-3I) \cdot \big((T-3I)(T-4I)\big) = \begin{pmatrix} 0 & 0 & 0 \\ 1 & 0 & 0 \\ 0 & 0 & 1 \end{pmatrix} \begin{pmatrix} 0 & 0 & 0 \\ -1 & 0 & 0 \\ 0 & 0 & 0 \end{pmatrix} = \begin{pmatrix} 0 & 0 & 0 \\ 0 & 0 & 0 \\ 0 & 0 & 0 \end{pmatrix}$$

(b) As in the prior item, the fact that the matrix is triangular makes computation of the characteristic polynomial easy.

$$c(x) = |T - xI| = \begin{vmatrix} 3-x & 0 & 0 \\ 1 & 3-x & 0 \\ 0 & 0 & 3-x \end{vmatrix} = (3-x)^3 = -1 \cdot (x-3)^3$$

There are three possibilities for the minimal polynomial $m_1(x) = (x-3)$, $m_2(x) = (x-3)^2$, and $m_3(x) = (x-3)^3$. We settle the question by computing $m_1(T)$

$$T - 3I = \begin{pmatrix} 0 & 0 & 0 \\ 1 & 0 & 0 \\ 0 & 0 & 0 \end{pmatrix}$$

and $m_2(T)$.

$$(T-3I)^2 = \begin{pmatrix} 0 & 0 & 0 \\ 1 & 0 & 0 \\ 0 & 0 & 0 \end{pmatrix} \begin{pmatrix} 0 & 0 & 0 \\ 1 & 0 & 0 \\ 0 & 0 & 0 \end{pmatrix} = \begin{pmatrix} 0 & 0 & 0 \\ 0 & 0 & 0 \\ 0 & 0 & 0 \end{pmatrix}$$

Because $m_2(T)$ is the zero matrix, $m_2(x)$ is the minimal polynomial.

(c) Again, the matrix is triangular.

$$c(x) = |T - xI| = \begin{vmatrix} 3-x & 0 & 0 \\ 1 & 3-x & 0 \\ 0 & 1 & 3-x \end{vmatrix} = (3-x)^3 = -1 \cdot (x-3)^3$$

Again, there are three possibilities for the minimal polynomial $m_1(x) = (x-3)$, $m_2(x) = (x-3)^2$, and $m_3(x) = (x-3)^3$. We compute $m_1(T)$

$$T - 3I = \begin{pmatrix} 0 & 0 & 0 \\ 1 & 0 & 0 \\ 0 & 1 & 0 \end{pmatrix}$$

and $m_2(T)$

$$(T-3I)^2 = \begin{pmatrix} 0 & 0 & 0 \\ 1 & 0 & 0 \\ 0 & 1 & 0 \end{pmatrix} \begin{pmatrix} 0 & 0 & 0 \\ 1 & 0 & 0 \\ 0 & 1 & 0 \end{pmatrix} = \begin{pmatrix} 0 & 0 & 0 \\ 0 & 0 & 0 \\ 1 & 0 & 0 \end{pmatrix}$$

and $m_3(T)$.

$$(T-3I)^3 = (T-3I)^2(T-3I) = \begin{pmatrix} 0 & 0 & 0 \\ 0 & 0 & 0 \\ 1 & 0 & 0 \end{pmatrix} \begin{pmatrix} 0 & 0 & 0 \\ 1 & 0 & 0 \\ 0 & 1 & 0 \end{pmatrix} = \begin{pmatrix} 0 & 0 & 0 \\ 0 & 0 & 0 \\ 0 & 0 & 0 \end{pmatrix}$$

Therefore, the minimal polynomial is $m(x) = m_3(x) = (x-3)^3$.

(d) This case is also triangular, here upper triangular.

$$c(x) = |T - xI| = \begin{vmatrix} 2-x & 0 & 1 \\ 0 & 6-x & 2 \\ 0 & 0 & 2-x \end{vmatrix} = (2-x)^2(6-x) = -(x-2)^2(x-6)$$

There are two possibilities for the minimal polynomial, $m_1(x) = (x-2)(x-6)$ and $m_2(x) = (x-2)^2(x-6)$. Computation shows that the minimal polynomial isn't $m_1(x)$.

$$(T-2I)(T-6I) = \begin{pmatrix} 0 & 0 & 1 \\ 0 & 4 & 2 \\ 0 & 0 & 0 \end{pmatrix} \begin{pmatrix} -4 & 0 & 1 \\ 0 & 0 & 2 \\ 0 & 0 & -4 \end{pmatrix} = \begin{pmatrix} 0 & 0 & -4 \\ 0 & 0 & 0 \\ 0 & 0 & 0 \end{pmatrix}$$

It therefore must be that $m(x) = m_2(x) = (x-2)^2(x-6)$. Here is a verification.

$$(T-2I)^2(T-6I) = (T-2I) \cdot \big((T-2I)(T-6I)\big) = \begin{pmatrix} 0 & 0 & 1 \\ 0 & 4 & 2 \\ 0 & 0 & 0 \end{pmatrix} \begin{pmatrix} 0 & 0 & -4 \\ 0 & 0 & 0 \\ 0 & 0 & 0 \end{pmatrix} = \begin{pmatrix} 0 & 0 & 0 \\ 0 & 0 & 0 \\ 0 & 0 & 0 \end{pmatrix}$$

(e) The characteristic polynomial is

$$c(x) = |T - xI| = \begin{vmatrix} 2-x & 2 & 1 \\ 0 & 6-x & 2 \\ 0 & 0 & 2-x \end{vmatrix} = (2-x)^2(6-x) = -(x-2)^2(x-6)$$

and there are two possibilities for the minimal polynomial, $m_1(x) = (x-2)(x-6)$ and $m_2(x) = (x-2)^2(x-6)$. Checking the first one

$$(T-2I)(T-6I) = \begin{pmatrix} 0 & 2 & 1 \\ 0 & 4 & 2 \\ 0 & 0 & 0 \end{pmatrix} \begin{pmatrix} -4 & 2 & 1 \\ 0 & 0 & 2 \\ 0 & 0 & -4 \end{pmatrix} = \begin{pmatrix} 0 & 0 & 0 \\ 0 & 0 & 0 \\ 0 & 0 & 0 \end{pmatrix}$$

shows that the minimal polynomial is $m(x) = m_1(x) = (x-2)(x-6)$.

(f) The characteristic polynomial is this.

$$c(x) = |T - xI| = \begin{vmatrix} -1-x & 4 & 0 & 0 & 0 \\ 0 & 3-x & 0 & 0 & 0 \\ 0 & -4 & -1-x & 0 & 0 \\ 3 & -9 & -4 & 2-x & -1 \\ 1 & 5 & 4 & 1 & 4-x \end{vmatrix} = (x-3)^3(x+1)^2$$

Here are the possibilities for the minimal polynomial, listed here by ascending degree: $m_1(x) = (x-3)(x+1)$, $m_1(x) = (x-3)^2(x+1)$, $m_1(x) = (x-3)(x+1)^2$, $m_1(x) = (x-3)^3(x+1)$, $m_1(x) = (x-3)^2(x+1)^2$, and $m_1(x) = (x-3)^3(x+1)^2$. The first one doesn't pan out

$$(T-3I)(T+1I) = \begin{pmatrix} -4 & 4 & 0 & 0 & 0 \\ 0 & 0 & 0 & 0 & 0 \\ 0 & -4 & -4 & 0 & 0 \\ 3 & -9 & -4 & -1 & -1 \\ 1 & 5 & 4 & 1 & 1 \end{pmatrix} \begin{pmatrix} 0 & 4 & 0 & 0 & 0 \\ 0 & 4 & 0 & 0 & 0 \\ 0 & -4 & 0 & 0 & 0 \\ 3 & -9 & -4 & 3 & -1 \\ 1 & 5 & 4 & 1 & 5 \end{pmatrix}$$

$$= \begin{pmatrix} 0 & 0 & 0 & 0 & 0 \\ 0 & 0 & 0 & 0 & 0 \\ 0 & 0 & 0 & 0 & 0 \\ -4 & -4 & 0 & -4 & -4 \\ 4 & 4 & 0 & 4 & 4 \end{pmatrix}$$

but the second one does.

$$(T-3I)^2(T+1I) = (T-3I)\big((T-3I)(T+1I)\big)$$
$$= \begin{pmatrix} -4 & 4 & 0 & 0 & 0 \\ 0 & 0 & 0 & 0 & 0 \\ 0 & -4 & -4 & 0 & 0 \\ 3 & -9 & -4 & -1 & -1 \\ 1 & 5 & 4 & 1 & 1 \end{pmatrix} \begin{pmatrix} 0 & 0 & 0 & 0 & 0 \\ 0 & 0 & 0 & 0 & 0 \\ 0 & 0 & 0 & 0 & 0 \\ -4 & -4 & 0 & -4 & -4 \\ 4 & 4 & 0 & 4 & 4 \end{pmatrix}$$
$$= \begin{pmatrix} 0 & 0 & 0 & 0 & 0 \\ 0 & 0 & 0 & 0 & 0 \\ 0 & 0 & 0 & 0 & 0 \\ 0 & 0 & 0 & 0 & 0 \\ 0 & 0 & 0 & 0 & 0 \end{pmatrix}$$

The minimal polynomial is $m(x) = (x-3)^2(x+1)$.

Five.IV.1.15 Its characteristic polynomial has complex roots.

$$\begin{vmatrix} -x & 1 & 0 \\ 0 & -x & 1 \\ 1 & 0 & -x \end{vmatrix} = (1-x)\cdot\Big(x-\big(-\frac{1}{2}+\frac{\sqrt{3}}{2}i\big)\Big)\cdot\Big(x-\big(-\frac{1}{2}-\frac{\sqrt{3}}{2}i\big)\Big)$$

As the roots are distinct, the characteristic polynomial equals the minimal polynomial.

Five.IV.1.16 We know that $\mathcal{P}_n$ is a dimension $n+1$ space and that the differentiation operator is nilpotent of index $n+1$ (for instance, taking $n=3$, $\mathcal{P}_3 = \{c_3x^3+c_2x^2+c_1x+c_0 \mid c_3,\ldots,c_0 \in \mathbb{C}\}$ and the fourth derivative of a cubic is the zero polynomial). Represent this operator using the canonical form for nilpotent transformations.

$$\begin{pmatrix} 0 & 0 & 0 & \ldots & & 0 \\ 1 & 0 & 0 & & & 0 \\ 0 & 1 & 0 & & & \\ & & & \ddots & & \\ 0 & 0 & 0 & & 1 & 0 \end{pmatrix}$$

This is an $(n+1)\times(n+1)$ matrix with an easy characteristic polynomial, $c(x) = x^{n+1}$. (*Remark:* this matrix is $\mathrm{Rep}_{B,B}(d/dx)$ where $B = \langle x^n, nx^{n-1}, n(n-1)x^{n-2}, \ldots, n!\rangle$.) To find the minimal polynomial as in Example 1.12 we consider the powers of $T-0I = T$. But, of course, the first power of T that is the zero matrix is the power $n+1$. So the minimal polynomial is also x^{n+1}.

Five.IV.1.17 Call the matrix T and suppose that it is $n\times n$. Because T is triangular, and so $T-xI$ is triangular, the characteristic polynomial is $c(x) = (x-\lambda)^n$. To see that the minimal polynomial is the same, consider $T-\lambda I$.

$$\begin{pmatrix} 0 & 0 & 0 & \ldots & 0 \\ 1 & 0 & 0 & \ldots & 0 \\ 0 & 1 & 0 & & \\ & & \ddots & & \\ 0 & 0 & \ldots & 1 & 0 \end{pmatrix}$$

Recognize it as the canonical form for a transformation that is nilpotent of degree n; the power $(T-\lambda I)^j$ is zero first when j is n.

Five.IV.1.18 The $n=3$ case provides a hint. A natural basis for $\mathcal{P}_3$ is $B = \langle 1, x, x^2, x^3\rangle$. The action of the transformation is

$$1 \mapsto 1 \quad x \mapsto x+1 \quad x^2 \mapsto x^2+2x+1 \quad x^3 \mapsto x^3+3x^2+3x+1$$

and so the representation $\mathrm{Rep}_{B,B}(t)$ is this upper triangular matrix.

$$\begin{pmatrix} 1 & 1 & 1 & 1 \\ 0 & 1 & 2 & 3 \\ 0 & 0 & 1 & 3 \\ 0 & 0 & 0 & 1 \end{pmatrix}$$

Because it is triangular, the fact that the characteristic polynomial is $c(x) = (x-1)^4$ is clear. For the minimal polynomial, the candidates are $m_1(x) = (x-1)$,

$$T - 1I = \begin{pmatrix} 0 & 1 & 1 & 1 \\ 0 & 0 & 2 & 3 \\ 0 & 0 & 0 & 3 \\ 0 & 0 & 0 & 0 \end{pmatrix}$$

$m_2(x) = (x-1)^2$,

$$(T - 1I)^2 = \begin{pmatrix} 0 & 0 & 2 & 6 \\ 0 & 0 & 0 & 6 \\ 0 & 0 & 0 & 0 \\ 0 & 0 & 0 & 0 \end{pmatrix}$$

$m_3(x) = (x-1)^3$,

$$(T - 1I)^3 = \begin{pmatrix} 0 & 0 & 0 & 6 \\ 0 & 0 & 0 & 0 \\ 0 & 0 & 0 & 0 \\ 0 & 0 & 0 & 0 \end{pmatrix}$$

and $m_4(x) = (x-1)^4$. Because m_1, m_2, and m_3 are not right, m_4 must be right, as is easily verified.

In the case of a general n, the representation is an upper triangular matrix with ones on the diagonal. Thus the characteristic polynomial is $c(x) = (x-1)^{n+1}$. One way to verify that the minimal polynomial equals the characteristic polynomial is argue something like this: say that an upper triangular matrix is 0-upper triangular if there are nonzero entries on the diagonal, that it is 1-upper triangular if the diagonal contains only zeroes and there are nonzero entries just above the diagonal, etc. As the above example illustrates, an induction argument will show that, where T has only nonnegative entries, T^j is j-upper triangular.

Five.IV.1.19 The map twice is the same as the map once: $\pi \circ \pi = \pi$, that is, $\pi^2 = \pi$ and so the minimal polynomial is of degree at most two since $m(x) = x^2 - x$ will do. The fact that no linear polynomial will do follows from applying the maps on the left and right side of $c_1 \cdot \pi + c_0 \cdot \mathrm{id} = z$ (where z is the zero map) to these two vectors.

$$\begin{pmatrix} 0 \\ 0 \\ 1 \end{pmatrix} \quad \begin{pmatrix} 1 \\ 0 \\ 0 \end{pmatrix}$$

Thus the minimal polynomial is m.

Five.IV.1.20 This is one answer.

$$\begin{pmatrix} 0 & 0 & 0 \\ 1 & 0 & 0 \\ 0 & 0 & 0 \end{pmatrix}$$

Five.IV.1.21 The x must be a scalar, not a matrix.

Five.IV.1.22 The characteristic polynomial of

$$T = \begin{pmatrix} a & b \\ c & d \end{pmatrix}$$

is $(a-x)(d-x)-bc=x^2-(a+d)x+(ad-bc)$. Substitute

$$\begin{pmatrix} a & b \\ c & d \end{pmatrix}^2 - (a+d)\begin{pmatrix} a & b \\ c & d \end{pmatrix} + (ad-bc)\begin{pmatrix} 1 & 0 \\ 0 & 1 \end{pmatrix}$$
$$= \begin{pmatrix} a^2+bc & ab+bd \\ ac+cd & bc+d^2 \end{pmatrix} - \begin{pmatrix} a^2+ad & ab+bd \\ ac+cd & ad+d^2 \end{pmatrix} + \begin{pmatrix} ad-bc & 0 \\ 0 & ad-bc \end{pmatrix}$$

and just check each entry sum to see that the result is the zero matrix.

Five.IV.1.23 By the Cayley-Hamilton theorem the degree of the minimal polynomial is less than or equal to the degree of the characteristic polynomial, n. Example 1.6 shows that n can happen.

Five.IV.1.24 Let the linear transformation be $t\colon V \to V$. If t is nilpotent then there is an n such that t^n is the zero map, so t satisfies the polynomial $p(x)=x^n=(x-0)^n$. By Lemma 1.10 the minimal polynomial of t divides p, so the minimal polynomial has only zero for a root. By Cayley-Hamilton, Theorem 1.8, the characteristic polynomial has only zero for a root. Thus the only eigenvalue of t is zero.

Conversely, if a transformation t on an n-dimensional space has only the single eigenvalue of zero then its characteristic polynomial is x^n. The Cayley-Hamilton Theorem says that a map satisfies its characteristic polynomial so t^n is the zero map. Thus t is nilpotent.

Five.IV.1.25 A minimal polynomial must have leading coefficient 1, and so if the minimal polynomial of a map or matrix were to be a degree zero polynomial then it would be $m(x)=1$. But the identity map or matrix equals the zero map or matrix only on a trivial vector space.

So in the nontrivial case the minimal polynomial must be of degree at least one. A zero map or matrix has minimal polynomial $m(x)=x$, and an identity map or matrix has minimal polynomial $m(x)=x-1$.

Five.IV.1.26 We can interpret the polynomial can geometrically as, "a 60° rotation minus two rotations of 30° equals the identity."

Five.IV.1.27 For a diagonal matrix

$$T=\begin{pmatrix} t_{1,1} & 0 & & \\ 0 & t_{2,2} & & \\ & & \ddots & \\ & & & t_{n,n} \end{pmatrix}$$

the characteristic polynomial is $(t_{1,1}-x)(t_{2,2}-x)\cdots(t_{n,n}-x)$. Of course, some of those factors may be repeated, e.g., the matrix might have $t_{1,1}=t_{2,2}$. For instance, the characteristic polynomial of

$$D=\begin{pmatrix} 3 & 0 & 0 \\ 0 & 3 & 0 \\ 0 & 0 & 1 \end{pmatrix}$$

is $(3-x)^2(1-x)=-1\cdot(x-3)^2(x-1)$.

To form the minimal polynomial, take the terms $x-t_{i,i}$, throw out repeats, and multiply them together. For instance, the minimal polynomial of D is $(x-3)(x-1)$. To check this, note first that Theorem 1.8, the Cayley-Hamilton theorem, requires that each linear factor in the characteristic polynomial appears at least once in the minimal polynomial. One way to check the other direction—that in the case of a diagonal matrix, each linear factor need appear at most once—is to use a matrix argument. A diagonal matrix, multiplying from the left, rescales rows by the entry on the diagonal. But in a product $(T-t_{1,1}I)\cdots$, even without any repeat factors, every row is zero in at least one of the factors.

For instance, in the product

$$(D-3I)(D-1I)=(D-3I)(D-1I)I=\begin{pmatrix} 0 & 0 & 0 \\ 0 & 0 & 0 \\ 0 & 0 & -2 \end{pmatrix}\begin{pmatrix} 2 & 0 & 0 \\ 0 & 2 & 0 \\ 0 & 0 & 0 \end{pmatrix}\begin{pmatrix} 1 & 0 & 0 \\ 0 & 1 & 0 \\ 0 & 0 & 1 \end{pmatrix}$$

because the first and second rows of the first matrix $D-3I$ are zero, the entire product will have a first row and second row that are zero. And because the third row of the middle matrix $D-1I$ is zero, the entire product has a third row of zero.

Five.IV.1.28 This subsection starts with the observation that the powers of a linear transformation cannot climb forever without a "repeat", that is, that for some power n there is a linear relationship $c_n \cdot t^n + \cdots + c_1 \cdot t + c_0 \cdot \mathrm{id} = z$ where z is the zero transformation. The definition of projection is that for such a map one linear relationship is quadratic, $t^2 - t = z$. To finish, we need only consider whether this relationship might not be minimal, that is, are there projections for which the minimal polynomial is constant or linear?

For the minimal polynomial to be constant, the map would have to satisfy that $c_0 \cdot \mathrm{id} = z$, where $c_0 = 1$ since the leading coefficient of a minimal polynomial is 1. This is only satisfied by the zero transformation on a trivial space. This is a projection, but not an interesting one.

For the minimal polynomial of a transformation to be linear would give $c_1 \cdot t + c_0 \cdot \mathrm{id} = z$ where $c_1 = 1$. This equation gives $t = -c_0 \cdot \mathrm{id}$. Coupling it with the requirement that $t^2 = t$ gives $t^2 = (-c_0)^2 \cdot \mathrm{id} = -c_0 \cdot \mathrm{id}$, which gives that $c_0 = 0$ and t is the zero transformation or that $c_0 = 1$ and t is the identity.

Thus, except in the cases where the projection is a zero map or an identity map, the minimal polynomial is $m(x) = x^2 - x$.

Five.IV.1.29 **(a)** *This is a property of functions in general, not just of linear functions.* Suppose that f and g are one-to-one functions such that $f \circ g$ is defined. Let $f \circ g(x_1) = f \circ g(x_2)$, so that $f(g(x_1)) = f(g(x_2))$. Because f is one-to-one this implies that $g(x_1) = g(x_2)$. Because g is also one-to-one, this in turn implies that $x_1 = x_2$. Thus, in summary, $f \circ g(x_1) = f \circ g(x_2)$ implies that $x_1 = x_2$ and so $f \circ g$ is one-to-one.

(b) If the linear map h is not one-to-one then there are unequal vectors $\vec{v}_1$, $\vec{v}_2$ that map to the same value $h(\vec{v}_1) = h(\vec{v}_2)$. Because h is linear, we have $\vec{0} = h(\vec{v}_1) - h(\vec{v}_2) = h(\vec{v}_1 - \vec{v}_2)$ and so $\vec{v}_1 - \vec{v}_2$ is a nonzero vector from the domain that h maps to the zero vector of the codomain ($\vec{v}_1 - \vec{v}_2$ does not equal the zero vector of the domain because $\vec{v}_1$ does not equal $\vec{v}_2$).

(c) The minimal polynomial $m(t)$ sends every vector in the domain to zero and so it is not one-to-one (except in a trivial space, which we ignore). By the first item of this question, since the composition $m(t)$ is not one-to-one, at least one of the components $t - \lambda_i$ is not one-to-one. By the second item, $t - \lambda_i$ has a nontrivial null space. Because $(t - \lambda_i)(\vec{v}) = \vec{0}$ holds if and only if $t(\vec{v}) = \lambda_i \cdot \vec{v}$, the prior sentence gives that λ_i is an eigenvalue (recall that the definition of eigenvalue requires that the relationship hold for at least one nonzero $\vec{v}$).

Five.IV.1.30 This is false. The natural example of a non-diagonalizable transformation works here. Consider the transformation of $\mathbb{C}^2$ represented with respect to the standard basis by this matrix.

$$N = \begin{pmatrix} 0 & 1 \\ 0 & 0 \end{pmatrix}$$

The characteristic polynomial is $c(x) = x^2$. Thus the minimal polynomial is either $m_1(x) = x$ or $m_2(x) = x^2$. The first is not right since $N - 0 \cdot I$ is not the zero matrix, thus in this example the minimal polynomial has degree equal to the dimension of the underlying space, and, as mentioned, we know this matrix is not diagonalizable because it is nilpotent.

Five.IV.1.31 Let A and B be similar $A = PBP^{-1}$. From the facts that

$$\begin{aligned} A^n &= (PBP^{-1})^n = (PBP^{-1})(PBP^{-1}) \cdots (PBP^{-1}) \\ &= PB(P^{-1}P)B(P^{-1}P) \cdots (P^{-1}P)BP^{-1} = PB^nP^{-1} \end{aligned}$$

and $c \cdot A = c \cdot (PBP^{-1}) = P(c \cdot B)P^{-1}$ follows the required fact that for any polynomial function f we have $f(A) = P\, f(B)\, P^{-1}$. For instance, if $f(x) = x^2 + 2x + 3$ then

$$\begin{aligned} A^2 + 2A + 3I &= (PBP^{-1})^2 + 2 \cdot PBP^{-1} + 3 \cdot I \\ &= (PBP^{-1})(PBP^{-1}) + P(2B)P^{-1} + 3 \cdot PP^{-1} = P(B^2 + 2B + 3I)P^{-1} \end{aligned}$$

shows that $f(A)$ is similar to $f(B)$.

(a) Taking f to be a linear polynomial we have that $A - xI$ is similar to $B - xI$. Similar matrices have equal determinants (since $|A| = |PBP^{-1}| = |P| \cdot |B| \cdot |P^{-1}| = 1 \cdot |B| \cdot 1 = |B|$). Thus the characteristic polynomials are equal.

(b) As P and P^{-1} are invertible, $f(A)$ is the zero matrix when, and only when, $f(B)$ is the zero matrix.

(c) They cannot be similar since they don't have the same characteristic polynomial. The characteristic polynomial of the first one is $x^2 - 4x - 3$ while the characteristic polynomial of the second is $x^2 - 5x + 5$.

Five.IV.1.32 Suppose that $m(x) = x^n + m_{n-1}x^{n-1} + \cdots + m_1x + m_0$ is minimal for T.

(a) For the 'if' argument, because $T^n + \cdots + m_1T + m_0I$ is the zero matrix we have that $I = (T^n + \cdots + m_1T)/(-m_0) = T \cdot (T^{n-1} + \cdots + m_1I)/(-m_0)$ and so the matrix $(-1/m_0) \cdot (T^{n-1} + \cdots + m_1I)$ is the inverse of T. For 'only if', suppose that $m_0 = 0$ (we put the $n = 1$ case aside but it is easy) so that $T^n + \cdots + m_1T = (T^{n-1} + \cdots + m_1I)T$ is the zero matrix. Note that $T^{n-1} + \cdots + m_1I$ is not the zero matrix because the degree of the minimal polynomial is n. If T^{-1} exists then multiplying both $(T^{n-1} + \cdots + m_1I)T$ and the zero matrix from the right by T^{-1} gives a contradiction.

(b) If T is not invertible then the constant term in its minimal polynomial is zero. Thus,

$$T^n + \cdots + m_1T = (T^{n-1} + \cdots + m_1I)T = T(T^{n-1} + \cdots + m_1I)$$

is the zero matrix.

Five.IV.1.33 **(a)** For the inductive step, assume that Lemma 1.7 is true for polynomials of degree $i, \ldots, k-1$ and consider a polynomial $f(x)$ of degree k. Factor $f(x) = k(x - \lambda_1)^{q_1} \cdots (x - \lambda_z)^{q_z}$ and let $k(x - \lambda_1)^{q_1 - 1} \cdots (x - \lambda_z)^{q_z}$ be $c_{n-1}x^{n-1} + \cdots + c_1x + c_0$. Substitute:

$$\begin{aligned} k(t - \lambda_1)^{q_1} \circ \cdots \circ (t - \lambda_z)^{q_z}(\vec{v}) &= (t - \lambda_1) \circ (t - \lambda_1)^{q_1} \circ \cdots \circ (t - \lambda_z)^{q_z}(\vec{v}) \\ &= (t - \lambda_1)\,(c_{n-1}t^{n-1}(\vec{v}) + \cdots + c_0\vec{v}) \\ &= f(t)(\vec{v}) \end{aligned}$$

(the second equality follows from the inductive hypothesis and the third from the linearity of t).

(b) One example is to consider the squaring map $s\colon \mathbb{R} \to \mathbb{R}$ given by $s(x) = x^2$. It is nonlinear. The action defined by the polynomial $f(t) = t^2 - 1$ changes s to $f(s) = s^2 - 1$, which is this map.

$$x \stackrel{s^2-1}{\longmapsto} s \circ s(x) - 1 = x^4 - 1$$

Observe that this map differs from the map $(s - 1) \circ (s + 1)$; for instance, the first map takes $x = 5$ to 624 while the second one takes $x = 5$ to 675.

Five.IV.1.34 Yes. Expand down the last column to check that $x^n + m_{n-1}x^{n-1} + \cdots + m_1x + m_0$ is plus or minus the determinant of this.

$$\begin{pmatrix} -x & 0 & 0 & & & m_0 \\ 0 & 1-x & 0 & & & m_1 \\ 0 & 0 & 1-x & & & m_2 \\ & & & \ddots & & \\ & & & & 1-x & m_{n-1} \end{pmatrix}$$

Five.IV.2: Jordan Canonical Form

Five.IV.2.18 We must check that

$$\begin{pmatrix} 3 & 0 \\ 1 & 3 \end{pmatrix} = N + 3I = PTP^{-1} = \begin{pmatrix} 1/2 & 1/2 \\ -1/4 & 1/4 \end{pmatrix} \begin{pmatrix} 2 & -1 \\ 1 & 4 \end{pmatrix} \begin{pmatrix} 1 & -2 \\ 1 & 2 \end{pmatrix}$$

That calculation is easy.

Five.IV.2.19 **(a)** The characteristic polynomial is $c(x) = (x-3)^2$ and the minimal polynomial is the same.
(b) The characteristic polynomial is $c(x) = (x+1)^2$. The minimal polynomial is $m(x) = x+1$.
(c) The characteristic polynomial is $c(x) = (x+(1/2))(x-2)^2$ and the minimal polynomial is the same.
(d) The characteristic polynomial is $c(x) = (x-3)^3$ The minimal polynomial is the same.
(e) The characteristic polynomial is $c(x) = (x-3)^4$. The minimal polynomial is $m(x) = (x-3)^2$.
(f) The characteristic polynomial is $c(x) = (x+4)^2(x-4)^2$ and the minimal polynomial is the same.
(g) The characteristic polynomial is $c(x) = (x-2)^2(x-3)(x-5)$ and the minimal polynomial is $m(x) = (x-2)(x-3)(x-5)$.
(h) The characteristic polynomial is $c(x) = (x-2)^2(x-3)(x-5)$ and the minimal polynomial is the same.

Five.IV.2.20 **(a)** The transformation $t-3$ is nilpotent (that is, $\mathscr{N}_\infty(t-3)$ is the entire space) and it acts on a string basis via two strings, $\vec{\beta}_1 \mapsto \vec{\beta}_2 \mapsto \vec{\beta}_3 \mapsto \vec{\beta}_4 \mapsto \vec{0}$ and $\vec{\beta}_5 \mapsto \vec{0}$. Consequently, $t-3$ can be represented in this canonical form.

$$N_3 = \begin{pmatrix} 0 & 0 & 0 & 0 & 0 \\ 1 & 0 & 0 & 0 & 0 \\ 0 & 1 & 0 & 0 & 0 \\ 0 & 0 & 1 & 0 & 0 \\ 0 & 0 & 0 & 0 & 0 \end{pmatrix}$$

and therefore T is similar to this canonical form matrix.

$$J_3 = N_3 + 3I = \begin{pmatrix} 3 & 0 & 0 & 0 & 0 \\ 1 & 3 & 0 & 0 & 0 \\ 0 & 1 & 3 & 0 & 0 \\ 0 & 0 & 1 & 3 & 0 \\ 0 & 0 & 0 & 0 & 3 \end{pmatrix}$$

(b) The restriction of the transformation $s+1$ is nilpotent on the subspace $\mathscr{N}_\infty(s+1)$, and the action on a string basis is $\vec{\beta}_1 \mapsto \vec{0}$. The restriction of the transformation $s-2$ is nilpotent on the subspace $\mathscr{N}_\infty(s-2)$, having the action on a string basis of $\vec{\beta}_2 \mapsto \vec{\beta}_3 \mapsto \vec{0}$ and $\vec{\beta}_4 \mapsto \vec{\beta}_5 \mapsto \vec{0}$. Consequently the Jordan form is this.

$$\begin{pmatrix} -1 & 0 & 0 & 0 & 0 \\ 0 & 2 & 0 & 0 & 0 \\ 0 & 1 & 2 & 0 & 0 \\ 0 & 0 & 0 & 2 & 0 \\ 0 & 0 & 0 & 1 & 2 \end{pmatrix}$$

Five.IV.2.21 For each, because many choices of basis are possible, many other answers are possible. Of course, the calculation to check if an answer gives that PTP^{-1} is in Jordan form is the arbiter of what's correct.

(a) Here is the arrow diagram.

$$\begin{array}{ccc} \mathbb{C}^3_{wrt\ \mathcal{E}_3} & \xrightarrow[T]{t} & \mathbb{C}^3_{wrt\ \mathcal{E}_3} \\ \text{id}\big\downarrow P & & \text{id}\big\downarrow P \\ \mathbb{C}^3_{wrt\ B} & \xrightarrow[J]{t} & \mathbb{C}^3_{wrt\ B} \end{array}$$

The matrix to move from the lower left to the upper left is this.

$$P^{-1} = \left(\text{Rep}_{\mathcal{E}_3,B}(\text{id})\right)^{-1} = \text{Rep}_{B,\mathcal{E}_3}(\text{id}) = \begin{pmatrix} 1 & -2 & 0 \\ 1 & 0 & 1 \\ -2 & 0 & 0 \end{pmatrix}$$

The matrix P to move from the upper right to the lower right is the inverse of P^{-1}.

(b) We want this matrix and its inverse.

$$P^{-1}=\begin{pmatrix}1&0&3\\0&1&4\\0&-2&0\end{pmatrix}$$

(c) The concatenation of these bases for the generalized null spaces will do for the basis for the entire space.

$$B_{-1}=\langle\begin{pmatrix}-1\\0\\0\\1\\0\end{pmatrix},\begin{pmatrix}-1\\0\\-1\\0\\1\end{pmatrix}\rangle\qquad B_3=\langle\begin{pmatrix}1\\1\\-1\\0\\0\end{pmatrix},\begin{pmatrix}0\\0\\0\\-2\\2\end{pmatrix},\begin{pmatrix}-1\\-1\\1\\2\\0\end{pmatrix}\rangle$$

The change of basis matrices are this one and its inverse.

$$P^{-1}=\begin{pmatrix}-1&-1&1&0&-1\\0&0&1&0&-1\\0&-1&-1&0&1\\1&0&0&-2&2\\0&1&0&2&0\end{pmatrix}$$

Five.IV.2.22 The general procedure is to factor the characteristic polynomial $c(x)=(x-\lambda_1)^{p_1}(x-\lambda_2)^{p_2}\cdots$ to get the eigenvalues λ_1,λ_2, etc. Then, for each λ_i we find a string basis for the action of the transformation $t-\lambda_i$ when restricted to $\mathscr{N}_\infty(t-\lambda_i)$, by computing the powers of the matrix $T-\lambda_i I$ and finding the associated null spaces, until these null spaces settle down (do not change), at which point we have the generalized null space. The dimensions of those null spaces (the nullities) tell us the action of $t-\lambda_i$ on a string basis for the generalized null space, and so we can write the pattern of subdiagonal ones to have N_{λ_i}. From this matrix, the Jordan block J_{λ_i} associated with λ_i is immediate $J_{\lambda_i}=N_{\lambda_i}+\lambda_i I$. Finally, after we have done this for each eigenvalue, we put them together into the canonical form.

(a) The characteristic polynomial of this matrix is $c(x)=(-10-x)(10-x)+100=x^2$, so it has only the single eigenvalue $\lambda=0$.

power p	$(T+0\cdot I)^p$	$\mathscr{N}((t-0)^p)$	*nullity*
1	$\begin{pmatrix}-10&4\\-25&10\end{pmatrix}$	$\{\begin{pmatrix}2y/5\\y\end{pmatrix}\mid y\in\mathbb{C}\}$	1
2	$\begin{pmatrix}0&0\\0&0\end{pmatrix}$	$\mathbb{C}^2$	2

(Thus, this transformation is nilpotent: $\mathscr{N}_\infty(t-0)$ is the entire space). From the nullities we know that t's action on a string basis is $\vec{\beta}_1\mapsto\vec{\beta}_2\mapsto\vec{0}$. This is the canonical form matrix for the action of $t-0$ on $\mathscr{N}_\infty(t-0)=\mathbb{C}^2$

$$N_0=\begin{pmatrix}0&0\\1&0\end{pmatrix}$$

and this is the Jordan form of the matrix.

$$J_0=N_0+0\cdot I=\begin{pmatrix}0&0\\1&0\end{pmatrix}$$

Note that if a matrix is nilpotent then its canonical form equals its Jordan form.

We can find such a string basis using the techniques of the prior section.

$$B=\langle\begin{pmatrix}1\\0\end{pmatrix},\begin{pmatrix}-10\\-25\end{pmatrix}\rangle$$

We took the first basis vector so that it is in the null space of t^2 but is not in the null space of t. The second basis vector is the image of the first under t.

(b) The characteristic polynomial of this matrix is $c(x) = (x+1)^2$, so it is a single-eigenvalue matrix. (That is, the generalized null space of $t+1$ is the entire space.) We have

$$\mathscr{N}(t+1) = \{\begin{pmatrix} 2y/3 \\ y \end{pmatrix} \mid y \in \mathbb{C}\} \qquad \mathscr{N}((t+1)^2) = \mathbb{C}^2$$

and so the action of $t+1$ on an associated string basis is $\vec{\beta}_1 \mapsto \vec{\beta}_2 \mapsto \vec{0}$. Thus,

$$N_{-1} = \begin{pmatrix} 0 & 0 \\ 1 & 0 \end{pmatrix}$$

the Jordan form of T is

$$J_{-1} = N_{-1} + -1 \cdot I = \begin{pmatrix} -1 & 0 \\ 1 & -1 \end{pmatrix}$$

and choosing vectors from the above null spaces gives this string basis (other choices are possible).

$$B = \langle \begin{pmatrix} 1 \\ 0 \end{pmatrix}, \begin{pmatrix} 6 \\ 9 \end{pmatrix} \rangle$$

(c) The characteristic polynomial $c(x) = (1-x)(4-x)^2 = -1 \cdot (x-1)(x-4)^2$ has two roots and they are the eigenvalues $\lambda_1 = 1$ and $\lambda_2 = 4$.

We handle the two eigenvalues separately. For λ_1, the calculation of the powers of $T - 1I$ yields

$$\mathscr{N}(t-1) = \{\begin{pmatrix} 0 \\ y \\ 0 \end{pmatrix} \mid y \in \mathbb{C}\}$$

and the null space of $(t-1)^2$ is the same. Thus this set is the generalized null space $\mathscr{N}_\infty(t-1)$. The nullities show that the action of the restriction of $t-1$ to the generalized null space on a string basis is $\vec{\beta}_1 \mapsto \vec{0}$.

A similar calculation for $\lambda_2 = 4$ gives these null spaces.

$$\mathscr{N}(t-4) = \{\begin{pmatrix} 0 \\ z \\ z \end{pmatrix} \mid z \in \mathbb{C}\} \qquad \mathscr{N}((t-4)^2) = \{\begin{pmatrix} y-z \\ y \\ z \end{pmatrix} \mid y, z \in \mathbb{C}\}$$

(The null space of $(t-4)^3$ is the same, as it must be because the power of the term associated with $\lambda_2 = 4$ in the characteristic polynomial is two, and so the restriction of $t-2$ to the generalized null space $\mathscr{N}_\infty(t-2)$ is nilpotent of index at most two—it takes at most two applications of $t-2$ for the null space to settle down.) The pattern of how the nullities rise tells us that the action of $t-4$ on an associated string basis for $\mathscr{N}_\infty(t-4)$ is $\vec{\beta}_2 \mapsto \vec{\beta}_3 \mapsto \vec{0}$.

Putting the information for the two eigenvalues together gives the Jordan form of the transformation t.

$$\begin{pmatrix} 1 & 0 & 0 \\ 0 & 4 & 0 \\ 0 & 1 & 4 \end{pmatrix}$$

We can take elements of the null spaces to get an appropriate basis.

$$B = B_1 {}^\frown B_4 = \langle \begin{pmatrix} 0 \\ 1 \\ 0 \end{pmatrix}, \begin{pmatrix} 1 \\ 0 \\ 1 \end{pmatrix}, \begin{pmatrix} 0 \\ 5 \\ 5 \end{pmatrix} \rangle$$

(d) The characteristic polynomial is $c(x) = (-2-x)(4-x)^2 = -1 \cdot (x+2)(x-4)^2$.

For the eigenvalue λ_{-2}, calculation of the powers of $T + 2I$ yields this.

$$\mathscr{N}(t+2) = \{\begin{pmatrix} z \\ z \\ z \end{pmatrix} \mid z \in \mathbb{C}\}$$

The null space of $(t+2)^2$ is the same, and so this is the generalized null space $\mathscr{N}_\infty(t+2)$. Thus the action of the restriction of $t+2$ to $\mathscr{N}_\infty(t+2)$ on an associated string basis is $\vec{\beta}_1 \mapsto \vec{0}$.

For $\lambda_2 = 4$, computing the powers of $T-4I$ yields

$$\mathscr{N}(t-4) = \{\begin{pmatrix} z \\ -z \\ z \end{pmatrix} \mid z \in \mathbb{C}\} \qquad \mathscr{N}((t-4)^2) = \{\begin{pmatrix} x \\ -z \\ z \end{pmatrix} \mid x, z \in \mathbb{C}\}$$

and so the action of $t-4$ on a string basis for $\mathscr{N}_\infty(t-4)$ is $\vec{\beta}_2 \mapsto \vec{\beta}_3 \mapsto \vec{0}$.

Therefore the Jordan form is

$$\begin{pmatrix} -2 & 0 & 0 \\ 0 & 4 & 0 \\ 0 & 1 & 4 \end{pmatrix}$$

and a suitable basis is this.

$$B = B_{-2}{}^\frown B_4 = \langle \begin{pmatrix} 1 \\ 1 \\ 1 \end{pmatrix}, \begin{pmatrix} 0 \\ -1 \\ 1 \end{pmatrix}, \begin{pmatrix} -1 \\ 1 \\ -1 \end{pmatrix} \rangle$$

(e) The characteristic polynomial of this matrix is $c(x) = (2-x)^3 = -1 \cdot (x-2)^3$. This matrix has only a single eigenvalue, $\lambda = 2$. By finding the powers of $T-2I$ we have

$$\mathscr{N}(t-2) = \{\begin{pmatrix} -y \\ y \\ 0 \end{pmatrix} \mid y \in \mathbb{C}\} \qquad \mathscr{N}((t-2)^2) = \{\begin{pmatrix} -y-(1/2)z \\ y \\ z \end{pmatrix} \mid y, z \in \mathbb{C}\} \qquad \mathscr{N}((t-2)^3) = \mathbb{C}^3$$

and so the action of $t-2$ on an associated string basis is $\vec{\beta}_1 \mapsto \vec{\beta}_2 \mapsto \vec{\beta}_3 \mapsto \vec{0}$. The Jordan form is this

$$\begin{pmatrix} 2 & 0 & 0 \\ 1 & 2 & 0 \\ 0 & 1 & 2 \end{pmatrix}$$

and one choice of basis is this.

$$B = \langle \begin{pmatrix} 0 \\ 1 \\ 0 \end{pmatrix}, \begin{pmatrix} 7 \\ -9 \\ 4 \end{pmatrix}, \begin{pmatrix} -2 \\ 2 \\ 0 \end{pmatrix} \rangle$$

(f) The characteristic polynomial $c(x) = (1-x)^3 = -(x-1)^3$ has only a single root, so the matrix has only a single eigenvalue $\lambda = 1$. Finding the powers of $T-1I$ and calculating the null spaces

$$\mathscr{N}(t-1) = \{\begin{pmatrix} -2y+z \\ y \\ z \end{pmatrix} \mid y, z \in \mathbb{C}\} \qquad \mathscr{N}((t-1)^2) = \mathbb{C}^3$$

shows that the action of the nilpotent map $t-1$ on a string basis is $\vec{\beta}_1 \mapsto \vec{\beta}_2 \mapsto \vec{0}$ and $\vec{\beta}_3 \mapsto \vec{0}$. Therefore the Jordan form is

$$J = \begin{pmatrix} 1 & 0 & 0 \\ 1 & 1 & 0 \\ 0 & 0 & 1 \end{pmatrix}$$

and an appropriate basis (a string basis associated with $t-1$) is this.

$$B = \langle \begin{pmatrix} 0 \\ 1 \\ 0 \end{pmatrix}, \begin{pmatrix} 2 \\ -2 \\ -2 \end{pmatrix}, \begin{pmatrix} 1 \\ 0 \\ 1 \end{pmatrix} \rangle$$

(g) The characteristic polynomial is a bit large for by-hand calculation, but just manageable $c(x) = x^4 - 24x^3 + 216x^2 - 864x + 1296 = (x-6)^4$. This is a single-eigenvalue map, so the transformation $t-6$

is nilpotent. The null spaces

$$\mathscr{N}(t-6)=\{\begin{pmatrix}-z-w\\-z-w\\z\\w\end{pmatrix} \mid z,w\in\mathbb{C}\}\quad \mathscr{N}((t-6)^2)=\{\begin{pmatrix}x\\-z-w\\z\\w\end{pmatrix} \mid x,z,w\in\mathbb{C}\}\quad \mathscr{N}((t-6)^3)=\mathbb{C}^4$$

and the nullities show that the action of $t-6$ on a string basis is $\vec{\beta}_1\mapsto\vec{\beta}_2\mapsto\vec{\beta}_3\mapsto\vec{0}$ and $\vec{\beta}_4\mapsto\vec{0}$. The Jordan form is

$$\begin{pmatrix}6&0&0&0\\1&6&0&0\\0&1&6&0\\0&0&0&6\end{pmatrix}$$

and finding a suitable string basis is routine.

$$B=\langle\begin{pmatrix}0\\0\\0\\1\end{pmatrix},\begin{pmatrix}2\\-1\\-1\\2\end{pmatrix},\begin{pmatrix}3\\3\\-6\\3\end{pmatrix},\begin{pmatrix}-1\\-1\\1\\0\end{pmatrix}\rangle$$

Five.IV.2.23 There are two eigenvalues, $\lambda_1=-2$ and $\lambda_2=1$. The restriction of $t+2$ to $\mathscr{N}_\infty(t+2)$ could have either of these actions on an associated string basis.

$$\vec{\beta}_1\mapsto\vec{\beta}_2\mapsto\vec{0}\qquad\begin{array}{l}\vec{\beta}_1\mapsto\vec{0}\\\vec{\beta}_2\mapsto\vec{0}\end{array}$$

The restriction of $t-1$ to $\mathscr{N}_\infty(t-1)$ could have either of these actions on an associated string basis.

$$\vec{\beta}_3\mapsto\vec{\beta}_4\mapsto\vec{0}\qquad\begin{array}{l}\vec{\beta}_3\mapsto\vec{0}\\\vec{\beta}_4\mapsto\vec{0}\end{array}$$

In combination, that makes four possible Jordan forms, the two first actions, the second and first, the first and second, and the two second actions.

$$\begin{pmatrix}-2&0&0&0\\1&-2&0&0\\0&0&1&0\\0&0&1&1\end{pmatrix}\quad\begin{pmatrix}-2&0&0&0\\0&-2&0&0\\0&0&1&0\\0&0&1&1\end{pmatrix}\quad\begin{pmatrix}-2&0&0&0\\1&-2&0&0\\0&0&1&0\\0&0&0&1\end{pmatrix}\quad\begin{pmatrix}-2&0&0&0\\0&-2&0&0\\0&0&1&0\\0&0&0&1\end{pmatrix}$$

Five.IV.2.24 The restriction of $t+2$ to $\mathscr{N}_\infty(t+2)$ can have only the action $\vec{\beta}_1\mapsto\vec{0}$. The restriction of $t-1$ to $\mathscr{N}_\infty(t-1)$ could have any of these three actions on an associated string basis.

$$\vec{\beta}_2\mapsto\vec{\beta}_3\mapsto\vec{\beta}_4\mapsto\vec{0}\qquad\begin{array}{l}\vec{\beta}_2\mapsto\vec{\beta}_3\mapsto\vec{0}\\\vec{\beta}_4\mapsto\vec{0}\end{array}\qquad\begin{array}{l}\vec{\beta}_2\mapsto\vec{0}\\\vec{\beta}_3\mapsto\vec{0}\\\vec{\beta}_4\mapsto\vec{0}\end{array}$$

Taken together there are three possible Jordan forms, the one arising from the first action by $t-1$ (along with the only action from $t+2$), the one arising from the second action, and the one arising from the third action.

$$\begin{pmatrix}-2&0&0&0\\0&1&0&0\\0&1&1&0\\0&0&1&1\end{pmatrix}\quad\begin{pmatrix}-2&0&0&0\\0&1&0&0\\0&1&1&0\\0&0&0&1\end{pmatrix}\quad\begin{pmatrix}-2&0&0&0\\0&1&0&0\\0&0&1&0\\0&0&0&1\end{pmatrix}$$

Five.IV.2.25 The action of $t+1$ on a string basis for $\mathscr{N}_\infty(t+1)$ must be $\vec{\beta}_1\mapsto\vec{0}$. Because of the power of $x-2$ in the minimal polynomial, a string basis for $t-2$ has length two and so the action of $t-2$ on $\mathscr{N}_\infty(t-2)$ must be of this form.

$$\begin{array}{l}\vec{\beta}_2\mapsto\vec{\beta}_3\mapsto\vec{0}\\\vec{\beta}_4\mapsto\vec{0}\end{array}$$

Therefore there is only one Jordan form that is possible.

$$\begin{pmatrix} -1 & 0 & 0 & 0 \\ 0 & 2 & 0 & 0 \\ 0 & 1 & 2 & 0 \\ 0 & 0 & 0 & 2 \end{pmatrix}$$

Five.IV.2.26 There are two possible Jordan forms. The action of $t+1$ on a string basis for $\mathscr{N}_\infty(t+1)$ must be $\vec{\beta}_1 \mapsto \vec{0}$. There are two actions for $t-2$ on a string basis for $\mathscr{N}_\infty(t-2)$ that are possible with this characteristic polynomial and minimal polynomial.

$$\begin{array}{ll} \vec{\beta}_2 \mapsto \vec{\beta}_3 \mapsto \vec{0} & \qquad \vec{\beta}_2 \mapsto \vec{\beta}_3 \mapsto \vec{0} \\ \vec{\beta}_4 \mapsto \vec{\beta}_5 \mapsto \vec{0} & \qquad \vec{\beta}_4 \mapsto \vec{0} \\ & \qquad \vec{\beta}_5 \mapsto \vec{0} \end{array}$$

The resulting Jordan form matrices are these.

$$\begin{pmatrix} -1 & 0 & 0 & 0 & 0 \\ 0 & 2 & 0 & 0 & 0 \\ 0 & 1 & 2 & 0 & 0 \\ 0 & 0 & 0 & 2 & 0 \\ 0 & 0 & 0 & 1 & 2 \end{pmatrix} \qquad \begin{pmatrix} -1 & 0 & 0 & 0 & 0 \\ 0 & 2 & 0 & 0 & 0 \\ 0 & 1 & 2 & 0 & 0 \\ 0 & 0 & 0 & 2 & 0 \\ 0 & 0 & 0 & 0 & 2 \end{pmatrix}$$

Five.IV.2.27 **(a)** The characteristic polynomial is $c(x) = x(x-1)$. For $\lambda_1 = 0$ we have

$$\mathscr{N}(t-0) = \{\begin{pmatrix} -y \\ y \end{pmatrix} \mid y \in \mathbb{C}\}$$

(of course, the null space of t^2 is the same). For $\lambda_2 = 1$,

$$\mathscr{N}(t-1) = \{\begin{pmatrix} x \\ 0 \end{pmatrix} \mid x \in \mathbb{C}\}$$

(and the null space of $(t-1)^2$ is the same). We can take this basis

$$B = \langle \begin{pmatrix} 1 \\ -1 \end{pmatrix}, \begin{pmatrix} 1 \\ 0 \end{pmatrix} \rangle$$

to get the diagonalization.

$$\begin{pmatrix} 1 & 1 \\ -1 & 0 \end{pmatrix}^{-1} \begin{pmatrix} 1 & 1 \\ 0 & 0 \end{pmatrix} \begin{pmatrix} 1 & 1 \\ -1 & 0 \end{pmatrix} = \begin{pmatrix} 0 & 0 \\ 0 & 1 \end{pmatrix}$$

(b) The characteristic polynomial is $c(x) = x^2 - 1 = (x+1)(x-1)$. For $\lambda_1 = -1$,

$$\mathscr{N}(t+1) = \{\begin{pmatrix} -y \\ y \end{pmatrix} \mid y \in \mathbb{C}\}$$

and the null space of $(t+1)^2$ is the same. For $\lambda_2 = 1$

$$\mathscr{N}(t-1) = \{\begin{pmatrix} y \\ y \end{pmatrix} \mid y \in \mathbb{C}\}$$

and the null space of $(t-1)^2$ is the same. We can take this basis

$$B = \langle \begin{pmatrix} 1 \\ -1 \end{pmatrix}, \begin{pmatrix} 1 \\ 1 \end{pmatrix} \rangle$$

to get a diagonalization.

$$\begin{pmatrix} 1 & 1 \\ 1 & -1 \end{pmatrix}^{-1} \begin{pmatrix} 0 & 1 \\ 1 & 0 \end{pmatrix} \begin{pmatrix} 1 & 1 \\ -1 & 1 \end{pmatrix} = \begin{pmatrix} -1 & 0 \\ 0 & 1 \end{pmatrix}$$

Five.IV.2.28 The transformation $d/dx\colon \mathcal{P}_3 \to \mathcal{P}_3$ is nilpotent. Its action on $B = \langle x^3, 3x^2, 6x, 6\rangle$ is $x^3 \mapsto 3x^2 \mapsto 6x \mapsto 6 \mapsto 0$. Its Jordan form is its canonical form as a nilpotent matrix.

$$J = \begin{pmatrix} 0 & 0 & 0 & 0 \\ 1 & 0 & 0 & 0 \\ 0 & 1 & 0 & 0 \\ 0 & 0 & 1 & 0 \end{pmatrix}$$

Five.IV.2.29 Yes. Each has the characteristic polynomial $(x+1)^2$. Calculations of the powers of $T_1 + 1 \cdot I$ and $T_2 + 1 \cdot I$ gives these two.

$$\mathscr{N}(t_1 + 1) = \{ \begin{pmatrix} y/2 \\ y \end{pmatrix} \mid y \in \mathbb{C} \} \qquad \mathscr{N}(t_2 + 1) = \{ \begin{pmatrix} 0 \\ y \end{pmatrix} \mid y \in \mathbb{C} \}$$

(Of course, for each the null space of the square is the entire space.) The way that the nullities rise shows that each is similar to this Jordan form matrix

$$\begin{pmatrix} -1 & 0 \\ 1 & -1 \end{pmatrix}$$

and they are therefore similar to each other.

Five.IV.2.30 Its characteristic polynomial is $c(x) = x^2 + 1$ which has complex roots $x^2 + 1 = (x+i)(x-i)$. Because the roots are distinct, the matrix is diagonalizable and its Jordan form is that diagonal matrix.

$$\begin{pmatrix} -i & 0 \\ 0 & i \end{pmatrix}$$

To find an associated basis we compute the null spaces.

$$\mathscr{N}(t + i) = \{ \begin{pmatrix} -iy \\ y \end{pmatrix} \mid y \in \mathbb{C} \} \qquad \mathscr{N}(t - i) = \{ \begin{pmatrix} iy \\ y \end{pmatrix} \mid y \in \mathbb{C} \}$$

For instance,

$$T + i \cdot I = \begin{pmatrix} i & -1 \\ 1 & i \end{pmatrix}$$

and so we get a description of the null space of $t + i$ by solving this linear system.

$$\begin{array}{rl} ix - y = 0 \\ x + iy = 0 \end{array} \xrightarrow{i\rho_1 + \rho_2} \begin{array}{rl} ix - y = 0 \\ 0 = 0 \end{array}$$

(To change the relation $ix = y$ so that the leading variable x is expressed in terms of the free variable y, we can multiply both sides by $-i$.)

As a result, one such basis is this.

$$B = \langle \begin{pmatrix} -i \\ 1 \end{pmatrix}, \begin{pmatrix} i \\ 1 \end{pmatrix} \rangle$$

Five.IV.2.31 We can count the possible classes by counting the possible canonical representatives, that is, the possible Jordan form matrices. The characteristic polynomial must be either $c_1(x) = (x+3)^2(x-4)$ or $c_2(x) = (x+3)(x-4)^2$. In the c_1 case there are two possible actions of $t + 3$ on a string basis for $\mathscr{N}_\infty(t+3)$.

$$\vec{\beta}_1 \mapsto \vec{\beta}_2 \mapsto \vec{0} \qquad \begin{array}{l} \vec{\beta}_1 \mapsto \vec{0} \\ \vec{\beta}_2 \mapsto \vec{0} \end{array}$$

There are two associated Jordan form matrices.

$$\begin{pmatrix} -3 & 0 & 0 \\ 1 & -3 & 0 \\ 0 & 0 & 4 \end{pmatrix} \qquad \begin{pmatrix} -3 & 0 & 0 \\ 0 & -3 & 0 \\ 0 & 0 & 4 \end{pmatrix}$$

Similarly there are two Jordan form matrices that could arise out of c_2.

$$\begin{pmatrix} -3 & 0 & 0 \\ 0 & 4 & 0 \\ 0 & 1 & 4 \end{pmatrix} \qquad \begin{pmatrix} -3 & 0 & 0 \\ 0 & 4 & 0 \\ 0 & 0 & 4 \end{pmatrix}$$

So in total there are four possible Jordan forms.

Five.IV.2.32 Jordan form is unique. A diagonal matrix is in Jordan form. Thus the Jordan form of a diagonalizable matrix is its diagonalization. If the minimal polynomial has factors to some power higher than one then the Jordan form has subdiagonal 1's, and so is not diagonal.

Five.IV.2.33 One example is the transformation of $\mathbb{C}$ that sends x to $-x$.

Five.IV.2.34 Apply Lemma 2.8 twice; the subspace is $t-\lambda_1$ invariant if and only if it is t invariant, which in turn holds if and only if it is $t-\lambda_2$ invariant.

Five.IV.2.35 False; these two 4×4 matrices each have $c(x)=(x-3)^4$ and $m(x)=(x-3)^2$.

$$\begin{pmatrix}3&0&0&0\\1&3&0&0\\0&0&3&0\\0&0&1&3\end{pmatrix}\quad\begin{pmatrix}3&0&0&0\\1&3&0&0\\0&0&3&0\\0&0&0&3\end{pmatrix}$$

Five.IV.2.36 (a) The characteristic polynomial is this.

$$\begin{vmatrix}a-x&b\\c&d-x\end{vmatrix}=(a-x)(d-x)-bc=ad-(a+d)x+x^2-bc=x^2-(a+d)x+(ad-bc)$$

Note that the determinant appears as the constant term.

(b) Recall that the characteristic polynomial $|T-xI|$ is invariant under similarity. Use the permutation expansion formula to show that the trace is the negative of the coefficient of x^{n-1}.

(c) No, there are matrices T and S that are equivalent $S=PTQ$ (for some nonsingular P and Q) but that have different traces. An easy example is this.

$$PTQ=\begin{pmatrix}2&0\\0&1\end{pmatrix}\begin{pmatrix}1&0\\0&1\end{pmatrix}\begin{pmatrix}1&0\\0&1\end{pmatrix}=\begin{pmatrix}2&0\\0&1\end{pmatrix}$$

Even easier examples using 1×1 matrices are possible.

(d) Put the matrix in Jordan form. By the first item, the trace is unchanged.

(e) The first part is easy; use the third item. The converse does not hold: this matrix

$$\begin{pmatrix}1&0\\0&-1\end{pmatrix}$$

has a trace of zero but is not nilpotent.

Five.IV.2.37 Suppose that B_M is a basis for a subspace M of some vector space. Implication one way is clear; if M is t invariant then in particular, if $\vec{m}\in B_M$ then $t(\vec{m})\in M$. For the other implication, let $B_M=\langle\vec{\beta}_1,\dots,\vec{\beta}_q\rangle$ and note that $t(\vec{m})=t(m_1\vec{\beta}_1+\cdots+m_q\vec{\beta}_q)=m_1t(\vec{\beta}_1)+\cdots+m_qt(\vec{\beta}_q)$ is in M as any subspace is closed under linear combinations.

Five.IV.2.38 Yes, the intersection of t invariant subspaces is t invariant. Assume that M and N are t invariant. If $\vec{v}\in M\cap N$ then $t(\vec{v})\in M$ by the invariance of M and $t(\vec{v})\in N$ by the invariance of N.

Of course, the union of two subspaces need not be a subspace (remember that the x- and y-axes are subspaces of the plane $\mathbb{R}^2$ but the union of the two axes fails to be closed under vector addition; for instance it does not contain $\vec{e}_1+\vec{e}_2$.) However, the union of invariant subsets is an invariant subset; if $\vec{v}\in M\cup N$ then $\vec{v}\in M$ or $\vec{v}\in N$ so $t(\vec{v})\in M$ or $t(\vec{v})\in N$.

No, the complement of an invariant subspace need not be invariant. Consider the subspace

$$\{\begin{pmatrix}x\\0\end{pmatrix}\mid x\in\mathbb{C}\}$$

of $\mathbb{C}^2$ under the zero transformation.

Yes, the sum of two invariant subspaces is invariant. The check is easy.

Five.IV.2.39 One such ordering is the *dictionary ordering*. Order by the real component first, then by the coefficient of i. For instance, $3+2i<4+1i$ but $4+1i<4+2i$.

Five.IV.2.40 The first half is easy — the derivative of any real polynomial is a real polynomial of lower degree. The answer to the second half is 'no'; any complement of $\mathcal{P}_j(\mathbb{R})$ must include a polynomial of degree $j+1$, and the derivative of that polynomial is in $\mathcal{P}_j(\mathbb{R})$.

Five.IV.2.41 For the first half, show that each is a subspace and then observe that any polynomial can be uniquely written as the sum of even-powered and odd-powered terms (the zero polynomial is both). The answer to the second half is 'no': x^2 is even while $2x$ is odd.

Five.IV.2.42 Yes. If $\text{Rep}_{B,B}(t)$ has the given block form, take B_M to be the first j vectors of B, where J is the $j\times j$ upper left submatrix. Take B_N to be the remaining k vectors in B. Let M and N be the spans of B_M and B_N. Clearly M and N are complementary. To see M is invariant (N works the same way), represent any $\vec{m}\in M$ with respect to B, note the last k components are zeroes, and multiply by the given block matrix. The final k components of the result are zeroes, so that result is again in M.

Five.IV.2.43 Put the matrix in Jordan form. By non-singularity, there are no zero eigenvalues on the diagonal. Ape this example:

$$\begin{pmatrix} 9 & 0 & 0 \\ 1 & 9 & 0 \\ 0 & 0 & 4 \end{pmatrix} = \begin{pmatrix} 3 & 0 & 0 \\ 1/6 & 3 & 0 \\ 0 & 0 & 2 \end{pmatrix}^2$$

to construct a square root. Show that it holds up under similarity: if $S^2 = T$ then $(PSP^{-1})(PSP^{-1}) = PTP^{-1}$.

Topic: Method of Powers

1 (a) The largest eigenvalue is 4.
(b) The largest eigenvalue is 2.

3 (a) The largest eigenvalue is 3.
(b) The largest eigenvalue is -3.

5 In theory, this method would produce λ_2. In practice, however, rounding errors in the computation introduce components in the direction of $\vec{v}_1$, and so the method will still produce λ_1, although it may take somewhat longer than it would have taken with a more fortunate choice of initial vector.

6 Instead of using $\vec{v}_k = T\vec{v}_{k-1}$, use $T^{-1}\vec{v}_k = \vec{v}_{k-1}$.

Topic: Stable Populations

Topic: Page Ranking

1 The sum of the entries in column j is $\sum_i \alpha h_{i,j} + (1-\alpha)s_{i,j} = \sum_i \alpha h_{i,j} + \sum_i (1-\alpha)s_{i,j} = \alpha\sum_i \alpha h_{i,j} + (1-\alpha)\sum_i s_{i,j} = \alpha\cdot 1 + (1-\alpha)\cdot 1$, which is one.

2 This *Sage* session gives equal values.

```
sage: H=matrix(QQ,[[0,0,0,1], [1,0,0,0], [0,1,0,0], [0,0,1,0]])
sage: S=matrix(QQ,[[1/4,1/4,1/4,1/4], [1/4,1/4,1/4,1/4], [1/4,1/4,1/4,1/4], [1/4,1/4,1/4,1/4]])
sage: alpha=0.85
sage: G=alpha*H+(1-alpha)*S
sage: I=matrix(QQ,[[1,0,0,0], [0,1,0,0], [0,0,1,0], [0,0,0,1]])
sage: N=G-I
```

```
sage: 1200*N
[-1155.00000000000  45.0000000000000  45.0000000000000  1065.00000000000]
[ 1065.00000000000 -1155.00000000000  45.0000000000000  45.0000000000000]
[ 45.0000000000000  1065.00000000000 -1155.00000000000  45.0000000000000]
[ 45.0000000000000  45.0000000000000  1065.00000000000 -1155.00000000000]
sage: M=matrix(QQ,[[-1155,45,45,1065], [1065,-1155,45,45], [45,1065,-1155,45], [45,45,1065,-1155]])
sage: M.echelon_form()
[ 1  0  0 -1]
[ 0  1  0 -1]
[ 0  0  1 -1]
[ 0  0  0  0]
sage: v=vector([1,1,1,1])
sage: (v/v.norm()).n()
(0.500000000000000, 0.500000000000000, 0.500000000000000, 0.500000000000000)
```

3 We have this.

$$H = \begin{pmatrix} 0 & 0 & 1 & 1/2 \\ 1/3 & 0 & 0 & 0 \\ 1/3 & 1/2 & 0 & 1/2 \\ 1/3 & 1/2 & 0 & 0 \end{pmatrix}$$

(a) This *Sage* session gives the answer.

```
sage: H=matrix(QQ,[[0,0,1,1/2], [1/3,0,0,0], [1/3,1/2,0,1/2], [1/3,1/2,0,0]])
sage: S=matrix(QQ,[[1/4,1/4,1/4,1/4], [1/4,1/4,1/4,1/4], [1/4,1/4,1/4,1/4], [1/4,1/4,1/4,1/4]])
sage: I=matrix(QQ,[[1,0,0,0], [0,1,0,0], [0,0,1,0], [0,0,0,1]])
sage: alpha=0.85
sage: G=alpha*H+(1-alpha)*S
sage: N=G-I
sage: 1200*N
[-1155.00000000000  45.0000000000000  1065.00000000000  555.000000000000]
[ 385.000000000000 -1155.00000000000  45.0000000000000  45.0000000000000]
[ 385.000000000000  555.000000000000 -1155.00000000000  555.000000000000]
[ 385.000000000000  555.000000000000  45.0000000000000 -1155.00000000000]
sage: M=matrix(QQ,[[-1155,45,1065,555], [385,-1155,45,45], [385,555,-1155,555], [385,555,45,-1155]])
sage: M.echelon_form()
[            1             0             0 -106613/58520]
[            0             1             0        -40/57]
[            0             0             1        -57/40]
[            0             0             0             0]
sage: v=vector([106613/58520,40/57,57/40,1])
sage: (v/v.norm()).n()
(0.696483066294572, 0.268280959381099, 0.544778023143244, 0.382300367118066)
```

(b) Continue the *Sage* to get this.

```
sage: alpha=0.95
sage: G=alpha*H+(1-alpha)*S
sage: N=G-I
sage: 1200*N
[-1185.00000000000  15.0000000000000  1155.00000000000  585.000000000000]
[ 395.000000000000 -1185.00000000000  15.0000000000000  15.0000000000000]
[ 395.000000000000  585.000000000000 -1185.00000000000  585.000000000000]
[ 395.000000000000  585.000000000000  15.0000000000000 -1185.00000000000]
sage: M=matrix(QQ,[[-1185,15,1155,585], [395,-1185,15,15], [395,585,-1185,585], [395,585,15,-1185]])
sage: M.echelon_form()
[             1              0              0 -361677/186440]
[             0              1              0         -40/59]
[             0              0              1         -59/40]
[             0              0              0              0]
sage: v=vector([361677/186440,40/59,59/40,1])
sage: (v/v.norm()).n()
(0.713196892748114, 0.249250262646952, 0.542275102671275, 0.367644137404254)
```

(c) Page p_3 is important, but it passes its importance on to only one page, p_1. So that page receives a large boost.

Topic: Linear Recurrences

1 **(a)** We express the relation in matrix form.

$$\begin{pmatrix} 5 & -6 \\ 1 & 0 \end{pmatrix} \begin{pmatrix} f(n) \\ f(n-1) \end{pmatrix} = \begin{pmatrix} f(n+1) \\ f(n) \end{pmatrix}$$

The characteristic equation of the matrix

$$\begin{vmatrix} 5-\lambda & -6 \\ 1 & -\lambda \end{vmatrix} = \lambda^2 - 5\lambda + 6$$

has roots of 2 and 3. Any function of the form $f(n) = c_1 2^n + c_2 3^n$ satisfies the recurrence.

(b) The matrix expression of the relation is

$$\begin{pmatrix} 0 & 4 \\ 1 & 0 \end{pmatrix} \begin{pmatrix} f(n) \\ f(n-1) \end{pmatrix} = \begin{pmatrix} f(n+1) \\ f(n) \end{pmatrix}$$

and the characteristic equation

$$\begin{vmatrix} \lambda^2 - 2 \end{vmatrix} = (\lambda - 2)(\lambda + 2)$$

has the two roots 2 and -2. Any function of the form $f(n) = c_1 2^n + c_2(-2)^n$ satisfies this recurrence.

(c) In matrix form the relation

$$\begin{pmatrix} 5 & -2 & -8 \\ 1 & 0 & 0 \\ 0 & 1 & 0 \end{pmatrix} \begin{pmatrix} f(n) \\ f(n-1) \\ f(n-2) \end{pmatrix} = \begin{pmatrix} f(n+1) \\ f(n) \\ f(n-1) \end{pmatrix}$$

has a characteristic equation with roots -1, 2, and 4. Any combination of the form $c_1(-1)^n + c_2 2^n + c_3 4^n$ solves the recurrence.

Printed in Great Britain
by Amazon